APPLIED
ANALYSIS

APPLIED ANALYSIS

Cornelius Lanczos

DOVER PUBLICATIONS, INC.

New York

This Dover edition, first published in 1988, is an unabridged and unaltered republication of the work first published by Prentice-Hall, Inc., Englewood Cliffs, New Jersey, in 1956.

Manufactured in the United States of America
Dover Publications, Inc., 31 East 2nd Street, Mineola, N.Y. 11501

Library of Congress Cataloging-in-Publication Data

Lanczos, Cornelius, 1893–
    Applied analysis / Cornelius Lanczos.
        p.   cm.
    Reprint. Originally published: Englewood Cliffs, N.J. :Prentice-Hall, 1956.
    Bibliography: p.
    Includes index.
    ISBN 0-486-65656-X (pbk.)
    1. Mathematical analysis.   2. Mathematical physics.   I. Title.
                                                88-3961
                                                CIP

To the Memory of the Six Million
Who Died for the Kiddush Hashem

PREFACE

FOR MANY YEARS the author has been engaged in studies of those fields of mathematical analysis that are of primary concern to the engineer and the physicist. That this area of "workable mathematics" did not receive the same attention during the 19th century as did the classical fields of analysis is perhaps the result of a historical misunderstanding. Until the time of Gauss and Legendre the "workable" methods of analysis received the closest attention of the best mathematicians. The brilliant discovery of the theory of limits changed the emphasis. Thenceforth it was considered satisfactory to design infinite approximation processes by which the validity of certain analytical results could be established, irrespective of whether the process used were feasible or not for a given problem.

It was then that the gradual separation of "pure" and "applied" mathematics occurred until we now have the "pure analyst," who pursues his ideas in a world of purely theoretical constructions, and the "numerical analyst," who translates the processes of analysis into machine operations.

In actual fact there is a large area between the two, which is not less analytical than the analysis of infinite processes but devoted to a different branch of analysis, namely, the analysis of *finite* algorithms. Here our objective is the analysis and design of finite processes which *approximate* the solution of an analytical problem. To design procedures which will effectively minimize the error in a small number of steps and which will estimate the error with sufficient accuracy is not a matter of *practical* interest only, but a matter of *scientific* interest as well. This book is largely devoted to such problems.

A few remarks concerning the manner of presentation may not be out of place. The author is not in favor of sacrificing rigor under the disguise that the applied scientist is interested only in the results and not in the more or less intricate procedures which lead to those results. The concepts and statements of mathematics are sharp and uncompromising and any "sloppy" presentation of a mathematical theorem disqualifies the formulation and throws doubt on the claimed result. It seems permissible, however, to state and prove a theorem under less exacting conditions than those in which the pure analyst is interested, if the gain achieved by this confinement is that the methods and results of mathematical investigations become presentable to the student of physics or scientific engineering in a language which is not overly strange to him. Furthermore, the author has the notion that mathematical formulas have their "secret life," behind their Golem-like appearance. To bring out the "secret life" of mathematical relations by an occasional narrative digression does not appear to him a profanation of the sacred rituals of formal analysis but merely an attempt to a more integrated way of understanding. The reader who has to struggle through a maze of "lemmas," "corollaries," and "theorems," can easily get lost in formalistic details, to the detriment of the essential elements of the results obtained. By keeping his mind on the principal points he gains in depth, although he may lose in details. The loss is not serious, however, since any reader equipped with the elementary tools of algebra and calculus can easily interpolate the missing details. It is a well-known experience that the only truly enjoyable and profitable way of studying mathematics is the method of "filling-in details" by one's own efforts. This additional work, the author hopes, will stir the reader's imagination and may easily lead to stimulating discussions and further explorations, on both the university and the research levels.

That a book of this nature cannot exhaust the subject without becoming unduly bulky, goes without saying. The broad subject of boundary-value problems, together with the theory of integral equations, had to be omitted, due to lack of space. But it is perhaps no exaggeration to say that the topics considered in each chapter are encountered almost daily by the engineer and physicist. A brief description of each chapter follows.

Chapter I. *Algebraic Equations*. The search for the roots of an algebraic equation is frequently encountered in vibration and flutter problems and in problems of static and dynamic stability. Some useful computing techniques, based on the "movable strip method," are discussed. The Bernoulli method with all its ramifications plays the central role, but the scanning of the unit circle for the separation of complex roots of nearly equal magnitude and the method of reciprocal radii for stability questions are likewise of interest.

Chapter II. *Matrices and Eigenvalue Problems*. This chapter is devoted to a systematic development of the properties of matrices, with particular emphasis on those features which are most frequently encountered in industrial research.

Chapter III. *Large-Scale Linear Systems.* The advent of the electronic computer brings the iterative techniques for the solution of complicated boundary value problems and vibration problems into the foreground. This leads at once to the investigation of polynomial operations with matrices. While the general case of complex eigenvalues could not be included, the "spectroscopic method" of finding the real eigenvalues of large matrices and the corresponding method of solving large-scale linear equations is of such general usefulness—and so naturally tied up with the later chapter on harmonic analysis—that their treatment was hardly out of place. An additional treatment of a perturbation problem gives at least a partial answer to the complex eigenvalue problem by showing how an arbitrary complex eigenvalue and eigenvector can be obtained if we can start with a fairly good *first approximation* of the desired eigenvalue.

Chapter IV. *Harmonic Analysis.* The length of this chapter may be excused by the extraordinary importance of the Fourier series and its corollaries, the Fourier integral and the Laplace transform, in all problems of analysis. One might be tempted to paraphrase the famous saying of Victor Hugo that if he were asked to destroy all literature but keep one single book, he would preserve the Book of Job. Similarly, if we were asked to abandon all mathematical discoveries save one, we would hardly fail to vote for the Fourier series as the candidate for survival. This series has influenced the entire course of analysis, in both its theoretical and practical aspects, most profoundly. Moreover, its interconnection with other parts of analysis is so intimate that if we said "the Fourier series with all its implications," a considerable part of our classical analysis would be preserved.

For the purposes of engineering the orthogonality of the Fourier functions with respect to equidistant data is perhaps the most important single item. Accordingly the present chapter deals primarily with the interpolation aspects of the Fourier series, its flexibility in representing empirically given equidistant data with great ease. An important artifice is needed here to make the series applicable, viz., the subtraction of a linear trend which reduces the two boundary values to zero and permits the use of a pure sine series.

The remaining Gibbs oscillations, caused by the discontinuity of the *second* derivative, are too small to cause any harm.

An additional artifice, the application of the "σ-factors," counteracts the divergence-producing Gibbs oscillations if the series is *differentiated*. The same method smooths out the unpleasant Gibbs oscillations in the neighborhood of a discontinuity and in the representation of the delta-function.

In electric network analysis the methods of the Fourier transform and the Laplace transform have gained enormous impetus during the last few years. These transforms are not only theoretical devices for proving certain basic theorems, but are of fundamental importance for the practical construction of the input-output relation of linear networks. The interpolatory solution of the filter problem and a variety of methods for the inversion of the Laplace transform are discussed as problems of applied analysis which have very real technical significance. Finally, the frequently encountered "search for hidden periodicities" is dealt with and a numerical scheme developed which achieves greater independence of the various frequencies—and thus higher resolution power and higher accuracy—than the traditional schemes.

Chapter V. *Data Analysis*. The problem of the reduction of data and the problem of obtaining the first and even second derivatives of an empirically given function are constantly encountered in tracking problems, but also occur in similar form in ordinary curve-fitting problems. Two methods of smoothing are discussed: smoothing in the small and smoothing in the large. In the first case, local least-square parabolas are employed which lead to a certain weighted average of every observation with a few of its left and right neighbors. In the second case, a Fourier analysis is performed and the smoothing achieved by merely *truncating* the series at a judiciously chosen point. The latter technique provides us simultaneously with an analytical expression which represents all our data and interpolates them at any desired point of the interval. If it so happens that a close *polynomial* approximation is desired, a method is described which transforms the Fourier series into a polynomial series of strongest convergence. We thus avoid the pitfalls of equidistant Lagrangian interpolation and obtain a polynomial which fits our data with small and practically uniformly distributed errors.

Chapter VI. *Quadrature Methods*. Since the dawn of science the problem of integration has fascinated the scholar. Each scientific period added its own share of knowledge to the problem of quadrature. While Archimedes, who first introduced integration as an exact limit process, used only trapezoids with constantly decreasing sides for the purpose of quadrature, later ages refined the technique by operating with interpolating polynomials of higher order. The present chapter gives a survey of a variety of these methods. The Gaussian quadrature looms large as the most advanced of all quadrature methods. A slight modification of the method is described which makes it numerically more palatable by avoiding interpolation at irrational points. Moreover, the fundamental idea of the Gaussian method is translated to the case when only boundary values are at our disposal, viz., the value of the function and its derivatives up to a certain order at both endpoints of the interval. The resulting quadrature formula has strong convergence and can be used for the solution of boundary value problems and eigenvalue problems associated with ordinary differential equations. The method is demonstrated with the help of a few numerical examples.

Chapter VII. *Power Expansions.* The representation of functions by polynomials is an old art. However, to represent a function within a given interval by a few powers but with small error goes beyond the realm of the Taylor expansion and requires the theory of orthogonal function systems. There is in particular one special class of polynomials, the "Chebyshev polynomials," which assures stronger convergence than any other class of polynomials. In actual fact we go once more back to the Fourier series, since an expansion into Chebyshev polynomials is in reality nothing but a cosine series in a modified variable.

We can put these polynomials to good use in the problem of solving ordinary differential equations with coefficients which are rational functions of x. Since most of the fundamental transcendentals encountered in mathematical physics are definable in terms of such differential equations, we obtain rapidly convergent expansions for a large variety of functions by a simple technique. We terminate our series from the beginning to a finite polynomial of a given order. This necessitates an error term on the right side of the given differential equation. This error term is put proportional to a Chebyshev polynomial of properly chosen order. We now obtain simple recurrence relations from which the coefficients of the approximating polynomial can be determined. We may have to repeat this process several times and obtain the final result as a linear superposition of the composing polynomials. This "τ-method" is very helpful in putting the powers to work with maximum efficiency, whether our aim is to approximate an elementary function for operational purposes, or a transcendental function for evaluation purposes. The gain in reducing the error in comparison to that of the Taylor series (if the Taylor series exists at all) is always very considerable.

Acknowledgments. This book has grown through years of scientific thinking. In these years the author had the good fortune of innumerable discussions with colleagues and friends, which gave him the basic background on which to build. The number of people who helped the author in his endeavors is thus very great and their enumeration impossible. It will be more adequate to enumerate the institutions with which he was connected during the slow ripening of his thoughts.

1. During the author's memorable years at Purdue University (1931–45) Dr. W. Marshall of the Department of Mathematics and Dr. K. Lark-Horowitz of the Department of Physics organized a lecture course in "Approximation Methods of Analysis," where the author first came in touch with approximation problems. It was during those years that the author discovered for himself the outstanding properties of the Chebyshev polynomials, which had a decisive influence on his later scientific development.

2. During the national emergency the author spent the year 1943–44 at the Mathematical Tables Project, New York City, headed by Dr. A. N. Lowan, Director. The associations with an outstanding staff of numerical analysts were most gratifying.

3. During 1944–45 the author gave two lecture courses: "Engineering Applications of Rapidly Converging Series" and "Approximation Mathematics Course," under the auspices of the Physical Research Unit of the Boeing Airplane Company in Seattle, Washington. The excellently prepared mimeographed notes of these courses— compiled with the able assistance of Marius Cohn—form the basic core of the present book.

In 1946 Dr. C. K. Stedman, Head of the Physical Research Unit of the Boeing Airplane Company, invited the author to join his unit as a mathematical consultant and research engineer. The benefits he derived from the daily discussions with an unusually select group of excellent physicists and electrical engineers cannot be measured.

4. At the invitation of Dr. J. H. Curtiss, then Chief, National Bureau of Standards, Washington, D. C. the author joined in 1949 the newly founded Institute for Numerical Analysis at the University of California, Los Angeles. Under the leadership of Dr. Curtiss, the Institute provided a scholarly atmosphere and colleagual associations on a level which had few parallels anywhere in the world. The generosity with which the Institute supported the author's scientific projects will remain in his memory with undiminished force.

5. At the invitation of the Dublin Institute for Advanced Studies, the author spent a memorable year (1952–53) in Dublin, Eire, in daily contact with the directors of the Institute, Professor E. Schroedinger and Professor J. L. Synge. The quadrature formula of § 22, Chapter VI, and its application to the solution of eigenvalue problems were developed during this year.

6. During the winter 1953–54 the author was affiliated with North American Aviation, Los Angeles, as a staff member of the Tabulating Department of Charles F. Davis, numerical analyst. A lecture course attended by a selected group of engineers led to memorable friendships and is reflected in Chapters II and III of this book. The "spectroscopic eigenvalue analysis" was developed during these months.

In enumerating the extraordinary opportunities with which his good fortune endowed him in his scientific life, the author should not fail to mention the splendid assistance he received in the numerical documentation of his mathematical endeavors. For the computation of the numerical examples of the book he is primarily indebted to Miss Mary Ellen Russell, Research Assistant of the Physical Research Unit, Boeing Airplane Company, Seattle, Washington, while the Appendix Tables were principally prepared by Miss Lillian Forthal, INA, National Bureau of Standards, Los Angeles, California.

C. L.

Dublin Institute for Advanced Studies
Dublin, Eire

CONTENTS

INTRODUCTION

Chapter I
ALGEBRAIC EQUATIONS

xiii

Contents

Chapter II

MATRICES AND EIGENVALUE PROBLEMS

Chapter III
LARGE-SCALE LINEAR SYSTEMS

Chapter IV
HARMONIC ANALYSIS

Chapter V

DATA ANALYSIS

Chapter VI

QUADRATURE METHODS

Contents

BIBLIOGRAPHY

The following books are recommended for collateral or more advanced reading. They represent a selective bibliography of books, written in English, whose topic is similar or related to that of the present book. If important sources have been overlooked, this did not occur intentionally. The Courant-Hilbert and Whittaker-Watson books are irreplaceable sources of information. In the field of numerical analysis the books of Milne, Scarborough, and Hartree are basic.

{1} COURANT, R. and D. HILBERT, *Methods of Mathematical Physics*, Vol. I (Interscience Publishers, New York, 1953).

{2} DOHERTY, R. E. and E. G. KELLER, *Mathematics of Modern Engineering*, Vols. I and II (John Wiley & Sons, Inc., New York, 1936 and 1942).

{3} DWYER, P. S., *Linear Computations* (John Wiley & Sons, Inc., New York, 1951).

{4} HARTREE, D. R., *Numerical Analysis* (Clarendon Press, Oxford, 1952).

{5} HOUSEHOLDER, A. S., *Principles of Numerical Analysis* (McGraw-Hill Book Company, Inc., New York, 1953).

{6} JAEGER, J. C., *Applied Mathematics* (Clarendon Press, Oxford, 1951).

{7} KARMAN, TH. V. and M. A. BIOT, *Mathematical Methods of Engineering* (McGraw-Hill Book Company, Inc., New York, 1940).

{8} MILNE, W. E., *Numerical Calculus* (Princeton University Press, Princeton, N. J., 1949).

{9} MURNAGHAN, T. D., *Introduction to Applied Mathematics* (John Wiley & Sons, Inc., New York, 1948).

{10} PIPES, L. A., *Applied Mathematics for Engineers and Physicists* (McGraw-Hill Book Company, Inc., New York, 1946).

{11} SCARBOROUGH, J. B., *Numerical Mathematical Analysis* (Johns Hopkins Press, Baltimore, 1950).

{12} SCHELKUNOFF, S. A., *Applied Mathematics for Engineers and Scientists* (D. Van Nostrand Company, Inc., New York, 1948).

{13} SMITH, L. P., *Mathematical Methods for Scientists and Engineers* (Prentice-Hall, Inc., Englewood Cliffs, N. J., 1953).

{14} WHITTAKER, E. T. and G. ROBINSON, *The Calculus of Observations* (Blackie & Son, Glasgow, 1940).

{15} WHITTAKER, E. T. and G. N. WATSON, *A Course of Modern Analysis* (Cambridge University Press, London, 1935).

Books of a more specific character are listed at the end of each chapter. References in braces { } refer to the books of the general Bibliography, those in brackets [ ] to the books and articles of the chapter bibliographies.

INTRODUCTION

1. Pure and applied mathematics. The history of mathematics reveals that the interest in the formal processes of mathematics was seldom divested of the desire to obtain an adequate picture of the physical universe. The postulates and operations of analysis are not chosen *arbitrarily*, but are postulates and operations which map the geometrical order of things in the abstract realm of numbers. The formal operations of analysis are thus merely one link in our desire to discover the inherent functional order of the physical universe. The entire process involves three phases:

1. A given physical situation is translated into the realm of numbers.

2. By purely formal operations with these numbers certain mathematical results are obtained.

3. These results are translated back into the world of physical reality.

During the development of mathematics, Phase 2 of this translation process becomes eventually an independent endeavor in itself. We can concentrate on the formal processes of analysis without asking ourselves whether the problem posed has necessarily a counterpart in nature. Similarly, we can stop with the analytical result obtained in Phase 2 without interpreting it back into the physical order of things.

The nineteenth century invented the terms "pure" and "applied" mathematics for the characterization of these two situations, a terminology which is far from being adequate and satisfactory. The term "pure mathematics" as an expression for isolation of the formal processes of mathematics has the connotation that this kind of intellectual endeavor is more "l'art pour l'art," more "pure" and thus more "idealistic" than the other endeavor which keeps in closer touch with the inherent order of the physical universe. This connotation is hardly tenable and could have been avoided by a more precise terminology. Yet we continue with the usage of words, even if they

Introduction

have been coined wrongly by conjuring up associations which are not warranted philosophically. The expressions "negative numbers" and "imaginary numbers" originate from a period which did not understand properly the true significance of these concepts. The names survive, however, and cause misunderstandings among those who do not pay close attention to the technical definition of these words. Similar is the situation relative to the misnomers "pure" and "applied mathematics."

2. Pure analysis, practical analysis, numerical analysis. There exists, however, a somewhat different situation which again necessitates a distinctive terminology and where again the words "pure" and "practical" came into use without proper justification. Already the old Greek mathematicians of the fourth and third centuries B.C. recognized the fundamental fact that certain mathematical situations require an *infinite sequence of never-ending approximations* which come nearer and nearer to the desired result but without ever reaching it exactly. Archimedes discovered in his fundamental treatise on the circle that the circumference of a circle is not only not *calculable* exactly but not even *definable* exactly. The exact processes of algebra are thus replaced by another kind of approach termed by the Greeks the "method of exhaustion," while nineteenth century mathematics adopted the name "infinite limit processes." In these limit processes, frequently characterized by an infinite expansion, we do not endeavor to *obtain* a quantity, but merely to *approach* it with an error which can be made as small as we wish. "Absolute accuracy" is thus replaced by "arbitrarily great accuracy." The first kind of accuracy, obtainable in algebra but *not* obtainable in higher mathematics, means the "error zero." The second kind of accuracy, typical for the processes of higher mathematics, means a "finite but arbitrarily small error."

The great perspectives opened by the theory of limits of Cauchy and Gauss led to an overwhelming emphasis of *infinite approximation processes*. In these processes the number of steps employed is of no further consequence. We are not interested in what happens during a *finite* number of steps. We are satisfied if we know what happens *eventually*, if the number of steps is increased unlimitedly.

Much less attention was paid to another kind of approximation process which is logically equally feasible. We may like to know what error is involved if we pursue a certain process a definite *finite* number

of times. In particular we may be interested in designing approximation processes which obtain a certain result with a minimum error
for a given number of steps. The accuracy here is neither of the
absolute nor of the arbitrarily great kind. We are reconciled to the
fact that an error is committed. We should like to have tools for
estimating this error. Moreover, we would like to investigate what
processes can be pursued for effective *reduction* of this error.

This branch of analysis, which received historically much less
attention than the previously discussed limit processes, has not been
designated by any adequate name. Since, however, the classical or
"pure" analysis deals with infinite approximation processes, this
other branch of analysis is sometimes referred to as "practical
analysis." The name comes from the fact that in physics and engineering we are not interested in reducing the error of a certain approximation process to zero or to an arbitrarily small amount, since our
observations are of limited accuracy anyway. Hence we are satisfied
with an answer which is "theoretically" imperfect but "practically"
acceptable. Moreover, the theoretically perfect answer may involve
tools which are too cumbersome to be available practically. Hence in
practical analysis we are concerned with finite processes which are
numerically accessible and which in a relatively few steps give an
accuracy deemed satisfactory for certain aims of physics and
engineering.

Here again the juxtaposition of the words "pure" and "practical"
conveys a wrong evaluation. One may be inclined to believe that only
the processes of "pure" analysis are of distinctly mathematical
interest, while the processes of "practical" analysis are developed
solely in view of the needs of the "practical" man, i.e., the physicist
and engineer. This, however, is equivalent to putting an accidental
feature of the historical development on an absolute pedestal. Let us
consider as an example an infinite expansion of the number π which
may require 100 terms for an accuracy of 1%. If another expansion
of "practical analysis" obtained the same accuracy with only five
terms, we are justified in assuming that the second expansion is,
even from the purely analytical angle, *more adequate* than the first
very slowly convergent series. The increased convergence of the
second series may be booked as a "practical" achievement, but it
may also be booked as a result of purely logical interest.

One can hardly deny that here is a branch of mathematical analysis which would deserve a more adequate name than the word "practical" with its utilitarian and pragmatic overtones. The present book makes a move in this direction by introducing a technical term for this class of analytical processes. The Greek word "parexic" (with the roots para = almost, quasi, and ek = out) means "nearby." Hence the term "parexic analysis" can well be adopted to mean that we do not want an exact but only a "nearby" determination of a certain quantity. We can then speak of parexic methods, parexic expansions, parexic viewpoints, in contradistinction to the corresponding methods, expansions, and viewpoints of "pure analysis" which aim at arbitrary accuracy with the help of infinite processes. The author found that such a terminology is quite rewarding by obtaining a brief notation for something which under the customary terminology is expressible only by the way of circumlocution.[1]

During the last few years a new word came into vogue: "numerical analysis." This term is well suited to the designation of certain aspects of mathematical analysis which deal with translation of mathematical processes into operations with numbers. The basic viewpoint here is the ease with which certain analytical methods can be handled if the mathematical quantities are replaced by actual numbers; moreover, this branch of analysis is concerned with the accumulation of rounding errors, caused not by the approximate nature of parexic processes but by the approximate nature of the arithmetic processes of multiplications and divisions, if we restrict ourselves to a definite number of decimal places.

We thus come to the conclusion that pure analysis, parexic analysis, and numerical analysis represent three well-circumscribed phases of mathematical investigations which have their own fields of interest and which pursue these interests with characteristic methods of their own. The present book is primarily devoted to the field of "parexic analysis," but without losing sight of the general aims of pure analysis and the more arithmetical aspects of numerical analysis.

---

[1] This is the only instance in which the present book deviates from the customary terminology (or lack of terminology) of mathematics. The author is well aware that the coining of new words is the privilege of much more advanced minds. But in this particular instance an obvious emergency existed, and if the author's suggestion is not accepted, at least he has called attention to the existence of this problem.

I

ALGEBRAIC EQUATIONS

1. Historical introduction. Algebraic equations of the first and second order aroused the interest of scientists from the earliest days. While the early Egyptians solved mostly equations of first order, the early Babylonians (about 2000 B.C.) were already familiar with the solution of the quadratic equation, and constructed tables for the solution of cubic equations by bringing the general cubic into a normal form.

The Hindus developed the systematic algebraic theory of the equations of first and second order (seventh century). The standard method of solving the general quadratic equation by completing the square is a Hindu invention. The Hindus were familiar with the operational viewpoint and were not afraid of the use of negative numbers, considering them as "debts." The clear insight into the nature of imaginary and complex numbers came much later, in the time of Euler (eighteenth century).

The solution of cubic equations was first discovered by the Italian Tartaglia (early sixteenth century); Cardano's pupil Ferrari added a few years later the solution of biquadratic equations. The essentially different character of equations of fifth and higher order was clearly recognized by Lagrange (late eighteenth century), but the first exact proof that general equations of fifth and higher order cannot be solved by purely algebraic tools is due to the Norwegian, Abel (1824), while a few years later (1832) the French Galois gave the general group-theoretical foundation of the entire problem.

The "fundamental theorem of algebra" states that every algebraic equation has at least one solution within the realm of complex numbers. If this is proved, we can immediately infer (by successive divisions by the root factors) that every polynomial of the order n can

5

be resolved into a product of n root factors. The first rigorous proof of the fundamental theorem of algebra was given by Gauss when only 22 years of age (1799). Later Cauchy's theory of the functions of a complex variable provided a deeper insight into the nature of the roots of an algebraic equation and yielded a simplified proof for the fundamental theorem.

The existence of n generally complex roots of an algebraic equation of nth order is in no contradiction to the unsolvability of an algebraic equation of fifth or higher order by algebraic means. The latter statement means that the roots of a general algebraic equation of higher than fourth order are not obtainable by purely algebraic operations on the coefficients (i.e., addition, subtraction, multiplication, division, raising to a power and taking the root). Such operations can *approximate*, however, the roots with any degree of accuracy.

2. Allied fields. (a) The problem of solving an algebraic equation of nth order is closely related to the theory of vibrations around a state of equilibrium. The frequencies (or the squares of the frequencies) of a mechanical system appear as the "characteristic roots" or "eigenvalues" of a matrix, obtainable by solving the "characteristic equation" of the matrix, which is an algebraic equation of the order n.

(b) In electrical engineering the response of an electric network is always a linear superposition of exponential functions. The exponents of these functions are obtainable as the roots of a certain polynomial which can be constructed if the elements of the network and the network diagram are given.

(c) Intricate algebraic and geometric relations frequently yield by elimination an algebraic equation of second or higher order for one of the unknowns.

3. Cubic equations. Equations of third and fourth order are still solvable by algebraic formulas. However, the numerical computations required by the formulas are usually so involved and time-absorbing that we prefer less cumbersome methods which give the roots *in approximation* only but still close enough for later refinement.

The solution of a cubic equation (with real coefficients) is particularly convenient since one of the roots must be real. After finding this root, the other two roots follow immediately by solving a quadratic equation.

A general cubic equation can be written in the form

$$f(\xi) = \xi^3 + a\xi^2 + b\xi - c = 0 \qquad (1\text{-}3.1)$$

The factor of ξ^3 can always be normalized to 1 since we can divide through by the highest coefficient. Moreover, the absolute term can always be made negative because, if it is originally positive, we put $\xi_1 = -\xi$ and operate with this ξ_1.

Now it is convenient to introduce a new scale factor which will normalize the absolute term to -1. We put

$$x = \alpha\xi, \quad a_1 = \alpha a, \quad b_1 = \alpha^2 b, \quad c_1 = \alpha^3 c \qquad (1\text{-}3.2)$$

and write the new equation

$$f(x) = x^3 + a_1 x^2 + b_1 x - c = 0 \qquad (1\text{-}3.3)$$

If we choose $\qquad\qquad\qquad \alpha = 1/\sqrt[3]{c} \qquad\qquad\qquad (1\text{-}3.4)$

we obtain $\qquad\qquad\qquad\qquad c_1 = 1 \qquad\qquad\qquad\qquad (1\text{-}3.5)$

Now, since $f(0)$ is negative and $f(\infty)$ is positive, we know that there must be at least one root between $x = 0$ and $x = \infty$. We put $x = 1$ and evaluate $f(1)$. If $f(1)$ is positive, the root must be between 0 and 1; if $f(1)$ is negative, the root must be between 1 and ∞. Moreover, since

$$x_1 \cdot x_2 \cdot x_3 = 1 \qquad (1\text{-}3.6)$$

we know in advance that we cannot have *three* roots between 0 and 1, or 1 and ∞. Hence, if $f(1) > 0$, we know that there must be one and *only one* real root in the interval [0,1], while if $f(1) < 0$, we know that there must be one and only one real root in the interval [1,∞]. The latter interval can be changed to the interval [1,0] by the transformation

$$\bar{x} = \frac{1}{x} \qquad (1\text{-}3.7)$$

which simply means that the coefficients of the equation change their sequence:

$$-c_1 \bar{x}^3 + b_1 \bar{x}^2 + a_1 \bar{x} + 1 = 0 \qquad (1\text{-}3.8)$$

Hence we have reduced our problem to the new problem: find the real root of a cubic equation in the range [0,1]. We solve this problem in good approximation by taking advantage of the remarkable properties of the Chebyshev polynomials (cf. VII, 9) which enable us to reduce a higher power to lower powers with a small error. In particular, the third Chebyshev polynomial

$$T_3^*(x) = 32x^3 - 48x^2 + 18x - 1 \qquad (1\text{-}3.9)$$

normalized to the range [0,1] gives

$$x^3 = \frac{48x^2 - 18x + 1}{32} = 1.5x^2 - 0.5625x + 0.03125 \qquad (1\text{-}3.10)$$

with a maximum error of $\pm\frac{1}{32}$. The original cubic is thus reducible to a quadratic, with an error not exceeding 3%.

We now solve this quadratic, retaining only the root between 0 and 1.

4. Numerical example. In actual practice α need not be taken with great accuracy but can be rounded off to two significant figures. Consider the solution of the following cubic:

$$\xi^3 + 1.2\xi^2 - 17.0\xi - 70 = 0 \qquad (1\text{-}4.1)$$

Barlow's Tables give the cube root of 70 as $4.1212\cdots$, the reciprocal of which gives $\alpha = 0.2426\cdots$. We conveniently choose

$$\alpha = 0.25 \qquad (1\text{-}4.2)$$

obtaining

$$f(x) = x^3 + 0.3x^2 - 1.0625x - 1.09375 = 0 \qquad (1\text{-}4.3)$$

At $x = 1$, $f(1) = -0.856$ is still negative. The root is thus between $x = 1$ and ∞. We invert the range by putting

$$\bar{x} = 1/x \qquad (1\text{-}4.4)$$

$$1.09375\bar{x}^3 + 1.0625\bar{x}^2 - 0.3\bar{x} - 1 = 0 \qquad (1\text{-}4.5)$$

The substitution (1-3.10) reduces this equation to the quadratic

$$2.703\bar{x}^2 - 0.915\bar{x} - 0.966 = 0 \qquad (1\text{-}4.6)$$

solution of which gives

$$\bar{x} = \frac{0.915 \pm 3.370}{5.406} \tag{1-4.7}$$

The negative sign of the square root yields a spurious result, since it falls outside the range considered. The positive sign gives

$$\bar{x} = 0.79 \tag{1-4.8}$$

and thus

$$x = \frac{1}{0.79} = 1.27$$

and

$$\xi = \frac{1.27}{0.25} = 5.08 \tag{1-4.9}$$

Substitution in the original equation shows that the left side gives the remainder 5.692, which in comparison with the absolute term 70 is an error of 8%.

The operation with large roots is numerically not advantageous. It is thus of considerable importance that we can always restrict ourselves to roots which are in absolute value less than 1, because if the absolute value of the root is greater than 1, the reciprocal transformation $\bar{x} = 1/x$, which merely inverts the polynomial, changes the root to its reciprocal. Hence in our example we will prefer to substitute the reciprocal of (9)[1], i.e.,

$$\bar{\xi} = 1/5.08 = 0.197 \tag{1-4.10}$$

into the inverted equation

$$70\bar{\xi}^3 + 17\bar{\xi}^2 - 1.2\bar{\xi} - 1 = 0 \tag{1-4.11}$$

The remainder is now -0.0395, an error of 4% compared with the absolute term 1.

---

[1] Equations encountered in the current section are quoted by the digits after the decimal point only. Hence (9) refers to (1-4.9) since the equation has to be found among the equations of the current section. An equation in the current chapter, but *not* in the current section is quoted by Section number and equation digit. Hence (3.9) refers to (1-3.9).

5. Newton's method. If we have a good approximation $x = x_0$ to a root of an algebraic equation, we can improve that approximation by a method known as "Newton's method." We put

$$x = x_0 + h \tag{1-5.1}$$

and expand $f(x_0 + h)$ into powers of h:

$$f(x_0 + h) = f(x_0) + hf'(x_0) + \frac{h^2}{2}f''(x_0) + \cdots \tag{1-5.2}$$

For small h the higher order terms will rapidly diminish. If we neglect everything beyond the second term, then the solution of the equation

$$f(x) = f(x_0 + h) = 0 \tag{1-5.3}$$

is obtained in good approximation by

$$h_0 = -\frac{f(x_0)}{f'(x_0)} \tag{1-5.4}$$

We can now consider

$$x_1 = x_0 + h_0 \tag{1-5.5}$$

as a new first approximation, replacing x_0 by x_1. Hence

$$h_1 = -\frac{f(x_1)}{f'(x_1)} \tag{1-5.6}$$

combined with

$$x_2 = x_0 + h_0 + h_1 \tag{1-5.7}$$

is a still closer approximation of the root, and generally we obtain the iterative scheme

$$h_n = -\frac{f(x_n)}{f'(x_n)} \tag{1-5.8}$$

$$x_{n+1} = x_0 + h_0 + \cdots + h_n \tag{1-5.9}$$

which converges rapidly to x, if x_0 is a sufficiently close first approximation.

Newton's scheme is not restricted to algebraic equations, but is equally applicable to transcendental equations.

An increase of convergence is obtainable if we stop only after the *third* term, considering the second-order term a small correction of the first-order term. Hence we write

$$f(x_0) + h(f'(x_0) + \frac{h}{2}f''(x_0)) = 0 \qquad (1\text{-}5.10)$$

and solve this equation in the form

$$h = - \frac{f(x_0)}{f'(x_0) + \frac{h}{2}f''(x_0)} \qquad (1\text{-}5.11)$$

replacing the h in the denominator by the first approximation (4). This yields a formula which can best be remembered in the following form:

$$\frac{1}{h_0} = -\frac{f'(x_0)}{f(x_0)} + \frac{1}{2}\frac{f''(x_0)}{f'(x_0)} \qquad (1\text{-}5.12)$$

6. Numerical example for Newton's method. In § 4[1] the cubic equation

$$f(x) = 1.09375x^3 + 1.0625x^2 - 0.3x - 1 = 0 \qquad (1\text{-}6.1)$$

was treated, and the approximation

$$x_0 = 0.79 \qquad (1\text{-}6.2)$$

was obtained. We substitute this value in $f(x)$ and likewise in

$$f'(x) = 3.28125x^2 + 2.125x - 0.297619 \qquad (1\text{-}6.3)$$

and
$$\tfrac{1}{2}f''(x) = 3.28125x + 1.0625 \qquad (1\text{-}6.4)$$
obtaining

$$f(x_0) = -0.034631, \quad f'(x_0) = 3.42658, \quad \tfrac{1}{2}f''(x_0) = 3.65468 \quad (1\text{-}6.5)$$

Substitution in the formula (1-5.12) gives

$$\frac{1}{h} = 98.945453 + 1.066568 = 100.012021 \qquad (1\text{-}6.6)$$

$$h = 0.009998798 \qquad (1\text{-}6.7)$$

$$x_1 = x_0 + h = 0.7999988 \qquad (1\text{-}6.8)$$

---

[1] Throughout the book the § sign refers to sections of the same chapter.

If this new x_1 is substituted in $f(x)$, we obtain

$$f(x_1) = -0.00000418 \qquad (1\text{-}6.9)$$

At this point we can stop, since the error is only 4 units in the 6th place; the coefficients of an algebraic equation are seldom given with more than 5 decimal place accuracy.

7. Horner's scheme. Direct substitution of a number into a polynomial is simple enough if the number is real and the polynomial is of low order. In view of later occasions, however, when polynomials of higher order have to be considered and the numbers to be substituted are complex, we will now discuss a numerically more elegant scheme, called "Horner's scheme," which obtains $f(x_0)$, $f'(x_0)$, ⋯ by a process of synthetic divisions.

We consider the algebraic division

$$\frac{f(x) - f(x_0)}{x - x_0} = f_1(x) \qquad (1\text{-}7.1)$$

If $f(x)$ is a polynomial of the order n, $f_1(x)$ is a polynomial of the order $n - 1$. We can continue the process and gradually decrease the order of the resulting polynomial to zero:

$$\frac{f_1(x) - f_1(x_0)}{x - x_0} = f_2(x) \qquad (1\text{-}7.2)$$

$$\frac{f_2(x) - f_2(x_0)}{x - x_0} = f_3(x) \qquad (1\text{-}7.3)$$

. . . .

What we accomplish by this successive decomposition of the given polynomial $f(x)$ is that we automatically obtain the successive coefficients of the Taylor expansion about the point $x = x_0$. Indeed, multiplying through our equations by $x - x_0$ and making successive substitutions, we obtain

$$f(x) = f(x_0) + f_1(x_0)(x - x_0) + f_2(x_0)(x - x_0)^2 + \cdots \quad (1\text{-}7.4)$$

The coefficients of the Taylor expansion thus appear as the successive *remainders* of a sequence of synthetic divisions. This process is known as "Horner's scheme" [W. G. Horner (1819), also P. Ruffini (1804)].

8. The movable strip technique. In numerical work frequently a great deal of time is lost by noting down partial results which could have been avoided by a more concise arrangement of the calculations. One particular device which is of great help in many numerical algorithms, involves a "movable strip." We formulate the algorithm of the movable strip in terms of desk calculations, but the technique can easily be coded for electronic computers, too.

A certain fixed set of numbers is written down on a vertical strip of paper. This movable column operates on another given column, the "fixed strip." The operation consists in multiplying two numbers facing each other, one of the movable strip and one of the fixed strip. The partial products are summed and the result written down on a third strip, the "nascent strip." Now the movable strip glides vertically downward by one step, the operation is repeated, and the next element of the "nascent strip" is obtained. Thus we continue until the movable strip arrives at the bottom of the fixed strip. Hence we obtain an arrangement which is demonstrated in the following numerical scheme:

Movable strip	Fixed strip	Nascent strip	
−3			
−2			
4			(1-8.1)
1	2	2	
	−3	5	
	−5	−21	
	0	−20	
	1	20	

In this arrangement the sequence of operations was not decisive, since the results written down on the "nascent strip" have no influence on the later operations. The nascent strip of the scheme (1) could have been obtained by starting the movable strip from the bottom and moving upwards or in any other sequence. Frequently, however, another kind of algorithm is encountered in which the nascent strip is put *between* the movable strip and the fixed strip. This is a "feed-back" arrangement which can be performed only in the right sequence. The movable strip now operates on the nascent strip, and only the lowest element of the movable strip (the 1 in the above example) reaches over to the fixed strip. The algorithm operates now as follows:

Movable strip	Nascent strip	Fixed strip	
−3			
−2			
4			(1-8.2)
1	2	2	
	5	−3	
	11	−5	
	28	0	
	76	1	

The first kind of arrangement will be encountered later in all procedures which involve the weighting of data, such as local smoothing, differentiation of an empirically given function, etc. (cf. V, 8, 10). But all problems associated with division of polynomials require the second kind of arrangement.

As a simple application of the movable strip technique, let us consider the synthetic division of a given polynomial by $x - x_0$. The movable strip will now contain but two elements, viz., x_0, 1 in vertical arrangement. The fixed strip will contain the consecutive coefficients $a_n, a_{n-1}, \cdots, a_0$ of the given polynomial, likewise in vertical arrangement. The results are written down in succession between these two strips.

Finally we arrive at the last element of our scheme, when the 1 of the movable strip reaches over to the a_0 of the fixed strip. If the result is 0, we know that $x - x_0$ is a root factor of the given polynomial, and the nascent strip contains the successive coefficients of the ratio $p_n(x)/(x - x_0)$. If the result is not 0, we obtain the remainder of the division process. We do *not* write this last element in the nascent column but transfer it to a separate "remainder" column, filling out the last element of the nascent strip by zero.

We demonstrate this technique by obtaining the result of substituting $x_0 = 0.79$ into the cubic (1-6.1) on the basis of synthetic division:

Movable strip	*Nascent strip*	*Fixed strip*	*Remainder:*
0.79			
1	1.09375	1.09375	(1-8.3)
	1.9265625	1.0625	
	1.2219844	−0.3	
	0	−1	−0.03463234

Horner's scheme is obtained by repeating this algorithm again and again. In each instance the "nascent strip" of the previous step becomes the "fixed strip" of the next step. For the sake of neater arrangement we will write the remainder in each instance at the bottom of the fixed strip in question. The complete synthetic division scheme of problem (3) now becomes

1.09375	1.09375	1.09375	1.09375
1.09375	2.790625	1.9265625	1.0625
	3.6546875	1.2219844	−0.3
		3.4265781	−1
			−0.0346323

Hence

$$f(0.79 + h) = 1.09375h^3 + 3.6546875h^2 + 3.4265781h - 0.0346323$$

In actual practice we can frequently stop after three divisions, since $f(x_0), f'(x_0)$ and $\frac{1}{2}f''(x_0)$ are sufficient for application of formula (5.12).

9. The remaining roots of the cubic. In the present example formula (5.12) gave the improved root (6.8). We repeat the synthetic division with this new root, but do not go beyond the first step:

1	1	
1.86909628	1.041667	(1-9.1)
1.24892599	−0.297619	
	−1.033399	
	−0.00000107	

Since the remainder is already practically negligible, we have reduced the given cubic to the quadratic

$$x^2 + 1.869096x + 1.248926 = 0 \tag{1-9.2}$$

which can be solved by the standard formula. We obtain

$$x_2 = -0.934548 + 0.612818i$$
$$x_3 = -0.934548 - 0.612818i \tag{1-9.3}$$

Going back to the original equation (4.1) by taking the reciprocals and dividing by 0.24, we finally obtain the three roots

$$\xi_1 = 5.035677$$
$$\xi_2 = -3.117839 + 2.044483i \tag{1-9.4}$$
$$\xi_3 = -3.117839 - 2.044483i$$

10. Substitution of a complex number into a polynomial. Multiplication of two complex numbers is much more cumbersome than multiplication of two real numbers. Hence the substitution of a complex number $x_0 = a + ib$ into a polynomial is cumbersome even if synthetic division is used. The following method of substitution has the advantage that it reduces the operation with complex numbers to a minimum.

In the ordinary synthetic division scheme we would divide $f(x)$ by the root factor $x - x_0 = x - a - ib$, which involves the products of complex numbers from the beginning. Let us divide, however, by the real quantity (the notation "asterisk" refers to "conjugate complex"):

$$(x - x_0)(x - x_0^*) = (x - a)^2 + b^2 = x^2 - 2ax + (a^2 + b^2) \tag{1-10.1}$$

The result of the division can be written as follows:

$$\frac{f(x)}{(x - x_0)(x - x_0^*)} = f_1(x) + \frac{Ax + B}{(x - x_0)(x - x_0^*)} \tag{1-10.2}$$

Now the remainder is not a mere constant but the linear term $Ax_0 + B$. Multiplying through by the denominator on both sides we find

$$f(x_0) = Ax_0 + B \tag{1-10.3}$$

This modification of the ordinary synthetic division scheme has thus the advantage that substitution of the complex number x_0 occurs *only at the end,* and only in the form Ax_0. Up to that point all operations are real.

The movable strip is now composed of the numbers $-(a^2 + b^2)$, $2a$, 1. The operations proceed once more in the previous fashion, with the only difference that the synthetic division is finished one step *before* the last coefficient of $f(x)$ has been reached. Hence the process gives *two* remainder coefficients; the first is A, the second B. They are transferred to a separate "remainder" column to the right of the fixed column, while the last two elements of the nascent strip become 0.

Example. Substitute the complex number $x = 0.3 - 0.5i$ in

$$f(x) = x^4 - 2x^3 + 4x^2 - 2x + 1 \qquad (1\text{-}10.4)$$

M.S.	Quotient	f(x)	Remainder
−0.34			
0.6			
1			
	1	1	(1-10.5)
	−1.4	−2	
	2.82	4	
	0	−2	0.168
	0	1	0.0412

The remainder is thus $0.168x + 0.0412$, and we obtain

$$f(0.3 - 0.5i) = 0.168(0.3 - 0.5i) + 0.0412$$
$$= 0.0916 - 0.084i \qquad (1\text{-}10.6)$$

However, Horner's scheme is *not* applicable if this technique is adopted. We now have to form $f'(x)$ and $f''(x)$ by actual differentiations and apply the process to these polynomials.

In the present example, if we consider the given complex number as a preliminary root x_0 which shall be corrected by Newton's method, the scheme continues as follows:

$f'(x)$			$\frac{1}{2}f''(x)$		
4	4				
−3.6	−6		6	6	(1-10.7)
0	8	4.48	0	−6	−2.4
0	−2	−0.776	0	4	1.96

Hence

$$f'(x_0) = 4.48(0.3 - 0.5i) - 0.776 = 0.568 - 2.24i$$
$$\tfrac{1}{2}f''(x_0) = -2.40(0.3 - 0.5i) + 1.96 = 1.24 + 1.2i$$

Application of formula (5.12) gives

$$\frac{1}{h} = \frac{1.24 + 1.2i}{0.568 - 2.24i} - \frac{0.568 - 2.24}{0.0916 - 0.084i} \qquad (1\text{-}10.8)$$

and we obtain $\qquad h = -0.042909 - 0.029222i$

Hence the corrected root becomes

$$x_1 = 0.257091 - 0.529222i \qquad (1\text{-}10.9)$$

Substituting once more in $f(x)$ yields

$$
\begin{array}{lll}
1 & 1 & \\
-1.485818 & -2 & \qquad (1\text{-}10.10) \\
\underline{2.889847} & \underline{4} & \\
0 & -2\ \big| & 0.000256 \\
0 & 1\ \big| & -0.000383
\end{array}
$$

The new remainder becomes

$$f(x_1) = 0.000256x_1 - 0.000383 = -0.000318 - 0.000135i$$

An estimation of the accuracy of the root x_1 can be obtained as follows. Since x_1 is very near to x_0, $f'(x_1)$ is only slightly different from $f'(x_0)$.

The next correction in Newton's method requires the evaluation of $-f(x_1)/f'(x_1)$, which for estimation purposes can be replaced by $-f(x_1)/f'(x_0)$. Since the absolute value of $f'(x_0)$ is more than 2, the remainder $f(x_1)$ shows that the error of x_1 cannot be more than 1.5 units in the fourth decimal. If we are satisfied with this accuracy, we can consider

$$x_1 = 0.2571 - 0.5292i$$

as the final root. Then the coefficients of the division process (10) yield the reduced equation

$$x^2 - 1.4858x + 2.8898 = 0 \qquad (1\text{-}10.11)$$

the solution of which gives the other pair of complex roots.

11. Equations of fourth order. Algebraic equations of fourth order with generally complex roots occur frequently in the stability analysis of airplanes and in problems involving servomechanisms. The historical method of solving algebraic equations of fourth order (also called biquadratic or quartic equations) involves the following steps. By a transformation of the form $x + \alpha$ the coefficient of the cubic term is annihilated. Then an auxiliary cubic equation is solved. The roots of the original equation are constructed with the help of the three roots of the auxiliary cubic. Numerically this method is lengthy and cumbersome. The following modification of the traditional procedure yields the four roots of an arbitrary quartic equation with real coefficients on the basis of a quick and numerically convenient scheme.

Every equation of the form

$$x^4 + c_1 x^3 + c_2 x^2 + c_3 x + c_4 = 0 \qquad (1\text{-}11.1)$$

can be rewritten as follows:

$$(x^2 + \alpha x + \beta)^2 = (ax + b)^2 \qquad (1\text{-}11.2)$$

If the original c_i are real, the new coefficients are also real. Hence the original equation becomes solvable in form of the quadratic equation

$$x^2 + \alpha x + \beta \pm (ax + b) = 0 \qquad (1\text{-}11.3)$$

which has four (generally complex) roots, obtainable by the standard formula. The new coefficients can be determined as follows. We evaluate in succession the following numerical constants:

$$\alpha = \frac{c_1}{2}, \quad A = c_2 - \alpha^2, \quad B = c_3 - \alpha A \qquad (1\text{-}11.4)$$

and form the cubic equation

$$\xi^3 + (2A - \alpha^2)\xi^2 + (A^2 + 2B\alpha - 4c_4)\xi - B^2 = 0 \quad (1\text{-}11.5)$$

Since the left side is negative at $\xi = 0$, a positive real root must exist. We determine this root according to the method of § 3. In order to avoid later corrections, it is advisable to add at this point Newton's correction (cf. § 5), obtaining ξ with great accuracy. The coefficients of the reduced equation (3) are then determined as follows:[1]

$$\alpha = \tfrac{1}{2}c_1, \qquad \beta = \tfrac{1}{2}(A + \xi) \qquad\qquad (1\text{-}11.6)$$

$$a = \sqrt{\xi} \qquad b = \frac{a}{2}\left(\alpha - \frac{B}{\xi}\right)$$

Numerical example. We will demonstrate the general procedure by solving the following quartic equation:

$$x^4 + 7.64x^3 + 23.6044x^2 + 38.91024x + 38.149496 = 0$$

Here

$$c_1 = 7.64, \quad c_2 = 23.6044, \quad c_3 = 38.91024, \quad c_4 = 38.149496$$

and we obtain by substituting in the general equations (4):

$$\alpha = 3.82, \quad A = 9.012, \quad B = 4.4844$$

The cubic equation (5) becomes:

$$\xi^3 + 3.431600\xi^2 - 37.121024\xi - 20.109843 = 0$$

We make the substitution

$$\xi = \frac{\eta}{0.37}$$

obtaining

$$\eta^3 + 1.269692\eta^2 - 5.081868\eta - 1.018624 = 0$$

Since the left side is still negative at $\eta = 1$, we invert the equation by going to the reciprocal root (we divide by 1.018624):

$$\bar{\eta}^3 + 4.988954\bar{\eta}^2 - 1.246478\bar{\eta} - 0.981717 = 0$$

$$\begin{array}{ccc} 1.5 & -0.5625 & +0.03125 \\ \hline \end{array}$$

$$6.488954\bar{\eta}_0^2 - 1.808978\bar{\eta}_0 - 0.950467 = 0$$

$$\bar{\eta}_0^2 - 0.2788\bar{\eta}_0 - 0.1465 = 0$$

$$\bar{\eta}_0 = 0.1394 + \sqrt{0.1658} = 0.5466$$

[1] The exceptional case $B = 0$ deserves special attention. Then (5) has the solution $\xi = 0$ and we obtain by a limit process:

$$\alpha = \tfrac{1}{2}c_1, \quad \beta = \tfrac{1}{4}A, \quad a = 0, \quad b = \tfrac{1}{2}\sqrt{A^2 - 4c_4}$$

If, however, $A^2 - 4c_4$ happens to be negative, it is preferable to divide (5) by ξ and solve the resulting quadratic for the positive root. This ξ is then substituted in the general formulas (6); (with $B = 0$).

We will correct this root by Newton's method, using the synthetic division scheme of Horner (cf. § 7 and § 8), here displayed in horizontal arrangement:

1	4.988954	−1.246478	−0.981717
1	5.535554	1.779256	−0.009176
1	6.082154	5.103761	
1	6.628754		

$$\frac{1}{h} = \frac{5.103761}{0.009176} + \frac{6.628754}{5.103761} = \begin{array}{r} 556.2076 \\ +1.2988 \\ \hline 557.5064 \end{array}$$

$$\begin{array}{r} h = 0.001794 \\ 0.5466 \\ \hline \bar{\eta} = 0.548394 \end{array}$$

$$\bar{\eta} = 0.548394, \quad \eta = 1.823506, \quad \xi = 4.928394$$

Substitution in the equations (6) yields

$$\alpha = 3.82, \quad \beta = 6.970197, \quad a = 2.219998, \quad b = 3.230196$$

Hence the reduced quadratic equation becomes

$$x^2 + 3.82x + 6.970197 \pm (2.219998x + 3.230196) = 0$$

(a) $x^2 + 6.039998x + 10.200393 = 0$

$\qquad x = -3.019999 \pm 1.039230i$

$\qquad$ (correct: $-3.02 \pm 1.03923048i$)

(b) $x^2 + 1.600002x + 3.740001 = 0$

$\qquad x = -0.800001 \pm 1.760681i$

$\qquad$ (correct: $-0.8 \pm 1.76068169i$)

Great accuracy results, however, even if we do not correct the preliminary value $\bar{\eta}_0$ but accept it as $\bar{\eta}$. Then,

$$\bar{\eta} = 0.5466, \quad \eta = 1.8295, \quad \xi = 4.9446$$

and substitution in (6) now gives

$$\alpha = 3.82, \quad \beta = 6.9783, \quad a = 2.2236, \quad b = 3.2388$$

which leads to the reduced equation

$$x^2 + 3.82x + 6.9783 \pm (2.2236x + 3.2388) = 0$$

The four roots of this equation are

$$x = -3.0218 \pm 1.0420i \quad \text{and} \quad x = -0.7982 \pm 1.7614i$$

A good check on the accuracy to be expected is available by forming the product of all four roots. This should be equal to c_4. In our case the first set of roots gives 38.149461 (correct: 38.149496). The second set of roots gives 38.2081.

12. Equations of higher order. Newton's correction scheme is an important tool in the gradual evaluation of the roots of an algebraic equation, if we can start with a crude approximation. Unfortunately, we are not in the possession of any direct methods for approximate localization of the roots of equations of higher than fourth order. The *real* roots of an algebraic equation can be found with relatively little labor. For this purpose we divide the interval between -1 and $+1$ in, let us say, 10 equal parts, evaluating $f(x)$ at intervals of 0.2. We observe the change of sign and localize the root more exactly by linear interpolation. The synthetic division scheme of § 7 will then refine this root to the desired accuracy. The roots outside the interval -1 to $+1$ are now transformed inside by the reciprocal transformation (3.7), i.e., by inverting the sequence of the coefficients, and once more we scan the same interval in units of 0.2. Hence with relatively little labor all the *real* roots of a polynomial can be located.

However, a general polynomial need not have any real roots, and the polynomials encountered in vibration and stability problems are usually exactly of this type. Hence it is of importance that there exist a method, called the "method of moments," first described by Daniel Bernoulli (1728), which puts us in the position to locate the absolutely largest root of an algebraic equation of any order with comparative ease.

13. The method of moments. If we resolve a polynomial of nth order into its root factors:

$$f(x) = a_0 x^n + a_1 x^{n-1} + \cdots + a_n$$
$$\equiv a_0 (x - x_1)(x - x_2) \cdots (x - x_n) \qquad (1\text{-}13.1)$$

and differentiate logarithmically, we obtain a formula which is particularly useful in the study of the algebraic behavior of a polynomial:

$$-\frac{f'(x)}{f(x)} = \frac{1}{x_1 - x} + \frac{1}{x_2 - x} + \cdots + \frac{1}{x_n - x} \qquad (1\text{-}13.2)$$

Let $x = x_0$ be a point which is not far from the root $x = x_1$. Then

$$-\frac{f'(x_0)}{f(x_0)} = \frac{1}{x_1 - x_0} + \frac{1}{x_2 - x_0} + \cdots + \frac{1}{x_n - x_0} \qquad (1\text{-}13.3)$$

and we see that Newton's method of evaluating the correction of x_0 from the formula

$$\frac{1}{h} = -\frac{f'(x_0)}{f(x_0)} \qquad (1\text{-}13.4)$$

amounts to reducing the right side to the first term. This will be the more justified, the nearer x_0 is to the true root x_1.

Let us put in particular $x_0 = 0$ and obtain

$$-\frac{f(0)}{f'(0)} = \frac{1}{x_1} + \frac{1}{x_2} + \cdots + \frac{1}{x_n} \qquad (1\text{-}13.5)$$

If it so happens that one root, say x_1, is much nearer to the origin than all the other roots, then

$$\frac{1}{h} = -\frac{f'(0)}{f(0)} \qquad (1\text{-}13.6)$$

will give a close approximation of that root. But this approximation loses increasingly in value as the closeness of x_1 to the origin becomes less pronounced.

Let us now differentiate the function (2) $m - 1$ times. We thus obtain

$$-\frac{1}{(m-1)!} \left[\frac{f'(x)}{f(x)}\right]^{(m-1)} = \frac{1}{(x_1 - x)^m} + \cdots + \frac{1}{(x_n - x)^m} \qquad (1\text{-}13.7)$$

and putting $x = 0$:

$$-\frac{1}{(m-1)!} \left[\frac{f'(x)}{f(x)}\right]^{(m-1)}_{x=0} = \frac{1}{x_1^m} + \cdots + \frac{1}{x_n^m} \qquad (1\text{-}13.8)$$

By this method we put the spotlight on the nearest root x_1 with increasing sharpness, even if the closeness of x_1 is not very pronounced in the beginning.

We can thus generalize formula (6) to

$$\frac{1}{h^m} = -\frac{1}{(m-1)!}\left[\frac{f'(x)}{f(x)}\right]_{x=0}^{(m)} \qquad (1\text{-}13.9)$$

and obtain a good approximation of the nearest root x_1 by choosing m sufficiently high.

This general idea can be elaborated to a valuable method for locating the nearest root of an algebraic equation. The actual differentiation of the function (2) would be a cumbersome task. We can obtain, however, the quantities (9) by a simple division scheme which generalizes the synthetic division method of § 8 and § 9.

It is preferable to think of the roots of the *inverted* polynomial which are the reciprocals of the original roots. In these terms the quantities (8) become

$$S_m = x_1^m + x_2^m + \cdots + x_n^m \qquad (1\text{-}13.10)$$

They are called the "symmetric moments of the roots" but the name "power sums" is also frequently used. In the reciprocal plane the previously nearest root changed to the most remote root.

Formula (9) is now converted into

$$h^m = x_1^m + \cdots + x_n^m = S_m \qquad (1\text{-}13.11)$$

The following paragraph gives a simple and elegant numerical scheme for successive generation of the moments S_m.

14. Synthetic division of two polynomials. The movable strip technique can be successfully employed for generation of the symmetric root functions S_m. Let us write the coefficients of the polynomial $A(x)$ in a vertical column, starting with the highest coefficient and ending with the lowest ("fixed strip"). Moreover, let $B(x)$ be a polynomial whose highest coefficient is normalized to 1. We write the coefficients of $B(x)$ on a movable strip, starting with 1 and moving *upward*. The sign of each coefficient is reversed, with the exception of the highest coefficient 1 which remains unchanged.

The movable strip technique now generates the ratio of the two polynomials $A(x)$ and $B(x)$.

Example. Divide $6x^6 - 3x^5 + x^3 - x^2$ by $x^3 + 2x^2 + 4x - 1$.

$B(x)$	Quotient	$A(x)$	Remainder	
1	6	6		
−4	−15	−3		(1-14.1)
−2	6	0		
1	55	1		
	0	−1	−150	
	0	0	−214	
	0	0	55	

Hence

$$\frac{6x^6 - 3x^5 + x^3 - x^2}{x^3 + 2x^2 + 4x - 1}$$

$$= 6x^3 - 15x^2 + 6x + 55 + \frac{-150x^2 - 214x + 55}{x^3 + 2x^2 + 4x - 1}$$

If this technique is applied to the ratio $f'(x)/f(x)$ we have the difficulty that the numerator is of the order $n - 1$, the denominator of the order n. We make the division possible by multiplying the numerator by an arbitrarily high power x^N. Then the scheme can continue indefinitely, giving a polynomial of the order $N - 1$. If now we divide by x^N, we obtain the quotient in the form

$$\frac{f'(x)}{f(x)} = \frac{c_0}{x} + \frac{c_1}{x^2} + \frac{c_2}{x^3} + \cdots \qquad (1\text{-}14.2)$$

where the coefficients $c_0, c_1, c_2, \cdots$ are the successive entries of the "quotient" column.

On the other hand, the expansion of (13.2) in reciprocal powers of x gives

$$\frac{f'(x)}{f(x)} = \frac{1}{x}\left(\frac{1}{1 - x_1/x} + \cdots + \frac{1}{1 - x_n/x}\right)$$

$$= \frac{n}{x} + \frac{S_1}{x^2} + \frac{S_2}{x^3} + \cdots + \frac{S_m}{x^{m+1}} \qquad (1\text{-}14.3)$$

The comparison of (2) and (3) shows that

$$c_m = S_m \qquad (1\text{-}14.4)$$

The movable strip technique thus provides the successive power sums, up to any order m.

Example. Obtain the successive power sums of the polynomial

$$x^4 + 7.60x^3 + 23.34x^2 + 38.44x + 37.40 \qquad (1\text{-}14.5)$$

$f(x)$

m		S_m	$f'(x)$
	-37.40		
	-38.44		
	-23.34		
	7.60		
0	1	4	4
1		-7.60	22.80
2		11.08	46.68
3		-22.144	38.44
4		52.2312	
5		-616.6382	
6		4015.4799	

$$(1\text{-}14.6)$$

· · · · ·

15. Power sums and the absolutely largest root. The behavior of the moments of high order allows conclusions concerning the absolute value of the most remote root. Let us write the generally complex roots in polar form:

$$x_k = r_k e^{i\theta k} \qquad (1\text{-}15.1)$$

Then

$$S_m = r_1^m e^{im\theta_1}\left[1 + \left(\frac{r_2}{r_1}\right)^m e^{im(\theta_2-\theta_1)} + \cdots + \left(\frac{r_n}{r_1}\right)^m e^{im(\theta_m-\theta_1)}\right] \qquad (1\text{-}15.2)$$

If r_1 is the most remote root and all the other r_i are smaller than r_1, then the ratios $r_i/r_1 (i = 2, \cdots n)$, raised to an increasingly high power converge to zero, and thus for very large m,

$$S_m = r_1^m e^{im\theta_1}$$

and

$$r_1 = \sqrt[m]{|S_m|} \qquad (1\text{-}15.3)$$

The successive power sums have thus the valuable property that *they single out the absolutely largest root with ever-increasing strength.*

If the roots of an algebraic equation are of slightly different orders of magnitude, this difference becomes greatly magnified in the power sums of high order. For example, the ratio 1.5 : 1 is increased in ten steps to the ratio $(1.5)^{10} : 1 = 57.7$. The largest root will thus greatly overshadow all the other roots. It is possible, however, that more than one root will have the same absolutely largest distance r_1 from the origin. In fact, if the algebraic equation has real coefficients, we know in advance that the complex roots always appear in *pairs* $a + ib$ and $a - ib$, and thus at least *two* roots lie on the maximum circle $r = r_1$. The successive power sums S_m then show preference for one *pair* of roots. Generally in the S_m of high order only the absolutely largest roots will be practically present, while the absolutely smaller roots are practically obliterated. If the absolutely largest root is real, it is quickly obtainable by the ratio of two successive S_m:

$$x = \frac{S_{m+1}}{S_m} \tag{1-15.4}$$

The mere observance of a few consecutive S_m helps to spot this situation. The sign of the S_m is then either constantly positive or constantly alternating. Moreover, the ratio (4) remains approximately the same if two neighboring ratios S_{m+1}/S_m and S_{m+2}/S_{m+1} are used.

In the more frequent case of complex roots such a regularity in the size and sign of the S_m cannot be observed. A large value may be followed by a very small value, and the sign may change capriciously. A behaviour of this kind indicates that the absolutely largest root is complex and of the form $a \pm ib$. The associated S_m is now of the form

$$S_m = 2r^m \cos m\theta \tag{1-15.5}$$

The last factor is responsible for the irregular changes in sign and size. However, if the maximum circle contains only *one* pair of complex roots, we can again succeed with approximate localization of the largest root. We consider the determinant equation

$$\begin{vmatrix} 1 & \lambda & \lambda^2 & \cdots & \lambda^n \\ 1 & x_1 & x_1^2 & \cdots & x_1^n \\ \vdots & & & & \\ 1 & x_n & x_n^2 & \cdots & x_n^n \end{vmatrix}' = 0 \tag{1-15.6}$$

which is valid for $\lambda = x_1, x_2, \cdots x_n$. If this determinant is pre-multiplied by the determinant

$$\begin{vmatrix} 1 & 0 & 0 & \cdots & 0 \\ 0 & w_1 & w_2 & \cdots & w_n \\ 0 & w_1 x_1 & w_2 x_2 & \cdots & w_n x_n \\ \vdots & \vdots & & & \\ 0 & w_1 x_1^{n-1} & w_2 x_2^{n-1} & \cdots & w_n x_n^{n-1} \end{vmatrix}$$

(1-15.7)

we obtain the equation

$$\begin{vmatrix} 1 & \lambda & \lambda^2 & \cdots & \lambda^n \\ \omega_0 & \omega_1 & \omega_2 & \cdots & \omega_n \\ \omega_1 & \omega_2 & \omega_3 & \cdots & \omega_{n+1} \\ \vdots & & & & \\ \omega_{n-1} & \omega_{n-2} & \omega_{n+1} & \cdots & \omega_{2n-1} \end{vmatrix} = 0 \qquad (1\text{-}15.8)$$

where the ω_k denote the "weighted power sums"

$$\omega_k = w_1 x_1^k + w_2 x_2^k + \cdots + w_n x_n^k \qquad (1\text{-}15.9)$$

Now we can generate the equation (8) in successive steps, starting with the quadratic equation

$$\begin{vmatrix} 1 & \lambda & \lambda^2 \\ \omega_0 & \omega_1 & \omega_2 \\ \omega_1 & \omega_2 & \omega_3 \end{vmatrix} = 0 \qquad (1\text{-}15.10)$$

and constantly adding two more rows and columns of the original determinant. In this successive procedure each new step generates one additional pair of roots and at the same time corrects the roots previously obtained. Generally we cannot expect that these corrections will remain small. But this is actually the case if it so happens that the weights $w_1, w_2, \cdots$ are strongly biased in favor of the absolutely largest roots. If the weights w_1, w_2 dominate in comparison to the other weights, then the solution of the simple quadratic equation (10) will already approximate the absolutely largest pair of roots and the later phases of the process merely add small corrections.

Let us now assume that the absolutely largest complex root outdistances the others by a reasonably large factor, e.g., 1.5 or more. We consider four consecutive power sums S_m, belonging to the subscripts m, $m + 1$, $m + 2$, $m + 3$, where m is not below 9. These four S_m can be considered the four ω_i of equation (10):

$$\omega_0 = w_1 \quad + w_2 \quad + \cdots + w_n$$
$$\omega_1 = w_1 x_1 + w_2 x_2 + \cdots + w_n x_n \qquad (1\text{-}15.11)$$
$$\omega_2 = w_1 x_1^2 + w_2 x_2^2 + \cdots + w_n x_n^2$$
$$\omega_3 = w_1 x_1^3 + w_2 x_2^3 + \cdots + w_n x_n^3$$

with $\qquad w_1 = x_1^m, \quad w_2 = x_2^m, \quad \cdots, \quad w_n = x_n^m \qquad (1\text{-}15.12)$

The conditions for applicability of the simple quadratic (10) are thus fulfilled. We form the ratios

$$s_1 = \frac{S_m}{S_{m+1}}, \qquad s_2 = \frac{S_{m+2}}{S_{m+1}}, \qquad s_3 = \frac{S_{m+3}}{S_{m+1}} \qquad (1\text{-}15.13)$$

and solve the quadratic equation

$$\begin{vmatrix} 1 & \lambda & \lambda^2 \\ s_1 & 1 & s_2 \\ 1 & s_2 & s_3 \end{vmatrix} = 0 \qquad (1\text{-}15.14)$$

which means

$$(1 - s_1 s_2)\lambda^2 + (s_1 s_3 - s_2)\lambda + (s_2^2 - s_3) = 0 \qquad (1\text{-}15.15)$$

The (usually complex) roots of this equation establish the absolutely largest root with fair accuracy. In the case of real roots we keep only the larger of the two roots.

We demonstrate the method numerically by applying it to the example of § 14. If the movable strip technique of the table (1-14.6) is continued up to $m = 12$, the last four power sums become

$S_9 =$	61829.6	$s_1 =$	-0.31103
$S_{10} =$	-198790.6	hence $\quad s_2 =$	-2.91902
$S_{11} =$	580274.3	$s_3 =$	7.55682
$S_{12} =$	-1502225		

and the quadratic equation (15) becomes

$$0.092097\lambda^2 + 0.56862\lambda + 0.96386 = 0$$

$$\lambda = -3.087 \pm 0.967i$$

The correct root of (14.5) is

$$x = -3 \pm i$$

and we see that the error is not more than 3%. Hence we obtained a root which is close enough to the true value to make Newton's method (cf. § 5) applicable.

16. Estimation of the largest absolute value. The assumption that the absolute value of the largest root will be at least 1.5 times as large as the next largest root will not always be true. We have to be prepared for the possibility that two or even more pairs of complex roots will lie nearly on the same circle of the complex plane. In the extreme case *all* the complex roots of an algebraic equation may have nearly the same absolute value. Even in such cases the power sums S_m contain valuable information concerning the location of the absolutely largest roots. The radius of the maximum circle can be ascertained with sufficient accuracy, even if two or more pairs of roots lie near to that circle. Hence we can proceed as follows. We first obtain a close value for the absolute value r of the largest pair or pairs of roots. Then we proceed to localization of the angle θ—or several angles θ_i—associated with the complex roots which lie near to the maximum circle. For this second half of the problem *two* procedures will be considered. One is based on the properties of "hidden periodicities." while the other uses a transformation known as the "transformation by reciprocal radii."

The factor $\cos m\theta$ in the expression (15.5) interferes with a simple determination of r on the basis of the power sums S_m. In the absence of this factor we could obtain r with the help of the equation

$$\log r = \frac{1}{m} \log \frac{|S_m|}{2} \tag{1-16.1}$$

With the proper precaution this equation can still be used, in spite of the disturbing factor. No matter what θ is, it will inevitably happen that $m\theta$ comes near to a certain multiple of π. If this happens, the factor $\cos m\theta$ will be near to ± 1. We can watch out for such opportunities by spotting the *peak values* of S_m. Starting with $m = 10$ we can take the logarithm of $|S_m|$ and divide by m. We go up until $m = 16$ and select the maximum of these ratios. For this maximum we form (1) and obtain r. This method is applicable even if more than one pair of complex roots happen to be near the maximum circle. Even an error of 100% in S_{16}, for example, caused by the presence of a second pair of complex roots, will not vitiate r by more than 4%, since the sixteenth root of 2 is 1.0443. Hence it will always be possible to obtain a reasonably accurate value for the radius of the maximum circle, without going to an unduly large order in the tabulation of the S_m.

A numerical example will help in demonstrating the general procedure. We choose an equation of sixth order whose root factors are known. The following sixtic equation:

$$\xi^6 - 2\xi^5 + 6\xi^4 - 58\xi^3 + 633\xi^2 - 200\xi + 2500 = 0 \quad (1\text{-}16.2)$$

has three pairs of conjugate complex roots, located as follows:

$$\xi = \pm 2i, \qquad -3 \pm 4i, \qquad 4 \pm 3i$$

In harmony with our general policies, we first normalize the absolute term to the order of magnitude 1. For this purpose we divide the coefficients of the equation by the successive powers of 4, according to the substitution

$$\xi = 4x \qquad\qquad (1\text{-}16.3)$$

This yields the new equation

$$x^6 - 0.5x^5 + 0.375x^4 - 0.90625x^3$$
$$+ \, 2.47266x^2 - 0.19351x + 0.61035 = 0 \qquad (1\text{-}16.4)$$

The movable strip technique displayed in § 14 yields in succession the following power sums:

$$
\begin{aligned}
&6, \quad 0.5, \quad -0.5, \qquad 2.28125, \qquad -8.10939, \qquad -5.623069,\\
&\quad -0.0312454, \quad -11.299692, \quad 10.068455, \quad 20.171028, \quad (1\text{-}16.5)\\
&\quad -0.001920, \quad \underline{32.925676,} \quad 7.659814,
\end{aligned}
$$

The peak value $S_{11} = 32.925676$ leads to a particularly large r and will thus be retained; the other S_m are discarded. We divide by 2 and take the logarithm.

$$\log r_0 = \tfrac{1}{11} \log 16.463 = 0.1106$$

This gives $\qquad\qquad\qquad r_0 = 1.290$

which is a satisfactory approximation of the correct $r = 1.25$.

17. Scanning of the unit circle. A complex number

$$z = re^{i\theta}$$

is characterized by magnitude and direction. Even if r is already in our possession, we still have to find the angle θ associated with a complex root. If several roots are located in the vicinity of the maximum circle, all the corresponding θ_i will have to be found. Moreover, we will try to improve on the preliminary value of r, found by the previous peak value method.

Our first move will be to change the maximum circle to the unit circle by the scale transformation

$$x = r_0 u \qquad (1\text{-}17.1)$$

This transformation has the consequence that the moments related to the new variable u are equal to the old S_m divided by r_0^m. In our numerical example we found for r_0 the value 1.29, which can be rounded off to 1.3. Hence we divide the S_m values of the table (16.5) by the successive powers of 1.3, taken from *Barlow's Tables*. This gives the following new table, if we terminate our sequence with S_{12}:

$$
\begin{array}{cccccc}
6, & 0.38462, & -0.29586, & 1.03835, & -2.83932, & -1.51445, \\
0.00647, & -1.8008, & 1.2343, & 1.90212, & -0.00014, & 1.83720, \\
0.32877 & & & & & (1\text{-}17.2)
\end{array}
$$

Now, assuming that the new maximum circle is exactly 1 (which is in fact true only in approximation), the successive moments associated with the maximum root become

$$2, \quad 2\cos\theta, \quad 2\cos 2\theta, \quad 2\cos 3\theta, \quad \cdots \quad 2\cos m\theta$$

If more than one root lies on the unit circle, each root will be associated with such a consequence, and an arbitrary S_m becomes

$$S_m = 2\cos m\theta_1 + 2\cos m\theta_2 + \cdots + 2\cos m\theta_p \quad (1\text{-}17.3)$$

In practice p will seldom exceed 2 or 3, since the order of the equation will seldom exceed 6. However, for the sake of discussion, p can be left arbitrary.

Now S_m can be conceived as the value of a certain function $S(t)$ of the continuous variable t, at the definite points $t = m$. We define this function as follows:

$$S(t) = 2 \cos \theta_1 t + 2 \cos \theta_2 t + \cdots + 2 \cos \theta_p t \qquad (1\text{-}17.4)$$

We notice that $S(t)$ is composed of purely *periodic* components. Hence the search for the roots of an algebraic equation can be reformulated as a search for "hidden periodicities" of a function which is given in equidistant intervals. We will deal with this problem later in detail (cf. IV, 22). In the general problem each periodic component has its own amplitude, phase, and frequency. In our case the amplitude is fixed to 2 in advance; moreover, the "phase" aspect of the problem is irrelevant, since only cosine functions enter our considerations. It is the *frequency* ω_i in which we are primarily interested. The various ω_i of the periodic components correspond to the unknown angles θ_1, θ_2, $\cdots$, θ_p of the roots.

We anticipate the results of the later investigation and apply them to our present problem. It will be shown later how beneficial the "σ smoothing" is in cutting down the otherwise cumbersome "Gibbs oscillations." The application of the σ factors causes extra labor by multiplying each S_m by a pretabulated factor. However, the focusing power of the method is so strongly increased because of this smoothing procedure—by diminishing the mutual interference of the various roots on the unit circle—that the additional work is well justified. We will not go beyond S_{12}. Moreover, the weight factor of S_{12} becomes 0. Hence only 11 multiplications are involved because of the σ factors which are tabulated as follows:

m	σ_m	m	σ_m
0	1.	7	0.52708
1	0.98862	8	0.41350
2	0.95493	9	0.30010
3	0.90032	10	0.19099
4	0.82699	11	0.08987
5	0.73791	12	0
6	0.63662		

$$(1\text{-}17.5)$$

The table (2) of the S_m values is thus once more modified. We multiply them in succession with the corresponding σ_m factors of the table (5). The results S'_m of the multiplication are arranged in the following order. We start with $S'_m = 6$ but we divide by 2, and thus write down 3. We write next in the same horizontal row S'_1, S'_2, $\cdots$ until we come to S'_6. Then we continue in the line below, but now going *backward*. Hence S'_7 is lined up with S'_5, S'_8 with S'_4, $\cdots$, finally $\frac{1}{2}S'_{12} = 0$ with $\frac{1}{2}S'_0 = 3$. Then we form the sums and differences of these two lines. In our numerical example the arrangement looks as follows:

```
3, 0.38024, −0.28252, 0.93484, −2.34812, −1.11650, −0.00412
0, 0.16511, −0.00003, 0.57084,  0.51037, −0.94828,  ←────────
```

Sum: 3, 0.54535, −0.28255, 1.50568, −1.50568, −2.06478, −0.00412
Diff.: 3, 0.21453, −0.28249, 0.36400, −2.85849, −0.16822,   (1–17.6)

Since the half circle is divided into 12 parts, we are going to scan the unit circle in intervals of 15°. First, however, we do the scanning in intervals of 30°. We multiply the "sum" row by a cosine matrix [cf. (4-13.3)] which in fact coincides with the A_6 matrix of IV, 13. This matrix has pretabulated coefficients and we obtain the following scheme.

k	3	0.54535	−0.28255	1.50568	−1.83775	−2.06478	−0.00412
0	1	1	1	1	1	1	1
2	1	0.866	0.5	0	−0.5	−0.866	−1
4	1	0.5	−0.5	−1	−0.5	0.5	1
6	1	0	−1	0	1	0	−1
8	1	−0.5	−0.5	1	−0.5	−0.5	1
10	1	−0.866	0.5	0	−0.5	0.866	−1
12	1	−1	1	−1	1	−1	1

0.86183	6.04209	1.79063	1.44892	6.32142	1.64511	0.88933

The last line gives the resultant product of the top row with the successive rows of the matrix. At once we notice two well-pronounced maxima which belong to $k = 2 = 60°$ and $k = 8 = 150°$, thus indicating the existence of two complex roots near the unit circle. We will now refine our scanning by reducing the fundamental interval to 15°. It is now the "difference" row of the table (6) which comes into action. The previous multiplication matrix is replaced by a new preassigned matrix. We need not carry out the complete multiplication scheme but only the multiplications which will give us the two neighbors of the maxima found before. Hence in our problem the two products for $k = 1,3$ will be needed, and likewise the two products for $k = 7,9$. The numerical scheme appears as follows:

	3	0.21453	−0.28249	0.36400	−2.85849	−0.16822
1	1	0.9659	0.866	0.7071	0.5	0.2588
3	1	0.7071	0	−0.7071	−1	−0.7071
5	1	0.2588	−0.866	−0.7071	0.5	0.9659
7	1	−0.2588	−0.866	0.7071	0.5	−0.9659
9	1	−0.7071	0	0.7071	−1	0.7071
11	1	−0.9659	0.866	−0.7071	0.5	−0.2588
	1.74718	5.87175		2.17974	5.84523	

The two maxima are treated quite independently. First we concentrate on the maximum at $k = 2$ and its two neighbors. The three consecutive ordinates are:

$$y_0 = 1.74718, \qquad y_1 = 6.04209, \qquad y_2 = 5.87175$$

In order to interpolate for the position of the correct maximum, we lay a parabola of second order through these three ordinates and determine the point where the maximum occurs. The solution of this simple algebraic problem gives the following result. The abscissa of the maximum occurs at

$$x_m = \frac{1}{2} \frac{y_2 - y_0}{2y_1 - (y_0 + y_2)} \tag{1-17.7}$$

while the associated maximum ordinate becomes

$$y_m = y_1 + \tfrac{1}{4} x_m (y_2 - y_0) \tag{1-17.8}$$

In our numerical problem

$$x_m = \frac{1}{2} \frac{4.12457}{4.46525} = 0.46185$$

$$y_m = 6.04209 + \tfrac{1}{4} 0.46185 \cdot 4.12457 = 6.51832$$

Formula (7) assumes the labeling $-1, 0, 1$ for the abscissas of the 3 consecutive data points. In actual fact the distance between two neighboring abscissas is $15°$, and the middle ordinate belongs to the angle $k = 2 = 30°$. Hence the angle value of x_m becomes

$$\theta_m = 30° + 0.46185 \cdot 15° = 36°.928$$

The same calculation performed for the second maximum at $k = 8$ gives the following results.

$$y_0 = 2.17974, \qquad y_1 = 6.32142, \qquad y_2 = 5.84523$$

$$x_m = \frac{1}{2}\frac{3.66549}{4.61787} = 0.39688, \qquad \theta_m = 120° + 0.39688 \cdot 15°$$

$$= 125°.95$$

$$y_m = 6.32142 + \tfrac{1}{4} \, 0.39688 \cdot 3.66549 = 6.68511$$

The maximum ordinate y_m can be used for closer determination of the absolute value r of the root. If the assumption $r = 1$ were correct, y_m would have to be

$$\tfrac{1}{2} + \sigma_1 + \sigma_2 + \cdots + \sigma_{11} = 7.0669$$

But more generally for arbitrary values of r we obtain

$$y_m = \tfrac{1}{2} + \sigma_1 r + \sigma_2 r^2 + \cdots + \sigma_{11} r^{11} = A(r) \qquad (1\text{-}17.9)$$

It is not difficult to prepare a numerical table of the function $A(r)$ which proceeds in sufficiently close intervals of r, for the range between $r = 0.8$ and 1.2. The proper r value can then be ascertained by linear interpolation. This table is given as Table I of the Appendix. With the help of the tabular values we find that the y_m of the first maximum is associated with

$$r = 0.9802$$

while the y_m of the second maximum is associated with

$$r = 0.9864$$

We go back to the original variable x, which according to (1) requires multiplication by $r_0 = 1.3$. We thus obtain the first pair of complex roots of the equation (15.4) in the form

$$x = 0.9802 \cdot 1.3 \, (\cos 36°.93 \pm i \sin 36°.93) = 1.018 \pm 0.765i$$

which compares favorably with the correct root $1 \pm 0.75i$, the error being only 1%.

The second pair of complex roots comes out as follows:

$$x = 0.9864 \cdot 1.3 \, (\cos 125°.95 \pm i \sin 125°.95) = -0.753 \pm 1.038i$$

while the correct value is $-0.75 \pm i$. The error here is 3%.

18. Transformation by reciprocal radii. We will now consider a second and numerically even quicker procedure for locating the roots along the maximum circle. We start out exactly as before by estimating the radius of the maximum circle according to the peak value method of § 16. Then we perform the transformation (17.1). This transformation is now applied to the algebraic equation itself and not to the power sums S_m. In fact, the power sums will not be used beyond the determination of r_0.

In § 16 we dealt with the sixtic equation (16.4) and found $r_0 = 1.29$, which could be rounded off to $r_0 = 1.3$. Dividing the successive coefficients by the successive powers of 1.3, we obtain the new equation

$$u^6 - 0.38461u^5 + 0.22189u^4 - 0.41429u^3$$

$$+ 0.86575u^2 - 0.05260u + 0.12645 = 0 \qquad (1\text{-}18.1)$$

and we know in advance that at least one pair of complex roots will lie near the unit circle $|u| = 1$.

At this point we introduce a remarkable transformation, widely used in the theory of analytical functions. The simple reciprocal transformation

$$\bar{z} = 1/z \qquad (1\text{-}18.2)$$

considered in the complex plane, has remarkable properties. It gives a "conformal mapping" of the plane on itself, transforming the outside of the unit circle to the inside, and vice versa. The outstanding property of this transformation is that *all circles remain circles*, although with modified radius. Since straight lines can be conceived as circles with an infinite radius, the straight lines are also transformed into circles.

For our purposes a shifting of the origin of our variables will be of advantage. We consider the transformation

$$u = \frac{1-v}{1+v}, \qquad v = \frac{1-u}{1+u} \qquad (1\text{-}18.3)$$

If the complex variable u moves along the unit circle, the corresponding variable v moves along the imaginary axis. Hence a root along the unit circle is transformed into a purely imaginary root.

The transformation (3) can easily be accomplished if $f(u)$ is a polynomial in u of the order n. We then get, apart from the factor

$(1 + v)^{-n}$ which is of no interest for our present purposes, a new polynomial in v. The coefficients of this polynomial are in a linear relation to the original coefficients a_k and are obtainable by multiplying the a_k by a definite matrix B_n of $n + 1$ rows and $n + 1$ columns.

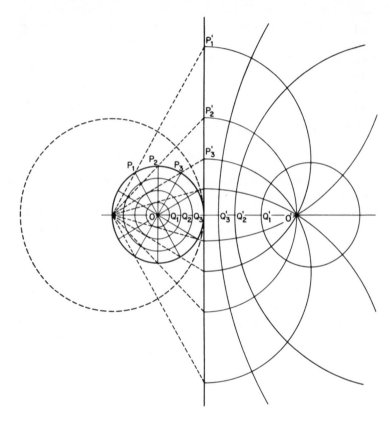

This matrix can be pretabulated for every n (cf. Table II of the Appendix); it is composed of *integers*. For example for $n = 1, 2,$ and 3 we get

$$\begin{bmatrix} -1 & 1 \\ 1 & 1 \end{bmatrix} \qquad \begin{bmatrix} 1 & -1 & 1 \\ -2 & 0 & 2 \\ 1 & 1 & 1 \end{bmatrix} \qquad \begin{bmatrix} -1 & 1 & -1 & 1 \\ 3 & -1 & -1 & 3 \\ -3 & -1 & 1 & 3 \\ 1 & 1 & 1 & 1 \end{bmatrix}$$

These matrices can easily be generated in succession, because of the recurrence relation

$$b_{ik}^{(n+1)} = b_{(i-1)k}^{(n)} - b_{ik}^{(n)} \qquad (1\text{-}18.4)$$

For example, the element (21) (i.e., second row, first column) of the third matrix is obtainable by taking the element (11) (first row, first column) of the second matrix and subtracting the element (21) (second row, first column): $1 - (-2) = 3$. The last column to be added always repeats the first column, but with uniformly positive signs.

An interesting property of the B matrices is that their square is always proportional to the unit matrix:

$$B_n^2 = 2^n \cdot I \qquad (1\text{-}18.5)$$

This means that multiplying any row of the B matrix by any column of the same matrix gives 2^n in the diagonal and 0 everywhere else. For example, in the third matrix the product of row 2 and column 2 gives

$$3 + 1 + 1 + 3 = 8$$

while the product of row 2 and column 3 gives

$$-3 + 1 - 1 + 3 = 0$$

We multiply the coefficients of our numerical problem (1)

$$1, \quad -0.38461, \quad 0.22189, \quad -0.41429, \quad 0.96575, \quad -0.05260, \quad 0.12645$$

by the matrix $n = 6$ of Table II. The multiplication occurs in row by row fashion. We thus get a transformed set of coefficients which belong to a polynomial of sixth order in v

$$3.06379v^6 - 5.28162v^5 + 16.75769v^4 - 20.04644v^3$$
$$+ 14.86053v^2 - 2.62554v + 1.36439 = 0 \qquad (1\text{-}18.6)$$

(Numerical check: sum of new coefficients $= 2^n$ times old absolute term. This check is absolute, since the multiplications by integers do not introduce any rounding errors. In our example: $\Sigma\, a_i' = 8.09280 = 64 \cdot 0.12645$).

19. Roots near the imaginary axis. The root which was near the unit circle in the variable u is now near the imaginary axis in the new variable v. But if the root of an algebraic equation with real coefficients is near the imaginary axis, such a root can easily be located. Assuming first that the root lies *exactly* on the imaginary axis, the given equation splits into two equations, since the even powers alone give one equation and the odd powers alone give another equation. We will put

$$v^2 = -\lambda \qquad (1\text{-}19.1)$$

and obtain for λ two algebraic equations whose order is not more than *one-half* of the previous order. Moreover, in the new problem we search for *real* roots only. Our task is thus greatly simplified.

The imaginary part of the equation provides a *linear* equation in the case of a quartic, and even in the case of a sixtic, the equation in λ is still only *quadratic* and can be solved by the standard formula. Only positive roots are of interest to us.

The imaginary part (odd powers) of the numerical example (1-18.6) gives for λ,

$$5.28162\lambda^2 - 20.04644\lambda + 2.62554 = 0$$

This equation has two positive roots:

$$\lambda_1 = 0.13583, \qquad \lambda_2 = 3.65967$$

Which of these two roots will be of interest to us cannot be told in advance. We have to substitute this λ in the real part of the equation and see whether the remainder is small or not. In our case both λ_i give small remainders, thus indicating that *two* separate pairs of complex roots are near the imaginary axis.

An added advantage of this method is that the correction of the preliminary root by Newton's scheme (cf. § 5) becomes greatly simplified. The "movable strip" used in evaluating $f(x_0)$, $f'(x_0)$ and $\frac{1}{2}f''(x_0)$ is now composed of the numbers

$$-\lambda, \quad 0, \quad 1$$

in vertical arrangement. In view of the middle zero, the number of multiplications is halved, and the calculating load considerably reduced.

We display the substitution scheme for the case of the smaller root λ_1.

-0.13583	*Quot.*	$f(v)$		*Quot.*	$f'(v)$	
0						
1	3.06379	3.06379				
	-5.28162	-5.28162		18.38274	18.38274	
	16.34153	16.75769		-26.40810	-26.40810	
	-19.32904	-20.04644		64.40810	67.03075	
	12.64086	14.86053	Remainder	-56.55231	-60.13932	Remainder
	0	-2.62554	-0.00008	0	29.72106	20.95543
	0	1.36439	-0.35262	0	-2.62554	5.05596

	$\frac{1}{2}f''(v)$		
45.95685	45.95685		
-52.81620	-52.81620		
94.30380	100.54612		
0	-60.13932	-52.96530	
0	14.86053	2.05124	

$v_0 = \sqrt{0.13583i} = 0.36855i$
$f(v_0) = -0.3526$
$f'(v_0) = 7.7231i + 5.0560$
$\frac{1}{2}f''(v_0) = -19.5204i + 2.0512$

$$\frac{1}{h} = \frac{5.0560 + 7.7231i}{0.3526} + \frac{2.0512 - 19.5204i}{5.0560 + 7.7231i}$$

$$= (14.338 + 21.902i) - (1.719 - 1.344i) = 12.619 + 20.558i$$

$$h = 0.02169 - 0.03533i$$

$$v = v_0 + h = 0.02169 + 0.33322i$$

Transforming back to the original variable u we obtain

$$u = \frac{1-v}{1+v} = -1 + \frac{2}{1+v} = -1 + \frac{2}{1.02169 + 0.33322i}$$
$$= 0.76933 - 0.57706i$$

Finally we return to the original x by multiplying by 1.3.

$$x = (0.76933 - 0.57706i)1.3 = 1.0001 - 0.7502i$$

$$[\text{correct}: \ 1 - 0.75i]$$

The correction scheme for the larger root is quite similar. It is advisable, however, to substitute the *reciprocal* of λ:

$$\bar{\lambda} = 0.27325, \qquad \bar{v}_0 = \sqrt{0.27325i} = 0.52213i$$

into the inverted polynomial

$$1.36439 - 2.62554\bar{v} + 14.86053\bar{v}^2 - \cdots + 3.06379\bar{v}^6$$

We thus obtain, if we perform the calculations,

$$\bar{h} = 0.02448 - 0.02291i$$

$$\bar{v} = 0.02448 + 0.49982i$$

$$u = \frac{\bar{v} - 1}{\bar{v} + 1} = 1 - \frac{2}{\bar{v} + 1} = 1 - \frac{2}{1.02448 + 0.49982i}$$

$$= -0.57688 + 0.76932i$$

$$x = 1.3(-0.57688 + 0.76932i)$$

$$= -0.7499 + 1.0001i$$

[correct: $-0.75 + i$]

20. Multiple roots. Whenever the remainder $f(x_0)$ is small, Newton's scheme will give a powerful correction, provided, however, that $f'(x_0)$ is not too small at the same time. Special precautions are demanded, however, if it so happens that not only $f(x_0)$ but also $f'(x_0)$ is small so that the ratio $f(x_0)/f'(x_0)$ is not small. In such a case, the smallness of $f'(x_0)$ indicates that we are near to a double root, or to two roots very close together. The separation of these roots cannot be obtained by a purely linear interpolation. We have to evaluate $f(x_0), f'(x_0)$, and $f''(x_0)$ and solve the quadratic equation

$$\tfrac{1}{2}f''(x_0)h^2 + f'(x_0)h + f(x_0) = 0 \qquad (1\text{-}20.1)$$

by the standard formula. The two roots, if added to x_0, will approximate the two roots closely. The x_1 thus obtained can now be used for further refinement, by the regular technique described in § 5. Going through the routine independently for both roots, we can obtain each root with any accuracy we want, after the first approximation brought us so close to each respective root that the damaging influence of the other root is eliminated.

A similar procedure is demanded if it so happens that not only $f(x_0)$ and $f'(x_0)$ but even $f''(x)$ is so small that the h calculated from (1) is not small. In that case we have to add $f'''(x_0)$ and solve the cubic

$$\tfrac{1}{6}f'''(x_0)h^3 + \tfrac{1}{2}f''(x_0)h^2 + f'(x_0)h + f(x_0) = 0 \qquad (1\text{-}20.2)$$

in order to separate the three roots, which are now very closely together. Subsequent application of Newton's scheme to every single root thus obtained will finally correct each root to the desired accuracy.

We see that the separation of closely bunched multiple roots is generally a cumbersome task which cannot be accomplished without a great deal of computational work.

21. Algebraic equations with complex coefficients. Let $F_n(x)$ be a polynomial with complex coefficients. The roots of such a polynomial are all complex but they do not appear in pairs of the form $\alpha \pm \beta i$. The method of the power sums is still available, but the evaluation of the higher moments is greatly slowed down by the complex character of the coefficients. The product of two complex numbers requires four multiplications, and thus the time of obtaining the radius of the maximum circle is greatly increased.

It is frequently preferable to avoid the operation with complex numbers altogether by multiplying the polynomial $F_n(x)$ by its complex conjugate $F_n^*(x)$, where the symbol * indicates that every i in $F_n(x)$ is changed to $-i$. We form the polynomial of the order $2n$.

$$f_{2n}(x) = F_n(x)\, F_n^*(x) \qquad (1\text{-}21.1)$$

which has real coefficients. This introduces n extraneous roots, since every $\alpha + i\beta$ is now complemented by the corresponding $\alpha - i\beta$. But we have the great advantage that the operation with complex coefficients is avoided. We operate solely with the real $f_{2n}(x)$ and obtain its n pairs of conjugate complex roots $x_k = \alpha_k \pm i\beta_k$.

Now the decision has to be made whether the sign of the imaginary part is plus or minus. We do that by substituting x_k into the original $F_n(x)$. The movable strip has still real coefficients. Hence we can take the real part of $F_n(x)$ alone and the imaginary part of $F_n(x)$ alone and go through the procedure independently, avoiding any complex numbers. The remainders combined appear in the form

$$(a + ib)x_k + (c + id) \qquad (1\text{-}21.2)$$

where the coefficients a, b, c, d are not small. By putting this remainder equal to zero we obtain a root which will be very near to either $\alpha_k + i\beta_k$ or $\alpha_k - i\beta_k$. We thus decide the right sign and get at the same time an additional check on the accuracy of the root.

22. Stability analysis. Whenever a mechanical or electrical system is in a state of equilibrium, small disturbing forces will tend to modify this equilibrium. Since the potential energy in the neighborhood of a state of equilibrium is always a positive definite form of the displacements, the system will perform small oscillations around the state of equilibrium. These oscillations can be conceived as a linear superposition of n "normal oscillations." The frequencies of these oscillations are determined by solving the "characteristic equation" associated with the dynamic system. While frictional forces tend to cause damping of these oscillations and thus return the system in its state of equilibrium, the situation is different if an energy source is present in the system which can counteract the effect of frictional forces and even build up the amplitudes to ever-increasing amounts. The general problem of stability thus involves an algebraic equation whose complex roots may reveal either positive or negative damping, depending on the sign of the real parts of the roots. The stability of the system demands that *all* the roots of the characteristic equation shall lie in the negative half plane of the complex plane. One single root with a positive real part is sufficient to throw the system out of gear.

Under these circumstances it is of interest to decide whether a given algebraic equation possesses roots in the positive half plane or not, without going through a detailed quantitative analysis of the roots. The English physicist, E. J. Routh discovered an elegant method for testing an algebraic equation as to its stability, without any calculation of the roots.[1] However, occasionally we want to go further and actually locate the unstable root if the system is found to be unstable. For this purpose the transformation by reciprocal radii is again of great advantage. Let us replace x by $x = -u$ and again transform to the variable v by the transformation (1-18.3). This transformation maps the entire left complex half plane $R(x) < 0$ into the unit circle. Hence the condition that all the roots x_i are in the negative half plane is equivalent to the condition that all the roots u_i are in absolute value smaller than 1. Hence the radius r_m of the maximum circle must come out as <1 in order to guarantee the stability of the system. Since the method of synthetic division makes the power sums and also r_m easily accessible, we obtain a stability

---

[1] For a description of Routh's method see {2}, Vol. I, p. 129; also Ref. [3] of Chapter II, p. 154.

criterion with relatively little labor, together with a localization of the unstable root if r_m turns out to be greater than 1.

In transforming the original equation it is important that the equation shall be given in a proper scale. We normalize the scale by demanding that the absolute term of the equation shall be nearly 1 (assuming that the coefficient of x^n is 1).

Example 1. The roots of the previously treated quartic had negative real parts, the polynomial is thus of the "stable" kind. Let us now see how we can establish this fact without going through the evaluation of the roots. The original equation is

$$x^4 + 760x^3 + 23.34x^2 + 38.44x + 37.40 = 0 \qquad (1\text{-}22.1)$$

We normalize the absolute term and change x to u by the transformation

$$x = -\frac{1}{0.4} u \qquad (1\text{-}22.2)$$

which means that the successive coefficients of (1) are multiplied by the powers of 0.4, and the sign of the odd powers is changed to the opposite.

$$u^4 - 3.04u^3 + 3.7344u^2 - 2.4602u + 0.9574 = 0 \qquad (1\text{-}22.3)$$

The new absolute term is nearly 1. Multiplication by the B_4 matrix gives the new equation in v.

$$11.1920v^4 - 1.3300v^3 + 4.2756v^2 + 0.9892v + 0.1916 = 0$$
$$(1\text{-}22.4)$$

The stability of this equation can be established by mere inspection, without further computation. Assuming that v is an arbitrary complex number, we can conceive the terms of an algebraic equation as vectors of the complex plane. The fact that a sum of vectors is zero means geometrically that the vector polygon closes, i.e., the end point returns to the origin. If now we separate the first vector from the rest, we have the proposition that the straight line from A to B is complemented by a polygon which starts from B and returns to A. We see from the minimum property of a straight line that the total length of the path from B back to A can never be less than the path from A to B. Therefore, if we find that for a certain v,

$$\left| a_1 v^{n-1} \right| + \left| a_2 v^{n-2} \right| + \cdots < \left| a_n v^n \right| \qquad (1\text{-}22.5)$$

we can be sure that equation (22.4) cannot hold. Now if we divide by v^n and write $\bar{v}$ for $1/v$, we see at once that

$$11.192 > |1.3300|\,|\bar{v}| + |4.2756|\,|\bar{v}|^2 + |0.9892|\,|\bar{v}|^3 + 0.1916|\bar{v}|^4$$

$$(1\text{--}22.6)$$

for any $|\bar{v}| \leq 1$, that is, for any $|v| \geq 1$. Generally we can say that if the coefficient of v^n is larger than the absolute sum of all the remaining coefficients, the equation in v can have no root outside or on the unit circle, which means that the system is stable. This simple criterion (which is sufficient *but not necessary* for stability) is satisfied in our case and thus the stability of the equation established.

Example 2. The following equation is also of the stable type.

$$x^6 + 4.24x^5 + 8.50x^4 + 10.27x^3 + 8.09x^2 + 3.96x + 0.98 = 0$$

The absolute term is here already properly normalized, and we can immediately proceed to the transformation to v, after changing the sign of the odd powers. Multiplication by the matrix B_6 gives

$$37.04v^6 - 2.06v^5 + 23.30v^4 + 1.24v^3 + 2.92v^2 + 0.18v + 0.10 = 0$$

Once more we see that the coefficient of v^6 outdistances the sum of the remaining coefficients, which shows directly the stability of the system.

Example 3. The previous polynomial becomes unstable if we multiply it by the root factor

$$x^2 - 0.6x + 1$$

This gives the eighth-order polynomial,

$$x^8 + 3.64x^7 + 6.956x^6 + 9.41x^5 + 10.428x^4$$
$$+ 9.376x^3 + 6.694x^2 + 3.372 + 0.98$$

The transformation to the v plane yields

$$51.856v^8 - 2.884v^7 + 128.924v^6 - 3.620v^5 + 64.668v^4$$
$$+ 3.476v^3 + 7.732v^2 + 0.468v + 0.260 = 0$$

The large coefficients of v^4, v^6, v^8 allow an immediate localization of the unstable root. Neglecting all the other terms we get the following quadratic in $v^2 = -y$.

$$51.86y^2 - 128.92y + 64.67 = 0$$

which gives

$$y = 1.243 \pm 0.546$$

The smaller root is <1 and thus stable. The unstable root yields

$$v = \sqrt{-1.79} = \pm 1.33i$$

and going back to the original variable x,

$$-x = \frac{1-v}{1+v} = -1 + \frac{2}{1+v} = -0.28 \pm 0.96i$$

$$x = 0.28 \pm 0.96i$$

This is a very close estimate of the true value of the unstable root, which is $x = 0.3 \pm 0.95i$

Example 4. Conditions are not always so favorable for spotting unstable roots. In the example given by Doherty-Keller, {2}, I, p. 129) the following sixtic shall be investigated for stability.

$$x^6 + 68.6x^5 + 785x^4 + 7213x^3 + 50700x^2 + 8200x + 435000 = 0$$

For an approximate normalization of the absolute term we put

$$x = -10u$$

and obtain

$$u^6 - 6.86u^5 + 7.85u^4 - 7.213u^3 + 5.07u^2 - 0.082u + 0.435 = 0$$

The transformation to v gives

$$28.51v^6 - 36.062v^5 + 21.676v^4 - 0.18v^3$$
$$- 4.466v^2 + 18.162v + 0.2 = 0$$

The previous criterion for stability fails to hold, since the absolute sum of the last six coefficients outweigh the first coefficient. Hence the stable or unstable nature of the equation is not yet decided, and we have to evaluate a few of the power sums. Dividing by the largest coefficient and truncating to two decimal places, we obtain

$$v^6 - 1.26v^5 + 0.76v^4 - 0.16v^2 + 0.64v = 0$$

The equation is thus reducible to the fifth order, since the root $v = 0$

(strictly speaking $v = $ very small) is of no interest. Synthetic division by the movable strip method yields the successive power sums:

$k = $	$S_k = $
0	5
1	1.26
2	0.0676
3	−0.87242
⋮	⋮
10	3.7941
11	5.5555
12	6.0846

As soon as a power sum larger than 5 is reached, we can stop, since this clearly establishes the instability of the system. Indeed, the successive powers constantly diminish the absolute value of the roots if all the roots lie inside the unit circle. Hence it is impossible that the sum $v_1^k + v_2^k + \cdots + v_n^k$ shall ever go beyond n if all the $|v_i|$ are smaller than 1. If our goal is merely to show the instability of the system, this goal is already accomplished by having obtained $S_{11} > 5$. But if we want to locate the unstable root, we can evaluate a few more S_m and obtain the unstable root by solving the quadratic equation (15.14).

This method of deciding the stability of a system will obviously fail if the critical r is very nearly 1. But this means that the root in the original variable u is very near to the imaginary axis. Since we have seen that roots near to the imaginary axis can easily be located (by looking for the real roots of a redundant system of two equations), a very slightly unstable system can be spotted anyway and needs no further consideration.

Bibliographical References

[1]   Cf. Ref. {11}, Chapter IX; [2] Cf. Ref. {14}, Chapter VI.

[3]   Bôcher, M., *Introduction to Higher Algebra* (Macmillan, New York, 1938).

Article:

[4]   Fry, Th. C., "Some Numerical Methods of Locating Roots of Polynomials," *Quart. Applied Math.*, **3**, 89 (1945).

II

MATRICES
AND EIGENVALUE
PROBLEMS

1. Historical survey. The solution of simultaneous linear algebraic equations has a very old history. The Hindus, who are the inventors of the decimal system and of the algebraic method, solved simultaneous linear algebraic equations from the sixth century on. In the early stages of the development only the unknowns of the problem were denoted by symbols, while the given coefficients of the linear system came in with their actual numerical values. The complete symbolization of algebra came in the time of the great French algebraist François Viète (1540–1603). This symbolization made it possible to develop general methods for the solution of systems of linear equations. The great philosopher and mathematician Leibniz invented the notion of a "determinant" and introduced a notation which is essentially the same we use today. He showed how the unknowns of any consistent linear algebraic system are obtainable by forming the ratio of two determinants.

During the nineteenth century the operational viewpoint of mathematics came into the focus of interest. In contrast to arithmetic, where only the final numerical answer is of importance, the interest of algebra centers around the *operations* involved in the evaluation of the results. Our interest is not what the final numerical answer will be but by what operations that answer will be obtained. Hence in a purely algebraic problem parentheses have to be removed, fractions cleared, multiplications performed, etc., in purely *symbolic* form. We do not care what the numerical value will be of the symbols which stand for quantities. Our conclusions are based on the fact that all numbers satisfy certain general laws, called the "postulates of algebra." This makes it possible for us to draw valid conclusions without knowing what the actual numerical parameters of the problem will be. We need not carry out the complicated algebraic operations with the help of the given numbers but with the help of *symbols* which imitate the behavior of numbers. The actual numbers enter only the final formula which tells us what operations are to be performed with the given numbers in order to get the answer.

The fundamental operations of arithmetic—addition and subtraction, multiplication and division—are first performed on simple numbers only. We start with the integers and gradually extend the realm of numbers by introducing the negative numbers and the fractions. In algebra two further fundamental operations are added, viz., raising to a power and taking the root. These operations give rise to a further enlargement of the number system by the discovery of the imaginary and complex numbers of the form $a + bi$. All these numbers satisfy the basic postulates of algebra, and thus the algebraic operations with all these numbers are equally satisfactory. But in 1859 the English mathematician Cayley greatly extended the realm of algebra by showing that a "matrix," although composed of a *system* of quantities, can be conceived as one single algebraic operator which satisfies all the postulates of ordinary algebra, with the single exception of the commutative law of multiplication. The algebra of matrices thus demonstrates the operation of a noncommutative algebra. The matrix algebra differs, however, in one further feature from ordinary algebra. It satisfies its own characteristic equation and thus leads to a polynomial identity which has no analogy in the algebra of ordinary real or complex numbers. This phase of the theory was developed by Sylvester (1851)—originator of the term "latent roots"—and later by Weierstrass (1868). The most complete algebraic theory of the characteristic equation was finally given by Frobenius (1879).

Fredholm (1900) introduced the notion of matrices of infinite order and extended the algebraic theory of the characteristic equation to the case of infinitely many variables. This became the foundation of the theory of orthogonal function systems (cf. V, 16) and the geometrical treatment of linear differential and integral operators.

Fields of application. The characteristic equation with the associated eigenvalues and eigenvectors plays a fundamental part in the theory of vibrations, whether these vibrations be of a mechanical or electric, macroscopic or microscopic kind. The elastic vibrations of a bridge or any other solid structure, the flutter vibrations of an airplane wing, the transient oscillations of an electric network, the wave-mechanical vibrations of atoms and molecules are all examples for the operation of the characteristic equation. The buckling of an elastic structure is likewise an eigenvalue problem, since buckling occurs if the smallest vibrational frequency of the elastic structure reaches the value zero.

In Schroedinger's wave mechanics the atomic and molecular oscillations of particles play a fundamental role in the description of the physical and chemical properties of matter. The eigenvalues of Schroedinger's wave equation are directly proportional to the energy value of the various quantum states.

In boundary value problems, expansion of the solution into the orthogonal functions associated with the given differential operator provides a powerful tool in the problem of constructing the solution in terms of the given boundary values. Since the Green's function associated with the given differential operator is frequently not available in explicit form, the bilinear expansion gives an indirect method of generating the Green's function.

The theory of linear differential and integral operators gains greatly in clarity and conciseness by introducing the associated eigenfunctions as an auxiliary frame of reference. This means in geometrical interpretation that the associated quadratic surface is transformed to its principal axes. The principal axis transformation of quadratic forms becomes thus a fundamental connecting link between widely different branches of mathematics. The solution of systems of linear algebraic equations, matrix calculus, the general theory of linear differential and integral operators, can all be conceived as various formulations of the same fundamental problem, viz., the principal axis transformation of a quadratic surface in a Euclidean space of either finite or infinitely many dimensions.

2. Vectors and tensors. Vectors are different from scalars by having magnitude *and directions*. But vectors are not the only quantities which go beyond the realm of scalars. A vector can be analyzed in a certain frame of reference and given with the help of its components. Thus we may say that a vector a has in three dimensions the components a_1, a_2, a_3, and we may write

$$a = a_1 i + a_2 j + a_3 k$$

but we may also write

$$a = (a_1, a_2, a_3)$$

which indicates that the three numbers a_1, a_2, a_3, called the "components" of the vector, uniquely characterize the vector a. During the development of the natural sciences in the nineteenth century the discovery was made that vectors represent only a very special class of directed quantities, called "tensors." Vectors are the simplest class of tensors, namely a set of quantities which can be characterized by one single subscript. But generally the number of indices may be two, three, or more: a_{ijk} ....

Of greatest importance among these tensors are the tensors of *second order*. A tensor of second order is characterized by two subscripts: a_{ik}. The components of such a tensor can be arranged in a two-dimensional scheme, in such a way that the first subscript gives the *row* and the second subscript the *column* to which a certain component belongs.

$$\begin{pmatrix} a_{11} & a_{12} & a_{13} & \cdots & a_{1n} \\ a_{21} & a_{22} & a_{23} & \cdots & a_{2n} \\ \vdots & \vdots & \vdots & & \vdots \\ a_{n1} & a_{n2} & a_{n3} & \cdots & a_{nn} \end{pmatrix}$$

By putting brackets around the scheme we indicate that we want to consider the *entire assembly* of these components as one single integrated unit. None of these components have existence in themselves. It is the entire assembly of all the components which constitutes a tensor, in a similar way as the entire assembly of components constitutes a vector.

3. Matrices as algebraic quantities. Cayley made the fundamental discovery that such an array of numbers can be conceived as one single algebraic quantity A, with which certain algebraic operations can be performed. We can add, subtract, multiply, and divide with such an array of numbers, just as if it were a single algebraic quantity. We will call such an array of numbers a "matrix" and denote it by

$$A = \begin{pmatrix} a_{11} & a_{12} & \cdots & a_{1n} \\ a_{21} & a_{22} & \cdots & a_{2n} \\ \vdots & \vdots & & \vdots \\ a_{n1} & a_{n2} & \cdots & a_{nn} \end{pmatrix} ; \quad also \quad A = \begin{bmatrix} a_{11} & a_{12} & \cdots & a_{1n} \\ a_{21} & a_{22} & \cdots & a_{2n} \\ \vdots & \vdots & & \vdots \\ a_{n1} & a_{n2} & \cdots & a_{nn} \end{bmatrix}$$

The quantities a_{ik} are called the "elements" of the matrix A. In contrast to the "determinant"

$$|A| = \begin{vmatrix} a_{11} & a_{12} & \cdots & a_{1n} \\ a_{21} & a_{22} & \cdots & a_{2n} \\ \vdots & \vdots & & \vdots \\ a_{n1} & a_{n2} & \cdots & a_{nn} \end{vmatrix} \qquad (2\text{-}3.1)$$

we do not think of a matrix A as a single number, obtained by a certain process out of the elements a_{ik} of the matrix. The matrix A refers merely to the *entire assembly* of the numbers a_{ik}, but arranged in a very definite manner in rows and columns. Every element a_{ik} has its definite place in the array and cannot be replaced by other elements. Hence a matrix is defined as an assembly of numbers, arrayed in a strictly prescribed geometric pattern.

The fact that we can operate with a matrix algebraically means that the fundamental operations of algebra, viz., addition, subtraction, multiplication, and division, raising to a power, and taking the root can be extended to the realm of matrices.

The two fundamental operations from which everything else is derivable are addition and multiplication. Subtraction is merely the inverse of addition, division merely the inverse of multiplication. Hence it is enough to know how to add and how to multiply matrices.

The addition of matrices is a simple operation. Given the two matrices

$$A = \begin{bmatrix} a_{11} & a_{12} & \cdots & a_{1n} \\ a_{21} & a_{22} & \cdots & a_{2n} \\ \vdots & \vdots & & \vdots \\ a_{n1} & a_{n2} & \cdots & b_{nn} \end{bmatrix} ; \quad B = \begin{bmatrix} b_{11} & b_{12} & \cdots & b_{1n} \\ b_{21} & b_{22} & \cdots & b_{2n} \\ \vdots & \vdots & & \vdots \\ b_{n1} & b_{n2} & \cdots & b_{nn} \end{bmatrix} \qquad (2\text{-}3.2)$$

we can add corresponding elements and obtain the new matrix

$$C = \begin{bmatrix} a_{11} + b_{11} & a_{12} + b_{12} & \cdots & a_{1n} + b_{1n} \\ a_{21} + b_{21} & a_{22} + b_{22} & \cdots & a_{2n} + b_{2n} \\ \vdots & \vdots & & \vdots \\ a_{n1} + b_{n1} & a_{n2} + b_{n2} & \cdots & a_{nn} + b_{nn} \end{bmatrix} \qquad (2\text{-}3.3)$$

We define this new matrix C as the sum of the matrices A and B.

$$C = A + B \qquad (2\text{-}3.4)$$

More complicated is the operation of *multiplication*. The clue to this operation is given to us by the fact that matrices appear primarily in connection with *linear equations*. A general system of simultaneous linear algebraic equations can be written in the following systematic manner.

$$\begin{aligned} a_{11}x_1 + a_{12}x_1 + a_{23}x_3 &+ \cdots & a_{1n}x_n &= b_1 \\ a_{21}x_1 + a_{22}x_2 + a_{23}x_3 &+ \cdots & a_{2n}x_n &= b_2 \\ \vdots \qquad \vdots \qquad \vdots & & \vdots & \quad \vdots \\ a_{n1}x_1 + a_{n2}x_2 + a_{n3}x_3 &+ \cdots & a_{nn}x_n &= b_n \end{aligned} \qquad (2\text{-}3.5)$$

This systematic way of writing a system of algebraic equations was absolutely essential for the development of matrix calculus. In earlier days the elements of a matrix were denoted by different letters: a, b, c, $\cdots$, without the use of subscripts. Similarly the unknowns of the problem fell apart into a system of disconnected quantities x, y, z, $\cdots$. The ingenious symbolism of the subscripts in both matrix elements and unknowns was absolutely essential for development of the theory of matrices, because it made it evident that the elements of a matrix are in reality the components of a single tensor of second order, while the unknowns of a set of linear equations are to be conceived as the components of one single vector x. Similarly the given right side of the equations has to be conceived as the components of another vector b. Accordingly we will write the given set of equations in the form

$$Ax = b \qquad (2\text{-}3.6)$$

and conceive the left side as the product of the matrix A and the vector x. Hence we know already what the operation "multiplication" means if applied to a matrix and a vector.

This equation brings out an important feature of matrices. Let us write the previous equation in reversed sequence:

$$b = Ax \tag{2-3.7}$$

Considering x as a given vector, we can say that the matrix A associates with a given vector x a new vector b. We can also say that the matrix A "transforms" the vector x into a new vector b. For example, we may rotate the vector x around a certain axis, by a certain angle. Such a rotation would be an example of multiplication by a certain matrix. But the matrix associated with a rigid rotation in space is a very *special* matrix. A general matrix transforms the vector x into a new position b, but this transformation cannot be pictured as a mere rotation.

Let us now start with a vector u and transform it into the vector v, with the help of the matrix A. Then we continue the process and transform v into a new vector w, with the help of another matrix B

$$v = Au, \qquad w = Bv \tag{2-3.8}$$

Now w was generated out of v, but v itself was generated out of u. Hence we can say that w was generated out of u, leaving out the intermediate vector v. If we substitute for v in the previous equation, we obtain

$$w = BAu \tag{2-3.9}$$

Hence if we consider the direct transition from u to w by writing

$$w = Cu \tag{2-3.10}$$

we see that the matrix C must be conceived as

$$C = BA \tag{2-3.11}$$

This gives the rule by which the product of two matrices B and A is obtained. If we perform the substitution, we find that an arbitrary element c_{ik} of the product matrix C must be constructed as follows. Select the ith row of the first factor B and the kth column of the second factor A. Multiply these two together. "Multiplication" here means to form the product of corresponding elements and take the sum. For example,

$$v_1 = 3u_1 - 2u_2, \qquad w_1 = v_1 + v_2$$

$$v_2 = -u_1 + 4u_2, \qquad w_2 = 2v_1 - 5v_2$$

Then

$$w_1 = 3u_1 - 2u_2 - u_1 + 4u_2 = 2u_1 + 2u_2$$

$$w_2 = 6u_1 - 4u_2 + 5u_1 - 20u_2 = 11u_1 - 24u_2$$

On the other hand,

$$\begin{bmatrix} 1 & 1 \\ 2 & -5 \end{bmatrix} \cdot \begin{bmatrix} 3 & -2 \\ -1 & 4 \end{bmatrix} =$$

$$\begin{bmatrix} 1 \cdot 3 + 1 \cdot (-1) = 2 & 1 \cdot (-2) + 1 \cdot 4 = 2 \\ 2 \cdot 3 + (-5) \cdot (-1) = 11 & 2 \cdot (-2) + (-5) \cdot 4 = -24 \end{bmatrix}$$

If the very same transformations are applied in the *opposite* sequence, we do *not* get the same result.

$$v_1 = u_1 + u_2, \qquad w_1 = 3v_1 - 2v_2$$

$$v_2 = 2u_1 - 5u_2, \qquad w_2 = -v_1 + 4v_2$$

$$w_1 = 3u_1 + 3u_2 - 4u_1 + 10u_2 = -u_1 + 13u_2$$

$$w_2 = -u_1 - u_2 + 8u_1 - 20u_2 = 7u_1 - 21u_2$$

$$\begin{bmatrix} 3 & -2 \\ -1 & 4 \end{bmatrix} \cdot \begin{bmatrix} 1 & 1 \\ 2 & -5 \end{bmatrix} =$$

$$\begin{bmatrix} 3 \cdot 1(-2) \cdot 2 = -1 & 3 \cdot 1(+2) \cdot 5 = 13 \\ -1 \cdot 1(+4) \cdot 2 = 7 & -1 \cdot 1(-4) \cdot 5 = -21 \end{bmatrix}$$

This shows that the products AB and BA are *not* the same. The ordinary commutative law of multiplication does *not* hold in the case of matrices.

On the other hand, the *associative* law of multiplication is satisfied.

$$A(BC) = (AB)C \qquad (2\text{-}3.12)$$

Indeed, let us transform u to v to w to z by the following operations.

$$v = Cu, \quad w = Bv, \quad z = Aw$$

Then

$$z = (ABC)u$$

But this transformation could have been obtained by going from u directly to w and then to z,

$$ABC = A(BC)$$

or by going from u to v and from v directly to z,

$$ABC = (AB)C$$

Our customary algebra is based on six fundamental postulates:

1. Commutative law of addition: $a + b = b + a$
2. Associative law of addition: $(a + b) + c = a + (b + c)$
3. Commutative law of multiplication: $ab = ba$
4. Associative law of multiplication: $(ab)c = a(bc)$
5. Distributive law of multiplication: $(a + b)\ c = ac + bc$, $c(a + b) = ca + cb$
6. The nonfactorability of zero: If $ab = 0$, then $a = 0$, or $b = 0$, or $a = b = 0$.

In matrix algebra the postulates 1, 2, 4, and 5 still hold, but the postulates 3 and 6 are violated. The postulate 3 does not hold, because in defining the product of two matrices the first and the second factor do not enter symmetrically, since the *rows* of the first factor are combined with the *columns* of the second factor. Hence, generally

$$AB \neq BA$$

4. Eigenvalue analysis. If we have one single matrix A alone, without any second matrix B, the noncommutative property of matrix multiplication is irrelevant and we would think that the algebra of the matrix A is equivalent to the algebra of any ordinary algebraic variable x. Yet this is *not* the case because, although at present all the first five postulates of algebra are fulfilled, the nonfulfillment of the *sixth* postulate causes a profound difference.

This postulate is closely related to a remarkable equation in matrix algebra which has no analogy in the realm of ordinary algebraic numbers. We have said that multiplication of a vector x by the matrix A generates a new vector b which can be conceived as a transformation of the original vector x. We will now ask the question whether or not it may happen that the new vector b has the *same* direction as the original vector x. In this case b is simply *proportional* to x and we obtain the condition

$$Ax = \lambda x \tag{2-4.1}$$

or written out in components,

$$\begin{aligned}
a_{11}x_1 + a_{12}x_2 + \cdots + a_{1n}x_n &= \lambda x_1 \\
a_{21}x_1 + a_{22}x_2 + \cdots + a_{2n}x_n &= \lambda x_2 \\
\vdots \\
a_{n1}x_1 + a_{n2}x_2 + \cdots + a_{nn}x_n &= \lambda x_n
\end{aligned} \tag{2-4.2}$$

The right side is not truly a "right side" in the sense of a given vector, and it is more logical to bring the right side over to the left side and write the entire equation as a *homogeneous* set of equations, without any right side.

$$\begin{aligned}
(a_{11} - \lambda)x_1 + a_{12}x_2 + \cdots + a_{1n}x_n &= 0 \\
a_{21}x_1 + (a_{22} - \lambda)x_2 + \cdots + a_{2n}x_n &= 0 \\
\vdots \\
a_{n1}x_1 + a_{n2}x_2 + \cdots + (a_{nn} - \lambda)x_n &= 0
\end{aligned} \tag{2-4.3}$$

The matrix of this system of equations is still the original matrix A, but after subtracting λ in all the diagonal terms.

Now we know that n homogeneous linear equations in n unknowns have no solution (outside of the vanishing of all the x_i, which means that the vector x does not exist), unless one very definite condition is fulfilled, viz., that the *determinant* of the system is zero.

$$\begin{vmatrix}
a_{11} - \lambda & a_{12} & \cdots & a_{1n} \\
a_{21} & a_{22} - \lambda & \cdots & a_{2n} \\
\vdots \\
a_{n1} & a_{n2} & \cdots & a_{nn} - \lambda
\end{vmatrix} = 0 \tag{2-4.4}$$

Now the actual expansion of a determinant according to the original definition is a very cumbersome task if n is larger than 4. There are other less direct but numerically simpler methods for the determination of λ. However, for theoretical purposes the existence of the determinant condition (4) is of greatest importance. The technique of expanding a determinant of the order n shows that on the left side a *polynomial of the order n* in λ appears. It will be convenient to multiply the determinant by $(-1)^n$. Then the largest power of the polynomial appears in the form λ^n.

$$(-1)^n \begin{vmatrix} a_{11} - \lambda & a_{12} & \cdots & a_{1n} \\ a_{21} & a_{22} - \lambda & \cdots & a_{2n} \\ \vdots & & & \\ a_{n1} & a_{n2} & \cdots & a_{nn} - \lambda \end{vmatrix} \tag{2-4.5}$$

$$= \lambda^n + c_{n-1}\lambda^{n-1} + c_{n-2}\lambda^{n-2} + \cdots + c_0$$

For example, let us choose $n = 3$ and expand the determinant (5).

$$- \begin{vmatrix} 2 - \lambda & 0 & 3 \\ 1 & -1 - \lambda & 5 \\ 0 & 4 & -2 - \lambda \end{vmatrix} = \lambda^3 + \lambda^2 - 24\lambda + 24$$

The determinant (5), if evaluated as a polynomial of λ, is called the "characteristic polynomial" of A, and if we set this polynomial equal to zero, we get the "characteristic equation":

$$\lambda^n + c_{n-1}\lambda^{n-1} + c_{n-2}\lambda^{n-2} + \cdots + c_0 = 0 \tag{2-4.6}$$

The condition (4) shows that λ has to fulfill this equation, i.e., λ has to be one of the roots of the algebraic equation (6).

We know that an algebraic equation of the order n has always n and only n generally complex roots. Some of the roots may collapse into one, but then they count as multiple roots. Hence we can say that there are definitely n and only n λ values, called the "characteristic values" or "eigenvalues" of the matrix A, for which the equation (1) is solvable.

$$\lambda = \lambda_1, \lambda_2, \lambda_3, \cdots \lambda_n \tag{2-4.7}$$

To every possible $\lambda = \lambda_i$ a solution of the homogeneous set (1) can be found. We tabulate these solutions as follows:

$$
\begin{aligned}
\lambda = \lambda_1: \quad & x = x_1{}^{(1)}, \quad x_2{}^{(1)}, \quad \cdots, \quad x_n{}^{(1)} \\
\lambda = \lambda_2: \quad & x = x_1{}^{(2)}, \quad x_2{}^{(2)}, \quad \cdots, \quad x_n{}^{(2)} \\
\lambda = \lambda_n: \quad & x = x_1{}^{(n)}, \quad x_2{}^{(n)}, \quad \cdots, \quad x_n{}^{(n)}
\end{aligned}
\qquad (2\text{-}4.8)
$$

The actual construction of these solutions is generally, if n goes beyond 4, a very cumbersome task. For the general understanding of the nature of matrices, however, it is enough to know that these solutions exist and are actually obtainable by solving a linear set of equations.

The solutions of the table (8) represent n distinct vectors of the n-dimensional space. In some exceptional cases some (or all) of these vectors may collapse into one, but then we conceive such cases as limits of the regular cases. These solutions are called the "eigenvectors" or "principal axes" of the matrix A. We will denote them by $u_1, u_2, \cdots, u_n$.

$$
\begin{aligned}
u_1 &= (x_1^{(1)}, x_2^{(1)}, \cdots, x_n^{(1)}) \\
u_2 &= (x_1^{(2)}, x_2^{(2)}, \cdots, x_n^{(2)}) \\
&\vdots \\
u_n &= (x_1^{(n)}, x_2^{(n)}, \cdots, x_n^{(n)})
\end{aligned}
\qquad (2\text{-}4.9)
$$

The entire eigenvector analysis of the matrix A can thus be summarized as follows:

$$\text{the } n \text{ eigenvalues:} \quad \lambda = \lambda_1, \lambda_2, \lambda_3, \cdots, \lambda_n$$

$$\text{the } n \text{ associated eigenvectors:} \quad u_1, u_2, u_3, \cdots, u_n$$

Since the eigenvectors are solutions of a homogeneous set of equations, the solution is determined only up to a universal factor. Every one of the u vectors can be multiplied by an arbitrary factor and still remain an eigenvector. The eigenvectors are thus uniquely determined in their *directions* only, but their *length* (absolute value) is arbitrary.

5. The Hamilton-Cayley equation. Let us consider the solution of the linear vector equation,

$$(A - \lambda_1)x = 0$$

This equation defines the first principal axis u, and thus the general solution of the equation is $x = \alpha_1 u_1$ where α_1 is arbitrary. We now consider the solution of another linear vector equation which is quadratic in A.

$$(A - \lambda_1)(A - \lambda_2)x = 0$$

This equation will be satisfied by an arbitrary linear combination of the first *two* eigenvectors.

$$x = \alpha_1 u_1 + \alpha_2 u_2$$

Similarly the equation

$$(A - \lambda_1)(A - \lambda_2)(A - \lambda_3)x = 0$$

will be satisfied by an arbitrary linear combination of the first *three* eigenvectors.

$$x = \alpha_1 u_1 + \alpha_2 u_2 + \alpha_3 u_3$$

Finally the full equation which contains all the root factors:

$$(A - \lambda_1)(A - \lambda_2)(A - \lambda_3) \cdots (A - \lambda_n)x = 0$$

will be satisfied by an arbitrary linear combination of *all* the n eigenvectors.

$$x = \alpha_1 u_1 + \alpha_2 u_2 + \cdots \alpha_n u_n$$

But the n eigenvectors $u_1, u_2, \cdots, u_n$ include the *entire space*[1] and thus the last x becomes an *arbitrary vector* of the n-dimensional space. This means that the matrix

$$H = (A - \lambda_1)(A - \lambda_2)(A - \lambda_3) \cdots (A - \lambda_n) \qquad (2\text{-}5.1)$$

operating on an *arbitrary* vector x, gives zero.

$$Hx = 0 \qquad (2\text{-}5.2)$$

This is possible only if the matrix H *vanishes identically*. Hence we find that an arbitrary matrix A satisfies the following polynomial identity.

$$(A - \lambda_1)(A - \lambda_2) \cdots (A - \lambda_n) \equiv 0 \qquad (2\text{-}5.3)$$

---

[1] This is not true for "defective" matrices (cf. § 11), but here the theorem is establishable by a limit process.

We have to write this equation somewhat more lucidly, since the $\lambda_1, \lambda_2, \cdots, \lambda_n$ are scalars (pure numbers), while A is a matrix. The equation which defines the eigenvectors: $Ax = \lambda x$, if written in homogeneous form

$$(A - \lambda)x = 0$$

should actually mean

$$(A - \lambda I)x = 0$$

where I is the so-called "unit matrix." This matrix transforms any vector in itself since it is the nature of unity that, if used as a multiplier, it does not change anything. If we require that

$$Iu = u$$

shall hold for any arbitrary vector u, it is necessary and sufficient that all the diagonal elements of I shall be 1, and all the other elements 0.

$$I = \begin{bmatrix} 1 & 0 & & 0 \\ 0 & 1 & & 0 \\ & & \vdots & \\ 0 & 0 & & 1 \end{bmatrix} \tag{2-5.4}$$

Hence equation (3) becomes in proper writing,

$$(A - \lambda_1 I)(A - \lambda_2 I) \cdots (A - \lambda_n I) \equiv 0 \tag{2-5.5}$$

We encountered this very same equation earlier in scalar form when we were interested in finding the eigenvalues of a matrix. The expansion of the determinant (4.5) gave the polynomial

$$\lambda^n + c_1 \lambda^{n-1} + \cdots + c_0$$

and since the roots of this polynomial were denoted by $\lambda_1, \lambda_2, \cdots, \lambda_n$, we have by the laws of algebra:

$$\lambda^n + c_1 \lambda^{n-1} + \cdots + c_0 = (\lambda - \lambda_1)(\lambda - \lambda_2) \cdots (\lambda - \lambda_n)$$

This shows that by multiplying together the root factors (5) we will get exactly the left side of (4.6), i.e., the characteristic polynomial, with the only difference that A takes the place of λ.

$$(A - \lambda_1 I) \cdots (A - \lambda_n I) \equiv A^n + c_1 A^{n-1} \cdots c_0 I$$

But the characteristic polynomial again is nothing but the determinant (2-4.5) associated with the characteristic equation. Hence the equation (5) may also be written in the form

$$(-1)^n \begin{vmatrix} a_{11} - A & a_{12} \cdots & a_{1n} \\ a_{21} & a_{22} - A \cdots & a_{2n} \\ \vdots & & \\ a_{n1} & a_{n2} & a_{nn} - A \end{vmatrix} \equiv 0 \qquad (2\text{-}5.6)$$

and we obtain the remarkable theorem, discovered independently by Hamilton and by Cayley, that *every matrix satisfies identically its own characteristic equation.* The characteristic equation, written in terms of the scalar λ, defines the characteristic values of the matrix, but written in terms of the matrix A expresses an *algebraic identity.*

Numerical example. In the numerical example of Section 4, the matrix A was defined as follows:

$$A = \begin{bmatrix} 2 & 0 & 3 \\ 1 & -1 & 5 \\ 0 & 4 & -2 \end{bmatrix}$$

The characteristic polynomial appeared in the form

$$\lambda^3 + \lambda^2 - 24\lambda + 24$$

If we square and cube the matrix A according to the rules of matrix multiplication, we obtain

$$A^2 = \begin{bmatrix} 4 & 12 & 0 \\ 1 & 21 & -12 \\ 4 & -12 & 24 \end{bmatrix}, \qquad A^3 = \begin{bmatrix} 20 & -12 & 72 \\ 23 & -69 & 132 \\ -4 & 108 & -96 \end{bmatrix}$$

We now form $A^3 + A^2 - 24A + 24I$:

$$\begin{bmatrix} 20+4-48+24=0 & -12+12-0+0=0 & 72+0-72+0=0 \\ 23+1-24+0=0 & -69+21+24+24=0 & 132-12-120+0=0 \\ -4+4-0+0=0 & 108-12-96+0=0 & -96+24+48+24=0 \end{bmatrix}$$

Hence we have demonstrated that

$$A^3 + A^2 - 24A + 24I \equiv 0$$

The existence of a polynomial relation of the form

$$A^n + c_{n-1}A^{n-1} + \cdots c_1 A + c_0 I = 0$$

distinguishes matrix algebra from ordinary algebra even in the case of one single matrix. If x is an ordinary algebraic quantity, we can form polynomials of first, second, third, $\cdots$ order, up to any order, because the powers of x are linearly independent of each other. No power is ever reducible to a linear combination of lower powers (although in a given finite range such a reduction is possible with a high degree of accuracy, cf. VII, 9). With a matrix of n rows and n columns the situation is different. Since A^n is reducible to a linear combination of lower powers, the same is true of A^{n+1}, A^{n+2}, $\cdots$; generally of A^{n+k} ($k = 0, 1, 2, \cdots$). Hence any polynomial of A which is of an order larger than $n - 1$ can always be exactly reduced to a polynomial of not more than $(n - 1)$st degree.

Another important difference concerns the process of *division*. In ordinary algebra the quantity

$$x^{-1} = 1/x$$

exists for any value of x except $x = 0$ since we cannot divide by zero. But if the zero has factors, the situation is different. In matrix algebra the zero has the factors

$$A - \lambda_i I$$

Hence the operation $X^{-1} = 1/X$

if X is a matrix, loses its significance not only if $X = 0$, i.e., if all the elements of the matrix vanish, but also if

$$X = A - \lambda_i I$$

Here X is not zero and in fact no element of X need be zero. Nevertheless, the reciprocal of X cannot be formed.

The reciprocal of A itself is involved in this difficulty if one of the eigenvalues of A happens to be $\lambda_i = 0$. The problem of finding the reciprocal or "inverse" of a matrix is of fundamental importance in solving systems of linear equations. We see that this problem has no solution if the matrix A has a zero eigenvalue. In the strict mathematical sense the inversion problem is impossible only if one of the eigenvalues of A is *exactly* zero. But from the practical standpoint we come into great numerical difficulties not only if A has a zero eigenvalue, but also if A has one or more *very small eigenvalues*. The mathematical analysis of such "nearly singular systems" deserves particularly close attention. In the strict sense, the inversion problem of a matrix can be pursued without knowing anything about the eigenvalue problem. But in actual fact we cannot understand the peculiar behavior of singular or nearly singular systems if we dissociate this problem from the eigenvalue problem of the matrix.

The eigenvalue problem is of profound importance in all flutter and vibration phenomena, since the frequency of elastic or electric vibrations is determined by the eigenvalues of a certain matrix, while the eigenvectors or principal axes of that matrix provide the vibrational modes. But even purely static phenomena, such as the stability analysis of an airplane, or the problem of buckling, are equivalent to an eigenvalue problem. The eigenvalue analysis of matrices became thus a leading item in the engineering research of our days.

6. Numerical example of a complete eigenvalue analysis. It will be of interest to carry through in an actual numerical example all the operations which lead to a complete eigenvalue analysis of a given matrix. Hence we will choose a particularly simple matrix of only 3 rows and columns, in order to reduce the numerical computations to a minimum and yet display all the characteristic features of the eigenvalues and eigenvectors.

Let the matrix A be given as follows:

$$A = \begin{bmatrix} 33 & 16 & 72 \\ -24 & -10 & -57 \\ -8 & -4 & -17 \end{bmatrix} \qquad (2\text{-}6.1)$$

Our aim will be to obtain all the three eigenvalues and eigenvectors associated with this matrix.

First we construct the characteristic equation by putting $-\lambda$ in the diagonal and setting the determinant equal to zero. The determinant is obtained by expanding in the elements of the first row. We multiply by $(-1)^3 = -1$ in order to obtain the characteristic polynomial with a plus sign in front of λ^3.

$$(-1)^3 \begin{vmatrix} 33-\lambda & 16 & 72 \\ -24 & -10-\lambda & -57 \\ -8 & -4 & -17-\lambda \end{vmatrix} = \begin{vmatrix} \lambda-33 & -16 & -72 \\ -24 & -10-\lambda & -57 \\ -8 & -4 & -17-\lambda \end{vmatrix}$$

$$= (\lambda - 33)[10 + \lambda)(17 + \lambda) - 228] + 16[24(17 + \lambda) - 456]$$

$$-72[96 - 8(10 + \lambda)]$$

$$= (\lambda - 33)(\lambda^2 + 27\lambda - 58) + 16(24\lambda - 48) - 72(-8\lambda + 16)$$

$$= \lambda^3 + 27\lambda^2 - 58\lambda$$
$$- 33\lambda^2 - 891\lambda + 1914$$
$$+ 384\lambda - 768$$
$$+ 576\lambda - 1152$$
$$\overline{\lambda^3 - 6\lambda^2 + 11\lambda - 6}$$

$$(2\text{-}6.2)$$

Hence the characteristic equation becomes

$$\lambda^3 - 6\lambda^2 + 11\lambda - 6 = 0 \qquad (2\text{-}6.3)$$

The roots of this equation are

$$\lambda_1 = 1, \qquad \lambda_2 = 2, \qquad \lambda_3 = 3 \qquad (2\text{-}6.4)$$

We will also check the Hamilton-Cayley identity which agrees with the characteristic equation, but with A taking the place of λ.

$$A^3 - 6A^2 + 11A - 6I = 0 \qquad (2\text{-}6.5)$$

which gives

$$A^3 = 6A^2 - 11A + 6I \qquad (2\text{-}6.6)$$

In order to check this equation, we have to form the square and the cube of the original matrix. We know that in multiplying two matrices together we have to combine the *rows* of the first matrix with the *columns* of the second matrix. To keep track of the corresponding elements is not easy under these conditions, and errors can easily be made. It is better to "transpose" the first matrix by changing rows to columns. Then we have to multiply *columns* by *columns* which is much less confusing and avoids the misplacing of elements.

The transposition (i.e., exchange of rows and columns of a matrix) is denoted by $\tilde{A}$. In our example

$$\tilde{A} = \begin{bmatrix} 33 & -24 & -8 \\ 16 & -10 & -4 \\ 72 & -57 & -17 \end{bmatrix}$$

Hence we will obtain A^2 by multiplying $\tilde{A}$ and A in column-by-column fashion. In order to indicate that we do not mean ordinary multiplication but column by column multiplication, we will use the symbol o.

$$A^2 = \begin{bmatrix} 33 & -24 & -8 \\ 16 & -10 & -4 \\ 72 & -57 & -17 \end{bmatrix} \circ \begin{bmatrix} 33 & 16 & 72 \\ -24 & -10 & -57 \\ -8 & -4 & -17 \end{bmatrix} = \begin{bmatrix} 129 & 80 & 240 \\ -96 & -56 & -189 \\ -32 & -20 & -59 \end{bmatrix}$$

We repeat the process once more and obtain A^3.

$$A^3 = \begin{bmatrix} 129 & -96 & -32 \\ 80 & -56 & -20 \\ 240 & -189 & -59 \end{bmatrix} \circ \begin{bmatrix} 33 & 16 & 72 \\ -24 & -10 & -57 \\ -8 & -4 & -17 \end{bmatrix} = \begin{bmatrix} 417 & 304 & 648 \\ -312 & -220 & -507 \\ -104 & -76 & -161 \end{bmatrix}$$

We now form the right side of equation (6).

$$6A^2 - 11A + 6I =$$

$$\begin{bmatrix} 6\cdot129 - 11\cdot33 + 6 = & 417 & 6\cdot80 - 11\cdot16 = & 304 \\ -6\cdot96 + 11\cdot24 = -312 & -6\cdot56 + 11\cdot10 + 6 = -220 \\ -6\cdot32 + 11\cdot8 = -104 & -6\cdot20 + 11\cdot4 = - 76 \end{bmatrix}$$

$$\begin{bmatrix} 6\cdot240 - 11\cdot72 = & 648 \\ -6\cdot189 + 11\cdot57 = -507 \\ -6\cdot59 + 11\cdot17 + 6 = -161 \end{bmatrix}$$

The elements thus obtained agree with the elements of A^3, and thus the Hamilton-Cayley equation is demonstrated.

We now come to determination of the eigenvectors or principal axes of our matrix. This means the solution of the homogeneous linear equations

$$(A - \lambda I)x = 0 \qquad (2\text{-}6.7)$$

The process has to be repeated for every λ_i, since every λ_i is associated with a definite principal axis. First we choose $\lambda = \lambda_i = 1$. We subtract 1 from the diagonal elements of the matrix and obtain the following homogeneous linear equations.

$$32x_1 + 16x_2 + 72x_3 = 0$$
$$-24x_1 - 11x_2 - 57x_3 = 0 \qquad (2\text{-}6.8)$$
$$- 8x_1 - 4x_2 - 18x_3 = 0$$

We know in advance that the determinant of this linear system is zero, since it was exactly this condition which led to the determination of the eigenvalues. Now the vanishing of the determinant of a homogeneous linear set of equations has a very definite significance. It means that these n equations are not independent of each other, but the last equation is a consequence of the previous equations. Hence we can *omit* the third equation of the set (8) and it suffices to solve the remaining *two* (generally $n - 1$) equations.

$$32x_1 + 16x_2 + 72x_3 = 0 \qquad -24x_1 - 11x_2 - 57x_3 = 0 \qquad (2\text{-}6.9)$$

If these two equations are satisfied, the last one is automatically satisfied.

But then the difficulty is that we have to obtain 3 unknowns from only 2 equations. On the other hand, we know from the homogeneous nature of the eigenvalue problem that the length of the principal axes must remain undetermined. This leaves a universal factor α undetermined. If x_1, x_2, x_3 is some solution of our problem, then $\alpha x_1, \alpha x_2, \alpha x_3$ (with any arbitrary α) is an equally valid solution. But then we can take advantage of the arbitrariness of α for normalizing x in any arbitrary fashion. For example we may choose $x_3 = 1$. Then equations (9) become

$$32x_1 + 16x_2 + 72 = 0, \qquad -24x_1 - 11x_2 - 57 = 0$$

and this can be written in the inhomogeneous form:

$$32x_1 + 16x_2 = -72, \qquad -24x_1 - 11x_2 = 57 \qquad (2\text{-}6.10)$$

The original set of n homogeneous equations in n unknowns is thus replaceable by an inhomogeneous set of $n - 1$ equations in $n - 1$ unknowns.

These equations, unless inconsistent or redundant, are now solvable by determinants if n does not go beyond 3 or 4, or by matrix inversion if n is larger. In our simple example, the solution is directly obtainable by the simple formula of solving two simultaneous linear equations.

$$a_{11}x_1 + a_{12}x_2 = b_1 \qquad \text{(2-6.11)}$$

$$a_{21}x_1 + a_{22}x_2 = b_2$$

$$x_1 = \frac{b_1 a_{22} - b_2 a_{12}}{a_{11}a_{22} - a_{21}a_{12}}$$

$$x_2 = \frac{a_{11}b_2 - a_{21}b_1}{a_{11}a_{22} - a_{21}a_{12}}$$

provided $a_{11}a_{12} - a_{21}a_{12} \neq 0$.

We include the process for all three values $\lambda = 1, 2, 3$ in the following table.

$\lambda = 1$	$\lambda = 2$	$\lambda = 3$
$32x_1 + 16x_2 = -72$ $-24x_1 - 11x_2 = 57$	$31x_1 + 16x_2 = -72$ $24x_1 - 12x_2 = 57$	$30x_1 + 16x_2 = -72$ $24x_1 - 13x_2 = 57$
$x_1 = \dfrac{-120}{32} = -\dfrac{15}{4}$ $x_2 = \dfrac{96}{32} = 3$ $x_3 = 1$	$x_1 = \dfrac{-48}{12} = -4$ $x_2 = \dfrac{39}{12} = \dfrac{13}{4}$ $x_3 = 1$	$x_1 = \dfrac{24}{-6} = -4$ $x_2 = \dfrac{-18}{-6} = 3$ $x_3 = 1$

We will now tabulate our results as follows. We have obtained 3 eigenvalues λ_i and 3 associated vectors which we will call u_1, u_2, u_3. We write the components of these vectors in 3 separate columns. Taking advantage of the arbitrariness of the lengths, we can multiply each of these vectors by any constant. Hence we can eliminate fractions, and the vector $(-15/4, 3, 1)$ may be replaced by $(-15, 12, 4)$. Our table then looks as follows:

$\lambda = 1$	$\lambda = 2$	$\lambda = 3$	
u_1	u_2	u_3	
-15	-16	-4	(2-6.12)
12	13	3	
4	4	1	

While this solution came about very easily in our simple example, we can imagine that in the case of matrices of high order the procedure is much more difficult, and methods have to be designed by which the eigenvalues and eigenvectors of such matrices become numerically accessible. We will discuss such methods later in greater detail.

We would think that our eigenvalue analysis is now complete. Yet this is not the case. There is still one feature of matrix algebra which we have not considered up to now and which is of fundamental importance. Any given matrix A is automatically associated with another matrix which is inseparably attached to it. This is the "transposed matrix" or briefly "transpose" of A which we will call $\tilde{A}$.

The arrangement of the matrix components into rows and columns is somewhat arbitrary. We have a square in front of us, but what is "row" and what is "column" depends on how we look at this square. A square displays complete right-left and up-down symmetry.

We look at our square in the normal way. We designate certain elements as being in a "row," certain others as being in a "column." But turning the square 90° and looking at it again, we find that the previous rows turn to columns and the columns to rows. We thus have a duality associated with every matrix. The matrix A is inevitably associated with its transpose $\tilde{A}$, and we cannot operate with A without simultaneously operating also with the transposed matrix $\tilde{A}$. In ordinary algebra we have a somewhat analogous phenomenon in the field of complex numbers. The complex number $a + bi$, is inevitably associated with the "conjugate complex" number $a - bi$, and both appear simultaneously in many algebraic problems. For example, in solving an algebraic equation with real coefficients, the roots $a + bi$ and $a - bi$ always appear together, and one is inseparable from the other. The operation of changing A into $\tilde{A}$ thus corresponds in ordinary algebra to the operation of changing i to $-i$.

Our eigenvalue analysis is thus incomplete if we do not extend it to the transposed matrix $\tilde{A}$. Here the defining equation of the eigenvalue problem becomes

$$\tilde{A}x = \lambda x \qquad (2\text{-}6.13)$$

In principle we could think that this problem is completely separated from the previous one. However, as far as the eigenvalues λ_i are concerned, the two problems coincide. We have seen that the eigenvalues λ of a matrix satisfy a determinant condition. We know that a determinant does not change its value if rows and columns are interchanged. Hence the characteristic polynomial associated with $\tilde{A}$ is exactly the same as that associated with A. Consequently the eigenvalues of $\tilde{A}$ are identical with the eigenvalues of A. But each eigenvalue has a dual aspect inasmuch as we can use the same λ for solving the equation $Ax = \lambda x$ and the equation $\tilde{A}y = \lambda y$. In the first case we determine the principal axes of A, in the second case the principal axes of $\tilde{A}$.

Hence we will go through our previous procedure once more, but now using the transposed matrix $\tilde{A}$ instead of A.

$$\tilde{A} = \begin{bmatrix} 33 & -24 & -8 \\ 16 & -10 & -4 \\ 72 & -57 & -17 \end{bmatrix}$$

We need not repeat the solution of the characteristic equation since it remains unchanged. Once more we obtain the 3 characteristic values $\lambda = 1, 2, 3$. However, the table of the equations which lead to the principal axes will now appear as follows:

$\lambda = 1$	$\lambda = 2$	$\lambda = 3$
$32x_1 - 24x_2 = 8$	$31x_1 - 24x_2 = 8$	$30x_1 - 24x_2 = 8$
$16x_1 - 11x_2 = 4$	$16x_1 - 12x_2 = 4$	$16x_1 - 13x_2 = 4$
$x_1 = \dfrac{8}{32} = \dfrac{1}{4}$	$x_1 = \dfrac{0}{12} = 0$	$x_1 = \dfrac{-8}{-6} = \dfrac{4}{3}$
$x_2 = \dfrac{0}{32} = 0$	$x_2 = \dfrac{-4}{12} = -\dfrac{1}{3}$	$x_2 = \dfrac{-8}{-6} = \dfrac{4}{3}$
$x_3 = 1$	$x_3 = 1$	$x_3 = 1$

Once more we tabulate our results, denoting the eigenvectors of the transposed matrix $\tilde{A}$ by v_1, v_2, v_3. Once more we can avoid the fractions by multiplying by a suitable factor, since the length of the principal axes is undetermined.

$\lambda = 1$	$\lambda = 2$	$\lambda = 3$	
v_1	v_2	v_3	
1	0	4	(2-6.14)
0	−1	4	
4	3	3	

The tables (12) and (14) together represent the complete eigenvalue analysis of our problem. It consists of n scalars, viz., the eigenvalues

$$\lambda = \lambda_1, \lambda_2, \cdots, \lambda_n$$

and $2n$ vectors, viz., the n eigenvectors or principal axes of the matrix A:

$$u_1, u_2, \cdots, u_n$$

and the n eigenvectors or principal axes of the matrix $\tilde{A}$:

$$v_1, v_2, \cdots, v_n$$

In our numerical example the complete eigenvalue analysis of the given matrix A can be tabulated as follows:

Complete eigenvalue analysis

$\lambda = 1$	$\lambda = 2$	$\lambda = 3$	$\lambda = 1$	$\lambda = 2$	$\lambda = 3$	
u_1	u_2	u_3	v_1	v_2	v_3	
−15	−16	−4	1	0	4	(2-6.15)
12	13	3	0	−1	4	
4	4	1	4	3	3	

These two sets of vectors are in a remarkable reciprocity relation to each other. The u vectors in themselves do not reveal any particular inner relations, nor do the v vectors in themselves. But let us form the dot product of one u and one v vector. For example, pairing u with all the v vectors, we obtain

$$u_1 \cdot v_1 = -15 \cdot 1 + 12 \cdot 0 + 4 \cdot 4 = 1$$
$$u_1 \cdot v_2 = -15 \cdot 0 - 12 \cdot 1 + 4 \cdot 3 = 0$$
$$u_1 \cdot v_3 = -15 \cdot 4 + 12 \cdot 4 + 4 \cdot 3 = 0$$

Similarly pairing u_2 with all the v vectors, we obtain

$$u_2 \cdot v_1 = -16 \cdot 1 + 13 \cdot 0 + 4 \cdot 4 = 0$$
$$u_2 \cdot v_2 = -16 \cdot 0 - 13 \cdot 1 + 4 \cdot 3 = -1$$
$$u_2 \cdot v_3 = -16 \cdot 4 + 13 \cdot 4 + 4 \cdot 3 = 0$$

Finally, pairing u_3 with all the v vectors, we obtain

$$u_3 \cdot v_1 = -4 \cdot 1 + 3 \cdot 0 + 1 \cdot 4 = 0$$
$$u_3 \cdot v_2 = -4 \cdot 0 - 3 \cdot 1 + 1 \cdot 3 = 0$$
$$u_3 \cdot v_3 = -4 \cdot 4 + 3 \cdot 4 + 1 \cdot 3 = -1$$

If the dot product of two vectors comes out as zero, this means, in the language of geometry, that these two vectors are *perpendicular* or *orthogonal* to each other. We see that *any vector of the u set is orthogonal to any vector of the v set, except its own pair.*

$$u_i \cdot v_k = 0 \quad (i \neq k) \tag{2-6.16}$$

The dot product $u_1 \cdot v_1$ came out as 1, the dot product $u_2 \cdot v_2$ as -1, and the dot product $u_3 \cdot v_3$ as -1. We can see without difficulty that these dot products could have come out as anything we like, since the lengths of the principal axes remained undetermined. We now obtain a practical normalization of the free lengths of the principal axes by demanding that the dot products of $u_1 \cdot v_1$ shall all become 1.

$$u_1 \cdot v_1 = 1 \tag{2-6.17}$$

This leaves the lengths of u_i still undetermined, but the lengths of the vectors v_i can always be adjusted in such a way that the condition (17) shall be satisfied (leaving apart the singular case, which occurs only in the case of "defective matrices," cf. § 11, that a certain $u_i \cdot v_i$ may come out as 0). If originally a certain $u_i \cdot v_i$ gives c_i, we change v_i to

$$\bar{v}_i = \frac{1}{c_i} v_i \tag{2-6.18}$$

and then in the new $\bar{v}_i$ system the condition (17) will already hold.

In our numerical scheme the vector v_1 is already properly normalized since $u_1 \cdot v_1$ was accidentally 1. The vectors v_2 and v_3 have to be divided by -1, and thus the final v vector scheme becomes

v_1	v_2	v_3
1	0	-4
0	1	-4
4	-3	-3

$$(2\text{-}6.19)$$

A dual set of vectors in which any vector of the one set is orthogonal to all vectors of the other set, except its own pair, is called a "biorthogonal" set. If in addition the dot product of every vector with its own pair is normalized to 1, we speak of a biorthogonal and normalized set of vectors.

We will now omit the vertical dividing lines which separate the vectors u_i from each other and the vectors v_i from each other. Then we obtain n columns of elements, n in each column, which together form an n by n matrix. We thus obtain one matrix U which includes all the vectors u_i, and one matrix V which includes all the vectors v_i. Hence the results of a complete eigenvalue analysis can be stated in still different fashion by giving the n eigenvalues $\lambda_1, \lambda_2, \cdots, \lambda_n$ and the two n by n matrices U and V. In our example,

$$\lambda = 1, 2, 3$$

$$U = \begin{bmatrix} -15 & -16 & -4 \\ 12 & 13 & 3 \\ 4 & 4 & 1 \end{bmatrix}, \quad V = \begin{bmatrix} 1 & 0 & -4 \\ 0 & 1 & -4 \\ 4 & -3 & -3 \end{bmatrix} \quad (2\text{-}6.20)$$

The inner relations between these two matrices can be expressed in the form of a single matrix equation.

$$\tilde{U} V = I \qquad (2\text{-}6.21)$$

Indeed, to multiply two matrices together means to form the dot products of rows and columns. The rows of the first matrix are multiplied by the columns of the second matrix. But the transposition of U has the effect that the *rows* of $\tilde{U}$ are actually the *columns* of U. And thus on the left side of (21) we have the column-by-column products of the matrices U and V. The fact that these products come out as 1 for corresponding pairs and 0 for all other combinations means that the diagonal elements of the product matrix become 1 and all other elements 0. But this is exactly the definition of the unit matrix I. An alternative way of stating (21) is the equation

$$\tilde{V}U = I \tag{2-6.22}$$

7. Algebraic treatment of the orthogonality of eigenvectors. What we have demonstrated purely numerically can be corroborated quite generally by the operations of matrix algebra. For this purpose we first reformulate the general definition of an eigenvalue problem.

$$Ax = \lambda x \tag{2-7.1}$$

This equation defines only *one* principal axis at a time, because λ is here chosen as one of the eigenvalues, and for each $\lambda = \lambda_i$ the equation has to be repeated. The equation (1) has thus to be used n times, for $\lambda = \lambda_1, \lambda_2, \cdots, \lambda_n$.

We will now include *all* the eigenvectors of A in a single matrix equation. For this purpose we introduce a matrix Λ, defined as follows:

$$\Lambda = \begin{bmatrix} \lambda_1 & 0 & & 0 \\ 0 & \lambda_2 & & 0 \\ & & \vdots & \\ 0 & 0 & & \lambda_n \end{bmatrix} \tag{2-7.2}$$

A matrix of this form, which has only diagonal elements—all the other elements being zero—is called a "diagonal matrix." The operation with such a matrix is particularly simple. Let us use Λ as a first factor (called "premultiplication"). By the general rule of matrix multiplication we get

$$\Lambda A = \begin{bmatrix} \lambda_1 a_{11} & \lambda_1 a_{12} & \cdots & \lambda_1 a_{1n} \\ \lambda_2 a_{21} & \lambda_2 a_{22} & & \lambda_2 a_{2n} \\ \vdots & & & \\ \lambda_n a_{n1} & \lambda_n a_{n2} & \cdots & \lambda_n a_{nn} \end{bmatrix} \tag{2-7.3}$$

We see that the premultiplication by Λ has the effect that the successive *rows* of A are multiplied by $\lambda_1, \lambda_2, \cdots, \lambda_n$.

Let us now multiply by Λ as a second factor (called "post-multiplication").

$$A\Lambda = \begin{bmatrix} \lambda_1 a_{11} & \lambda_2 a_{12} & \cdots & \lambda_n a_{1n} \\ \lambda_1 a_{21} & \lambda_2 a_{22} & & \lambda_n a_{2n} \\ \vdots & & & \\ \lambda_1 a_{n1} & \lambda_2 a_{n2} & \cdots & \lambda_n a_{nn} \end{bmatrix} \qquad (2\text{-}7.4)$$

Hence the postmultiplication by Λ has the effect that the successive *columns* of A are multiplied by $\lambda_1, \lambda_2, \cdots, \lambda_n$.

We now write the matrix equation

$$AU = U\Lambda \qquad (2\text{-}7.5)$$

If we write out this equation in components, we find that the first column of the resultant equation scheme defines the first principal axis u_1 (i.e., the first column of the matrix U), the second column defines the second principal axis u_2, and so on. The entire principal axis problem is now included in a single matrix equation.

The principal axis problem of the transposed matrix $\tilde{A}$ is similarly included in the matrix equation

$$\tilde{A}V = V\Lambda \qquad (2\text{-}7.6)$$

which defines all the "adjoint" axes $v_1, v_2, \cdots, v_n$. They are the successive columns of the matrix V.

Now the following fundamental rule of matrix algebra is directly provable on the basis of the definition of matrix multiplication.

$$\widetilde{AB} = \tilde{B}\tilde{A} \qquad (2\text{-}7.7)$$

"The transpose of a product is equal to the product of the transposed matrices, but reversing their sequence." The same rule holds for any number of factors, e.g.,[1]

$$(\widetilde{ABC}) = \tilde{C}\tilde{B}\tilde{A} \qquad (2\text{-}7.8)$$

We must know, furthermore, that the transposition of a diagonal matrix Λ leaves that matrix unchanged, since the diagonal terms are not affected by the process of changing rows to columns.

$$\tilde{\Lambda} = \Lambda \qquad (2\text{-}7.9)$$

---

[1] Exactly the same rule holds for the operation "inverse", i.e., to raise a product of matrices to the power -1:

$$(ABC)^{-1} = C^{-1}B^{-1}A^{-1}$$

Finally, "the transposition of the transposition restores the original matrix":

$$\tilde{\tilde{A}} = A \qquad (2\text{-}7.10)$$

which is the direct consequence of the fact that changing rows to columns interchanges the sequence of the subscripts:

$$\tilde{a}_{ik} = a_{ki}$$

but two such interchanges restore the original sequence.

$$\tilde{\tilde{a}}_{ik} = \tilde{a}_{ki} = a_{ik}$$

Let us transpose on both sides of (6):

$$\tilde{V}A = \Lambda \tilde{V} \qquad (2\text{-}7.11)$$

and let us postmultiply this equation by U

$$\tilde{V}AU = \Lambda \tilde{V}U \qquad (2\text{-}7.12)$$

On the other hand, let us premultiply (7.5) by $\tilde{V}$

$$\tilde{V}AU = \tilde{V}U\Lambda \qquad (2\text{-}7.13)$$

Since the left sides are equal, we obtain

$$\Lambda \tilde{V}U = \tilde{V}U\Lambda \qquad (2\text{-}7.14)$$

or, denoting the product $\tilde{V}U$ by W,

$$\tilde{V}U = W \qquad (2\text{-}7.15)$$

we have

$$\Lambda W = W\Lambda \qquad (2\text{-}7.16)$$

This means that the matrix W is *commutative* with the diagonal matrix Λ. We may also write

$$\Lambda W - W\Lambda = 0 \qquad (2\text{-}7.17)$$

which means, in view of (3) and (4),

$$\begin{bmatrix} 0 & (\lambda_1 - \lambda_2)w_{12} & \cdots & (\lambda_1 - \lambda_n)w_{1n} \\ (\lambda_2 - \lambda_1)w_{21} & 0 & & (\lambda_2 - \lambda_n)w_{2n} \\ \vdots & & & \\ (\lambda_n - \lambda_1)w_{n1} & (\lambda_n - \lambda_2)w_{n2} & \cdots & 0 \end{bmatrix} = 0 \qquad (2\text{-}7.18)$$

But then, assuming that the roots λ_i of the characteristic equation are all different, we find that all the w_{ik} $(i \neq k)$ must be zero, i.e., W must be a pure *diagonal matrix*

$$W = \begin{bmatrix} w_1 & 0 & 0 \\ 0 & w_2 & 0 \\ \vdots & & \\ 0 & 0 & w_n \end{bmatrix} \tag{2-7.19}$$

This proves already the biorthogonality of the vectors U_i and v_i, since (19) expresses the fact the the dot product of any u_i with any v_k $(i \neq k)$ must be zero. The further condition that all the w_i become 1 cannot be proved, since this is a matter of definition. The equations defining the matrix U and the matrix V are such that each column of the matrices U and V can be multiplied by an arbitrary factor, and now we dispose of half of these factors in such a way that the dot products u_iv_i become 1. Then W becomes the unit matrix I, and we obtain

$$\tilde{V}U = \tilde{U}V = I \tag{2-7.20}$$

Here we have the fundamental relation between the principal axes u_i and the adjoint axes v_i (the principal axes of the transposed matrix) expressed as a matrix equation. In consequence of (20) we can say that the matrices U and $\tilde{V}$ are in a reciprocity relation to each other

$$\begin{aligned} \tilde{V} &= U^{-1}, & V &= \tilde{U}^{-1} \\ U &= \tilde{V}^{-1}, & \tilde{U} &= V^{-1} \end{aligned} \tag{2-7.21}$$

One matrix is the "inverse transpose" of the other.

A further remarkable relation is obtainable by postmultiplying (5) by $\tilde{V}$:

$$AU\tilde{V} = U\Lambda\tilde{V}$$

which gives, in view of (20),

$$A = U\Lambda\tilde{V} \tag{2-7.22}$$

Similarly,

$$\tilde{A} = V\Lambda\tilde{U} \tag{2-7.23}$$

Equation (22) shows that the original matrix A is obtainable by multiplying three matrices together, viz., the matrix U, the diagonal matrix Λ, and the matrix V.

However, complete solution of the eigenvalue problem solves also the problem of inverting the matrix. Let us write once more the fundamental defining equation of the principal axis problem of the matrix A

$$Ax = \lambda x$$

Premultiplying by A^{-1}, we obtain

$$x = \lambda A^{-1}x \quad \text{or} \quad A^{-1}x = \lambda^{-1}x$$

This equation defines the principal axes A^{-1}, and we come to the conclusion that the same vector u_i which was a principal axis of A is also a principal axis of A^{-1}, while the associated eigenvalue is the *reciprocal* of the original eigenvalue λ_i. Hence the solution of the eigenvalue problem of A^{-1} is given as follows:

$$\lambda = \frac{1}{\lambda_1}, \frac{1}{\lambda_2}, \quad \cdots, \quad \frac{1}{\lambda_n} \tag{2-7.24}$$

$$U' = U, \qquad V' = V$$

The diagonal matrix Λ contains now the *reciprocals* of the previous λ and is thus equal to the inverse of the previous matrix Λ. If we apply the relation (22) to the new eigenvalue problem, we obtain

$$A^{-1} = U\Lambda^{-1}\tilde{V}$$

The inverse of A is thus generated purely in terms of the principal axes and eigenvalues of A.

Going back to our previous numerical problem, we have

$$U = \begin{bmatrix} -15 & -16 & -4 \\ 12 & 13 & 3 \\ 4 & 4 & 1 \end{bmatrix}, \qquad \Lambda = \begin{bmatrix} 1 & 0 & 0 \\ 0 & 2 & 0 \\ 0 & 0 & 3 \end{bmatrix},$$

$$V = \begin{bmatrix} 1 & 0 & -4 \\ 0 & 1 & -4 \\ 4 & -3 & -3 \end{bmatrix}$$

Hence:

$$\tilde{V} = \begin{bmatrix} 1 & 0 & 4 \\ 0 & 1 & -3 \\ -4 & -4 & -3 \end{bmatrix}, \qquad \Lambda\tilde{V} = \begin{bmatrix} 1 & 0 & 4 \\ 0 & 2 & -6 \\ -12 & -12 & -9 \end{bmatrix}$$

and

$$U\Lambda\tilde{V}=\begin{bmatrix} -15 & -16 & -4 \\ 12 & 13 & 3 \\ 4 & 4 & 1 \end{bmatrix}\cdot\begin{bmatrix} 1 & 0 & 4 \\ 0 & 2 & -6 \\ -12 & -12 & -9 \end{bmatrix}$$

$$=\begin{bmatrix} -15 & 12 & 4 \\ -16 & 13 & 4 \\ -4 & 3 & 1 \end{bmatrix}\circ\begin{bmatrix} 1 & 0 & 4 \\ 0 & 2 & -6 \\ -12 & -12 & -9 \end{bmatrix}=\begin{bmatrix} 33 & 16 & 72 \\ -24 & -10 & -57 \\ -8 & -4 & -17 \end{bmatrix}$$

The resultant product coincides with the given matrix A.

We now repeat the procedure, but changing the λ_i to their reciprocals,

$$\tilde{V}=\begin{bmatrix} 1 & 0 & 4 \\ 0 & 1 & -3 \\ -4 & -4 & -3 \end{bmatrix}, \qquad \Lambda^{-1}\tilde{V}=\begin{bmatrix} 1 & 0 & 4 \\ 0 & \frac{1}{2} & -\frac{3}{2} \\ -\frac{4}{3} & -\frac{4}{3} & -1 \end{bmatrix}$$

$$U\Lambda^{-1}\tilde{V}=\begin{bmatrix} -15 & 12 & 4 \\ -16 & 13 & 4 \\ -4 & 3 & 1 \end{bmatrix}\circ\begin{bmatrix} 1 & 0 & 4 \\ 0 & \frac{1}{2} & -\frac{3}{2} \\ -\frac{4}{3} & -\frac{4}{3} & -1 \end{bmatrix}=\begin{bmatrix} -\frac{29}{3} & -\frac{8}{3} & -32 \\ 8 & \frac{5}{2} & \frac{51}{2} \\ \frac{8}{3} & \frac{2}{3} & 9 \end{bmatrix}$$

In order to demonstrate that the final result is actually the reciprocal (or inverse) matrix A^{-1}, we form the product AA^{-1}:

$$AA^{-1}=\tilde{A}\circ A^{-1}=\begin{bmatrix} 33 & -24 & -8 \\ 16 & -10 & -4 \\ 72 & -57 & -17 \end{bmatrix}\circ\begin{bmatrix} -\frac{29}{3} & -\frac{8}{3} & -32 \\ 8 & \frac{5}{2} & \frac{51}{2} \\ \frac{8}{3} & \frac{2}{3} & 9 \end{bmatrix}=\begin{bmatrix} 1 & 0 & 0 \\ 0 & 1 & 0 \\ 0 & 0 & 1 \end{bmatrix}$$

The product of the two matrices gives the unit matrix I, which demonstrates that one matrix is the reciprocal (or inverse) of the other.

We will seldom take recourse to this method of inverting a matrix, since inversion of a matrix is generally a much simpler task than solution of the complete eigenvalue problem. But if we have a problem in which the complete exploration of the properties of a given matrix A is demanded, it may be necessary to display all the eigenvectors and eigenvalues of A. In that case it can easily be of value that the inversion problem of the matrix is already included in the eigenvalue analysis. Moreover, for the critical study of "nearly singular systems," whose inversion offers great practical difficulties, the relation between the eigenvalue problem and the inversion problem is of inestimable importance.

8. The eigenvalue problem in geometrical interpretation. The algebraic treatment of the eigenvalue problem puts the spotlight on certain characteristic features of matrix algebra. In ordinary algebra the products ab and ba are interchangeable. In matrix algebra this is not the case, except under restricted conditions. We have found, for example, that the interchangeability of the products $W\Lambda$ and ΛW, where Λ is a diagonal matrix with noncollapsing elements, led to the conclusion that W itself has to be a diagonal matrix. We have also found that $AB = 0$ does not lead to the conclusion that either A or B has to be zero. In our numerical example the product

$$(A - I)(A - 2I)(A - 3I) = (A - I)(A^2 - 5A + 6I)$$

vanished. And yet neither $A - I$ nor $A^2 - 5A + 6I$ was zero. These are characteristic differences for which we have to watch if we operate with matrices.

The purely formalistic algebraic operations with matrices, while they produce spectacular results, can easily blind us to certain deeper implications of the matrix problem. Hence we will enlarge the picture and make it much more meaningful if we associate with it a certain space structure which gives a *geometrical* interpretation to the operation with matrices. The operation with matrices is in closest relation to the analytical geometry of second-order surfaces. The entire eigenvalue and principal axis problem is intimately connected with the geometrical properties of the second-order surfaces, called in our ordinary three-dimensional space ellipsoids and hyperboloids.

The theory of curves and surfaces of second order has a long and inspiring history. The Greeks spent a tremendous amount of ingenuity on the geometrical investigation of the conic sections, which they have called ellipses, parabolas, and hyperbolas. Apollonius of Perga, called "the great geometer," was in possession of all the basic properties of conic sections. He obtained his results by partly analytical and partly projective methods. When almost two thousand years later Descartes invented analytic geometry and showed how the problems of geometry can be solved with the help of algebra, he introduced a new and powerful mechanism to the systematic exploration of geometrical problems. Yet there was no basic theorem in the realm of the conic sections which the Greeks did not discover earlier by sheer ingenuity, without the help of algebra.

The Greeks did not know to what practical use the theory of conic sections may be put. Pragmatic evaluation of things was completely foreign to them, and they conceived the occupation with geometry as a privilege of the intellect, enjoyed for its own sake and not for any pragmatic gains. But in later ages this purely esthetic occupation with geometry paid rich dividends. When Kepler evaluated the observations of Tycho Brahe and came to the conclusion that the planets revolve not in circles but in ellipses around the sun, he based his calculations directly on the results of Apollonius. Without Greek geometry, Kepler could not have obtained his results. On the other hand, Newton's theory of gravity and theory of motion could not have come into being without the preliminary results of Kepler. And thus we see the direct line from the Greek investigation of the conic sections to Kepler, Newton, the foundation of physics and engineering in the eighteenth century, and their development to the present standards.

However, astronomy and physics are not the only instances in which the theory of conic sections plays a fundamental role. The analytic geometry of second-order curves and surfaces can be expanded from spaces of two and three dimensions to spaces of *any* dimensions. And then the discovery was made that the entire theory of linear operators—whether they appear as systems of ordinary linear algebraic equations, or as ordinary or partial differential equations, or as integral equations—can be formulated as a *geometrical* problem, associated with a certain second-order surface. The conic sections have been lifted out of the plane and put in a much more elaborate background. The space with which we operate is no longer a space of two or three dimensions, but a space of many dimensions and possibly even a space of infinitely many dimensions. But the quadratic surfaces, which find their place in these poly-dimensional spaces, still reflect the same fundamental properties that the Greeks discovered in their studies of the conic sections.

In the analytic geometry of conic sections the equation of an ellipse is usually given in the following form:

$$\frac{x^2}{a^2} + \frac{y^2}{b^2} = 1$$

The equation of a hyperbola is usually written in the form

$$\frac{x^2}{a^2} - \frac{y^2}{b^2} = 1$$

Then in solid analytic geometry—i.e., the analytic geometry of the three-dimensional space—the variable z is added, and we write the equation of an ellipsoid in the form

$$\frac{x^2}{a^2} + \frac{y^2}{b^2} + \frac{z^2}{c^2} = 1$$

If we want to eliminate the accidental numbers 2 and 3 from the study of spaces of arbitrary dimensions and formulate our equations in a way which leaves it free to choose *any* number of dimensions, the customary notations of analytic geometry have to be profoundly modified. We cannot use different letters x, y, z, $\cdots$ for the variables, since we would run out of letters. Moreover, we could never discover the inherent laws which govern our equations if we did not introduce a more systematic notation which takes into account the homogeneous nature of space in a more adequate fashion. The variables x, y, z have to be changed to x_1, x_2, x_3. The subscript notation has immediately the consequence that the number of dimensions can be systematically extended to any number we want.

Hence we will write the equation of an ellipse in the form

$$\lambda_1 x_1^2 + \lambda_2 x_2^2 = 1 \tag{2-8.1}$$

and the equation of an ellipsoid in the form

$$\lambda_1 x_1^2 + \lambda_2 x_2^2 + \lambda_3 x_3^2 = 1 \tag{2-8.2}$$

Then, if we want the equation of an ellipsoid in n dimensions, we can immediately generalize the previous equations to

$$\lambda_1 x_1^2 + \lambda_2 x_2^2 + \cdots + \lambda_n x_n^2 = 1 \tag{2-8.3}$$

The fact that n is not given is no handicap since we know the *law* according to which the equation is formed.

However, in physics and engineering, the problem of an ellipse or ellipsoid is usually not encountered in this fashion. If our aim is to study the properties of an ellipse, it is justifiable to put our frame of reference immediately in a definite relation to that ellipse. The major and the minor axes of that ellipse can immediately be chosen as the x and y axes of our frame of reference. But usually this is not the given situation. We have a given frame of reference, chosen by some other considerations, and a certain ellipse or ellipsoid may appear in this frame of reference, but in arbitrarily slanted position. In this

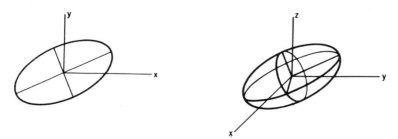

position the equation of the ellipse or the ellipsoid is less simple than before. In the case of an ellipse, an xy term is present, while in the case of an ellipsoid, the products xy, yz, and zx will occur. Hence the equation of an ellipse has now three instead of two terms, the equation of an ellipsoid six instead of three terms, and in the general n-dimensional case the appearance of the product terms has the consequence that the equation is composed of $\frac{1}{2}n(n+1)$ instead of n terms.

We have to learn how to handle these terms symbolically in order to make them operationally available. For this purpose we use a very definite method. Although the product xy appears only once, since xy and yx are equal by the commutative law of multiplication, we prefer to keep these two terms separated. Hence we will actually complicate the situation by writing n^2 instead of $\frac{1}{2}n(n+1)$ terms, but now we can trace the inner law of the terms more effectively. The equation of an ellipse will be written as follows:

$$(a_{11}x_1 + a_{12}x_2)x_1 + (a_{21}x_1 + a_{22}x_2)x_2 = 1 \qquad (2\text{-}8.4)$$

The equation of an ellipsoid becomes

$$(a_{11}x_1 + a_{12}x_2 + a_{13}x_3)x_1 + (a_{21}x_1 + a_{22}x_2 + a_{23}x)x_2$$
$$+ (a_{31}x_1 + a_{32}x_2 + a_{33}x_3)x_3 = 1 \qquad (2\text{-}8.5)$$

In arbitrary n dimensions, the same equation may be written as follows:

$$(a_{11}x_1 + a_{12}x_2 + \cdots + a_{1n}x_n)x_1$$
$$+ (a_{21}x_1 + a_{22}x_2 + \cdots + a_{2n}x_n)x_2$$
$$+ \cdots$$
$$+ (a_{n1}x_1 + a_{n2}x_2 + \cdots + a_{nn}x_n)x_n = 1 \qquad (2\text{-}8.6)$$

Since the terms x_ix_k and x_kx_i can be combined into one, it is the sum $a_{ik} + a_{ki}$ only which influences the equation, and since $a_{ik} + a_{ki}$ is symmetric with respect to an exchange of the indices i and k, we can assume from the very beginning that

$$a_{ik} = a_{ki} \qquad (2\text{-}8.7)$$

A matrix of this kind, which is insensitive with respect to an exchange of indices, is called a "symmetric matrix." For example, the matrix

$$A = \begin{bmatrix} 3 & 4 & -7 \\ 4 & -5 & 0 \\ -7 & 0 & 2 \end{bmatrix}$$

is symmetric, because $a_{12} = a_{21} = 4$, $a_{13} = a_{31} = -7$, $a_{23} = a_{32} = 0$. A symmetric matrix has the property that a transposition of rows and columns does not change anything on the matrix, and thus

$$\tilde{A} = A \qquad (2\text{-}8.8)$$

Matrices of this kind have particularly important properties which distinguish them within the wider class of arbitrary matrices.

In matrix notation the equation of a general second-order central surface becomes

$$xAx = 1 \qquad (2\text{-}8.9)$$

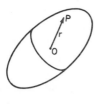

The vector $x = (x_1, x_2, x_3)$ in three dimensions and $x = (x_1, x_2, \cdots, x_n)$ in n dimensions has the significance of the "radius vector" r: which connects an arbitrary point P of the surface with the origin O.

It will now be our task to find the principal axes of this quadratic surface. This task does not exist if the surface is already given in the form

$$\lambda_1 x_1^2 + \lambda_2 x_2^2 + \cdots + \lambda_n x_n^2 = 1 \qquad (2\text{-}8.10)$$

since then the existence of the principal axes is taken for granted, and they are chosen as the rectangular axes of a Cartesian frame of reference. What characterizes specifically the principal axes of a quadratic surface? At every point of a surface the "normal" *n* can be constructed, i.e., a vector which is orthogonal to the tangential plane of the surface at the point *P*. The vectors *r* and *n* are generally not parallel to each other. Only in very exceptional directions, namely in those directions which we usually choose as coordinate axes, does it happen that normal and radius vectors become parallel to each other. Hence the name "principal axes" for this particular set of directions.

Now the direction cosines of the normal *n* are proportional to Ax. Thus the parallelism of the radius vector x and the normal Ax is expressed in the equation

$$Ax = \lambda x \qquad (2\text{-}8.11)$$

This equation was the fundamental equation of eigenvalue analysis. We obtained it here by asking for the principal axes of a quadratic surface.

In our previous discussion we obtained the complete solution of the eigenvalue problem by finding the n values of λ for which the equation is solvable, and the associated n vectors $u_1, u_2, \cdots, u_n$, for which the equation is solvable. The eigenvalues $\lambda_1, \lambda_2, \cdots, \lambda_n$ have a definite geometrical significance. Let us find the point P on the quadratic surface in which the principal axis intersects the surface. By the equation of the surface,

$$xAx = 1 \qquad (2\text{-}8.12)$$

By the equation of the principal axes,

$$Ax = \lambda x \qquad (2\text{-}8.13)$$

Multiplying this equation by x, we obtain

$$xAx = \lambda x^2 = 1 \qquad (2\text{-}8.14)$$

which gives

$$x^2 = 1/\lambda \qquad (2\text{-}8.15)$$

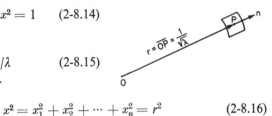

The significance of

$$x^2 = x_1^2 + x_2^2 + \cdots + x_n^2 = r^2 \qquad (2\text{-}8.16)$$

is the square of the distance of the point P in which the principal axis intersects the surface. Hence λ_i is the reciprocal of the square of the distance of this point P from the center. A large eigenvalue means that in the direction of a certain principal axis the quadratic surface comes near to the center. A small eigenvalue means that in the direction of a certain principal axis the surface stays far from the center.

The general principal axis problem is greatly simplified in the present case because of the fact that A is a *symmetric* matrix and thus $\tilde{A} = A$. In the general case we have to find separately the principal axes of A and of $\tilde{A}$. But if A is symmetric, and thus $\tilde{A} = A$, then the equation

$$Ax = \lambda x$$

solves simultaneously the equation

$$\tilde{A}x = \lambda x$$

The principal axes of A and $\tilde{A}$ now *coincide*. This has the consequence that the two matrices U and V, which characterize the general principal axis problem, become equal.

$$V = U \qquad (2\text{-}8.17)$$

But then the fundamental equation between these two matrices

$$\tilde{U}V = I$$

is reduced to the equation

$$\tilde{U}U = I \qquad (2\text{-}8.18)$$

This equation has the following significance. Let us form the dot products of the various columns of the U matrix with each other.

The dot product of any column with itself gives 1, the dot product of any column with another column gives 0. This means that the principal axes are mutually orthogonal, while their length is normalized to 1. A set of n vectors of this property, placed in an n-dimensional space, is called an "ortho-normal set."

We now have the proof that a "general" quadratic surface has n (and only n) principal axes and that these axes are orthogonal to each other. Hence we can introduce these axes as a new rectangular frame of reference and then we arrive at the point where the customary analytical geometry starts its investigations, by assuming from the beginning that the principal axes of the quadratic surface coincide with the rectangular axes of analytic geometry.

In actual fact we have not proved yet the *reality* of the principal axes, since the eigenvalues λ are the roots of an algebraic equation of nth order, and generally these roots may be complex numbers. The fundamental fact holds, however, that *all the eigenvalues of a symmetric matrix of real numbers are real.* And if the eigenvalue is real, then the associated solution

$$Ax = \lambda x \qquad (2\text{-}8.19)$$

must also be real. The reality of λ can be proved by assuming that λ is complex and showing that this assumption leads to a contradiction. Indeed, let us multiply (19) by x^*, where the notation $*$ refers to the operation "complex conjugate," i.e., the change of i to $-i$.

$$x^*Ax = \lambda x^*x \qquad (2\text{-}8.20)$$

Since in any algebraic relation involving complex numbers we know that such a relation remains true if every i is changed to $-i$, equation (19) has the consequence:

$$A^*x^* = \lambda^*x^* \qquad (2\text{-}8.21)$$

and premultiplying by x, we obtain

$$xA^*x^* = \lambda^*xx^* \qquad (2\text{-}8.22)$$

now we make use of the following fundamental transposition law.

$$x \cdot Ay = y \cdot \tilde{A}x \qquad (2\text{-}8.23)$$

This law, applied to the left side of (22), gives

$$x^*\tilde{A}^*x = \lambda^*x^*x \qquad (2\text{-}8.24)$$

Now, if A is real, then $A^* = A$, since the imaginary unit does not appear in A, and thus the change of i to $-i$ leaves the matrix unaltered. Moreover $\tilde{A} = A$, on account of the symmetry of the matrix. But then

$$x^* A x = \lambda^* x^* x \qquad (2\text{-}8.25)$$

and, taking the difference of (20) and (24), we obtain (the left sides being equal)

$$(\lambda - \lambda^*) x^* x = 0 \qquad (2\text{-}8.26)$$

The second factor cannot be zero, since it is the sum of positive quantities

$$x^* x = x_1^* x_1 + x_2^* x_2 + \cdots + x_n^* x_n = |x_1|^2 + |x_2|^2 + \cdots + |x_n|^2$$

We thus find
$$\lambda - \lambda^* = 0 \qquad (2\text{-}8.27)$$

or
$$\lambda = \lambda^* \qquad (2\text{-}8.28)$$

which means that λ must be *real*.

The reality of the principal axes of a quadratic surface is thus ascertained. It is important to know that the proof of the reality of λ depends solely on the condition

$$\tilde{A}^* = A \qquad (2\text{-}8.29)$$

Sometimes we have to solve the principal axis problem associated with a matrix with *complex* coefficients. If this matrix satisfies the condition (29), i.e., if the transposition of rows and columns *and* the simultaneous change of i to $-i$ restores the original matrix, we call such a matrix "Hermitian." For example, the following matrix is Hermitian.

$$A = \begin{bmatrix} 3 & 4 + 2i & -7 + 5i \\ 4 - 2i & -5 & -3i \\ -7 - 5i & 3i & 2 \end{bmatrix}$$

The transposition does not leave the matrix unchanged, but a transposition *and simultaneous change of every i to $-i$ restores the original matrix*. Although the eigenvectors of such a matrix are no longer real but complex vectors, the eigen*values* are still real, and the exceptional conditions which hold for real symmetric matrices carry over to the realm of Hermitian matrices. These matrices correspond in the complex realm to the symmetric matrices in the real realm. If a matrix has complex coefficients, the symmetry of the matrix is frequently of small advantage, although it is still true that the matrix V coincides with U. If A is Hermitian, the relation between V and U becomes

$$V = U^* \tag{2-8.30}$$

and thus $$\tilde{U}U^* = I \tag{2-8.31}$$

This, together with the reality of the eigenvalues, preserves the outstanding properties of ortho-normal vector systems.

9. The principal axis transformation of a matrix A. The geometrical approach to the problem of principal axes puts the emphasis on one particular phase of the theory which the purely algebraic theory does not reveal so conspicuously. If we picture a matrix in association with a quadratic surface, we see at once what the meaning of a principal axis is. But beyond that we see immediately the possibility of changing our frame of reference. Since the principal axes are mutually orthogonal and of the length 1, we can conceive them as a natural frame of reference associated with the given matrix. It will thus be advisable to study the problem of the *transformation of coordinates.*

Our original coordinates are

$$x = (x_1, x_2, \cdots , x_n) \tag{2-9.1}$$

Let us introduce a new set of coordinates:

$$\bar{x} = (\bar{x}_1, \bar{x}_2, \cdots , \bar{x}_n)$$

by changing our original axes to a new set of axes and generating the radius vector $\bar{x}$ as a linear combination of these new axes.

The general problem of coordinate transformations appears in the following form. Let us give a set of "base vectors":

$$u_1, u_2, \cdots , u_n \tag{2-9.2}$$

not necessarily orthogonal to each other and not normalized in length. Hence in general we want to assume that these vectors u_i are of arbitrary length and arbitrary direction, satisfying only *one* condition, viz., that they are *linearly independent.* Then the radius vector x can be generated as a linear superposition of these vectors.

$$x = x_1 u_1 + x_2 u_2 + \cdots + x_n u_n \tag{2-9.3}$$

In algebra the same vector is written with its components only:

$$x = (x_1, x_2, \cdots, x_n) \tag{2-9.4}$$

considering it an assembly of quantities x_i, characterized by a single subscript.

Now the choice of the base vectors $u_1, u_2, \cdots, u_n$ is more or less accidental. With equal right we could have chosen another set of vectors

$$\bar{u}_1, \bar{u}_2, \cdots, \bar{u}_n \tag{2-9.5}$$

and obtain the same vector x as a linear superposition of the new base vectors $\bar{u}_i$.

$$x = \bar{x}_1\bar{u}_1 + \bar{x}_2\bar{u}_2 + \cdots + \bar{x}_n\bar{u}_n \tag{2-9.6}$$

The relation between the two representations can be found as follows. Let us analyze the new vectors $\bar{u}_1$ in terms of the original vectors u_i. Since *any* vector can be obtained as a linear superposition of the base vectors, we can put

$$\begin{aligned}
\bar{u}_1 &= a_{11}u_1 + a_{21}u_2 + \cdots + a_{n1}u_n \\
\bar{u}_2 &= a_{12}u_1 + a_{22}u_2 + \cdots + a_{n2}u_n \\
&\vdots \\
\bar{u}_n &= a_{1n}u_1 + a_{2n}u_2 + \cdots + a_{nn}u_n
\end{aligned} \tag{2-9.7}$$

Writing the same equations in components only, we can put the components of $\bar{u}_1, \bar{u}_2, \cdots, \bar{u}_n$ in successive columns and thus obtain the following algebraic representation of the given coordinate transformation.

$$\tag{2-9.8}$$

Omitting the vertical bars, we obtain the matrix A:

$$A = \begin{bmatrix}
a_{11} & a_{12} & \cdots & a_{1n} \\
a_{21} & a_{22} & \cdots & a_{n2} \\
\vdots & & & \\
a_{nl} & a_{n2} & \cdots & a_{nn}
\end{bmatrix} \tag{2-9.9}$$

which uniquely characterizes the entire coordinate transformation. The successive columns of this matrix have the significance of giving the components of the first, second, $\cdots$, nth new base vector, analyzed in the original reference system. If we introduce in (6) the transformation (7), we obtain

$$
\begin{aligned}
x &= \bar{x}_1(a_{11}u_1 + a_{21}u_2 + \cdots + a_{n1}u_n) \\
&\quad + \bar{x}_2(a_{12}u_1 + a_{22}u_2 + \cdots + a_{n2}u_n) \\
&\quad + \cdots \\
&\quad + \bar{x}_n(a_{1n}u_1 + a_{2n}u_2 + \cdots + a_{nn}u_n) \\
&= (a_{11}\bar{x}_1 + a_{12}\bar{x}_2 + \cdots + a_{1n}\bar{x}_n)u_1 \\
&\quad + (a_{21}\bar{x}_1 + a_{22}\bar{x}_2 + \cdots + a_{2n}\bar{x}_n)u_2 \\
&\quad + \vdots \\
&\quad + (a_{n1}\bar{x}_1 + a_{n2}\bar{x}_2 + \cdots + a_{nn}\bar{x}_n)u_n
\end{aligned}
\tag{2-9.10}
$$

Originally the vector x was expressed in the form

$$
x = x_1 u_1 + x_2 u_2 + \cdots + x_n u_n \tag{2-9.11}
$$

The new form (10) must coincide with (11), since a vector cannot have two different representations in one and the same set of axes, if the axes are linearly independent. This yields

$$
\begin{aligned}
x_1 &= a_{11}\bar{x}_1 + a_{12}\bar{x}_2 + \cdots + a_{1n}\bar{x}_n \\
x_2 &= a_{21}\bar{x}_1 + a_{22}\bar{x}_2 + \cdots + a_{2n}\bar{x}_n \\
&\vdots \\
x_n &= a_{n1}\bar{x}_1 + a_{n2}\bar{x}_2 + \cdots + a_{nn}\bar{x}_n
\end{aligned}
\tag{2-9.12}
$$

or in matrix notation,

$$
x = A\bar{x} \tag{2-9.13}
$$

It will be advisable to change our notations slightly. Since the original vectors $u_1, u_2, \cdots, u_n$ do not appear in the final formulation, we may omit the bar over the $\bar{u}_1, \bar{u}_2, \cdots, \bar{u}_n$ and call the new base vectors simply $u_1, u_2, \cdots, u_n$. Moreover, the components of these vectors, analyzed in the original frame of reference, can be conveniently denoted by the symbol u_{ik} instead of a_{ik}. Hence the transformation matrix should be denoted by U, with the understanding that the subsequent columns of this matrix represent the components of $u_1, u_2, \cdots, u_n$. The transformation equation now becomes

$$x = U\bar{x} \qquad (2\text{-}9.14)$$

If the original base vectors of our coordinate system represent a customary rectangular set of axes, this will generally not be true any more of the new vectors $u_1, u_2, \cdots, u_n$. We may want, however, to preserve the rectangular character of our reference system. Then we have to set a condition on the transformation equation (14). The new vectors $u_1, u_2, \cdots, u_n$ have to be mutually orthogonal to each other and their length must be normalized to 1. This means that the dot product of any u vectors must come out as 0, while the dot product of any u vector with itself must come out as 1. All these conditions are included in the single matrix equation

$$\tilde{U}U = I \qquad (2\text{-}9.15)$$

This equation coincides with the equation (8.18) established earlier in the problem of finding the principal axes of a quadratic surface. The principal axes of a quadratic surface represent a set of base vectors which are automatically rectangular and which can thus be introduced as a new reference system. Let us see what happens to the quadratic surface as a result of this transformation. We put

$$x = U\bar{x} \qquad (2\text{-}9.16)$$

and introduce this transformation in the defining equation of the quadratic surface:

$$(U\bar{x}) \cdot (AU\bar{x}) = 1 \qquad (2\text{-}9.17)$$

We make use of the transposition law (8.23):

$$x \cdot Ay = y \cdot \tilde{A}x$$

which, applied to our problem, gives

$$\bar{x} \cdot (\widetilde{AU})U\bar{x} = 1 \qquad (2\text{-}9.18)$$

Since, however, (cf. 7.7)

$$(\widetilde{AU}) = \tilde{U}\tilde{A} = \tilde{U}A \qquad (2\text{-}9.19)$$

we obtain

$$\bar{x} \cdot \tilde{U}AU\bar{x} = 1 \qquad (2\text{-}9.20)$$

Now by the definition of the principal axes

$$AU = U\Lambda \qquad (2\text{-}9.21)$$

Premultiplying by $\tilde{U}$, we obtain

$$\tilde{U}AU = \tilde{U}U\Lambda = \Lambda \qquad (2\text{-}9.22)$$

and thus

$$\bar{x}\Lambda\bar{x} = 1 \qquad (2\text{-}9.23)$$

which means

$$\lambda_1\bar{x}_1^2 + \lambda_2\bar{x}_2^2 + \cdots + \lambda_n\bar{x}_n^2 = 1 \qquad (2\text{-}9.24)$$

We thus obtain the traditional form of a quadratic surface as it appears in analytic geometry, where we assume in advance that the principal axes of the given quadratic surface are chosen as the axes of a rectangular reference system.

The fact that in the new reference system the matrix A is replaced by the diagonal matrix Λ shows that the transformation to the principal axes had a profound influence on A by transforming the matrix into a particularly desirable form, viz., a purely *diagonal* form. The operation with diagonal matrices is infinitely simpler than the operation with arbitrary matrices. The equations are separated in the unknowns and immediately solvable. On the other hand, this diagonalization of the matrix requires the knowledge of the principal axes, which is generally not easily accomplished. However, the method of coordinate transformation is nevertheless a fundamentally important tool of matrix analysis. We may have a method of obtaining the principal axes of a matrix in rather *crude* approximation. Then the operation $\tilde{U}AU$ will not turn A into a diagonal form, but it will *boost up* the diagonal terms in comparison to the other terms. Such a matrix is still of great numerical advantage because matrix problems associated with such a matrix can be solved numerically in a series of quickly convergent iterations.

A special discussion is demanded in connection with *multiple roots*. The eigenvalues λ_i were obtained by solving an algebraic equation of the order n. Such an equation has always n roots but it may happen that some of the roots collapse into one. In that case we speak of a "multiple root" because that root stands for several distinct roots. The case of collapsing roots can be conceived as a limit, starting out with distinct roots and letting them approach each other indefinitely. Here again the geometrical picture helps us understand the nature of multiple roots. In the equation of an ellipse the equality of λ_1 and λ_2 brings about the equation of a *circle*:

$$\lambda_1(x_1^2 + x_2^2) = 1$$

In space the equality of two λ values generates a rotational ellipsoid:

$$\lambda_1(x_1^2 + x_2^2) + \lambda_3 x_3^2 = 1$$

and a *sphere* is generated if all the three λ values are equal:

$$\lambda_1(x_1^2 + x_2^2 + x_3^2) = 1$$

Now, if an ellipse becomes gradually a circle, its principal axes do not cease to exist. They become two mutually perpendicular diameters of the circle. However, the same circle may be the limiting position of an infinity of ellipses and thus any two mutually perpendicular diameters of the circle may serve as principal axes. In a similar way in the case of a sphere, any three mutually perpendicular diameters may serve as principal axes of that sphere. The same holds in higher dimensions. If in the general n-dimensional case m eigenvalues collapse into one, this means that in a certain m-dimensional "subspace" spherical conditions prevail. Any m mutually orthogonal axes can be chosen within that subspace as principal axes of the quadratic surface. The existence of multiple roots does not invalidate the existence of n distinct and mutually perpendicular axes. What happens is only that some of these axes are no longer uniquely determined but can be replaced by other equally valid axes. The collapse of certain eigenvalues into one is not connected with a corresponding collapse of the associated axes. The mutual orthogonality of the principal axes prevents them from ever collapsing into one.

Numerically the multiplicity of roots is always the cause of certain difficulties. If certain eigenvalues come very near together without collapsing into one, the associated principal axes are theoretically still uniquely determined. But to find these axes with any degree of accuracy becomes increasingly difficult as the difference between two eigenvalues decreases to smaller and smaller amounts.

10. Skew-angular reference systems. In our previous discussions we have assumed that the quadratic surface is in a slanted position relative to the axes of our reference system. Then the principal axes of the quadratic surface were obtained and these axes introduced as a new frame of reference. This transformation of the coordinates had the effect that the symmetric matrix A was transformed into a pure diagonal matrix Λ. We will now go one step further. Up to now we have assumed that we operate with a *rectangular* set of coordinates. The axes of our reference system did not coincide with the axes of the quadratic surface and this necessitated a transformation of the coordinates. But both the original coordinates axes and the principal axes of the quadratic surface were rectangular axes. The transformation involved was a mere *rotation* of the axes, i.e., an orthogonal transformation. Such an orthogonal transformation is characterized by the matrix equation

$$\tilde{U}U = I$$

The principal axes of a quadratic surface automatically satisfied this equation.

It can happen, however, that we encounter a still more general situation. The base vectors u_1, u_2, $\cdots$, u_n of our original reference system may not be orthogonal to each other. We then have a "skew-angular" set of axes. In this case the quadratic surface must

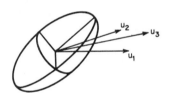

of necessity be in a slanted position relative to our axes, because the principal axes of the quadratic surface remain orthogonal to each other and thus we are sure that our skew-angular axes cannot coincide with the principal axes of the quadratic surface.

Since in ordinary analytical geometry we use almost exclusively a rectangular set of axes, our first task will be to investigate quite generally the operation with skew-angular axes. We assume that the base vectors u_1, u_2, $\cdots$, u_n are given as a set of n vectors of arbitrary direction and arbitrary length, except for one condition, viz., that these vectors shall be *linearly independent*. This means that no vector can be obtained as a linear combination of the other vectors. In other words a linear relation of the form (other than for each α_i equal to zero),

$$\alpha_1 u_1 + \alpha_2 u_2 + \cdots + \alpha_n u_n = 0 \qquad (2\text{-}10.1)$$

is *excluded*, since such a relation would imply that one vector—e.g., u_n if α_n is different from zero—could be obtained as a linear superposition of the other vectors.

As a consequence of this linear independence, we see at once that an arbitrary vector x, obtained as a linear superposition of the base vectors

$$x = x_1 u_1 + x_2 u_2 + \cdots + x_n u_n \qquad (2\text{-}10.2)$$

cannot have two different representations. If the same vector x could be obtained in a different way:

$$x = \xi_1 u_1 + \xi_2 u_2 + \cdots + \xi_n u_n \qquad (2\text{-}10.3)$$

then the difference of (2) and (3) gives

$$(x_1 - \xi_1)u_1 + (x_2 - \xi_2)u_2 + \cdots + (x_n - \xi_n)u_n = 0 \qquad (2\text{-}10.4)$$

This, however, would establish a linear relation of the form (1), which was excluded in advance. The representation (2) of the vector x is thus unique.

We can interpret (2) in the sense of a *synthesis*. Given the base vectors $u_1, u_2, \cdots, u_n$, we obtain a vector by multiplying each one of the base vectors by $x_1, x_2, \cdots, x_n$ and adding them up in the sense of vector addition. But frequently the *inverse* of the problem is encountered. The vector x is given, and we have to find out what linear superposition of the base vectors will generate that particular vector. We say that we *analyze* the vector x in the reference system of the base vectors $u_1, u_2, \cdots, u_n$. Here x is given, and we have to find the coefficients $x_1, x_2, \cdots, x_n$ of the linear superposition problem (2). These coefficients are called the "components" of the vector x in the reference system of the base vectors $u_1, u_2, \cdots, u_n$.

From the standpoint of this analysis, the rectangular systems are vastly superior to the skew-angular systems. If the vectors $u_1, u_2, \cdots, u_n$ form an orthogonal and normalized set of vectors, they satisfy the equations

$$\begin{aligned} u_1 \cdot u_k &= 0 \qquad (i \neq k) \\ u_i^2 &= 1 \end{aligned} \qquad (2\text{-}10.5)$$

These equations express in algebraic form the geometrical facts that any two vectors of the set are mutually orthogonal and that the length of any vector of the set is 1. In this reference system the "dotting" of the vector x with the unit vector u_i gives

$$x \cdot u_i = (x_1 u_1 + x_2 u_2 + \cdots + x_n u_n)u_i = x_i$$

and thus
$$x_i = x \cdot u_i \qquad (2\text{-}10.6)$$

The mere multiplication of the vector x by the vector u_i gives the scalar x_i, which is directly the component of x in the direction of the base vector u_1.

The same equation is *not* true, however, in a skew-angular reference system. Here the independence with which each one of the base vectors u_i operates is lost, and we have to find some other tools for replacing the simple equation (6). We do that by constructing a new set of vectors $v_1, v_2, \cdots, v_n$, called the "adjoint" set. Everything we do with rectangular axes can be duplicated with the help of skew-angular axes, provided only that we enlarge the given set of vectors to double capacity. Instead of operating with the *single* set of vectors $u_1, u_2, \cdots, u_n$, we operate with a *double* set of vectors.

$$\begin{pmatrix} u_1, u_2, \cdots, u_n \\ v_1, v_2 \cdots, v_n \end{pmatrix} \tag{2-10.7}$$

Every vector u_i is associated with a corresponding vector v_i, called the "conjugate" of u_i. The vectors v_i and u_i are in a definite duality relation to each other, which holds symmetrically in both directions. If the vectors u_i are given, their "adjoints" are the vectors v_i. If, on the other hand, we start with the vectors $v_1, v_2, \cdots, v_n$ as base vectors and construct their adjoints, we obtain the vectors $u_1, u_2, \cdots, u_n$. Hence the adjoint of the adjoint set is the original set.

The adjoint vectors $v_1, v_2, \cdots, v_n$ are uniquely defined. They are in a biorthogonality relation to the original vectors $u_1, u_2, \cdots, u_n$, in the sense that any vector u_i of the one set and any vector of the other set v_k (excluding its own conjugate v_i) are mutually orthogonal. Moreover, the length of the adjoint vectors v_i are normalized by the condition that the dot product $u_i \cdot v_i$ of any vector with its own conjugate shall be 1

$$u_i \cdot v_k = 0 \qquad (i \neq k)$$
$$u_i \cdot v_i = 1 \tag{2-10.8}$$

We see that the advantage of a rectangular set of vectors lies in the fact that the necessity of constructing a second set of adjoint vectors is obviated because there the vectors u_i themselves satisfy the orthogonality relations (5), and this means that the vectors v_i coincide with the vectors u_i

$$v_i = u_i \tag{2-10.9}$$

A rectangular and normalized set of vectors is therefore called "self-adjoint." For any other set of vectors the set v_i has to be specifically constructed. However, after this construction is done, we can operate with a skew-angular reference system just as easily and effectively as with an orthogonal system.

For this reason we will assume that whenever a skew-angular reference system $u_1, u_2, \cdots, u_n$ is given, we have automatically associated with it the adjoint system $v_1, v_2, \cdots, v_n$ and both vector systems are always considered together and not the one or the other system alone.

$$\begin{pmatrix} u_1, u_2, \cdots, u_n \\ v_1, v_2, \cdots, v_n \end{pmatrix}, \qquad u_i \cdot v_k = 0 \quad (i \neq k), \qquad u_i \cdot v_i = 1 \quad (2\text{-}10.10)$$

The problem of analyzing a vector x in the reference system of the base vectors u_i has now a simple solution. We multiply the equation

$$x = x_i u_1 + x_2 u_2 + \cdots + x_n u_n$$

by the vector v_i and obtain, on the basis of the biorthogonality relations (8),

$$x \cdot v_i = x_i$$

and thus
$$x_i = x \cdot v_i \qquad (2\text{-}10.11)$$

The dotting of x with conjugate vector v_i gives the factor x_i, i.e., the component of x in the direction of the base vector u_i.

Now the vectors u_i and v_i are so closely associated that one vector system has no meaning without the other. If we used the u_i as a base system for analyzing the vector x, we can use with equal justification the vectors v_i.

$$x = \xi_1 v_1 + \xi_2 v_2 + \cdots + \xi_n v_n \qquad (2\text{-}10.12)$$

Hence every vector has two representations in a skew-angular reference system, once denoted by Latin and once denoted by Greek letters. If (12) is dotted by u_i, we obtain

$$\xi_i = x \cdot u_i$$

The vector x is the same in both cases; only the *reference system* has changed in which the analysis was made. We do not have a dual pair of vectors, but two conjugate representations of one and the same vector, analyzed once in the original and once in the adjoint system.

$$x = x_1 u_1 + x_2 u_2 + \cdots + x_n u_n = \xi_1 v_1 + \xi_2 v_2 + \cdots + \xi_n v_n \qquad (2\text{-}10.13)$$

However, for algebraic operations we *omit* the unit vectors u_1, u_2, $\cdots$, u_n and retain the components only. We thus write

$$x = (x_1, x_2, \cdots, x_n) \tag{2-10.14}$$

Similarly, if we want to operate with the adjoint set of axes, we will omit the base vectors v_1, v_2, $\cdots$, v_n and write solely the components ξ_1, ξ_2, $\cdots$, ξ_n. But these components are completely different from the previous components, and thus the resultant vector cannot be called x any more, but ξ.

$$\xi = (\xi_1, \xi_2, \cdots, \xi_n) \tag{2-10.15}$$

While this notation is entirely logical and consistent as an algebraic notation, it is necessary to keep in mind that the two algebraic vectors x and ξ are not more than *two parallel representations* of one *single* geometrical entity, viz., the vector $\vec{x}$, on account of the two dual reference systems with which we have to operate.

In the algebraic sense the vector $\vec{x}$ has to be written down as the *pair* of vectors (x, ξ). In the customary rectangular reference systems of analytic geometry this doubling of the components does not occur, since the vector ξ and the vector x coincide. But in a skew-angular system the vector x is necessarily associated with the dual vector ξ.

We will now form the dot product of the two vectors x and y.

$$\vec{x} = x_1 u_1 + x_2 u_2 + \cdots + x_n u_n = \xi_1 v_1 + \xi_2 v_2 + \cdots + \xi_n v_n$$

$$\vec{y} = y_1 u_1 + y_2 u_2 + \cdots + y_n u_n = \eta_1 v_1 + \eta_2 v_2 + \cdots + \eta_n v_n$$

In view of the duality relations between the u_i and v_i axes, it is inevitable that the one vector must be used in the one, and the other in the other representation. We cannot form the products $x \cdot y$ or $\xi \cdot \eta$, but we can form the products $x \cdot \eta$ or $\xi \cdot y$.

$$x \cdot \eta = (x_1 u_1 + x_2 u_2 + \cdots + x_n u_n)(\eta_1 v_1 + \eta_2 v_2 + \cdots + \eta_n v_n)$$
$$= x_1 \eta_1 + x_2 \eta_2 + \cdots + x_n \eta_n$$
$$\xi \cdot y = (\xi_1 v_1 + \xi_2 v_2 + \cdots + \xi_n v_n)(y_1 u_1 + y_2 v_2 + \cdots + y_n v_n)$$
$$= \xi_1 y_1 + \xi_2 y_2 + \cdots + \xi_n y_n$$

In both cases we obtain the same dot product of the vector x and the vector y and thus:

$$x \cdot \eta = \xi \cdot y \tag{2-10.16}$$

The ordinary definition of the dot product of two vectors in the sense of multiplying corresponding components and forming the sum

$$a = (a_1, a_2, \cdots, a_n)$$
$$b = (b_1, b_2, \cdots, b_n)$$
$$a \cdot b = a_1 b_1 + a_2 b_2 + \cdots, a_n b_n$$

remains valid without any alteration. But we have to know that the dot product of two vectors can only be formed if the two factors are given in *dual representations*. The dot products

$$x \cdot y = x_1 y_1 + x_2 y_2 + \cdots + x_n y_n$$
$$\xi \cdot \eta = \xi_1 \eta_1 + \xi_2 \eta_2 + \cdots + \xi_n \eta_n$$

have no meaning whatsoever. One factor has to be analyzed in the u system, the other factor in the v system. (In tensor calculus the components $x_1, x_2, \cdots, x_n$ of a vector are called the "covariant components," and the components $\xi_1, \xi_2, \cdots, \xi_n$ of the same vector are called the "contravariant components" of the vector x. The distinction between the two sets of components is not made by using Latin and Greek letters, but by putting the subscripts once in an upper and once in a lower position.

$$x = (x_1, x_2, \cdots, x_n) \qquad \text{(covariant vector)}$$
$$x = (x^1, x^2, \cdots, x^n) \qquad \text{(contravariant vector)}$$

The dot product of the vectors x and y becomes

$$x \cdot y = x^1 y_1 + x^2 y_2 + \cdots + x^n y_n = x_1 y^1 + x_2 y^2 + \cdots + x_n y^n$$

This notation is less suited, however, for matrix operations.)

11. Principal axis transformation in skew-angular systems. We assume that a quadratic surface is given in a skew-angular frame of reference. The base vectors are

$$\begin{pmatrix} u_1, u_2, \cdots, u_n \\ v_1, v_2, \cdots, v_n \end{pmatrix} \qquad (2\text{-}11.1)$$

The radius vector x is given in dual fashion:

$$x = (x_1, x_2, \cdots, x_n)$$
$$\xi = (\xi_1, \xi_2, \cdots, \xi_n) \qquad (2\text{-}11.2)$$

The equation of a quadratic surface in ordinary rectangular co-ordinates has been

$$x \cdot Ax = 1 \tag{2-11.3}$$

In a skew-angular reference system the dot product can be formed only in dual representation, giving one vector in the base system and the other vector in the adjoint system. Hence now we have

$$\xi \cdot Ax = 1 \tag{2-11.4}$$

or written out in detail,

$$
\begin{aligned}
&\xi_1(a_{11}x_1 + a_{12}x_2 + \cdots + a_{1n}x_n) \\
&\xi_2(a_{21}x_1 + a_{22}x_2 + \cdots + a_{2n}x_n) \\
&\vdots \\
&\xi_n(a_{n1}x_1 + a_{n2}x_2 + \cdots + a_{nn}x_n) = 1
\end{aligned}
\tag{2-11.5}
$$

These are truly n^2 different terms which cannot be reduced to $n(n+1)/2$ terms, because the terms $a_{ik}\xi_i x_k$ and $a_{ki}\xi_k x_i$ do not combine any more. The matrix A is no longer symmetric but can be given as an arbitrary matrix with n^2 different elements.

The principal axes of the quadratic surface are once more those directions in which normal and radius vector become parallel. The direction cosines of the normal are once more proportional to Ax, and the proportionality of normal and radius vector is once more expressed in the equation

$$Ax = \lambda x \tag{2-11.6}$$

However, every vector has a dual aspect by being representable in both the original base vectors u_i and the adjoint vectors v_i. Hence we want to obtain the equations of the principal axes not only in the original system but also in the adjoint system. This means that we want to add an equation which will determine ξ instead of x. Now by the algebraic identity

$$x \cdot Ay = y \cdot \tilde{A}x$$

we can write the basic equation (3) in the alternate form

$$x \cdot \tilde{A}\xi = 1 \tag{2-11.7}$$

and, applying the equation (6) to the new form, we now obtain

$$\tilde{A}\xi = \lambda \xi \tag{2-11.8}$$

We have encountered the equations (6) and (8) earlier, in connection with the eigenvalue problem of the matrix A and its transpose $\tilde{A}$. In the purely algebraic treatment, the principal axes of A and $\tilde{A}$ separate into $n + n$ independent vectors. We have altogether $2n$ vectors, and we have seen that these two sets of vectors are in a biorthogonality relation to each other. In the geometrical treatment we have in fact only n vectors, namely, the n principal axes of a quadratic surface. But these n vectors have to be analyzed in two reference systems, viz., the u system and the v system. And thus every one of the principal axes has two sets of components, viz., the components $x_1, x_2, \cdots, x_n$ in the u system and the components $\xi_1, \xi_2, \cdots, \xi_n$ in the v system. Instead of $2n$ vectors we have merely n vectors, but each vector represented in two ways. If these vectors are once more called

$$\bar{u}_1, \bar{u}_2, \cdots, \bar{u}_n$$

and we use once more the notations of the previous section, we can put the components of these vectors in successive columns of a matrix.

$\bar{u}_1$	$\bar{u}_2$	$\cdots$	$\bar{u}_n$
u_{11}	u_{12}	$\cdots$	u_{1n}
u_{21}	u_{22}	$\cdots$	u_{2n}
$\vdots$			
u_{n1}	u_{n2}	$\cdots$	u_{nn}

$$(2\text{-}11.9)$$

This means in terms of vectors,

$$\vec{\bar{u}}_1 = u_{11}\vec{u}_1 + u_{21}\vec{u}_2 + \cdots + u_{n1}\vec{u}_n$$
$$\vdots \qquad\qquad (2\text{-}11.10)$$
$$\vec{\bar{u}}_n = u_{1n}\vec{u}_1 + u_{2n}\vec{u}_2 + \cdots + u_{nn}\vec{u}_n$$

But the very same vectors can likewise be analyzed in the v system.

$$\vec{\bar{u}}_1 = v_{11}\vec{v}_1 + v_{21}\vec{v}_2 + \cdots + v_{n1}\vec{v}_n$$
$$\vdots \qquad\qquad (2\text{-}11.11)$$
$$\vec{\bar{u}}_n = v_{1n}\vec{v}_1 + v_{2n}\vec{v}_2 + \cdots + v_{nn}\vec{v}_n$$

Now we abandon the reference to the original base vectors altogether. Then we can omit the bars and call the new base vectors simply

$u_1, u_2, \cdots, u_n$. These vectors are algebraically characterized by the matrix U:

$$
U = \begin{array}{c}
\begin{array}{cccc} u_1 & u_2 & \cdots & u_n \end{array} \\
\left[\begin{array}{cccc}
u_{11} & u_{12} & \cdots & u_{1n} \\
u_{21} & u_{22} & & u_{2n} \\
\vdots & & & \\
u_{n1} & u_{n2} & \cdots & u_{nn}
\end{array} \right]
\end{array}
\qquad (2\text{-}11.12)
$$

But the very same vectors, analyzed in the adjoint system, represent algebraically a second set of vectors, to be denoted by $v_1, v_2, \cdots, v_n$:

$$
V = \begin{array}{c}
\begin{array}{cccc} v_1 & v_2 & \cdots & v_n \end{array} \\
\left[\begin{array}{cccc}
v_{11} & v_{12} & \cdots & v_{1n} \\
v_{21} & v_{22} & & v_{2n} \\
\vdots & & & \\
v_{n1} & v_{n2} & \cdots & v_{nn}
\end{array} \right]
\end{array}
\qquad (2\text{-}11.13)
$$

The columns of U and the columns of V belong to the same set of n vectors. The fact that any two of the principal axes are orthogonal to each other means that the dot product of any U column with any V column except its own gives zero. Moreover, the fact that the length of the principal axes is normalized to 1 means that the dot product of any U column with the corresponding V column gives 1. In matrix equation,

$$
\tilde{U} V = I \qquad (2\text{-}11.14)
$$

We have encountered this equation before as the result of algebraic operations. The new insight we gain by the geometric treatment is that this equation expresses the orthogonality of the principal axes of a quadratic surface. In the earlier section this orthogonality appeared in the form

$$
\tilde{U} U = I
$$

and the matrix A was symmetric. The more general equation (14) comes about because the base vector system of our analysis is no longer rectangular but skew-angular.

We will now introduce the principal axes of the quadratic surface as a new reference system. Here the duality of the original and the adjoint system can be abandoned because the new reference system is "ortho-normal" (orthogonal and normalized in length) and therefore self-adjoint. The transformation to the new axes is given by the equations

$$x = U\tilde{x}, \qquad \xi = V\tilde{x} \qquad (2\text{-}11.15)$$

Introducing this transformation in (4), we obtain

$$V\tilde{x}\cdot AU\tilde{x} = 1 \qquad (2\text{-}11.16)$$

and by transposition,

$$\tilde{x}(\widetilde{AU})V\tilde{x} = \tilde{x}\tilde{U}\tilde{A}V\tilde{x} = 1 \qquad (2\text{-}11.17)$$

Since, however, by the definition of the matrix V,

$$\tilde{A}V = V\Lambda \qquad (2\text{-}11.18)$$

premultiplication by $\tilde{U}$ gives

$$\tilde{U}\tilde{A} = \tilde{U}V\Lambda = \Lambda \qquad (2\text{-}11.19)$$

and thus, according to (17), the equation of the quadratic surface in the new reference system becomes

$$\tilde{x}\Lambda\tilde{x} = 1 \qquad (2\text{-}11.20)$$

This equation coincides with the previous result (9.23) of transforming a quadratic surface to its principal axes. In the previous problem we started with a symmetric matrix, while now our starting point was an arbitrary matrix. The final result must be the same since in both cases the quadratic surface involved was the same and in the final transformation, after introducing the principal axes of the surface itself, all traces of the previous set of axes disappear, and thus it can make no difference whether we have started with a rectangular or a skew-angular set of axes.

In final analysis we can say that the principal axis problem of quadratic surfaces, analyzed in a skew-angular frame of reference, demonstrates that even nonsymmetric matrices can be transformed into a purely diagonal form. This transformation appears in the form

$$\tilde{V}AU = \Lambda \qquad \text{or also} \qquad \tilde{U}\tilde{A}V = \Lambda$$

There is, however, one fundamental difference which distinguishes the skew-angular case from the rectangular one. In the rectangular case ($\tilde{A} = A$), we could prove that all the eigenvalues and eigenvectors are *real*. Hence the rotation from one frame of reference to the other was always a real rotation within the real n-dimensional space. In the general case, this is no longer so. The eigenvalues λ_i are generally *complex* numbers, and thus the elements of the matrix Λ are generally complex elements. Similarly the elements of the matrices U and V are generally likewise complex numbers.

A second fundamental difference has something to do with the distinctness of the eigenvectors. As long as the eigenvalues λ_i are distinct, the corresponding eigenvectors are also distinct. In the case of multiple eigenvalues, however, special conditions prevail. In the symmetric case we have seen that the collapse of the eigenvalues did *not* mean a corresponding collapse of the eigenvectors. It meant only a partial indistinctness of the eigenvectors, inasmuch as within a certain subspace of m dimensions *any* m mutually orthogonal axes could be chosen as eigenvectors. This is different in the non-symmetric case. The original skew-symmetric reference system may be very far from an orthogonal system, and it may happen that in the limit some of the base vectors collapse into one.

The final transformation gives us a clue concerning the original orientation of the base vectors. The transformation from x to $\bar{x}$ occurred with the help of the equation

$$x = U\bar{x}$$

Multiplying by $\tilde{V}$, we obtain the inverse of this equation:

$$\bar{x} = \tilde{V}x \qquad (2\text{-}11.21)$$

This tranformation equation shows that the original base vectors, analyzed in the orthogonal reference system of the principal axes, are given by the successive columns of the matrix $\tilde{V}$; i.e., the successive rows of the matrix V. If we could guarantee that our original base vectors represented a linearly independent set of vectors, the determinant of the matrix V would of necessity be different from zero, and we would know that the principal axes of A form a system of n linearly independent vectors. It can happen, however, that during the limit process, when two near eigenvalues eventually collapse into one, also the original system of base vectors eventually becomes linearly dependent. In that case, the determinant of V becomes zero, which means that the number of linearly independent eigenvectors of A is less than n.

An interesting example is provided by the three matrices:

$$\begin{bmatrix} 1 & 0 & 0 \\ 0 & 1 & 0 \\ 0 & 0 & 1 \end{bmatrix}, \quad \begin{bmatrix} 1 & 1 & 2 \\ 0 & 1 & 0 \\ 0 & 0 & 1 \end{bmatrix}, \quad \begin{bmatrix} 1 & -1 & 3 \\ 0 & 1 & 2 \\ 0 & 0 & 1 \end{bmatrix} \quad \text{(2-11.22)}$$

In all three cases the determinant equation which gives the eigenvalues λ_i is reduced to

$$(\lambda - 1)^3 = 0 \quad \text{(2-11.23)}$$

which shows that $\lambda = 1$ is the only possible eigenvalue, with the multiplicity 3. The three generally distinct eigenvalues of a 3×3 matrix collapse in our case into one.

Now in the first case the given matrix is the unit matrix. The associated quadratic surface is a sphere. Here we can choose any three mutually orthogonal axes as the principal axes of the given matrix. The three principal axes exist and they are well separated, although their direction is not uniquely determined, and an infinity of orthogonal systems can be chosen as principal axes.

Let us now examine the second matrix. The equations of the principal axes become

$$\begin{aligned} x_1 + x_2 + 2x_3 &= x_1 \\ x_2 &= x_2 \\ x_3 &= x_3 \end{aligned}$$

This is reducible to the single equation

$$x_2 + 2x_3 = 0$$

which has two independent solutions:

$$\begin{aligned} x_2 &= x_3 = 0 \\ x_2 &= -2, \qquad x_3 = 1 \end{aligned}$$

In the first case x_1 cannot vanish, since otherwise the entire vector would vanish. Hence we can choose

$$u_1 = (1, 0, 0)$$

as our first solution. The second solution leaves x_1 arbitrary, and we may choose

$$u_2 = (0, -2, 1)$$

as our second solution. The other possibility,

$$u_3 = (a, -2, 1)$$

with an arbitrary a, is only a linear combination of u_1 and u_2.

$$u_3 = au_1 + u_2$$

Hence we see that the multiple eigenvalue $\lambda = 1$ in our case is associated with only *two* linearly independent eigenvectors.

Let us now examine the third matrix. Here the equations of the principal axes become

$$\begin{aligned} x_1 - x_2 + 3x_3 &= x_1 \\ x_2 + 2x_3 &= x_2 \\ x_3 &= x_3 \end{aligned}$$

This system is equivalent to

$$\begin{aligned} x_3 &= 0 \\ -x_2 + 3x_3 &= 0 \end{aligned}$$

The solution of this system is

$$u_1 = (1, 0, 0)$$

Hence in this case the number of eigenvectors is only *one*. The three generally independent eigenvectors collapsed into one.

We see that the collapse of eigenvalues can at the same time cause a collapse of the associated eigenvectors, a phenomenon which has no analogy in the realm of real symmetric matrices. The question can be asked whether this peculiar behavior of the eigenvectors could be foreseen without going through the explicit solution of the eigenvalue problem. This is indeed the case as we can see if we pay our attention to the Hamilton-Cayley equation. In the general case we know that

$$(A - \lambda_1 I)(A - \lambda_2 I) \cdots (A - \lambda_n I) = 0 \qquad (2\text{-}11.24)$$

This identity makes it possible to express the nth power of A in terms of the lower powers. Now, if all the λ_i are different, there is no identity of lower order in existence; i.e., it is not possible that already a power of A less than n is reducible to the lower powers. But in the case of multiple roots, the situation is different. It is possible that a multiple root may occur only once in the Hamilton-Cayley equation and not m times, if m is the multiplicity of the root.

In the case of our three matrices, the eigenvalue $\lambda = 1$ has the multiplicity 3. Hence we know in advance that the full Hamilton-Cayley equation will be

$$(A - I)^3 = 0 \qquad (2\text{-}11.25)$$

and this equation will certainly be satisfied in all three cases. But it may be that already

$$(A - I)^2 = 0 \qquad (2\text{-}11.26)$$

or even that

$$A - I = 0 \qquad (2\text{-}11.27)$$

In our first example the given matrix was actually the unit matrix, and thus the case (27) is actually realized. Here the matrix identity of lowest order is not of third but only of first order. Already A itself is reducible to a mere constant. The matrix identity of lowest order contains the multiple root not more than *once*. This is an indication that, while we lose uniqueness in the determination of the eigenvectors, we do not lose dimensions. The eigenvectors still include the entire n-dimensional space.

Let us see now what happens in the *second* case. We form $A - I$ and then $(A - I)^2$:

$$A-I = \begin{bmatrix} 0 & 1 & 2 \\ 0 & 0 & 0 \\ 0 & 0 & 0 \end{bmatrix}, \quad (A-I)^2 = \begin{bmatrix} 0 & 0 & 0 \\ 1 & 0 & 0 \\ 2 & 0 & 0 \end{bmatrix} \circ \begin{bmatrix} 0 & 1 & 2 \\ 0 & 0 & 0 \\ 0 & 0 & 0 \end{bmatrix} = \begin{bmatrix} 0 & 0 & 0 \\ 0 & 0 & 0 \\ 0 & 0 & 0 \end{bmatrix} = 0$$

Here $A - I$ is not zero but

$$(A - I)^2 = 0$$

The multiple root $\lambda = 1$ entered the identity of lowest order with the multiplicity 2, and this indicates that *one of the space dimensions is lost*. The eigenvectors of A include a two-dimensional space only. This is what we have actually found. There were only two linearly independent eigenvectors in existence.

We now come to the examination of the third case.

$$A - I = \begin{bmatrix} 0 & -1 & 3 \\ 0 & 0 & 2 \\ 0 & 0 & 0 \end{bmatrix}$$

$$(A - I)^2 = \begin{bmatrix} 0 & 0 & 0 \\ -1 & 0 & 0 \\ 3 & 2 & 0 \end{bmatrix} \circ \begin{bmatrix} 0 & -1 & 3 \\ 0 & 0 & 2 \\ 0 & 0 & 0 \end{bmatrix} = \begin{bmatrix} 0 & 0 & -2 \\ 0 & 0 & 0 \\ 0 & 0 & 0 \end{bmatrix}$$

$$(A - I)^3 = \begin{bmatrix} 0 & 0 & 0 \\ 0 & 0 & 0 \\ -2 & 0 & 0 \end{bmatrix} \begin{bmatrix} 0 & -1 & 3 \\ 0 & 0 & 2 \\ 0 & 0 & 0 \end{bmatrix} = \begin{bmatrix} 0 & 0 & 0 \\ 0 & 0 & 0 \\ 0 & 0 & 0 \end{bmatrix} = 0$$

Here the multiple root $\lambda = 1$ entered the matrix identity of lowest order with triple multiplicity. Hence we lose *two* dimensions and the eigenvectors of the matrix include a $3 - 2 = 1$-dimensional subspace only. Indeed, we have seen that the given matrix did not have more than *one* single eigenvector. (Matrices of this kind whose principal axes avoid certain dimensions of space are called "defective matrices.")

Hence we see that the degree of the defectiveness of a matrix can always be established by constructing the identity of lowest order which connects the powers of a matrix and examining the multiplicity with which the collapsing roots enter this identity.

12. The invariance of matrix equations under orthogonal transformations. The method of coordinate transformations is of profound importance in the study of matrices. Coordinates have no absolute significance and can be discarded in favor of other coordinates. We can view a matrix from a certain frame of reference, but we may also introduce a new frame of reference which may be more suited to the nature of the given problem. The question of the "most adequate reference system" is therefore frequently of profound significance. We may start with a given frame of reference which does not do justice to the general symmetry properties of the given problem. We will obtain a better basis for our analysis if we abandon the original system in favor of a more adequate system. In principle, however, *all* reference systems are equally admissible. Reference systems can be compared with a scaffold in erecting a building. The scaffold does not belong to the building; it serves merely the purpose of getting access to all possible points of the building. But after the building has been constructed, the scaffold can be removed. This fundamental difference between scaffold and building was the departure point of Einstein's celebrated "theory of relativity." It is not permissible to intermingle something which belongs to the scaffold with something which belongs to the building. The nature of the frame of reference has to stand out clearly as an *auxiliary construction* which is part of the *description* of nature but does not belong to the inner *essence* of nature. In Newtonian physics scaffold and building were so strongly cemented together that the entire building collapsed if the scaffold was removed. This difficulty was finally solved when Einstein's theory showed how to take into account the relative nature of coordinates.

Matrix algebra is likewise an example of the operation of the principle of relativity. A matrix is associated with a certain space structure and the analysis of that space structure demands the use of coordinates. These coordinates have no absolute significance and can be abandoned in favor of other coordinates. But the inner laws expressed by the equations between matrices cannot be affected by the accidental frame of reference in which matrices are analyzed. We say that the equations of matrix algebra "remain invariant" with respect to a transformation of the coordinates.

We will first restrict ourselves to a rectangular frame of axes. We then have a set of base vectors:

$$\vec{u}_1, \vec{u}_2, \cdots, \vec{u}_n \tag{2-12.1}$$

which are mutually orthogonal and whose length is 1.

$$\vec{u}_i \vec{u}_k = 0 \quad (i \neq k) \tag{2-12.2}$$

$$\vec{u}_i^2 = 1$$

We will now introduce a new set of orthogonal axes:

$$\vec{\bar{u}}_1, \vec{\bar{u}}_2, \cdots, \vec{\bar{u}} \tag{2-12.3}$$

by the transformation

$$\vec{\bar{u}}_1 = u_{11}\vec{u}_1 + u_{21}\vec{u}_2 + \cdots, u_{n1}\vec{u}_n$$
$$\vdots \tag{2-12.4}$$
$$\vec{\bar{u}}_n = u_{1n}\vec{u}_1 + u_{2n}\vec{u}_2 + \cdots, u_{nn}\vec{u}_n$$

This transformation is characterized by the matrix U.

$$U = \begin{bmatrix} u_{11} & u_{12} & \cdots & u_{1n} \\ u_{21} & u_{22} & & u_{2n} \\ \vdots & & & \\ u_{n1} & u_{n2} & \cdots & u_{nn} \end{bmatrix} \tag{2-12.5}$$

The successive columns of this matrix give us the components of the first, second, $\cdots$, nth new base vectors, analyzed in the original system. The fact that the new base vectors again satisfy the orthogonality conditions (2) finds expression in the matrix equation

$$\tilde{U}U = I \tag{2-12.6}$$

The relation between the original coordinates,

$$x = (x_1, x_2, \cdots, x_n)$$

and the new coordinates, $\bar{x} = (\bar{x}_1, \bar{x}_2, \cdots, \bar{x}_n)$

is expressed in the matrix equation

$$x = U\bar{x} \tag{2-12.7}$$

Let us now consider the matrix equation

$$Ax = b \tag{2-12.8}$$

In the new reference system the same equation will take the form

$$\bar{A}\bar{x} = \bar{b} \tag{2-12.9}$$

Now x and b are both vectors and thus they both follow the same transformation law (7).

$$x = U\bar{x}, \qquad b = U\bar{b} \tag{2-12.10}$$

Introducing these equations in (8), we obtain

$$AU\bar{x} = U\bar{b} \tag{2-12.11}$$

We will now premultiply by $\tilde{U}$ and take advantage of the orthogonality relation (6). Then

$$\tilde{U}AU\bar{x} = \bar{b}$$

Hence we have to put

$$\bar{A} = \tilde{U}AU \tag{2-12.12}$$

in order to obtain the form (9) of the equation.

We see that the transformation of a matrix differs from the transformation of a vector by the appearance of U in front of A and in the back of A. If the vector transformation (7) is premultiplied by $\tilde{U}$, we obtain

$$\bar{x} = \tilde{U}x \tag{2-12.13}$$

while

$$\bar{A} = \tilde{U}AU$$

In the vector transformation only premultiplication occurs; in the matrix transformation pre- and postmultiplication.

Let us now consider arbitrary combinations of matrices and see whether the transformation law (12) will hold again.

$$\bar{A} + \bar{B} = \tilde{U}AU + \tilde{U}BU = \tilde{U}(A + B)U$$

Hence the sum of matrices follows the same transformation law as a single matrix. The same is true of the products of matrices:

$$\bar{A}\bar{B} = \tilde{U}AU\tilde{U}BU = \tilde{U}ABU$$

in view of the fact that

$$\tilde{U}U = I$$

We can now say that *any* combination of matrices obtained by the fundamental operations addition and multiplication transforms exactly in the same way as one single matrix. This means the following. Let

$$F(A, B, \cdots, P)$$

be an arbitrary algebraic function of the arbitrary matrices $A, B, \cdots, P$. Then

$$F(\bar{A}, \bar{B}, \cdots, \bar{P}) = \tilde{U}F(A, B, \cdots, P)U$$

If now we have an arbitrary algebraic equation between matrices:

$$F(A, B, \cdots, P) = 0 \qquad (2\text{-}12.14)$$

then also

$$F(\bar{A}, \bar{B}, \cdots, \bar{P}) = 0 \qquad (2\text{-}12.15)$$

This shows that *matrix equations remain invariant under arbitrary orthogonal transformations.*

We can go one step further and include the operation "transposition" in our results. If

$$\bar{A} = \tilde{U}AU$$

we obtain by the law of transposition:

$$\tilde{\bar{A}} = \tilde{U}\tilde{A}U$$

We see that the transpose of a matrix is transformed in the same way as the matrix itself. Hence matrix equations which are formed by algebraic operations *and transposition* remain invariant under orthogonal transformations. The equation

$$F(A, B, \cdots, P, \tilde{A}, \tilde{B}, \cdots, \tilde{P}) = 0$$

implies

$$F(\bar{A}, \bar{B}, \cdots, \bar{P}, \tilde{\bar{A}}, \tilde{\bar{B}}, \cdots, \tilde{\bar{P}}) = 0$$

For example, if $A - \tilde{A} = 0$ then also $\bar{A} - \tilde{\bar{A}} = 0$

which means that the *symmetry of a matrix is preserved under orthogonal transformations.* Similarly, if

$$A = -\tilde{A} \qquad \text{then also} \qquad \bar{A} = -\bar{A}$$

which means that if a matrix is "antisymmetric" in one frame of reference, it remains antisymmetric in *all* orthogonal frames of reference.

Two fundamental quantities remain unchanged under an orthogonal transformation. The one is the unit matrix:

$$\bar{I} = \tilde{U}IU = \tilde{U}U = I$$

The other is the "dot product," or "scalar product" (inner product) of two vectors:

$$\bar{x}\cdot\bar{y} = \tilde{U}x\cdot\tilde{U}y = y\cdot U\tilde{U}x = y\cdot x = x\cdot y$$

13. The invariance of matrix equations under arbitrary linear transformations. We will now drop the restriction to rectangular coordinates and consider the general case of arbitrarily oriented skew-angular frames of reference. The base vectors $u_1, u_2, \cdots, u_n$ are no longer orthogonal to each other, nor is their length normalized to 1. *Any n* vectors which are linearly independent are admitted. We have seen that under these circumstances we have to *double* the given set by introducing the *adjoint* vectors $v_1, v_2, \cdots, v_n$.

$$\begin{pmatrix} u_1, & u_2, & \cdots & u_n \\ v_1, & v_2, & \cdots & v_n \end{pmatrix}, \qquad \begin{array}{l} u_i\cdot v_k = 0 \quad (i \neq k) \\ u_i\cdot v_i = 1 \end{array} \qquad (2\text{-}13.1)$$

Any vector can now be analyzed in either the one or the other frame of reference. Once more we will introduce a linear transformation by choosing a new set of base vectors, and their adjoints.

$$\begin{pmatrix} \bar{u}_1, & \bar{u}_2, & \cdots & \bar{u}_n \\ \bar{v}_1, & \bar{v}_2, & \cdots & \bar{v}_n \end{pmatrix}, \qquad \begin{array}{l} \bar{u}_i\cdot \bar{v}_k = 0 \quad (i \neq k) \\ \bar{u}_i\cdot \bar{v}_i = 1 \end{array} \qquad (2\text{-}13.2)$$

These vectors, if analyzed in the original frame of reference, can be characterized as follows:

$$\left. \begin{array}{l} \vec{\bar{u}}_1 = u_{11}\vec{u}_1 + \cdots + u_{n1}\vec{u}_n \\ \vdots \\ \vec{\bar{u}}_n = u_{1n}\vec{u}_1 + \cdots + u_{nn}\vec{u}_n \end{array} \right\} \left. \begin{array}{l} \vec{\bar{v}}_1 = v_{11}\vec{v}_1 + \cdots + v_{n1}\vec{v}_n \\ \vdots \\ \vec{\bar{v}}_n = v_{1n}\vec{v}_1 + \cdots + v_{nn}\vec{v}_n \end{array} \right\} (2\text{-}13.3)$$

These two transformations can be included in the two matrices

$$U = \begin{bmatrix} u_{11} & u_{12} & \cdots & u_{1n} \\ u_{21} & u_{22} & & u_{2n} \\ \vdots & \vdots & & \vdots \\ u_{n1} & u_{n2} & & u_{nn} \end{bmatrix}, \quad V = \begin{bmatrix} v_{11} & v_{12} & \cdots & v_{1n} \\ v_{21} & v_{22} & & v_{2n} \\ \vdots & \vdots & & \vdots \\ v_{n1} & v_{n2} & & v_{nn} \end{bmatrix} \quad (2\text{-}13.4)$$

The biorthogonality relations between the $\bar{u}_i$, $\bar{v}_k$ find expression in the matrix equation

$$\tilde{U}V = I \tag{2-13.5}$$

which can be written in the alternate forms

$$\tilde{U}V = V\tilde{U} = \tilde{V}U = U\tilde{V} = I \tag{2-13.6}$$

(since a matrix is always commutative with its inverse).

The transformation of coordinates is expressed by the following matrix equations.

$$\left. \begin{aligned} x &= U\bar{x} \\ \xi &= V\bar{\xi} \end{aligned} \right\} \tag{2-13.7}$$

Once more we consider the matrix equation

$$Ax = b \tag{2-13.8}$$

and introduce the transformation of the coordinates

$$x = U\bar{x}, \qquad b = U\bar{b}$$

We obtain

$$AU\bar{x} = U\bar{b}$$

and premultiplying by $\tilde{V}$ we get

$$\tilde{V}AU\bar{x} = \bar{b}$$

which shows that the transformation of the matrix A has to occur according to the equation

$$\bar{A} = \tilde{V}AU \tag{2-13.9}$$

On the other hand, if we premultiply the first equation of (7) by $\tilde{V}$ we obtain

$$\bar{x} = \tilde{V}x$$

Once more we observe that the transformation of a matrix differs from the transformation of a vector by the second factor by which it is postmultiplied.

Since $\hat{V}$ is the reciprocal of U, we can also put

$$\bar{A} = U^{-1}AU \tag{2-13.10}$$

Moreover, we could have started our deductions with the "adjoint equation"

$$\tilde{A}\xi = \beta \tag{2-13.11}$$

which expresses the equation (8) in the adjoint frame of reference. Then

$$\xi = V\bar{\xi}, \qquad \beta = V\bar{\beta} \tag{2-13.12}$$

leads to

$$\tilde{\bar{A}} = \tilde{U}\tilde{A}V \tag{2-13.13}$$

which may also be written in the form

$$\tilde{\bar{A}} = V^{-1}\tilde{A}V \tag{2-13.14}$$

Equation (13) is in harmony with (9) and can be obtained by transposing the previous equation.

Once more we can extend the transformation law (9) to the sum of matrices and to the product of matrices, and thus to any combination of the two fundamental operations "addition" and "multiplication" of matrix algebra. Once more we find that any algebraic equation between matrices,

$$F(A, B, \cdots, P) = 0 \tag{2-13.15}$$

which holds in one frame of reference, remains true in any other frame of reference

$$F(\bar{A}, \bar{B}, \cdots, \bar{P}) = 0 \tag{2-13.16}$$

The invariance of matrix equations with respect to coordinate transformations is once more demonstrated, but now extended from orthogonal to arbitrary linear transformations.

The only difference compared with the orthogonal case is that the relation (15) must not include the operation of transposition. The transposed matrix $\bar{A}$ follows a transformation law which is different from the transformation law of A [cf. (10) versus (14)]. Matrix equations which involve the transpose of a matrix do *not* remain invariant under arbitrary coordinate transformations.[1] For example, the equation

$$\tilde{A} = A$$

[1] The inverse A^{-1} of A transforms in the same way as A itself. Hence the equation (15) may include the inverses $A^{-1}, B^{-1}, \cdots, P^{-1}$ (if they exist).

(which expresses the symmetry of A) is lost under the impact of an arbitrary linear transformation, although it is preserved under strictly orthogonal transformations.

The unit matrix I and the scalar product of two vectors are again invariants of an arbitrary linear transformation:

$$\bar{I} = U^{-1}IU = U^{-1}U = I \qquad (2\text{-}13.17)$$

In the case of a scalar product, it is imperative that the two vectors shall be given in *adjoint representations:*

$$\bar{x}\cdot\bar{\eta} = \tilde{V}x\cdot\tilde{U}\eta = \eta\cdot U\tilde{V}x = \eta\cdot x = x\cdot\eta \qquad (2\text{-}13.18)$$

14. Commutative and noncommutative matrices. Matrix multiplication is generally noncommutative: $AB \neq BA$. We may have, however, two matrices for which the commutative law of ordinary multiplication is satisfied:

$$AB = BA \qquad (2\text{-}14.1)$$

By the principle of the invariance of matrix equations with respect to coordinate transformations, equation (1) must remain valid in every frame of reference and must thus express some inherent property of the two matrices A and B. What is this property?

We have seen that a matrix, by introducing its principal axes as a new frame of reference, can be transformed into a purely diagonal form. Then in the new frame of reference $\bar{A} = \Lambda$. Now it may happen that a second matrix B possesses the same principal axes— although different eigenvalues—and thus becomes likewise diagonal in the new reference system. Then

$$\bar{A} = \Lambda, \qquad \bar{B} = \Lambda'$$

Now the product of two diagonal matrices is again a diagonal matrix, with the diagonal elements:

$$\Lambda\Lambda' = \begin{bmatrix} \lambda_1\lambda_1' & & & \\ & \lambda_2\lambda_2' & & \\ & & \ddots & \\ & & & \lambda_n\lambda_n' \end{bmatrix}$$

These elements are symmetric with respect to the first and the second factor, and hence

$$\Lambda\Lambda' = \Lambda'\Lambda$$

Diagonal matrices are always commutative. But then A and B are commutative in the new frame of reference, and equation (1) proved in that particular frame. Since, however, the independence of that equation from any particular frame is proved in advance, we have found the necessary and sufficient condition for the commutability of two matrices; viz., *that their principal axes must be parallel.*

For example, the following two matrices are commutative and have therefore the same principal axes:

$$A = \begin{bmatrix} 33 & 16 & 72 \\ -24 & -10 & -57 \\ -8 & -4 & -17 \end{bmatrix}, \qquad B = \begin{bmatrix} -47 & -32 & -84 \\ 36 & 24 & 66 \\ 12 & 8 & 22 \end{bmatrix}$$

The unit matrix is geometrically represented by a sphere. *Any* axis can be chosen as a principal axis. Hence the unit matrix commutes with any matrix:

$$I \cdot A = A \cdot I = A$$

Moreover, the inverse A^{-1} of the matrix A has the same principal axes as A itself. Hence A and A^{-1} are commutative

$$A \cdot A^{-1} = A^{-1} A = I$$

Matrices which differ in their principal axes cannot be commutative. Matrix multiplication is generally noncommutative because two matrices have generally arbitrarily oriented principal axes and arbitrary eigenvalues. If the principal axes agree, although the eigenvalues are still arbitrary, the multiplication becomes commutative.

15. Inversion of a matrix. The Gaussian elimination method. A simultaneous system of linear algebraic equations can be written down in matrix form as follows:

$$Ax = b \tag{2-15.1}$$

where A is a given matrix, b is a given vector, and x is the unknown vector. By premultiplying by A^{-1}, the equation can be formally solved and we obtain

$$x = A^{-1}b \tag{2-15.2}$$

The matrix A^{-1} is called the "reciprocal" or the "inverse" of A, and any method by which A^{-1} can be calculated is called the inversion of A.

It is not always necessary, however, actually to invert the matrix for solving the linear equation (1). It depends on what the role of the right side b is, relative to the given problem. It is possible that the right side b is just as much an inherent part of our problem as the matrix A. In that case, we are satisfied if we can solve equation (1) for that particular right side. Frequently, however, the right side of the equation plays a more *accidental* role. The matrix A is inherently associated with the given physical situation, while the right side has frequently the significance of a "forcing function." We may want to obtain the response of the given structure to a variety of forcing functions. Hence we may want to change the right side freely, although the left side of the equation remains unchanged. In this case, we actually want the possession of A^{-1}; because, having A^{-1}, we let A^{-1} operate on b, and our solution is obtained. If we do not possess A^{-1}, we have to go through a larger number of algebraic operations for each separate b. On the other hand, the construction of the inverse of A requires much more labor than the solution of a set of linear equations for a given right side.

The fundamental method of inverting a matrix was introduced by Gauss and is called the Gaussian elimination method. It works equally for solving a set of linear equations and for inverting a matrix. Its underlying principle is simple and the operations involved are easily accomplished. It belongs to that class of numerical procedures which make use of a large number of very simple operations, instead of a smaller number of more involved operations. What we accomplish is that gradually the zero elements of the unit matrix shift over to the left side, while the right side fills up more and more with elements. Eventually the unit matrix is on the left side and the right side is completely filled up with elements. This terminates our task.

We can analyze the process by applying it first to the case of a linear system with a given right side. We write the equation (1) in the form

$$Ax - b = 0 \qquad (2\text{-}15.3)$$

and indicate the system symbolically by writing only the coefficients of the system, while the variables appear on the top of the scheme

$$
\begin{array}{ccccc}
x_1 & x_2 & \cdots & x_n & -1 \\
\hline
\begin{bmatrix}
a_{11} & a_{12} & \cdots & a_{1n} & b_1 \\
a_{21} & a_{22} & & a_{2n} & b_2 \\
\cdots & & & & \\
a_{n1} & a_{n2} & \cdots & a_{nn} & b_n
\end{bmatrix} & & & & = 0
\end{array}
\qquad (2\text{-}15.4)
$$

Since any equation can be multiplied by a factor, it is permissible to multiply any row by an arbitrary constant. Moreover, it is permissible to multiply any equation by an arbitrary factor and add it to another equation. Correspondingly, we can multiply any row of the above scheme by an arbitrary factor and add it to any other row. These are the two fundamental operations on which the elimination method is based.

First we look for the absolutely largest element of the matrix. Let this element be a_{ik}. Then we divide the entire ith row by a_{ik}, with the result that the new a_{ik} becomes 1. Our aim is now to make all the other elements of the kth column equal to 0. The kth column is then composed of zeros and one solitary 1. The zeroing of an element occurs by multiplying the ith row by that particular element and subtracting it from the row which that element occupies. For example,

$$
\begin{array}{cccc}
x_1 & x_2 & x_3 & -1 \\
\hline
\begin{bmatrix}
33 & 16 & 72 & 359 \\
-24 & -10 & -57 & -281 \\
-8 & -4 & -17 & -85
\end{bmatrix} & & & = 0
\end{array}
$$

The largest element of the matrix is 72, found in the first row. Dividing the first row by 72, the new first row becomes

$$
0.458333 \quad 0.222222 \quad 1 \quad 4.98611
$$

The 1 appears in the third column. We zero out the elements -57 and -17 of the same column by multiplying the new row by 57 and adding the result to the second row, and similarly by 17 and adding the result to the third row. The matrix appears at this stage as follows:

$$
\begin{array}{cccc}
x_1 & x_2 & x_3 & -1 \\
\hline
\begin{bmatrix}
0.458333 & 0.222222 & 1 & 4.98611 \\
2.12498 & 2.66665 & 0 & 3.20827 \\
-0.208339 & -0.222226 & 0 & -0.23613
\end{bmatrix}
\end{array}
$$

The variable x_3 is now eliminated, since the second and the third equations do not contain it any more. The original 3 by 3 problem is thus reduced to a 2 by 2 problem. Generally the zeroing of one column has the effect that the original n by n problem is reduced to an $(n-1)$ by $(n-1)$ problem.

We continue our scheme by hunting for the largest element in the remaining 2 by 2 matrix. We find 2.66665. Dividing the second row by this element, the new second row becomes

$$0.796872 \quad 1 \quad 0 \quad 1.203108$$

We zero the second column by multiplying this row by 0.222222 and -0.22226 and subtracting it from the first and the third rows. The new shape of the matrix becomes

$$
\begin{array}{rrrr}
0.281250 & 0 & 1 & 4.71875 \\
0.796872 & 1 & 0 & 1.20311 \\
-0.031253 & 0 & 0 & 0.031232
\end{array}
$$

Finally we divide the third row by -0.031253 and obtain

$$1 \quad 0 \quad 0 \quad -1.00067$$

We zero the first column by multiplying the new third row by 0.281250 and 0.796872 and subtracting it from the first and second rows. This gives

$$
\begin{bmatrix}
0 & 0 & 1 & 5.00019 \\
0 & 1 & 0 & 2.00052 \\
1 & 0 & 0 & -1.00067
\end{bmatrix}
$$

The transformed set of equations are thus

$$
\begin{aligned}
x_3 &-- 5.00019 = 0 \\
x_2 &- 2.00052 = 0 \\
x_1 &+ 1.00067 = 0
\end{aligned}
$$

This gives the solution

$$
\begin{array}{ll}
x_1 = -1.00067 & \text{[correct: } -1\text{]} \\
x_2 = 2.00052 & \text{[correct: } 2\text{]} \\
x_3 = 5.00019 & \text{[correct: } 5\text{]}
\end{array}
$$

Quite similar is the procedure of inverting the matrix. The only difference is that the right side is now taken by the unit matrix I. Hence we have to operate with n columns instead of one column.

But the successive operations are identical with those of the previous algorithm.

$$\begin{bmatrix} 33 & 16 & 72 & 1 & 0 & 0 \\ -24 & -10 & -57 & 0 & 1 & 0 \\ -8 & -4 & -17 & 0 & 0 & 1 \end{bmatrix}$$

First transformation:

$$\begin{bmatrix} -0.45833 & 0.222222 & 1 & 0.0138888 & 0 & 0 \\ 2.12498 & 2.66665 & 0 & 0.791662 & 1 & 0 \\ -0.208339 & -0.222226 & 0 & 0.236110 & 0 & 1 \end{bmatrix}$$

Second transformation:

$$\begin{bmatrix} 0.281250 & 0 & 1 & -0.052083 & -0.083333 & 0 \\ 0.796872 & 1 & 0 & 0.296875 & 0.375002 & 0 \\ -0.031253 & 0 & 0 & 0.302083 & 0.083335 & 1 \end{bmatrix}$$

Third transformation:

$$\begin{bmatrix} 0 & 0 & 1 & 2.66640 & 0.66661 & -8.9991 \\ 0 & 1 & 0 & 7.9992 & 2.49983 & 25.4974 \\ 1 & 0 & 0 & -9.66572 & -2.66646 & -31.9969 \end{bmatrix}$$

The position of the 1 in the first part of the final matrix indicates that the rows of the final result have to be rearranged in the sequence 3, 2, 1. The inverse matrix thus becomes

$$A^{-1} = \begin{bmatrix} -9.66572 & -2.66646 & -31.9969 \\ 7.9992 & 2.49983 & 25.4974 \\ 2.66640 & 0.66661 & 8.9991 \end{bmatrix}$$

The correct inverse, given to five decimal places, is

$$A^{-1} = \begin{bmatrix} -9.66667 & -2.66667 & -32.00000 \\ 8.00000 & 2.50000 & 25.50000 \\ 2.66667 & 0.66667 & 9.00000 \end{bmatrix}$$

The discrepancies are caused by rounding errors. How quickly the rounding errors can accumulate is demonstrated even by this simple 3 by 3 example. Although the calculations were made with an accuracy of 10^{-6}, the results have an accuracy of only 10^{-4}. In large matrices the accumulation of rounding errors can become quite serious.

16. Successive orthogonalization of a matrix. The numerical aspects of inverting a matrix are distinctly different from the purely mathematical aspects. Our calculations are always of only limited accuracy. The rounding errors constantly accumulate, and considering the large number of operations involved in the inversion process, they may eventually obliterate the desired results. The Gaussian elimination scheme insures proper results if the effect of rounding errors can be neglected. But when inverting large matrices this is almost never the case. The numerical inversion of such matrices is of paramount interest, considering the fact that so many problems of contemporary physics and engineering lead to the solution of large linear systems. What procedure shall we follow under these circumstances?

The establishment of electronic digital computers started a new chapter in the history of numerical analysis. The extraordinary rapidity with which these machines perform the fundamental operations of arithmetic leads to a shift in our general philosophy of numerical operations. The emphasis is no longer on procedures which obtain a result in the smallest number of operations. More important is the viewpoint of *simple codibility*, together with the demand on high accuracy. In the problem of inverting a matrix we will be interested in a procedure which, in spite of the limited accuracy of arithmetical operations, cannot come to grief, no matter how extensive the matrix is to which it is applied.

In discussing the question of accuracy we must realize that under no circumstances can we expect *absolute* accuracy. Nor is the question of accuracy a matter of arbitrary decisions. Rarely are the elements of the given matrix A absolute mathematical numbers. In most cases the elements a_{ik} of the matrix are obtained on the basis of some *measurements*, which are automatically of limited accuracy only. Let us now assume that as the result of some inversion process we have obtained an approximate inverse of A. This $\bar{A}^{-1}$ is certainly the exact inverse of a certain $\bar{A}$ which differs from A by the small amount αA_1:

$$\bar{A} = A + \alpha A_1$$

Let it be true that all the elements of αA_1 are smaller than the possible errors of a_{ik}. Under these circumstances we can describe $\bar{A}$ as being "numerically equivalent" to A, and the obtained $\bar{A}^{-1}$ can be accepted as the correct answer to our problem. It would be entirely superfluous to try to correct $\bar{A}^{-1}$ in order to arrive at the correct A^{-1}. The mathematically correct answer has no significance, in view of the limited accuracy of the given A. In the case of a "mathematical matrix" A, the substitution of a "numerically equivalent" $\bar{A}$ loses, of course, all significance. But even then it will be advisable to replace the exact elements of A by elements limited to a definite number of decimal places, carry through the inversion process and, finally, if necessary, correct $\bar{A}^{-1}$ by a perturbation process (cf. § 23).

The inversion method described in this section has the advantage that we can keep the rounding errors under control, no matter how far the process continues. The accuracy of our calculations is determined by the accuracy of the elements of A and can be adjusted, if the need arises, even to variable accuracy. We shall first discuss the mathematical aspects of the method, and then come to the question of the rounding errors.

We once more consider the equation

$$Ax = b \qquad (2\text{-}16.1)$$

but interpret it in a slightly different manner. Writing out these equations in detail, we have

$$\left.\begin{array}{l} a_{11}x_1 + a_{12}x_2 + \cdots + a_{1n}x_n = b_1 \\ a_{21}x_1 + a_{22}x_2 + \quad\;\; a_{2n}x_n = b_2 \\ \vdots \\ a_{n1}x_1 + a_{n2}x_2 + \cdots + a_{nn}x_n = b_n \end{array}\right\} \qquad (2\text{-}16.2)$$

We will now consider each one of the columns of the matrix an independent vector, and do the same with the right side of the set. Hence we put

$$\begin{array}{l} u_1 = (a_{11}, a_{21}, \cdots, a_{n1}) \\ u_2 = (a_{12}, a_{22}, \cdots, a_{n2}) \\ \vdots \\ u_n = (a_{1n}, a_{2n}, \cdots, a_{nn}) \\ b = (b_1, b_2, \cdots, b_n) \end{array} \qquad (2\text{-}16.3)$$

Our problem can now be interpreted as follows. Find a linear combination of the vectors $u_1, u_2, \cdots, u_n$ such that the resultant vector shall become the given vector b.

$$x_1u_1 + x_2u_2 + \cdots + x_nu_n = b \qquad (2\text{-}16.4)$$

In this interpretation we have a given skew-angular reference system, characterized by the base vectors $u_1, u_2, \cdots, u_n$, and we want to analyze the given vector b in this reference system. Hence $(x_1, x_2, \cdots, x_n)$ are the skew-angular components of the vector b in the given set of base vectors. The condition that these base vectors are linearly independent, is equivalent to the condition that the determinant of the given linear set is different from zero.

$$\begin{vmatrix} a_{11} & a_{12} & \cdots & a_{1n} \\ a_{21} & a_{22} & & a_{2n} \\ \vdots & & & \\ a_{n1} & a_{n2} & \cdots & a_{nn} \end{vmatrix} \neq 0 \qquad (2\text{-}16.5)$$

In discussing the operation with skew-angular reference systems we have seen that a skew-angular set of base vectors demands construction of the adjoint set $v_1, v_2, \cdots, v_n$. But this is equivalent to construction of the inverse matrix, since the rows of the inverse matrix represent the vectors $v_1, v_2, \cdots, v_n$.

In one particular case the problem is easily solvable. It may happen that the base vectors are *orthogonal* to each other and their length equals 1:

$$u_i \cdot u_k = 0 \quad (i \neq k), \qquad u_i^2 = 1 \qquad (2\text{-}16.6)$$

In this case the system is self-adjoint, and we obtain

$$v_i = u_i \qquad (2\text{-}16.7)$$

The inverse matrix is then simply the transpose of the original matrix.

$$A^{-1} = \tilde{A} \qquad (2\text{-}16.8)$$

Orthogonal reference systems are thus of particular advantage in dealing with linear systems. It will be our aim to introduce an auxiliary reference system which has the property of being orthogonal. We solve the problem in the new reference system and then return to the original frame of reference.

We choose the first vector u_1 as the first vector of the new system. However, we normalize its length to 1; i.e., we divide by the length of the vector:

$$w_1 = \frac{u_1}{\sqrt{u_1^2}} \qquad (2\text{-}16.9)$$

Our next vector w_2 is chosen in the plane of the first two vectors u_1 and u_2. In this plane we find a vector which is orthogonal to w_1 and whose length is 1. We call this vector w_2. Except for a $\pm$ sign, this vector is uniquely determined. Next we go into the space included by the first *three* vectors u_1, u_2, and u_3. In this space we find a particular vector w_3 (except its orientation up or down) which is orthogonal to both vectors w_1 and w_2, and whose length is 1. Next we go into the space included by the first *four* vectors u_1, u_2, u_3, u_4 and find a vector w of the length 1 which is orthogonal to the three vectors w_1, w_2, w_3. This construction can be continued until finally the space of all the vectors u_1, u_2, $\cdots$, u_n is exhausted. We have thus constructed a system of n mutually orthogonal vectors w_1, w_2, $\cdots$, w_n of the length 1.

Generally each new vector is obtained in two consecutive steps. First we construct

$$w_i' = u_i - p_{i1}w_1 - p_{i2}w_2 - \cdots - p_{ii}w_i \qquad (2\text{-}16.10)$$

where the coefficients p_{ik} are available since the dot-product of w_i' with the previous vectors w_k gives

$$p_{ik} = u_i w_k \qquad (2\text{-}16.11)$$

Then we normalize the length of w_i to 1[1]:

$$w_i = \frac{w_i'}{|w_i'|} \qquad (2\text{-}16.12)$$

We will write

$$p_{ii} = |w_i'| \qquad (2\text{-}16.13)$$

and obtain

$$u_1 = p_{i1}w_1 + p_{i2}w_2 + \cdots + p_{ii}w_i \qquad (2\text{-}16.14)$$

The elements $p_{ik}, (k \leq i)$, can be included in a "triangular matrix," i.e., a matrix which has elements in the diagonal and below the diagonal, while all the elements above the diagonal are put equal to zero.

The equation (14) can now be written in the form of a matrix equation:

$$A = W\tilde{P} \qquad (2\text{-}16.15)$$

---

[1] Division by zero cannot occur because w_i cannot vanish in all its components if the determinant of A is not zero.

where W is a matrix whose columns are in succession the vectors $w_1, w_2, \cdots, w_n$. By construction W is an orthogonal matrix:

$$\tilde{W}W = I \tag{2-16.16}$$

and thus

$$W^{-1} = \tilde{W} \tag{2-16.17}$$

Moreover, a triangular matrix can be inverted by successive eliminations (cf. § 17), thus giving the new triangular matrix

$$Q = P^{-1} \tag{2-16.18}$$

But then we have, in consequence of (15):

$$A^{-1} = \tilde{Q}\tilde{W} \tag{2-16.19}$$

and the problem of inverting A is accomplished.

In actual fact the limited accuracy of our arithmetical operations demand a slight modification of the method. If we proceed in the previous fashion, not paying attention to the rounding errors, we will gradually lose in the orthogonality of the vectors w_i. They would remain automatically orthogonal to each other if we could count on the exact orthogonality of the previous vectors w_i and the exactness of the coefficients p_{ik}. But neither of these conditions is fulfilled and the result is that the vectors w_i become gradually deorthogonalized. Hence we cannot count on the accuracy of (17), and the end result (19) becomes more and more unreliable as the order of A increases.

In view of this difficulty we shall now define the vector (10) by a process which does not rely on the exact orthogonality of the previous vectors w_k but insures the orthogonality of the new w_i' to all the previous w_k even if these w_k are not orthogonal. For this purpose we determine the p_{ik} of equation (10) by the least-square principle that the square of the length of w_i' shall be minimized:

$$(u_i - p_{i1}w_1 - \cdots - p_{i,i-1}w_{i-1})^2 = \text{minimum}$$

This gives the following linear set of equations:

$$\begin{aligned} p_{i1}w_1^2 + \cdots + p_{i1i-1}w_1w_{i-1} &= u_iw_1 \\ &\vdots \\ p_{i1}w_{i-1}w_1 + \cdots + p_{i1i-1}w_{i-1}^2 &= u_iw_{i-1} \end{aligned} \tag{2-16.20}$$

We need not solve these equations with absolute accuracy. We can count on the *approximate* orthogonality of the w_k vectors. As soon as a new w_i is obtained, we immediately dot it with the previous w_k, thus testing its orthogonality. In view of the rounding errors we do not get zero, but the components of a symmetric matrix:

$$w_i w_k = \varepsilon_{ik} \qquad (k = 1, 2, \cdots i - 1) \qquad (2\text{-}16.21)$$

$$w_i^2 = 1 + \varepsilon_{ii}$$

These ε_{ik}, however, are small, and thus it suffices to obtain the solution of the system (20) in two steps. First we evaluate the preliminary quantities (11):

$$p'_{ik} = u_i w_k \qquad (k < i) \qquad (2\text{-}16.22)$$

and then we correct them according to the following scheme:

$$p_{ik} = p'_{ik} - (\varepsilon_{ki} p'_{i1} + \cdots + \varepsilon_{k, i-1} p'_{i, i-1}) \qquad (2\text{-}16.23)$$

This correction scheme is the only change compared with the previous procedure. Having obtained p_{ik} we again generate the vector w'_i according to (10) and normalize its length according to (12). Also p_{ii} is again defined by (13), without further correction.

The orthogonality of the vector system w_i within the chosen number of decimal places is now insured by the construction of each new vector. It is important, however, that in the evaluation of the p_{ik} we shall keep a sufficient number of decimal places. In highly skew-angular systems the length of w'_i may become very small because we may lose several decimal places in the construction of w'_i. We have to make up for the loss of significant figures by adding a corresponding number of significant figures in writing down the p_{ik}, *without* demanding, however, an increased accuracy of the previously obtained w_k. Hence it cannot happen that the entire calculation has to be repeated with increased accuracy, because of a surprise encountered during the course of our calculations. The worst that can happen (excepting systems which are so nearly singular that an *exact* solution of (20) is demanded, in which case the system is void of any physical significance) is that the quantities (22) and (23) may be demanded with maximum accuracy. In the anticipation of this possibility it is advisable to obtain and store the ε_{ik} with *full accuracy*.

The number of decimal places to which the vectors w_i may be truncated can be decided as follows. We interpret the truncation of the vector w_i', determined according to (10) and (12), as a modification of the vector u_i, which is the ith column of the given matrix. If w_i is truncated to μ decimal places, the rounding errors in any component cannot exceed $\frac{1}{2} \cdot 10^{-\mu}$. The length of w_i' cannot exceed the length of u_i. Consequently the vector $\bar{u}_i$, which takes the place of u_i in the numerical inversion process, cannot differ from u_i by an amount which is more than $0.5 \cdot 10^{-\mu}$ times the length of u_i. This shows that if the elements of the ith column of A, after dividing by the length of that column, can be guaranteed to μ decimal places, it suffices to truncate w_i to μ decimal places. We have then succeeded in replacing the given matrix A by another, but numerically equivalent, matrix $\bar{A}$, which is split into the product of two matrices, according to (15). Now it is true that the matrix W is not exactly orthogonal, and thus its inverse is not exactly $\tilde{W}$. Since, however, we possess the matrix ε [cf. (21)], we can correct $\tilde{W}$ according to

$$W^{-1} = \tilde{W} - \varepsilon \tilde{W} \qquad (2\text{-}16.24)$$

The matrix P is exactly triangular and its inverse is obtainable to any degree of accuracy, according to the numerical scheme of § 17.

In actual fact it will seldom be necessary to apply the correction (24) to the simple inverse $\tilde{W}$, nor will it be necessary to invert P with excessive accuracy. The numerical certainty of A^{-1} is usually much smaller than the numerical certainty of A itself. Let the true matrix A_0 be equal to the given truncated matrix A plus a small but unknown correction:

$$A_0 = A - \alpha A_1$$

We will assume that the columns of A have been normalized to the length 1 so that the elements of A are all between ± 1. Now

$$A_0^{-1} = A^{-1} + \alpha A^{-1} A_1 A^{-1}$$

The relative error of A_0^{-1} on account of the presence of A_1 would be of the same order of magnitude as αA_1 itself, were it not for the post-multiplication by A^{-1}. Since the average order of the inverted matrix is usually much larger than 1, this multiplication will cause a considerable increase in the uncertainty of A^{-1}. Hence it is generally not necessary to ascertain the inverse of a matrix to the same number of significant figures to which the original matrix was given. This is not so, however, if A happens to be nearly orthogonal. But even then the relative accuracy of A^{-1} cannot be *greater* than that of the original matrix.

17. Inversion of a triangular matrix. A triangular matrix can be inverted by a simple numerical algorithm. Algebraically we proceed as follows. The equations

$$\begin{aligned}
u_1 &= p_{11}w_1 \\
u_2 &= p_{21}w_1 + p_{22}w_2 \\
&\vdots \\
u_n &= p_{n1}w_1 + p_{n2}w_2 + \cdots + p_{nn}w_n
\end{aligned} \tag{2-17.1}$$

have the property that every new equation brings in but one new unknown. Hence the first equation can be solved for w_1, by dividing by p_{11}. Then we solve the second equation for w_2, substituting for w_1 the previously obtained value. Then the third equation is solved for w_3, substituting for w_1 and w_2 their values. And so we proceed, until we arrive at w_n, which becomes a linear combination of $u_1, u_2, \cdots, u_n$:

$$\begin{aligned}
w_1 &= q_{11}u_1 \\
w_2 &= q_{21}u_1 + q_{22}u_2 \\
&\vdots \\
w_n &= q_{n1}u_1 + q_{n2}u_2 + \cdots + q_{nn}u_n
\end{aligned} \tag{2-17.2}$$

We see that the inverse of a triangular matrix is once more a triangular matrix.

In a numerical algorithm the operation with symbols has to be abandoned and a scheme developed which operates with numbers only. Our ordinary arithmetical operations—such as longhand division for example—are based on complicated algebraic operations, but in the actual numerical algorithm the algebraic operations are eliminated and replaced by purely numerical processes. Often we are not even aware what algebraic process underlies a certain numerical algorithm.

The numerical inversion of a triangular matrix proceeds as follows. We write down the given triangular matrix P in transposed form. Hence the *upper* and not the lower triangle is now filled up with elements. In the lower triangle we are going to put the elements of the inverse matrix $Q = P^{-1}$. The dividing line between the two matrices is the diagonal. Here we will immediately write down the q_{ii}

as the *reciprocals* of the p_{ii} and not as the original p_{ii}. The final construction will thus look as follows.

$$
\begin{array}{ccccc}
\underline{q_{11}} & p_{21} & p_{31} & \cdots & p_{n1} \\
q_{21} & \underline{q_{22}} & p_{32} & & p_{n2} \\
q_{31} & q_{32} & \underline{q_{33}} & & p_{n3} \\
& & \vdots & & \\
q_{n1} & q_{n2} & q_{n3} & \cdots & \underline{q_{nn}}
\end{array}
$$

Now an arbitrary q_{ik} $(i > k)$ can be constructed as follows. Two rows are involved in the construction; viz., the row i and the row k. The row i involves the elements of the *lower* triangle, the row k the elements of the *upper* triangle. For example, the element q_{53} will be constructed as follows. We will use the fifth row below and the third row above. We start from the underlined pivotal element q_{55} and go backward. We multiply the third row and the fifth row element by element, but with the understanding that we stop at a point where the pivotal element of the upper row is reached. Hence in the case of q_{53} only two terms are left:

$$
q_{55}p_{53} + q_{54}p_{43}
$$

because the next product would involve q_{53} which we do not have. The resulting sum is multiplied by p_{33} (the upper pivotal element), and the final answer is reversed in sign.

$$
q_{53} = -(q_{55}p_{53} + q_{54}p_{43})q_{33}
$$

We show the operation of this algorithm with the help of a simple example. Let us invert the following triangular matrix:

$$
Q = \begin{bmatrix}
2 & & & \\
3 & 5 & & \\
0 & -7 & 10 & \\
6 & 1 & 8 & 1
\end{bmatrix}
$$

Numerical construction of the inverse matrix:

$$
\begin{array}{cccc}
\underline{0.5} & 3 & 0 & 6 \\
-0.3 & \underline{0.2} & -7 & 1 \\
-0.21 & 0.14 & \underline{0.1} & 8 \\
-1.02 & -1.32 & -0.8 & \underline{1}
\end{array}
$$

Element 21: Multiply $0.2 \cdot 3$ and still multiply by 0.5, changing the sign to minus: $-0.2 \cdot 3 \cdot 0.5 = -0.3$

Element 42: Multiply $1 \cdot 1 + (-0.8)(-7) = 6.6$ and then multiply by 0.2, changing the sign: $-6.6 \cdot 0.2 = -1.32$

and so on.

We have thus obtained our Q matrix:

$$Q = \begin{bmatrix} 0.5 & & & \\ -0.3 & 0.2 & & \\ -0.21 & 0.14 & 0.1 & \\ -2.532 & -0.312 & -0.3 & 1 \end{bmatrix}$$

We check our calculations by forming the product PQ which has to come out as the unit matrix I. In order to multiply column by column, we transpose the first factor:

$$\begin{bmatrix} 2 & 3 & 0 & 6 \\ 0 & 5 & -7 & 1 \\ 0 & 0 & 10 & 8 \\ 0 & 0 & 0 & 0 \end{bmatrix} \circ \begin{bmatrix} 0.5 & 0 & 0 & 0 \\ -0.3 & 0.2 & 0 & 0 \\ -0.21 & 0.14 & 0.1 & 0 \\ -1.02 & -1.32 & -0.8 & 1 \end{bmatrix} = \begin{bmatrix} 1 & 0 & 0 & 0 \\ 0 & 1 & 0 & 0 \\ 0 & 0 & 1 & 0 \\ 0 & 0 & 0 & 1 \end{bmatrix}$$

The construction of the inverse of the given triangular matrix is thus completed.

18. Numerical example for the successive orthogonalization of a matrix. The following 5 by 5 matrix was obtained as the result of an industrial least square problem. Matrices of this kind are always symmetric and they have a further property: they are "positive definite." This means that the quadratic form associated with the matrix: $\Sigma a_{ik} x_i x_k$, cannot become zero or negative for any choice of the variables; (except the trivial case that all the x_i vanish).

Such matrices can be properly normalized by the following device. We transform the variables x_i by changing their scale:

$$x_i' = \sqrt{a_{ii}} x_i \tag{2-18.1}$$

and we divide the ith equation by $\sqrt{a_{ii}}$. This amounts to the following transformation of the elements:

$$a_{ik}' = \frac{a_{ik}}{\sqrt{a_{ii}} \sqrt{a_{kk}}} \tag{2-18.2}$$

The result of this transformation is that all the diagonal elements of the matrix become 1. Moreover, all the nondiagonal elements must lie between ± 1. In our problem the initial normalization (2) has been performed in advance and we will start immediately with the normalized matrix. The elements of this matrix are given, in harmony with the accuracy of the observations from which they were derived, to five decimal places:

$$A = \begin{bmatrix} 1 & 0.99877 & 0.99297 & 0.98341 & 0.99580 \\ 0.99877 & 1 & 0.99325 & 0.98044 & 0.99363 \\ 0.99297 & 0.99325 & 1 & 0.98233 & 0.98776 \\ 0.98341 & 0.98044 & 0.98233 & 1 & 0.98749 \\ 0.99580 & 0.99363 & & 0.98749 & 1 \end{bmatrix}$$

The fact that all the elements are so near to 1 indicates that the system is very skew-angular.

In order to ensure that the "equivalent matrix" does not deviate from the given matrix by more than $\frac{1}{2}$ unit of the fifth decimal place, we will keep six decimal places in the formation of the matrix W. The evaluation of the p_{ik} occurred with an accuracy of nine decimal places because the calculation of the w'_i showed that two decimal places were lost in the process of forming the differences (16.10). The corrections (16.23) were thus constantly demanded. The following table lists in its columns the vectors w_i as they emerged in succession during the orthogonalization and normalization process (decimal point in front):

$$W = \begin{bmatrix} 449819 & -074358 & -321824 & -476554 & -679308 \\ 449266 & 636326 & -372075 & -502115 & 051862 \\ 446657 & 360073 & 744406 & -317396 & 126348 \\ 442357 & -582393 & 297446 & 582606 & -192965 \\ 447930 & -347456 & -339669 & -283912 & 694732 \end{bmatrix}$$

The column-by-column product of this matrix by itself—obtained gradually but here shown in its final appearance—demonstrated that the deviations from orthogonality never reached 10^{-6}. The following table gives this product matrix (the ε-matrix of the text), multiplied by 10^{12}:

$$10^{12}\varepsilon = \begin{bmatrix} 547515 & -464906 & 338988 & 148894 & 292931 \\ -464906 & -265846 & -742034 & -026372 & 290133 \\ 338988 & -742034 & -062086 & 518499 & -317668 \\ 148874 & -026372 & 518499 & -816063 & -035420 \\ 292931 & 290133 & -317668 & -035420 & 886061 \end{bmatrix}$$

The elements of the P-matrix emerged from row to row. Apart from the diagonal elements—which were obtained by taking the square root of the sum of the squares of the elements of the w_i' vectors—every other element is listed as the sum of two elements, viz., the preliminary p_{ik}', obtained according to (16.22) and the correction, contained in the second term of (16.23), and listed directly below the corresponding p_{ik}':

$p_{11} = 2.223116$				
$p_{21} = 2.220954971$				
corr.: -1216				
$p_{2k} = \underline{2.220953755}$	0.003458902			
2.216535115	0.002963282			
corr.: -1212	$+1031$			
$p_{3k} = \underline{2.216533903}$	$\underline{0.002964313}$	0.01195886		
2.206282826	0.021036753	0.011995689		
corr.: -1222	$+1029$	-763		
$p_{4k} = \underline{2.206281604}$	$\underline{-0.021035724}$	$\underline{0.011995689}$	0.01410346	
2.220276968	-0.008670650	-0.000826800	0.002258580	
corr.: -1220	$+1029$	-760	-328	
$p_{5k} = \underline{2.220275748}$	$\underline{-0.008669621}$	$\underline{-0.000827560}$	$\underline{0.002258252}$	0.004058572

An interesting conclusion can be drawn concerning the *determinant* of the given matrix A. The equation (16.15) shows that the determinant of A is equal to the product of the determinants of W and P. But the determinant of the orthogonal matrix W can only be ± 1, the sign being decided by taking the product of the signs of the diagonal elements. In our case all the diagonal terms of W are positive and thus the sign is plus. The determinant of P, being a diagonal matrix, is simply the product of the diagonal terms. This product is in our problem $\Delta = 5.26367 \cdot 10^{-9}$, thus showing the extraordinarily skew-angular character of the given linear system.

We now come to the inversion of the P-matrix, according to the numerical scheme of § 17. This yields the Q-matrix:

$$
\begin{bmatrix}
0.449819082 & & & & \\
-288.827893 & 289.109087 & & & \\
-11.7789627 & -71.6631707 & 83.6200106 & & \\
-491.144346 & 492.1677422 & -71.1229472 & 70.9045865 & \\
-592.171546 & 329.1112848 & 56.6243775 & -39.4524044 & 246.392080
\end{bmatrix}
$$

(The unmarked elements are all zero.) Finally (16.19) shows that the column by row product of the Q matrix by the W matrix gives the desired inverse A^{-1}:

$$A^{-1} = \begin{bmatrix} 661.793 & -456.547 & -31.482 & -6.990 & -167.376 \\ -456.526 & 474.825 & -63.888 & 33.556 & 12.778 \\ -31.499 & -63.878 & 91.976 & -27.490 & 31.131 \\ -6.970 & 33.542 & -27.491 & 48.922 & -47.545 \\ -167.401 & 12.800 & 31.128 & -47.539 & 171.176 \end{bmatrix}$$

The exact inverse of A should be a symmetric matrix. Our inverse shows slight deviations from symmetry. It is unnecessary, however, to try to eliminate these deviations since we know that our A^{-1} is the exact inverse of a matrix which differs from the given matrix by less than $\frac{1}{2}$ unit of the fifth decimal place which is outside the reach of our measurements. Any conclusions involving quantities of such smallness would be automatically spurious. That the inaccuracies occur already in the second decimal, is not the fault of the inversion process but a consequence of the strongly skew-angular character of A. The columns of A had a relative accuracy of about $1 \cdot 10^{-6}$. In the columns of A^{-1} we cannot expect a relative accuracy exceeding $5 \cdot 10^{-4}$. In actual fact the deviations from symmetry are considerably smaller than the permissible maximum errors.

19. Triangularization of a matrix. The orthogonalization process which we have discussed in § 16 has still another aspect. If we transpose (16.15) and premultiply the old equation by the new one, we obtain

$$\tilde{A}A = P\tilde{P} \qquad (2\text{-}19.1)$$

In this equation the orthogonal matrix W has *dropped out completely*. This shows that we can obtain the triangular matrix P even *without* explicitly generating the orthogonal vectors w_i. While this procedure is much shorter than the previous one, it has the disadvantage that we cannot keep the rounding errors so simply under control as in the previous case. However, the method has its advantages if A is not too skew-angular and we can operate with a sufficient number of extra decimal places to keep the rounding errors below the danger point.

The equation (1), written out in its elements, is solvable by going from row to row and in each row systematically from the left to the right, until the diagonal element is reached. Every equation brings in one new unknown which can thus be eliminated. Finally the entire P matrix is constructed. Then again we invert P and obtain the new triangular matrix Q. Now the equation (16.15) shows that the matrix W can be eliminated by postmultiplying by $\tilde{Q}$:

$$W = A\tilde{Q} \qquad (2\text{-}19.2)$$

Substitution in (16.19) gives

$$A^{-1} = \tilde{Q}QA \qquad (2\text{-}19.3)$$

The matrix $\tilde{A}A$ is automatically a symmetric and positive definite matrix. We can conceive the equation (1) as an expression of the fact that any symmetric and positive definite matrix can be split into the row-by-row product of a triangular matrix by itself. Now it so happens that in many problems of applied mathematics the given matrix $\tilde{A}$ is already in itself a symmetric and positive definite matrix. In this case it is not necessary to premultiply by $\tilde{A}$, but we can put directly

$$A = P\tilde{P} \qquad (2\text{-}19.4)$$

Not only do we save now the column-by-column multiplication of the given matrix by itself—in the case of large matrices an elaborate operation—but we have the further great advantage that the matrix P will be much less skew-angular than the original matrix has been. The product of the diagonal elements of P is now only the *square root* of the original determinant. Hence the loss in significant figures is now much less pronounced than before. This reduction of a symmetric matrix to a triangular matrix is thus strongly advocated if the matrix is the result of a least-square problem, provided that we have enough decimal places at our disposal to counteract the accumulation of rounding errors. If the given matrix A is not excessively skew-angular, we will succeed with the triangularization without constant corrections. Finally we invert P and obtain A^{-1} in the form

$$A^{-1} = \tilde{Q}Q \qquad (2\text{-}19.5)$$

If A is not symmetric, the preliminary symmetrization according to equation (1)—premultiplication by $\tilde{A}$—cannot be avoided. Even then the scheme is remarkably simple, compared with the more elaborate scheme of § 18. However, the complete scheme is always safe and will always be preferable if the given matrix is strongly skew-angular, or if its order is so high that the extra number of decimal places is still unable to counteract the accumulation of rounding errors.

20. Inversion of a complex matrix. Occasionally linear systems with complex elements have to be solved, and the question arises of how to invert a matrix whose elements are complex numbers. In principle this problem needs no special attention, since complex numbers satisfy all the rules of ordinary algebra, and all operations with real numbers carry over to the realm of complex numbers with the added information that $i^2 = -1$. However, the numerical operations with complex numbers are often prone to errors and we often prefer to translate everything into the realm of real numbers. We can do that with complex matrices, at the cost of increasing the size of a matrix of n rows and columns to a matrix of $2n$ rows and columns. The amount of work in inverting such a matrix is almost 8 times as great. However, the algebra of complex numbers shows that this factor is reduced to only 4; this corresponds to the fact that multiplication of two complex numbers amounts to 4 times the work of a simple multiplication. We gain greatly, however, in the simplicity of the resulting work scheme.

Let C be a complex matrix which can be split into a real and an imaginary part.

$$C = A + iB \tag{2-20.1}$$

Consider the solution of the equation

$$Cz = c \tag{2-20.2}$$

Both vectors z and c are complex. We will write them in the form

$$z = x + iy, \qquad c = a + ib \tag{2-20.3}$$

Then the algebraic equation

$$(A + iB)(x + iy) = a + ib \tag{2-20.4}$$

splits into the two real equations

$$Ax - By = a, \qquad Bx + Ay = b \tag{2-20.5}$$

The components of the vectors x and y can be combined into one vector of $2n$ components. The same can be done with the two vectors a and b:

$$\bar{x} = (x_1, x_2, \cdots, x_n, y_1, y_2, \cdots, y_n)$$

$$\bar{a} = (a_1, a_2, \cdots, a_n, b_1, b_2, \cdots, b_n)$$

Our linear system can now be written in the form

$$\bar{C}\bar{x} = \bar{a} \tag{2-20.6}$$

Where the $2n$ by $2n$ matrix $\bar{C}$ is defined as

$$\bar{C} = \begin{bmatrix} A & -B \\ B & A \end{bmatrix} \tag{2-20.7}$$

We can avoid, however, the inversion of this large matrix, if we are in the possession of the inverse of A, i.e., A^{-1}. We premultiply the first equation of (5) by A^{-1} and thus liberate x.

$$x - A^{-1}By = A^{-1}a \tag{2-20.8}$$

Hence x is now expressible in terms of y.

$$x = A^{-1}By + A^{-1}a \tag{2-20.9}$$

Substituting in the second equation, we obtain

$$(A + BA^{-1}B)y = b - BA^{-1}a \tag{2-20.10}$$

We will now invert the n by n matrix

$$\bar{A} = A + BA^{-1}B$$

Let the inverse of this matrix be A_1:

$$A_1 = (A + BA^{-1}B)^{-1}$$

After the proper simplification the final result of our inversion scheme can be expressed as follows. The inverse C^{-1} of the complex matrix $C = A + iB$ may be written as

$$C^{-1} = A_1 - iB_1 \tag{2-20.11}$$

where

$$B_1 = A_1BA^{-1} = A^{-1}BA_1 \tag{2-20.12}$$

21. Solution of codiagonal systems. In many problems of analysis a special type of equation occurs as the result of changing a given linear differential equation to a difference equation. The matrix of the resulting linear system has many zeros. The only nonzero elements of the matrix are in the diagonal and a few "codiagonals" (these are lines parallel to the main diagonal, on both sides of the main diagonal). Hence the nonvanishing elements of the matrix are contained in a relatively narrow band around the main diagonal. Such systems can be solved without going through the regular inversion technique. If the number of codiagonals occupied by elements is small, the original large set of equations is reducible to a small set of equations. The remaining simultaneous set contains only as many unknowns as there are diagonals occupied by elements.

The accompanying figure demonstrates the situation graphically, before going into analytical details. A system with three codiagonals is pictured. The codiagonals *below* the main diagonal are not indicated specifically, since for the general procedure it makes no difference whether the matrix below the main diagonal is solid (i.e., completely occupied by elements) or composed of a few codiagonals and the rest of the elements vanishing.

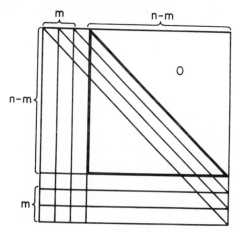

We reduce the original matrix of n rows and n columns to the heavily drawn matrix of $n - m$ rows and $n - m$ columns which is *purely triangular*. Since we know how to invert a triangular matrix, the reduced problem can be solved at once.

The original operation has the form

$$Ax = b \qquad (2\text{-}21.1)$$

We will agree that the notation $\bar{A}$ shall refer to the new triangular matrix, obtained by omitting the first m columns and the last m rows. Similarly, $\bar{x}$ shall denote the vector $x = (x_1, x_2, \cdots, x_n)$ but omitting the first m elements.

$$\bar{x} = (x_{m+1}, x_{m+2}, \cdots, x_n) \tag{2-21.2}$$

The omission of the last m equations is permissible since it merely means that for the time being we will not make use of the information contained in the last m equations. Later we will satisfy the remaining equations too. The first m columns cannot be omitted, but it is permissible to carry over these columns to the right side if the original columns of A are denoted by $u_1, u_2, \cdots, u_n$. Similarly the notation b will indicate that the last m elements of the vector b have been omitted. Hence the given system, disregarding the last m equations, may be written

$$\bar{A}\bar{x} = \bar{b} - (x_1\bar{u}_1 + x_2\bar{u}_2 + \cdots + x_m\bar{u}_m) \tag{2-21.3}$$

In the figure we have chosen $m = 3$.

Now $\bar{A}$ is a triangular matrix P whose inverse Q can be constructed according to the algorithm described in Section 17. We thus obtain

$$\bar{x} = Q\bar{b} - x_1 Q\bar{u}_1 - x_2 Q\bar{u}_2 - \cdots - x_m Q\bar{u}_m \tag{2-21.4}$$

This equation expresses $x_{m+1}, x_{m+2}, \cdots, x_n$ in terms of the first m unknowns $x_1, x_2, \cdots, x_m$.

We now come to the solution of the last m equations. This is now an easy task, since all the x_k beyond $k = m$ are expressed in terms of the first m unknowns. Hence we have a simultaneous set of m linear equations in m unknowns, which causes no particular difficulties, since m is small. The originally large set of n equations is reduced to the much smaller set of only m equations. Finally, after obtaining the $x_1, x_2, \cdots, x_m$ by solving the reduced set, we substitute these values in (4) and thus obtain the complete solution of our problems.

The inversion of a triangular matrix P demands that none of the diagonal elements of P shall vanish. If some of these elements vanish, the method is still applicable with the following modification. We replace the zero by 1 and put a compensating term on the right side of the equation. Let us assume, for example, that in the example of the figure the third codiagonal has a zero in the fourth and fifth equations. These zeros are replaced by 1 and correspondingly x_7 is put on the right side of the fourth, and x_8 on the right side of the fifth equation. Inverting the system we now obtain an apparent redundancy, since the x_i are expressed not only in terms of x_1, x_2, x_3, but also in terms of x_7 and x_8, while the last 3 equations cannot determine more than 3 unknowns. In actual fact, however, the fourth and the fifth equations give us two additional conditions which are not yet satisfied. And thus we obtain $3 + 2 = 5$ equations for $3 + 2 = 5$ unknowns which can be solved, provided that the system is not singular.

22. Matrix inversion by partitioning. The direct inversion of a large matrix is frequently not advisable, because of the eventual accumulation of rounding errors. Hence it is of importance to know that inversion of large matrices can be avoided by partitioning the matrix into smaller components. Consider a system of 20 equations with 20 unknowns, requiring the inversion of a 20 by 20 matrix. We might have coded the inversion of a 10 by 10 matrix. Then we can obtain the solution of our larger problem by two inversions of the 10 by 10 type. Generally this method of reducing a large matrix to a combination of smaller matrices can be characterized as follows. Consider the matrix equation.

$$Az = c \qquad (2\text{-}22.1)$$

Let us partition the given large matrix A into four smaller parts according to the following scheme. The partitioning divides our original matrix A into four parts, viz., the two square matrices A_1 and B_2 and the two nonsquare matrices B_1 and A_2. We partition also the unknown vector z and the given right side c.

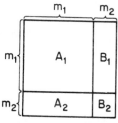

$$z = (x_1, x_2, \cdots x_{m_1}), \quad (y_1, y_2, \cdots y_{m_2}); \quad \begin{aligned} x &= (x_1, x_2, \cdots, x_{m_1}) \\ y &= (y_1, y_2, \cdots, y_{m_2}) \end{aligned}$$

$$c = (a_1, a_2, \cdots a_{m_1}), \quad (b_1, b_2, \cdots b_{m_2}); \quad \begin{aligned} a &= (a_1, a_2, \cdots, a_{m_1}) \\ b &= (b_1, b_2, \cdots, b_{m_2}) \end{aligned}$$

Now the first m_1 equations of the given system may be written

$$A_1 x + B_1 y = a$$

The last $m_2 = n - m_1$ equations give

$$A_2 x + B_2 y = b$$

and thus the entire system of equations is now equivalent to the two simultaneous vector equations

$$A_1 x + B_1 y = a, \qquad A_2 x + B_2 y = b \qquad (2\text{-}22.2)$$

Let us now assume that we are able to invert the matrix A_1, obtaining A_1^{-1}. We premultiply the first equation of (2) by A_1^{-1} and thus liberate x.

$$x = A_1^{-1} a - A_1^{-1} B_1 y \qquad (2\text{-}22.3)$$

Substitution in the second equation gives

$$(B_2 - A_2 A_1^{-1} B_1) y = b - A_2 A_1^{-1} a \qquad (2\text{-}22.4)$$

We encounter here the product of matrices which are not of the square type; that is, the numbers of rows and columns are not necessarily the same. Actually the law of multiplying two matrices together does not necessitate equality of the numbers of rows and columns. Since the rows of the first matrix are multiplied by the columns of the second matrix, it is necessary that the length of the *rows* of the first factor shall correspond to the length of the *columns* of the second factor. But this is the only condition, and generally we can multiply a matrix of n rows and l columns with a matrix of l rows and m columns. The product will be a matrix of n rows and m columns, as indicated in the figure below.

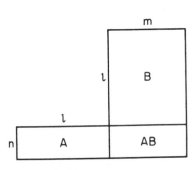

Now A_2 is a matrix of m_2 rows and m_1 columns, A_1^{-1} is a matrix of m_1 rows and m_1 columns. Hence $(m_2,m_1)(m_1,m_1) = (m_2,m_1)$ and thus the product $A_2 A_1^{-1}$ is an m_2 by m_1 matrix. Multiplying by B_1, which is an m_1 by m_2 matrix, we obtain $(m_2,m_1)\cdot(m_1,m_2) = (m_2,m_2)$. Hence the product $A_2 A_1^{-1} B_1$ is a square matrix of m_2 rows and m_2 columns which can be added to B_2.

We will now invert the matrix

$$\bar{B}_2 = B_2 - A_2 A_1^{-1} B_1 \qquad (2\text{-}22.5)$$

and thus solve equation (4):

$$y = \bar{B}_2^{-1}b - \bar{B}_2^{-1}A_2A_1^{-1}a \qquad (2\text{-}22.6)$$

We need the following matrix products:

$$C_2 = \bar{B}_2^{-1}A_2A_1^{-1}, \qquad C_1 = A_1^{-1}B_1\bar{B}_2^{-1}$$

Then $\quad x = (A_1^{-1} + A_1^{-1}B_1C_2)a - C_1b, \qquad y = -C_2a + \bar{B}_2^{-1}b$

The inverse of the system (1) becomes

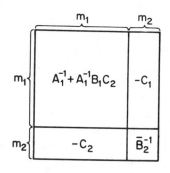

This method of inverting a large matrix by partitioning it to smaller matrices can be conceived as a generalization of the Gaussian elimination scheme. There we consider each equation separately and eliminate one unknown at a time. The process has thus to be repeated n times. By combining a certain number of equations into one system and inverting a larger matrix, we can simultaneously eliminate a whole *group of unknowns* and thus reduce the system in larger steps. For example, if we have coded the inversion of a 5 by 5 matrix, a system of twenty equations with twenty unknowns can be reduced to a 15 by 15 system by eliminating five of the unknowns. Then again we eliminate five unknowns and reduce the system to 10 by 10. The next elimination reduces the system to 5 by 5, which can now be solved directly. The process had to be repeated 4 times, while the original Gaussian scheme would have required 20 successive transformations. The partition method has the further advantage that accumulation of rounding errors will be greatly retarded if each single inversion is performed with great precision by back-checking and subsequent corrections (cf. the next section).

23. Perturbation methods. The word "perturbation" is taken from astronomy. When Newton discovered the law of gravity and laid the foundation to exact mathematical calculations, the orbits of the planets could be predicted with a high degree of mathematical accuracy. The predicted orbits, however, did not fit the actual observations with sufficient accuracy. The discrepancy was caused by the fact that the planets are subject not only to the gravity of the sun but to a minor degree to the gravity between planet and planet. This influence is much smaller than the influence of the sun, on account of the much smaller masses of the planets. The influence is nevertheless not negligible, and has to be added to the primary influence of the sun as a *correction*. The orbits of the planets are thus "perturbed" because of the added influence of the planetary masses and the "perturbation" can be mathematically calculated. If the masses of the planets were of the same order of magnitude as the mass of the sun, the mathematical problem of determining the orbits of the planets would be a practically unsolvable problem. The solution becomes possible by the fact that a good first approximation is available by neglecting the mutual influence of the planets and then adding it as a small correction. The discovery of the planet Neptune, based on the elaborate perturbation calculations of the planet Uranus by the eminent French astronomer Leverrier, and in our day the discovery of the planet Pluto, based on the perturbation of the orbit of Neptune, are impressive examples of the ingenuity of perturbation methods.

In matrix algebra, perturbation methods can be used as a powerful tool for counteracting the disturbing influence of rounding errors. We design methods for solving linear equations or inverting a matrix. We would get exact results, were it not for rounding errors which interfere with the exactitude of our calculations and cause small errors whose cumulative effect is not negligible. We can always check our results by substituting in the defining equation and examining how closely the right sides of the equations are satisfied. Since we are not interested in absolute accuracy, we may consider a certain result as sufficiently accurate. But it is equally possible that the equations did not check with the desired accuracy. Shall we now throw away our entire computation and start over again, using double or triple precision?

This is in fact seldom necessary. Even an inaccurate result can be very valuable because the exact result may be calculable on the basis of a perturbation procedure, without starting our computations over again. All perturbation methods are characterized by a certain "perturbation parameter" ε which has to be sufficiently small. We expand the exact solution into powers of ε. We thus obtain an infinite expansion, but in actual fact only a few terms of this expansion will be needed. If ε is of the order of magnitude of 10^{-2}, the second power of ε is of the order of magnitude 10^{-4}, the third power is of the order of magnitude 10^{-6}. Hence it will hardly be necessary to go beyond the third-order term of our expansion. Quite generally the convergence of our expansion will always be sufficiently rapid if we start with a first approximation which is not too far from the true solution. The corrections are then available on the basis of a successive iteration scheme, starting with a first crude result which in itself is not of sufficient accuracy but yet sufficiently close to serve as the basis of a perturbation procedure.

We will apply this perturbation method to three important cases. First we consider the inversion of a matrix. Let B be the exact inverse of a given matrix A, i.e.,

$$BA = I \tag{2-23.1}$$

In actual fact we have obtained an approximate inverse $\bar{B}$ which, if multiplied by A, gives the unit matrix I only approximately. The elements in the diagonal are not exactly 1 and the elements outside the diagonal not exactly zero. Generally we can put

$$\bar{B}A = I + \varepsilon C \tag{2-23.2}$$

The magnitude of ε can be normalized by the requirement that the absolutely largest element of the matrix C shall become 1. The smaller ε is, the quicker will be the convergence of the procedure.

We will now expand B in an infinite series, starting with $\bar{B}$.

$$B = \bar{B} + \varepsilon B_1 + \varepsilon^2 B_2 + \varepsilon^3 B_3 + \cdots \tag{2-23.3}$$

Since B is the exact inverse of A, we must have

$$(\bar{B} + \varepsilon B_1 + \varepsilon^2 B_2 + \varepsilon^3 B_3 + \cdots)A = I \tag{2-23.4}$$

and carrying over the first term to the right side and dividing by ε, we obtain

$$(B_1 + \varepsilon B_2 + \varepsilon B_3 + \cdots) A = -C$$

Now we postmultiply on both sides by the expansion (3). On the left side the second factor disappears since $AB = I$, and we get

$$B_1 + \varepsilon B_2 + \varepsilon^2 B_3 + \cdots = C(\bar{B} + \varepsilon B_1 + \varepsilon^2 B_2 + \cdots) \tag{2-23.5}$$

Equating equal powers of ε on the two sides of the equation, we obtain the following sequence of equations which permit us to determine the B_1, B_2, B_3, $\cdots$ by successive recurrences.

$$B_1 = C\bar{B}$$
$$B_2 = CB_1 \tag{2-23.6}$$
$$B_3 = CB_2$$
$$\vdots$$

A few steps will give sufficient accuracy. In fact, a large number of these successive matrix multiplications would miss their aim by bringing in new rounding errors which eventually become larger than the remaining error of the process. We can omit the parameter ε from our resulting formulas by putting

$$\bar{B}A = I + C \tag{2-23.7}$$
$$B = \bar{B} + B_1 + B_2 + B_3 + \cdots$$

with

$$B_1 = C\bar{B}$$
$$B_2 = CB_1 \tag{2-23.8}$$
$$B_3 = CB_2$$
$$\vdots$$

Wherever we stop we can take the resulting B, and multiplying it by A, see how near the resulting product comes to the unit matrix I. If the premises for the successful application of a perturbation method were satisfied at all, the remaining errors will now be negligibly small.

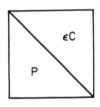

As a second application of the perturbation method we consider the case of a matrix which is *nearly triangular.* We have elements below the main diagonal which form the triangular matrix P. Above the diagonal we do not have zero, but elements which are small in comparison with the diagonal terms of P. We include all these elements in the matrix εC and put

$$A = P + \varepsilon C \tag{2-23.9}$$

Now we have no difficulty with the inversion of the triangular matrix P, and thus we obtain the new triangular matrix Q.

$$QP = I \qquad (2\text{-}23.10)$$

We can assume that this equation is satisfied with an error which is small in comparison with εC.

Let us now expand the true inverse of A in an infinite series, starting with Q.

$$B = Q + \varepsilon B_1 + \varepsilon^2 B_2 + \varepsilon^3 B_3 + \cdots \qquad (2\text{-}23.11)$$

Hence by the definition of the inverse,

$$(Q + \varepsilon B_1 + \varepsilon^2 B_2 + \varepsilon^3 B_3 + \cdots)(P + \varepsilon C) = I$$
$$QP + \varepsilon B_1 P + \varepsilon^2 B_2 P + \varepsilon^3 B_3 P + \cdots \qquad (2\text{-}23.12)$$
$$+ \varepsilon QC + \varepsilon^2 B_1 C + \varepsilon^3 B_2 C + \cdots = I$$

Since, however, $QP = I$, we obtain, by equating every coefficient of this expansion to 0,

$$B_1 P = -QC$$
$$B_2 P = -B_1 C$$
$$B_3 P = B_2 C$$
$$\vdots$$

and postmultiplying by Q,

$$B_1 = -QCQ$$
$$B_2 = -B_1 CQ \qquad (2\text{-}23.13)$$
$$B_3 = -B_2 CQ$$
$$\vdots$$

Once more we can formulate our results without the use of the expansion parameter.

$$A = P + C$$
$$B = Q + B_1 + B_2 + B_3 + \cdots$$
$$B_1 = -QCQ \qquad (2\text{-}23.14)$$
$$B_2 = -B_1 CQ$$
$$B_3 = -B_2 CQ$$
$$\vdots$$

Once more we can stop with B_k, where k is suitably chosen (it will seldom go beyond $k = 4$), and once more we check the accuracy of our results by forming the product of the resulting B with A. If the conditions were suitable to successful application of a perturbation method, the deviations from the unit matrix will be negligibly small.

As a third application of a perturbation procedure, we consider the solution of the linear set of equations

$$Ax = b \tag{2-23.15}$$

We assume that we have obtained a solution of this equation by some method, without inverting the matrix A. However, by substituting our solution in (15) we will not get b exactly, but only approximately. Hence we cannot call the obtained solution x, but $\bar{x}$. The difference

$$b - A\bar{x} = b_1 \tag{2-23.16}$$

is called the "residual vector" of our solution. Our problem is now reduced to the solution of the new set

$$Ax_1 = b_1 \tag{2-23.17}$$

Again we do not succeed exactly but obtain an approximate $\bar{x}_1$ which gives rise to a new residual vector b_2:

$$b_1 - A\bar{x}_1 = b_2 \tag{2-23.18}$$

together with the new equation,

$$Ax_2 = b_2 \tag{2-23.19}$$

$$x = \bar{x} + \bar{x}_1 + x_2 \tag{2-23.20}$$

This procedure can be continued until we come to a residual vector b_k which is so small that the associated x_k becomes negligible. Then x is equal to the sum of the partial solutions $\bar{x} + \bar{x}_1 + \bar{x}_2 + \cdots + \bar{x}_{k-1}$. This method of successive approximations is operative whenever some analog machine is employed to the solution of a linear system of equations. Such a machine of a mechanical or electrical type will seldom give an accuracy of more than 1%. This means that the original right side can be reduced by two significant figures if we substitute the obtained solution in the given set.

Now a new solution is obtained, and substituting again in the residual set, the right side is reduced by another two significant figures. Hence in four successive steps, 8 significant figures can be gained. The original solution of moderate accuracy has been corrected to a precise solution of high accuracy. The same principle is applicable, however, if we are in possession of some digital method (elimination or iterative techniques) by which a moderately accurate solution of a given linear set of equations can be obtained.

24. The compatibility of linear equations. The mere mathematical solution of a set of linear equations frequently blinds us to the dangers which arise in connection with large linear systems. The temptation is to use the large-scale computing facilities of the big electronic digital calculators for solution of extensive linear systems, without realizing that the exact mathematical solution obtained in this manner may have no physical significance whatever. The question concerning the physical significance of a mathematically correct solution has to be raised and the problem of "noise" has to be discussed. The "noise" here in question does not refer to the "arithmetical noise" caused by the rounding errors of our calculations, but to the "physical noise" caused by the inexactitude of our measurements.

The following example is well suited to characterize on a simplified scale the difficulties inherent in the solution of linear systems of a strongly skew-angular type. Consider the two equations:

$$x + y = 2.00001, \qquad x + 1.00001y = 2.00002 \qquad (2\text{-}24.1)$$

The solution of this system is

$$x = 1.00001, \qquad y = 1$$

From the purely mathematical standpoint we have two equations in two unknowns, and the system is not singular since the determinant of the system is not zero. Hence the system allows a unique solution, and after finding that solution our task is done.

From the standpoint of a physical system the situation is quite different. The right side of a system of equations is usually the result of physical measurements, and these measurements are of limited accuracy. Hence the right side of the above system may not be given with five, but only with two-decimal-place accuracy. But then we see at once that our system cannot be solved for x and y. The reason is that the unknowns x and y of our problem enter the equations practically with their *sum* only. The value of $x + y$ is well obtainable from the given equations. But in order to separate x and y, we need the combination $x - y$ too. Now, if we put

$$\tfrac{1}{2}(x + y) = \xi, \qquad \tfrac{1}{2}(x - y) = \eta$$

our equations become

$$2\xi = 2.00001, \qquad 2.00001\xi - 0.00001\eta = 2.00002$$

$$\text{(2-24.2)}$$

Hence the quantity η is only exceedingly *weakly* represented in our system, and thus requires *excessive accuracy* for its evaluation. We obtain η by a division by 10^{-5}; i.e., multiplication by 10^5. This requires that the right side shall be known with excessive accuracy, which is usually not possible for physical reasons. If, for example, the physical noise of the observations would cause an error of 0.001 in the second equation, then ξ would still be practically 1, while η would become 100, thus giving the entirely erroneous solution $x = 101$, $y = -99$. We should have recognized that under the given physical circumstances we can obtain $x + y$ with sufficient accuracy, but separate determination of x and y is out of the question.

While in this simple example we can follow each detail of the situation and demonstrate explicitly the unsatisfactory nature of nearly singular systems, we frequently accept the results deduced from strongly skew-angular systems without realizing that in view of the physical noise of the problem the mathematical solution may have little relation to the true values of the quantities that our solution is supposed to yield.

In order to develop the proper critical faculty for realistic appraisal of strongly skew-angular systems, we have first to develop the mathematical theory of the compatibility of linear systems and then properly modify it in order to apply it to the question of the physical feasibility of a given set of linear equations.

The mathematical compatibility problem of linear systems arises from the following consideration. If a problem contains n unknowns, our first thought is to get n equations for the determination of these unknowns. If the number of equations is less than n, we know in advance that the given information will not suffice for unique determination of all the unknowns. If the number of equations is more than n, we have an abundance of information which will generally lead to contradictions. Hence we discard underdetermined systems because they contain too little information and cannot lead to a unique solution of the problem, and we discard overdetermined systems because they contain too much information and serve no useful purpose.

It is important to realize, however, that the compatibility of a given set of equations bears no relation to the underdetermined or overdetermined or evendetermined ("balanced") nature of the problem. The following two equations are underdetermined, since two equations are given for five unknowns.

$$x_1 + x_2 + x_3 + x_4 + x_5 = 3 \qquad (2\text{-}24.3)$$
$$2x_1 + 2x_2 + 2x_3 + 2x_4 + 2x_5 = 8$$

But these two equations are self-contradictory and cannot be solved for any values of the five unknowns. On the other hand, the following five equations are given for only two unknowns and thus the system is overdetermined.

$$x_1 + x_2 = 0$$
$$2x_1 + 3x_2 = -1$$
$$3x_1 + 2x_2 = 1 \qquad (2\text{-}24.4)$$
$$x_1 - x_2 = 2$$
$$3x_1 + 5x_2 = -2$$

These five equations are not contradictory, but have the solution $x_1 = 1$, $x_2 = -1$.

The question arises whether there is a systematic way of deciding the compatible or incompatible nature of a given set of equations. Such a systematic method exists indeed and can be described as follows. Let us consider the linear equation

$$Ax = b \qquad (2\text{-}24.5)$$

and let us augment it by the adjoint equation

$$\tilde{A}y = c \qquad (2\text{-}24.6)$$

We form the scalar product of the first equation with y, the second equation with x:

$$y \cdot Ax = y \cdot b, \qquad x \cdot \tilde{A}y = x \cdot c \qquad (2\text{-}24.7)$$

Now by the fundamental transposition rule (8.23) we have

$$y \cdot Ax = x \cdot \tilde{A}y \qquad (2\text{-}24.8)$$

This shows that the left sides of equations (7) are equal for any choice of x and y. Hence the right sides must also be equal and we obtain

$$y \cdot b - x \cdot c = 0 \qquad (2\text{-}24.9)$$

This equation is of no particular use, however, in deciding the compatibility of the system $Ax = b$, since it demands the knowledge of x (i.e., the solution of the given system), while our aim is to decide whether the system is solvable at all. In one particular case, however, the vector x will drop out from our relation; viz., if c happens to be zero. Then we get

$$y \cdot b = 0 \qquad (2\text{-}24.10)$$

where y is the solution of the equation

$$\tilde{A}y = 0 \qquad (2\text{-}24.11)$$

One can show that this condition is not only necessary but also *sufficient*. We thus obtain the following general principle which answers all compatibility problems of linear systems: "The right side of a given set of linear equations has to be orthogonal to any solution of the adjoint homogeneous equation."

This general principle operates in a given case in a variety of ways. First, it is possible that the adjoint homogeneous equation $\tilde{A}y = 0$ has *no solution* outside of the identical vanishing of y. In that case we do not get any compatibility condition, which means that the given set (5) is compatible with *any* given right side. Second, it is possible that the adjoint homogeneous equation $\tilde{A}y = 0$ has one and *only one* solution (not counting the trivial solution $y = 0$ and not counting the freedom of an arbitrary factor by which y can be multiplied, because of the homogeneity of the equation). In this case the given right side has to satisfy one compatibility condition by being orthogonal to the homogeneous adjoint solution. Third, it is possible that the adjoint homogeneous equation $\tilde{A}y = 0$ has a number of independent solutions. In that case the given right side has to be orthogonal to *every one* of these independent solutions.

The question of overdetermination or underdetermination or evendetermination does not enter specifically the application of this general principle except for the fact that in the case of an overdetermined system the adjoint set has *always* nontrivial solutions, and thus the given right side must always satisfy one or more conditions.

Examples. 1. We consider the linear system

$$x_1 + x_2 + x_3 = 1$$
$$2x_1 + 2x_2 + 3x_3 = 3$$

Here the matrix A is $A = \begin{bmatrix} 1 & 1 & 1 \\ 2 & 2 & 3 \end{bmatrix}$

The adjoint matrix becomes $\tilde{A} = \begin{bmatrix} 1 & 2 \\ 1 & 2 \\ 1 & 3 \end{bmatrix}$

This gives rise to the adjoint equations

$$\left. \begin{aligned} y_1 + 2y_2 &= 0 \\ y_1 + 2y_2 &= 0 \\ y_1 + 3y_3 &= 0 \end{aligned} \right\}$$

These equations have no nonvanishing solution, and thus the given equations are compatible, irrespective of what the right side is.

2. We now consider the system

$$x_1 + x_2 + x_3 = 1$$
$$2x_1 + 2x_2 + 2x_3 = 3$$

The adjoint system is $y_1 + 2y_2 = 0$

$$y_1 + 2y_2 = 0$$
$$y_1 + 2y_2 = 0$$

This system has the solution

$$y_1 = -2$$

$$y_2 = 1$$

Hence the right side has to satisfy the condition

$$-2c_1 + c_2 = 0 \quad \text{which means} \quad c_2 = 2c_1$$

The given right side does not satisfy this condition, and thus the equations are incompatible and thus unsolvable.

3. We consider the overdetermined system

$$x_1 + x_2 = 0$$

$$2x_1 + 3x_2 = -1$$

$$3x_1 + 2x_2 = 1$$

$$x_1 + x_2 = 2$$

$$3x_1 + 5x_2 = -2$$

The matrix A has now two columns and five rows.

$$A = \begin{bmatrix} 1 & 1 \\ 2 & 3 \\ 3 & 2 \\ 1 & -1 \\ 3 & 5 \end{bmatrix}$$

Hence the adjoint matrix has two rows and five columns.

$$\tilde{A} = \begin{bmatrix} 1 & 2 & 3 & 1 & 3 \\ 1 & 3 & 2 & -1 & 5 \end{bmatrix}$$

The adjoint system is composed of two equations in five unknowns.

$$y_1 + 2y_2 + 3y_3 + y_4 + 3y_5 = 0$$

$$y_1 + 3y_2 + 2y_3 - y_4 + 5y_5 = 0$$

This system has three independent solutions:

$$y_1 = -5, \quad y_2 = 1, \quad y_3 = 1, \quad\quad y_4 = 0, \quad y_5 = 0$$
$$y_1 = -5, \quad y_2 = 2, \quad y_3 = 0, \quad\quad y_4 = 1, \quad y_5 = 0$$
$$y_1 = 0, \quad\quad y_2 = 9, \quad y_3 = -1, \quad y_4 = 0, \quad y_5 = -5$$

Correspondingly the right sides have to satisfy three conditions:

$$-0 \cdot 5 - 1 \cdot 1 + 1 \cdot 1 + 2 \cdot 0 - 2 \cdot 0 = 0$$
$$-0 \cdot 5 - 1 \cdot 2 + 1 \cdot 0 + 2 \cdot 1 - 2 \cdot 0 = 0$$
$$0 \cdot 0 - 1 \cdot 9 + 1 \cdot (-1) + 2 \cdot 0 - 2 \cdot (-5) = 0$$

Since these conditions are actually satisfied, the given overdetermined system is compatible.

The decision that a certain system of equations is compatible does not necessarily imply that the solution of the system is *unique*. A compatible system of equations may have one or more solutions. In our first example, we had an underdetermined incompatible system which had no solutions. In our second example we had an underdetermined compatible system which has an infinity of solutions. In our third example we have an overdetermined compatible system which has a unique solution. Generally we can say that if a compatibility condition is satisfied, we can drop one of the equations as superfluous. An overdetermined system will thus turn to an evendetermined system. But it is also possible that it may turn into an underdetermined system. Consider, for example, the following four equations in three unknowns:

$$x_1 + 3x_2 - 2x_3 = 11$$
$$2x_1 - 5x_2 + 7x_3 = -11$$
$$-x_1 + 2x_2 - 3x_3 = 4$$
$$x_1 + 2x_2 - x_3 = 8$$

The adjoint equations become

$$y_1 + 2y_2 - y_3 + y_4 = 0$$
$$3y_1 - 5y_2 + 2y_3 + 2y_4 = 0$$
$$-2y_1 + 7y_2 - 3y_3 - y_4 = 0$$

They have two independent solutions:

$$y_1 = 1, \quad y_2 = 5, \quad y_3 = 11, \quad y_4 = 0$$
$$y_1 = 9, \quad y_2 = 1, \quad y_3 = 0, \quad y_4 = -11$$

Correspondingly the given right side has to satisfy two conditions:

$$11 \cdot 1 - 11 \cdot 5 + 4 \cdot 11 + 8 \cdot 0 = 0$$
$$11 \cdot 9 - 11 \cdot 1 + 4 \cdot 0 + 8 \cdot (-11) = 0$$

These conditions are actually satisfied and the compatibility of the given system is thus guaranteed. In accordance with the general principle we can drop as many equations as we have independent compatibility conditions. This reduces the number of independent equations to two, and the system becomes underdetermined, since we have only two equations left for three unknowns.

25. Overdetermination and the principle of least squares. The difficulties inherent in many large-scale linear systems are comparable to the difficulties of an orator who in his speech tries to cover too large a variety of items. In the beginning his speech goes on rather fluently. However, as he checks off more and more items on his list, he becomes more and more tired and occasionally loses track of his thoughts. He does not remember all the incidents he wanted to tell at the right moment and thus omits certain items and repeats instead with different words the things he has previously said. Since he does not stick very closely to the truth, he comes into contradictions by forgetting in what direction he slanted the story in his earlier remarks (violating the old oratorical principle *Mendacem oportet esse memorem*—The liar should have a good memory). In the last ten minutes his mind goes completely blank, he garbles everything, and finally sits down to thunderous applause.

The liar of bad memory of our story is the "noise" which interferes with the accuracy of our measurements and distorts the true course of events. Since noise is of a random nature, the distortion is not consistent but occurs once in one, once in the other direction. This is *one* danger encountered in large-scale recordings of physical events. The *other* danger is that the information we have at our disposal is *insufficient* for actual determination of all the unknowns of the problem. In our story the speaker omitted to comment on certain items of his journey and replaced these comments by retelling with different words certain episodes on which he commented before. In analogy to this situation it can happen (and it frequently does happen) that the statements of our system of equations are insufficient for complete determination of all the unknowns of our problem. We count the number of equations and find that we have just as many equations as unknowns. Hence we think that our system is balanced and allows a unique solution. Yet it can happen that certain equations merely repeat in different words the statements made before, without adding anything essentially new to the previous statements. In this case our system is underdetermined and not in the position to yield a complete solution of our problem.

The two difficulties are interconnected. If the left sides of the equations are interdependent, the right sides have to satisfy certain compatibility conditions. But these conditions may not be satisfied in view of the noise of the measurements, which renders our equations incompatible in the strict mathematical sense. Coupled with this incompatibility is the fact that our equations are not sufficient for detei mination of all the unknowns of our problem, since in the case of compatibility we could drop a certain number of equations as superfluous, in which case we have not enough equations for the complete solution.

The problems of underdetermination and overdetermination are thus interlocked. Our system is in fact underdetermined because of absence of certain linear combinations of the unknowns, which makes it impossible to obtain *all* the unknowns of the system. A corresponding number of equations becomes superfluous and could be dropped. Hence an n by n set of equations which omits m linear combinations of the unknowns represents in reality a set of $n - m$ equations in $n - m$ unknowns, and is thus overdetermined by m equations.

One of the two difficulties is solvable; viz., the problem of over-determination. The ingenious method of least squares makes it possible to adjust an arbitrarily overdetermined and incompatible set of equations. In fact, we make an asset out of a liability and try to overdetermine a set of equations as much as possible by making an arbitrary number of surplus observations beyond the minimum number demanded by the number of unknowns. We now have a linear set of equations, characterized by

$$Ax = b \qquad (2\text{-}25.1)$$

where A is not a square matrix, but a matrix which has many more rows than columns. In view of the errors of our measurements, the equations become mathematically incompatible. This means that we cannot make all the components of the residual vector $Ax - b$ equal to zero.

But now we can ask for the "best" solution which is still available under the given circumstances. For this purpose we form the residual vector

$$Ax - b = r \qquad (2\text{-}25.2)$$

and now we take the square of the length of the residual vector and determine x by the condition that r^2 shall become a minimum. The problem of minimizing $(Ax - b)^2$ has always a definite solution, no matter how compatible or incompatible the given system is. If the given system is compatible, the least square solution x, if substituted in $Ax - b$, will automatically give 0. If the given system is incompatible, the residual vector $Ax - b$ will not give zero but the solution of smallest error in the sense that the sum of squares of the residuals will be smaller than that for any other choice of x.

The least-square solution thus completely dispenses with the investigation of the compatibility of a given system since we are reconciled to the fact that we do not get the exact solution of our problem, but the best solution possible under the given circumstances. If the given system is compatible, the residual vector of the least-square solution becomes automatically zero, thus proving that the system is compatible.

The least-square solution of the equation $Ax = b$ becomes

$$\tilde{A}Ax = \tilde{A}b \qquad (2\text{-}25.3)$$

The remarkable fact about this equation is that it always gives an evendetermined system of just as many equations as we have unknowns, no matter how strongly overdetermined the original system has been. We see at once from the figure that the product of an

m row, n column matrix multiplied by an n row, m column matrix gives an m by m square matrix, and thus in the final set the number of equations balances the number of unknowns, no matter how many equations the original system contained. Moreover, the matrix $\tilde{A}A$ is always a symmetric matrix,

$$(\tilde{A}\widetilde{A}) = \tilde{A}A \qquad (2\text{-}25.4)$$

and its eigenvalues are not only real but *positive* or in the limit zero.

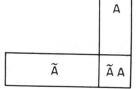

Example. In our 5 by 2 overdetermined system (24.4), we obtain

$$\begin{bmatrix} 1 & 1 \\ 2 & 3 \\ 3 & 2 \\ 1 & -1 \\ 3 & 5 \end{bmatrix} \circ \begin{bmatrix} 1 & 1 \\ 2 & 3 \\ 3 & 2 \\ 1 & -1 \\ 3 & 5 \end{bmatrix} = \begin{bmatrix} 24 & 27 \\ 27 & 40 \end{bmatrix}$$

Moreover $\tilde{A}b$ becomes

$$\begin{matrix} 0 & -1 & 1 & 2 & -2 \\ \end{matrix}$$
$$\begin{bmatrix} 1 & 2 & 3 & 1 & 3 \\ 1 & 3 & 2 & -1 & 5 \end{bmatrix} = \begin{matrix} -3 \\ -13 \end{matrix}$$

Hence $24x_1 + 27x_2 = -3, \qquad 27x_1 + 40x_2 = -13 \qquad (2\text{-}25.5)$

which has the solution

$$x_1 = 1, \qquad x_2 = -1$$

Substituting in the original set (24.4), we find that all equations are satisfied. The compatibility of the given set is thus demonstrated.

On the other hand, even the least-square formulation of an underdetermined system cannot help us in getting a unique solution. If certain combinations of the unknowns are missing in our equations, there is no magic by which they could be conjured up. An underdetermined system remains thus underdetermined, even in the least-square formulation. However, an incompatible underdetermined system is transformed into a compatible underdetermined system.

The following four equations in three unknowns represent an apparently overdetermined but in fact underdetermined and incompatible set of equations.

$$\begin{aligned}
x_1 + 3x_2 - 2x_3 &= 11 \\
2x_1 - 5x_2 + 7x_3 &= -10 \\
-x_1 + 2x_2 - 3x_3 &= 4 \\
x_1 + 2x_2 - x_3 &= 8
\end{aligned} \tag{2-25.6}$$

The least-square formulation of these equations becomes

$$\begin{aligned}
7x_1 - 7x_2 + 14x_3 &= -5 \\
-7x_1 + 42x_2 - 49x_3 &= 107 \\
14x_1 - 49x_2 + 63x_3 &= -112
\end{aligned} \tag{2-25.7}$$

The transposed matrix is now identical with the original one, and we have to look for the solutions of the homogeneous set:

$$\begin{aligned}
7y_1 - 7y_2 + 14y_3 &= 0 \\
-7y_1 + 42y_2 - 49y_3 &= 0 \\
14y_1 - 49y_2 + 63y_3 &= 0
\end{aligned} \tag{2-25.8}$$

The solution is

$$y_1 = 1, \qquad y_2 = -1, \qquad y_3 = -1 \tag{2-25.9}$$

The right side of (7) satisfies the condition of being orthogonal to the homogeneous solution. We can now drop the third equation of the set (7) and reduce it to the set of two equations

$$\begin{aligned}
7x_1 - 7x_2 + 14x_3 &= -5 \\
-7x_1 + 42x_2 - 49x_3 &= 107
\end{aligned}$$

which has an infinity of solutions. The choice $x_3 = 0$ leads to the solution

$$x_1 = 2.2, \qquad x_2 = 2.9143, \qquad x_3 = 0$$

which gives, if substituted back in (6), the right sides

$$10.943, \qquad -10.171, \qquad 3.629, \qquad 8.029$$

instead of the required

$$11, \quad -10, \quad 4, \quad 8$$

which are not attainable because of the incompatibility of the given equations.

26. Natural and artificial skewness of a linear set of equations. The difficulty of inverting the matrix of a strongly skew-angular set of equations is sometimes unnecessarily increased by an improper scaling of the variables. We conceived the equation

$$Ax = b \qquad (2\text{-}26.1)$$

as the problem of analyzing the given vector b in a given skew-angular reference system, determined by the columns of A:

$$x_1 u_1 + x_2 u_2 + \cdots + x_n u_n = b \qquad (2\text{-}26.2)$$

The determinant $|A|$ of the linear set (1) has the following striking geometrical significance. It represents the volume included by the skew-angular base vectors $u_1, u_2, \cdots, u_n$. This determinant can become very small for two reasons. One reason is that the vectors u_i are at a very small angle to each other. The other is that the length of some of these vectors becomes very small. The latter fact is caused by an inadequate scaling of the unknowns $x_1, x_2, \cdots, x_n$, and can be avoided. By a mere rescaling of the variables according to the equations

$$x_1 = \alpha_1 \bar{x}_1, \quad x_2 = \alpha_2 \bar{x}_2, \quad \cdots, \quad x_n = \alpha_n \bar{x}_n \qquad (2\text{-}26.3)$$

we can multiply the vectors $u_1, u_2, \cdots, u_n$ by arbitrary constants $\alpha_1, \alpha_2, \cdots, \alpha_n$ and can thus avoid any strong discrepancies in their length. In fact, we can normalize the length of every column exactly to 1. In that case the determinant of the new set is truly a fair measure of the skewness of the system. This determinant is always smaller in absolute value than 1. The maximum value 1 belongs to the case of orthogonal axes. The smaller the determinant, the smaller is the mutual independence of the system. On the lower end of the scale we find the determinant zero which occurs if one of the base vectors becomes a linear combination of the others.

A further advantage of this normalization is that all the elements of the matrix become numbers between ± 1. The operation with excessively large or excessively small numbers is avoided.

There is, however, a second process of scaling to which we have to pay attention. We have seen that in the least-square formulation of a linear set of equations the sum of the squares of the residuals was formed and this quantity was minimized. Now it is possible that the equations themselves are not properly balanced in the process of forming the sum of the squares. An arbitrary equation

$$a_{i1}x_1 + a_{i2}x_2 + \cdots + a_{in}x_n - b_i = 0$$

could first be multiplied by an arbitrary factor β_i before it is squared and added to the sum of the squares of the other equations. We can use these factors β_i to avoid the danger that certain equations are stated with a too small or a too large weight factor. If the weight factor of an individual equation is too small, that equation is practically nonexistent in comparison with the other equations, which means that we lose one of the important pieces of information and reduce our system to a practically underdetermined system. If the weight factor is too large, we overemphasize that particular piece of information at the cost of all the others, which leads to an even worse case of underdetermination, excluding all the other valuable pieces of information. The best balance is obtained if we choose our β_i factors in such a way that the sum of the squares of the elements of each row, excluding the right side, shall become approximately 1. With this normalization we have done equal justice to every equation and have avoided the two extremes of either overemphasizing or underemphasizing one particular equation.

The two normalizations—the one pertaining to the rows, the other to the columns—cannot be performed simultaneously (except in the case of symmetric matrices). The balancing of the length of the columns will upset the balancing of the length of the rows. Several alternating readjustments of rows and columns will be demanded before we can be satisfied that the sums of the squares of the coefficients in both rows and columns are sufficiently equalized. The equalization need not be carried out with great accuracy as long as the deviation from an average length is not serious in either rows or columns. The double equalization guarantees that the natural skewness of the system is not made worse artificially by inadequate scaling of either equations or unknowns.

27. Orthogonalization of an arbitrary linear system. We have seen that the difficulty with strongly skew-angular systems is that certain linear combinations of the unknowns enter the equations with excessively small weight. Hence it is not easy to disentangle each unknown for itself. This would require an accuracy on the right side of the equations, which is frequently not at our disposal. For a general analysis of our system the original skew-angular system of base vectors, provided by the successive columns of the matrix A, is not particularly suited, since we cannot operate effectively with skew-angular axes. The fortunate circumstance holds that by mere rotation of the reference system, our strongly skew-angular base vectors open up more and more. We can introduce a new reference system by *mere rotation*, without any other artifices, in which the originally arbitrarily skew-angular axes become completely *orthogonal*. This construction is entirely general and is applicable even to singular systems. *Any arbitrary matrix can be orthogonalized by a proper rotation of the frame of reference.*

This result seems at first sight paradoxical since we would think that a mere rotation of the reference system cannot alter the mutual orientation of the base vectors. However, in this transformation the matrix A participates as a *matrix* and not as an assembly of vectors. We have looked upon the matrix A as if its columns represented a succession of vectors. While this viewpoint is justified for a geo- metrical interpretation of a set of linear equations, we have to realize that the transformation of a matrix does *not* follow the transforma- tion law of a vector. A vector is transformed according to the law.

$$\bar{b} = \tilde{U}b \qquad (2\text{-}27.1)$$

and a matrix according to the law

$$\bar{A} = \tilde{U}AU \qquad (2\text{-}27.2)$$

where U is an orthogonal matrix

$$\tilde{U}U = I \qquad (2\text{-}27.3)$$

The columns of a matrix A are thus not invariant in either length or orientation if an orthogonal transformation is performed. If we conceive the columns of A as base vectors, these vectors change their length and orientation under the impact of an orthogonal transformation. There exists one particular coordinate transformation which reorients these axes in a particularly desirable manner by making them mutually orthogonal and thus changing an arbitrarily skew-angular system to an orthogonal system.

We consider the equation

$$Ax = b \qquad (2\text{-}27.4)$$

in its least-square formulation:

$$\tilde{A}Ax = \tilde{A}b \qquad (2\text{-}27.5)$$

Now the matrix $\tilde{A}A$ is symmetric. Its eigenvalues are all real and positive (or zero) and its principal axes are orthogonal. The eigenvalues of $\tilde{A}A$ are in no direct relation to the eigenvalues of A itself. The latter eigenvalues λ_i may be arbitrary complex quantities, while the eigenvalues of $\tilde{A}A$ are real. We will call the latter eigenvalues μ_i, in distinction to the eigenvalues of A itself, but it will be still more suitable to call them μ_i^2. This brings into evidence that the eigenvalues of $\tilde{A}A$ are not only real but even positive. Moreover, if A happens to be symmetric in itself, then $\tilde{A}A = A^2$, and the eigenvalues of $\tilde{A}A$ become λ_i^2, in which case $\mu_i = \lambda_i$.

Now the principal axes of $\tilde{A}A$ form a real orthogonal reference system in the n-dimensional space which can be introduced as a new reference system. In the new system,

$$\widetilde{\tilde{A}A}\bar{x} = \widetilde{\tilde{A}}b \qquad (2\text{-}27.6)$$

Moreover, in the new system $\tilde{A}A$ becomes a diagonal matrix.

$$\widetilde{\bar{A}}\bar{A} = \begin{bmatrix} \mu_1^2 & & & \\ & \mu_2^2 & & \\ & & \vdots & \\ & & & \mu_n^2 \end{bmatrix} \qquad (2\text{-}27.7)$$

Since, however, the matrix product $\widetilde{\bar{A}}\bar{A}$ means the column-by-column multiplication of the matrix by itself, we obtain for the base vectors $\bar{a}_i$ of the new linear system:

$$\bar{a}_i \cdot \bar{a}_k = 0, \qquad (i \neq k) \qquad (2\text{-}27.8)$$
$$\bar{a}_i^2 = \mu_i^2$$

The orthogonality of the new base vectors is thus demonstrated, although the length of these vectors is by no means equal to 1. The eigenvalues μ_i^2 of $\tilde{A}A$ can be interpreted as the squares of the lengths of the new base vectors $\bar{a}_i$.

The orthogonal transformation here required never fails to exist. Some of the μ_i^2 may become equal, in which case the orthogonal matrix U is not uniquely determined, since in certain directions circular conditions hold, and any two mutually perpendicular axes of the length 1 may be chosen as principal axes. But this means only that in such cases the orthogonal transformation which can accomplish our final goal of transforming our original skew-angular axes to a rectangular system is not unique. In the case of singular systems one or more of the μ_i values become zero but otherwise the general laws remain valid. However, since the sum of squares of the elements of a column can vanish only if every element vanishes, we see that in the case of a singular matrix *an entire column of $\bar{A}$ becomes zero.*

An interesting example of such a transformation is provided by an extreme case which is instructive because it demonstrates the behavior of strongly skew-angular systems. We consider the matrix

$$A = \begin{bmatrix} a_1 & a_1 & \cdots & a_1 \\ a_2 & a_2 & & a_2 \\ \vdots & \vdots & & \vdots \\ a_n & a_n & \cdots & a_n \end{bmatrix}$$

Here all the axes of the original skew-angular system collapse into one, thus realizing the most extreme case of linear dependence. The matrix $\tilde{A}A$ has now only a single element $\Sigma\, a_i^2$, repeated n^2 times. We can normalize it to 1.

$$a_1^2 + a_2^2 + \cdots + a_n^2 = 1$$

The solution of the eigenvalue problem of $\tilde{A}A$ becomes

either: $\lambda = 1;$ $x_1 = x_2 = \cdots = x_n = \dfrac{1}{\sqrt{n}}$

or: $\lambda = 0;$ $x_1 + x_2 + \cdots + x_n = 0$

The second solution splits into $n - 1$ orthogonal axes, since the eigenvalue $\lambda = 0$ has the multiplicity $n - 1$. The orthogonal matrix U whose columns give the eigenvectors of $\tilde{A}A$, can be written as follows:

$$U = \begin{bmatrix} \dfrac{1}{\sqrt{n}} & \dfrac{1}{\sqrt{2}} & \dfrac{1}{\sqrt{2\cdot 3}} & \dfrac{1}{\sqrt{(n-1)n}} \\[2ex] \dfrac{1}{\sqrt{n}} & -\dfrac{1}{\sqrt{2}} & \dfrac{1}{\sqrt{2\cdot 3}} & \dfrac{1}{\sqrt{(n-1)n}} \\[2ex] \vdots & 0 & -\dfrac{2}{\sqrt{2\cdot 3}} & \vdots \\[2ex] \cdot & \cdot \quad \cdot \quad \cdot \quad \cdot \quad \cdot \quad & \cdot \quad & \cdot \\[2ex] \dfrac{1}{\sqrt{n}} & 0 & \vdots & \vdots \quad \vdots \\[2ex] \dfrac{1}{\sqrt{n}} & 0 & 0 & -\dfrac{n-1}{\sqrt{(n-1)n}} \end{bmatrix}$$

and, forming the product $\tilde{U}AU$, we obtain the transformed matrix $\bar{A}$ in the new reference system:

$$\bar{A} = \begin{bmatrix} a_1 + a_2 + \cdots a_n & 0 & \cdots & 0 \\[1ex] \sqrt{n}\dfrac{a_1 - a_2}{\sqrt{2}} & 0 & & 0 \\[2ex] \sqrt{n}\dfrac{a_1 + a_2 - 2a_3}{\sqrt{2\cdot 3}} & 0 & & 0 \\[2ex] \vdots & & & \vdots \\[1ex] \sqrt{n}\dfrac{a_1 + a_2 + \cdots - (n-1)a_n}{\sqrt{(n-1)n}} & 0 & \cdots & 0 \end{bmatrix}$$

The orthogonality conditions are trivial in this case, since the last $n-1$ vectors vanish identically. But our example demonstrates what happens if the original matrix A is less extreme by being composed of columns which do not collapse into one, although they differ from each other by only small amounts. In this case the zero field of the above matrix $\bar{A}$ fills up with small elements, but the columns remain orthogonal to each other. The sum of the squares of the elements give in succession $\mu_1^2, \mu_2^2, \cdots, \mu_n^2$. While μ_1 can again be normalized to 1, by applying a universal scale factor to A, the consecutive $\mu_2, \mu_3, \cdots, \mu_n$ will be small numbers, if the columns of the matrix A are without exception at very small angles to each other. But it is equally possible that we have only a few "bad" axes which are at a very small angle to the subspace included by the previous axes. The product

$$\Delta = \mu_1\mu_2 \cdots \mu_n \qquad (2\text{-}27.9)$$

is equal to the absolute value of the determinant of A. If n is large and the determinant of A is very small, this can be brought about in two ways. We might have a relatively large number of μ_i of moderate smallness, or we might have a relatively large number of μ_i of the order of magnitude 1 and a few excessively small μ_i. At all events the new reference system is exceedingly well suited to the numerical appraisal as well as theoretical study of large linear systems.

28. The effect of noise on the solution of large linear systems. Inversion of large matrices is a numerically cumbersome procedure. If with the proper circumspection we have succeeded in obtaining a mathematically satisfactory solution, it is still a question to what extent that solution has a bearing on the given physical problem. Very accurate calculations require very accurate data. But the data of physical systems are frequently far from that accuracy which is required by the mathematical calculations. In particular the right side of linear systems is frequently the result of physical measurements and cannot be guaranteed to more than two or three significant figures. It is thus imperative that we should investigate what effect small but random changes of the elements of the right side have on the solution. This investigation is not simple in the original skew-angular reference system established by the successive columns of A, but we get ideally suited conditions if we introduce that rotated reference system in which $\tilde{A}A$ is transformed to a diagonal matrix.

The transformation is characterized by the equations

$$x = U\bar{x}, \qquad \bar{x} = \tilde{U}x \quad (2\text{-}28.1)$$

Now in the new reference system the matrix A became orthogonal with columns a_i, whose lengths are $\mu_1, \mu_2, \cdots, \mu_n$. The solution of the system $\tilde{A}A\bar{x} = \tilde{A}b$ can now be given as follows. On the left side the unknowns are separated.

$$\mu_1^2\bar{x}_1, \mu_2^2\bar{x}_2, \cdots, \mu_n^2\bar{x}_n \qquad (2\text{-}28.2)$$

On the right side we have to perform the operation

$$a_i \cdot b = |a_i| \ |b| \cos \theta_i = \mu_i |b| \cos \theta_i \qquad (2\text{-}28.3)$$

if θ_i denotes the angle between the vectors b and a_i. Hence

$$\mu_i^2 \bar{x}_i = \mu_i |b| \cos \theta_i \qquad (2\text{-}28.4)$$

which gives

$$\bar{x}_i = \frac{|b| \cos \theta_i}{\mu_i} \qquad (2\text{-}28.5)$$

The difficulty of solving nearly singular systems is caused by the division by μ_i if μ_i is very small. The result of this division is that the solution becomes very sensitive to small errors of the vector b. The true position of the vector b is not known. It is masked by an added error vector which is small compared with the length of b. This error vector, however, is of a random nature, and will have components in the direction of the large *and small* vectors a_i. If now the smallest vector a_n has the length 10^{-3} compared with the largest vector a_1—i.e., if the ratio of the smallest to the largest eigenvalue of $\tilde{A}A$ is 10^{-6}—then in x_n the noise will be magnified by the factor 1000. This can easily cause an error in the solution which renders it physically meaningless. The solution $x_1, x_2, \cdots, x_n$ of a linear system is frequently not biased in favor of the very small eigenvectors of $\tilde{A}A$, but favors all the eigenvectors with the same order of magnitude. This means that all the $\bar{x}_i$ are of the same order of magnitude. But then the vector b is strongly slanted against the small vectors a_i, because of multiplication by the small factors μ_i. If the noise were of the same character, the damage would be slight. The ratio (5) would come out as a number of average order of magnitude because the numerator becomes small together with the denominator. But in actual fact this is not so because, while b itself is strongly slanted in disfavor of the small eigenvalues, the noise has no bias against these directions and thus causes spurious components x_i which overpower the true components by a large factor. The result is a solution which may be wrong by 100% or more.

This analysis shows that the critical quantity which decides the physical reliability of a strictly mathematical solution is not the determinant of the system, but the *ratio of the largest to the smallest eigenvalue of the symmetrized matrix* $\tilde{A}A$[1]. It is the *square root* of this

---

[1] If A has complex coefficients, the symmetric matrix $\tilde{A}A$ is replaced by the Hermitian matrix $\tilde{A}^*A$.

ratio which measures the magnification of the noise in the direction of the smallest eigenvalue. As long as this ratio does not increase above a certain danger point, the problem of noise is not critical. But if that ratio becomes 10^4 and more, magnification of the noise in the direction of the smallest eigenvalue becomes 100 and more. The accuracy of our physical measurements is seldom sufficient to tolerate such an increase of the noise in certain directions. Any linear system whose critical ratio surpasses 10^4 can hardly be considered adequate for full determination of the unknowns of the problem.

The reference system of the principal axes of $\tilde{A}A$ has still another advantage. It brings out in purified form those particular combinations of the variables which are too weakly represented in the given system. The variables $\bar{x}_i$ are completely separated in this frame of reference. Moreover, certain $\bar{x}_i$ enter the equations with a too small factor. These $\bar{x}_i$ can be singled out at once as those quantities which practically drop out of the system. But now we can go back to our original variables x_i, remembering that the variables with which we have operated, had the following relation to the original x_i:

$$\bar{x} = \tilde{U}x$$

Since the matrix U is given, we can find those linear combinations of the unknowns in which the given linear system is deficient.

In our first oversimplified problem we considered a 2 by 2 matrix of the following form:

$$A = \begin{bmatrix} 1 & 1 \\ 1 & 1 + \varepsilon \end{bmatrix}$$

where ε was a small quantity. We have no difficulty in solving the associated eigenvalue problem for $\tilde{A}A$. The resulting solution becomes, considering ε a small quantity,

$$\underset{\mu = 2 \quad \varepsilon/2}{U = \begin{bmatrix} \dfrac{1}{\sqrt{2}} & -\dfrac{1}{\sqrt{2}} \\[2ex] \dfrac{1}{\sqrt{2}} & \dfrac{1}{\sqrt{2}} \end{bmatrix}}$$

We form $\bar{x} = \tilde{U}x$:

$$\bar{x}_1 = \frac{1}{\sqrt{2}}(x_1 + x_2), \qquad \bar{x}_2 = \frac{1}{\sqrt{2}}(-x_1 + x_2)$$

It is $\bar{x}_2$ which is connected with the small eigenvalue. Hence it is the particular linear combination

$$-x_1 + x_2$$

which is too weakly represented in our system. In the given simple example we were able to establish this fact by mere inspection. But in more involved systems we have no easy way of telling which particular linear relation (or relations) of the variables is practically absent in the system. The construction of the matrix U gives a systematic answer to the problem, and the knowledge of the eigenvalues μ_i^2 presents a quantitative measure for the weakness with which these combinations enter the given system.

If we do not go into the detailed analysis of the noise problem by finding the eigenvalues and eigenvectors of the matrix $\tilde{A}A$, it is still imperative that we should convince ourselves that the physical noise will not drown out our alleged solution. For this purpose we modify the given right side by random quantities of the order of magnitude of the errors of the measurements and observe the influence of this modification on our solution. If the solution changes by too large amounts as the result of this perturbation, we must come to the conclusion that our solution, although mathematically correct, cannot be considered an adequate solution of the given physical problem.

Bibliographical References

[1] AITKEN, A. C., *Determinants and Matrices* (Interscience Publishers, New York, 1944).

[2] FERRAR, W. L., *Algebra* (Oxford Press, New York, 1941).

[3] FRAZER, R. A., DUNCAN, W. I., and COLLAR, A. R., *Elementary Matrices* (Cambridge University Press, London, 1938; Macmillan, New York, 1947).

[4] TURNBULL, H. W., *The Theory of Determinants, Matrices and Invariants* (Blackie & Son, Glasgow, 1929).

III

LARGE-SCALE LINEAR SYSTEMS

1. Historical introduction. The early masters of infinitesimal calculus ventured out in a new direction which seemed to differ radically from the more conservative concepts of algebra. The operations with "infinitesimals" and quantities which "vanish in first, second, $\cdots$, nth order," although of obvious merits, had its grave logical difficulties, as pointed out by George Berkeley in his "Analyst" (1734). Before the exact limit concept of Cauchy and Gauss emerged in the early nineteenth century, Lagrange attempted a different solution. He showed how the concepts of algebra need not be given up even if working with the problems of higher mathematics. Derivatives cannot be distinguished from ordinary difference coefficients if the Δx is made small enough. Integration could be replaced by summation.

This tendency to "algebraize" calculus and all processes of higher mathematics became of increasing importance and eventually revolutionized the entire edifice of higher mathematics. Fredholm (in 1900) put the entire theory on a rigorous basis when he showed how a certain class of integral equations could be conceived as the limit of an ordinary simultaneous set of linear algebraic equations whose order gradually increases to infinity. Moreover, the error remained small even if the order of the associated algebraic system was far from becoming infinite.

While this development was in the beginning purely theoretical, it became in our own day of eminent practical importance in view of the construction of the large electronic digital calculators which made numerical solution of large sets of algebraic equations possible. The resolution of a boundary value problem into a large set of algebraic equations is thus no longer a purely theoretical but a very *practical* tool for solving partial differential equations.

The actual inversion of a large matrix, however, is even now a formidable task. Instead of getting an exact solution by matrix inversion, it is preferable to apply simpler operations which in many steps come gradually nearer and nearer to the solution desired. These are "iteration techniques," based on the constant repetition of the same algorithm. Procedures of this kind are particularly adapted to the coding for the big machines. The simplest and most natural iterations operate with the successive *powers* of a matrix. And thus we come to the investigation of *polynomials*, formed with the help of matrices.

2. Polynomial operations with matrices. If x is an algebraic quantity, the repeated operations of multiplication and addition lead to a linear superposition of the powers of x, called a "polynomial of x."

$$P_n(x) = p_n x^n + p_{n-1} x^{n-1} + \cdots + p_1 x + p_0 \qquad (3\text{-}2.1)$$

Polynomials of x have a tremendous variety of applications, because of their flexible nature and the ease with which analytical operations such as differentiation and integration can be performed with their help. If A is a matrix, we can investigate the possibility of using matrix polynomials for solution of certain matrix problems. Matrix algebra is a complete counterpart of ordinary algebra, although the commutative law of multiplication has to be sacrificed. But in the presence of one single matrix A, assuming that the coefficients of the polynomial $P_n(A)$ are ordinary numbers, the noncommutative nature of matrices does not enter the picture, since every matrix is commutable with itself. And thus we can investigate the possible operational advantages of a matrix polynomial.

$$P_n(A) = p_n A^n + p_{n-1} A^{n-1} + p_{n-2} A^{n-2} + \cdots + p_1 A + p_0 \quad (3\text{-}2.2)$$

However, from the viewpoint of applications, we immediately encounter the following difficulty. The successive powers of x are generated by simply multiplying the previous power x^{k-1} by x.

$$x^k = x x^{k-1}$$

We can similarly generate the successive powers of a matrix.

$$A^k = AA^{k-1}$$

But this operation involves a tremendous number of single operations, since the multiplication by the matrix A means row-by-column composition of two matrices. This amounts to n multiplications for every element, and since we have to find n^2 new elements, we need altogether n^3 multiplications and $n^3 - n^2$ additions for every new power of A. This is a prohibitive number of operations.

This unfavorable balance is greatly improved, however, if we do not generate the matrix polynomial (2) as a matrix but let this polynomial operate on a given vector b. Hence we consider the polynomial operation

$$P_n(A) = (p_n A^n + p_{n-1} A^{n-1} + \cdots + p_1 A + p_0)b \qquad (3\text{-}2.3)$$

This is no longer a matrix but a vector, obtainable by successive substitutions in the scheme:

$$Ab = \begin{matrix} a_{11}b_1 + a_{12}b_2 + \cdots + a_{1n}b_n \\ a_{21}b_1 + a_{22}b_2 + \quad + a_{2n}b_n \\ \vdots \\ a_{n1}b_1 + a_{n2}b_2 + \cdots + a_{nn}b_n \end{matrix} \qquad (3\text{-}2.4)$$

The number of multiplications is reduced to n^2, and the number of additions to $n^2 - n$. The operation $A^k b$ is obtainable by k successive substitutions of this kind, according to the associative law of multiplication.

$$b_k = A^k b = A(A^{k-1}b) = Ab_{k-1} \qquad (3\text{-}2.5)$$

Moreover, certain polynomials satisfy some simple "recurrence relations," by means of which the polynomial $P_n(A)b$ may be obtained in a simple way in terms of $P_{n-1}(A)b$ and $P_{n-2}(A)b$. All the important polynomials of applied mathematics satisfy a recurrence relation of the following type (cf. V, 21):

$$P_{n+1}(x) = (\alpha_n x - \beta_n)P_n(x) - \gamma_n P_{n-1}(x) \qquad (3\text{-}2.6)$$

If a polynomial of this kind is applied to matrices, we obtain

$$P_{n+1}(A)b_0 = (\alpha_n A - \beta_n)P_n(A)b_0 - \gamma_n P_{n-1}(A)b_0 \qquad (3\text{-}2.7)$$

Let us agree that the vector generated by the polynomial $P_n(A)$, operating on a given "trial vector" b_0, shall be denoted by b_n.

$$b_n = P_n(A)b_0 \qquad (3\text{-}2.8)$$

Then the relation (7) implies the following successive generation of the vectors b_k.

$$b_{n+1} = (\alpha_n A - \beta_n)b_n - \gamma_n b_{n-1} \qquad (3\text{-}2.9)$$

The nature of the operation $P_n(A)b_0$ can best be analyzed if we introduce the principal axes associated with the matrix A (cf. Sections 6 and 7).

$$Au = \lambda u, \qquad \tilde{A}v = \lambda v \qquad (3\text{-}2.10)$$

The first equation has the n solutions (omitting the "defective" case)

$$u_1, u_2, \cdots, u_n$$

The second equation has the n solutions

$$v_1, v_2, \cdots, v_n$$

This double set of vectors is in a biorthogonality relation to each other.

$$\begin{aligned} u_i \cdot v_k &= 0 \qquad (i \neq k) \\ u_i \cdot v_i &= 1 \end{aligned} \qquad (3\text{-}2.11)$$

Let us analyze the vector b_0 in the reference system of the base vectors u_i:

$$b_0 = \beta_1 u_1 + \beta_2 u_2 + \cdots + \beta_n u_n \qquad (3\text{-}2.12)$$

where
$$\beta_i = b_0 \cdot v_i$$

Then by the definition of the principal axes we obtain

$$Ab_0 = \beta_1 \lambda_1 u_1 + \beta_2 \lambda_2 u_2 + \cdots + \beta_n \lambda_n u_n$$

$$A^k b_0 = \beta_1 \lambda_1^k u_1 + \beta_2 \lambda_2^k u_2 + \cdots + \beta_n \lambda_n^k u_n$$

and for an arbitrary polynomial operator $P_n(A)$,

$$P_n(A)b_0 = \beta_1 P_n(\lambda_1)u_1 + \beta_2 P_n(\lambda_2)u_2 + \cdots + \beta_n P_n(\lambda_n)u_n \qquad (3\text{-}2.13)$$

We see that by introduction of the principal axes of A as a new frame of reference, operation of the matrix polynomial $P_n(A)$ is reducible to operation of the purely algebraic polynomial $P_n(\lambda)$, replacing λ by the successive eigenvalues $\lambda_1, \lambda_2, \cdots, \lambda_n$ of A.

3. The p,q algorithm. A particularly useful set of polynomials is established by the matrix A itself if we add to it an arbitrary vector b which plays the role of the "right side" of the linear equation $Ax = b$ associated with the given matrix. For our iteration technique, b will be identified with an arbitrary "trial vector" b_0, formed out of random numbers, on which the matrix A is going to operate. We can construct the successive powers $A^k b_0 = b_k$ by generating in succession,

$$b_0, \ b_1 = Ab_0, \ b_2 = Ab_1 = A^2 b_0, \ \cdots, \ b_n = Ab_{n-1} = A^n b_0 \quad (3\text{-}3.1)$$

In view of the Hamilton-Cayley identity, this last vector must become a linear combination of the previous vectors. Although generally the previous vectors are linearly independent, we can at every stage of the procedure ask for the "best identity" which can be established between them. This means that while the linear combination

$$p_k = b_k + \gamma_{k-1} b_{k-1} + \cdots + \gamma_0 b_0 \quad (3\text{-}3.2)$$

cannot be made zero for any choice of the γ_k, we can nevertheless *minimize* the length of this vector. If A is symmetric, we ask for the minimum of the scalar

$$p_k^2 = (b_k + \gamma_{k-1} b_{k-1} + \cdots + \gamma_0 b_0)^2 \quad (3\text{-}3.3)$$

If A is not symmetric, we have to distinguish between the operations with A itself and with the adjoint matrix A. We then have to complement the vector set (3.1) by the adjoint set,

$$\tilde{b}_0, \ \tilde{b}_1 = \tilde{A}\tilde{b}_0, \ \cdots, \ \tilde{b}_n = \tilde{A}\tilde{b}_{n-1} \quad (3\text{-}3.4)$$

The length to be minimized is now defined by the scalar

$$p_k^2 = (b_k + \gamma_{k-1} b_{k-1} + \cdots + \gamma_0 b_0)(\tilde{b}_k + \gamma_{k-1}\tilde{b}_{k-1} + \cdots + \gamma_0\tilde{b}_0)$$
$$(3\text{-}3.5)$$

The solution of this minimum problem leads to a set of linear equations for the γ_k, characterized by a special kind of matrix, called "recurrent." However, instead of solving this set independently for every k, we can develop an easily coded *progressive algorithm*[1] which generates the polynomials

$$p_k(A)b_0 = (A^k + \gamma_{k-1}A^{k-1} + \cdots + \gamma_0)b_0 \quad (3\text{-}3.6)$$

---

[1] Cf. the paper [5] of the bibliography, which contains an exhaustive treatment of the numerous mathematically interesting properties of these polynomials.

and an interlocked second set of polynomials $q_k(A)b_0$ *gradually*, in successive steps. Finally we reach the stage $k = n$, and $p_n(A) b_0$ becomes identically zero. The polynomial $p_n(A)$ coincides with the characteristic equation of the matrix; its roots give all the eigenvalues of A.

The polynomials $p_k(A)$ [and also $q_k(A)$] are of the type of the classical orthogonal polynomials. They satisfy a recurrence relation of the form (cf. V, 21)

$$p_{k+1}(A) = (A - \alpha_k)p_k(A) - \beta_k p_{k-1}(A) \qquad (3\text{-}3.7)$$

The numerical constants α_k and β_k of these relations are not given in advance; they are determined by the matrix itself, in conjunction with the right side b_0, and they come into evidence gradually, as the algorithm unfolds itself.

The p,q algorithm gives a *complete solution* of the eigenvalue problem. It yields all the eigenvalues and eigenvectors of the matrix in successive approximations which constantly improve and in the end become accurate (apart from the rounding errors). Of particular importance are the *beginning stages* of the process. This phase of the algorithm deserves attention of its own since it has many useful applications.

We construct the three vectors

$$b_0, \quad b_1, \quad b_2$$
$$\tilde{b}_0, \quad \tilde{b}_1, \quad \tilde{b}_2$$

and form their scalar products:

$$
\begin{aligned}
c_0 &= b_0 b_0 \\
c_1 &= b_0 \tilde{b}_1 = b_1 \tilde{b}_0 \\
c_2 &= b_0 \tilde{b}_2 = b_1 \tilde{b}_1 = b_2 \tilde{b}_0 \\
c_3 &= b_1 \tilde{b}_2 = b_2 \tilde{b}_1
\end{aligned}
\qquad (3\text{-}3.8)
$$

(In the case of a symmetric matrix, $\tilde{b}_k = b_k$, while in the case of a Hermitian matrix, $\tilde{b}_k = b_k^*$). Then the determinant condition

$$
\begin{vmatrix}
1 & \lambda & \lambda^2 \\
c_0 & c_1 & c_2 \\
c_1 & c_2 & c_3
\end{vmatrix} = 0
\qquad (3\text{-}3.9)
$$

will give us two roots which are near to the two largest eigenvalues of the matrix A if we use a trial vector b_0 which is strongly biased in favor of the large eigenvalues. This can be done by multiplying an arbitrary random vector $v_0, \tilde{v}_0$ several times by $A, \tilde{A}$; i.e., by forming in succession

$$v_1 = Av_0, \qquad v_2 = Av_1, \qquad \cdots \qquad v_m = Av_{m-1}$$
$$\tilde{v}_1 = \tilde{A}\tilde{v}_0, \qquad \tilde{v}_2 = \tilde{A}\tilde{v}_1, \qquad \cdots \qquad \tilde{v}_m = \tilde{A}\tilde{v}_{m-1}$$

and then choosing $b_0 = v_m$, $\tilde{b}_0 = \tilde{v}_m$. If m is as high as 5 or 6, the largest eigenvalues are already predominant. In this fashion the largest eigenvalue of a symmetric matrix or the largest pair of conjugate complex roots of an arbitrary matrix can be obtained in good approximation. The method is a natural extension of the Bernoulli method of finding the largest root of an algebraic equation (cf. 1-15.10). The same method is applicable to the frequently even more important problem of finding the *smallest* eigenvalues of a matrix, if we use a transformation which reverses the order of magnitude of the eigenvalues (cf. §§ 9 and 10).

The eigen*vectors* associated with the two largest roots are also obtainable in good approximation. Since equation (9) expresses the approximate validity of the relation,

$$(A - \lambda_1 I)(A - \lambda_2 I)b_0 = 0$$

we obtain for the eigenvector u_1 (associated with λ_1),

$$u_1 = Ab_0 - \lambda_2 b_0 = b_1 - \lambda_2 b_0$$

or in better approximation, multiplying by A,

$$u_1 = b_2 - \lambda_2 b_1 \qquad (3\text{-}3.10)$$

and similarly for the eigenvector u_2 (associated with λ_2),

$$u_2 = b_2 - \lambda_1 b_1 \qquad (3\text{-}3.11)$$

Likewise for the adjoint vectors,

$$\tilde{u}_1 = \tilde{b}_2 - \lambda_2 \tilde{b}_1, \qquad \tilde{u}_2 = \tilde{b}_2 - \lambda_1 \tilde{b}_1 \qquad (3\text{-}3.12)$$

(The accuracy of $u_2, \tilde{u}_2$ is greatly diminished compared with $u_1, \tilde{u}_1$, if the absolute value of λ_1 strongly overshadows that of λ_2; in the complex conjugate case, however, both vectors have the same degree of accuracy.)

4. The Chebyshev polynomials. The previous section dealt with polynomials which were generated by the matrix A itself. The α_k, β_k of the recurrence relation (3.7) had to be obtained in successive steps, in the course of a progressive minimization process. Much simpler are processes which employ *universal* polynomials, whose α_k, β_k are given constants. Amongst these polynomials the simplest and most useful ones are the Chebyshev polynomials, named after the Russian mathematician P. F. Chebyshev (1821–1894). They have the advantage that for them the α_k, β_k are constants which are *independent of k*. The coding of these polynomials is thus particularly simple.

We will encounter this remarkable set of polynomials in a great many garbs; (cf. IV, 16; V, 20; Chapter VII). At present we are interested in their applicability to matrix operations, on account of the particularly simple recurrence relation by which they can be generated.

We take our start from the simple trigonometric identity

$$\cos (n + 1)\theta + \cos (n - 1)\theta = 2 \cos \theta \cos n\theta \qquad (3\text{-}4.1)$$

Similarly

$$\sin (n + 1)\theta + \sin (n - 1)\theta = 2 \cos \theta \sin n\theta \qquad (3\text{-}4.2)$$

These identities make it possible that knowing the values of $\cos \theta$ and $\sin \theta$, all the later values of $\cos n\theta$ and $\sin n\theta$ can be evaluated in successive steps. Indeed, having obtained $\cos n\theta$, the identity (1) immediately gives $\cos (n + 1)\theta$ and thus the scheme continues. The same is true of the sine functions.[1] Now, if we put $\cos \theta = x$ and start with $\cos 0 = 1$, $\cos \theta = x$, the recurrence relation (1) gives for $\cos 2\theta$ a quadratic expression in x, then $\cos 3\theta$ follows as a cubic expression in x, and so on. We thus see that the recurrence relation (1) may be rewritten in the following form:

$$T_{n+1}(x) = 2xT_n(x) - T_{n-1}(x) \qquad (3\text{-}4.3)$$

Thus defining a set of *polynomials*. What we obtain is still $\cos n\theta$ in

---

[1] It is of interest to observe that the first trigonometric tables constructed by the Hindus actually calculated the values of $\sin x$ from degree to degree from this recurrence relation, starting with $\sin 1°$ as key value which was obtained from other considerations.

value but this $\cos n\theta$ is expressed as a polynomial of the order n in the variable $x = \cos \theta$:

$$\cos n\theta = T_n(x) \tag{3-4.4}$$

Quite similar is the situation with respect to the recurrence relation (2) if we divide on both sides by $\sin \theta$. We can now start with

$$\frac{\sin \theta}{\sin \theta} = 1, \quad \frac{\sin 2\theta}{\sin \theta} = 2 \cos \theta$$

and obtain $\sin 3\theta / \sin \theta$ as a quadratic expression in x, then $\sin 4\theta / \sin \theta$ as a cubic expression in x, and so on. Generally we now obtain the value of $\sin (n + 1)\theta / \sin \theta$ but expressed as a polynomial of the order n in the variable $x = \cos \theta$:

$$\frac{\sin (n + 1)\theta}{\sin \theta} = U_n(x) \tag{3-4.5}$$

The recurrence relation is exactly the same as before:

$$U_{n+1}(x) = 2xU_n(x) - U_{n-1}(x) \tag{3-4.6}$$

only the starting point is different because now

$$U_0(x) = 1, \quad U_1(x) = 2x \tag{3-4.7}$$

against

$$T_0(x) = 1, \quad T_1(x) = x \tag{3-4.8}$$

The relation

$$x = \cos \theta \tag{3-4.9}$$

has a simple significance. The interval of x extends from -1 to $+1$. We erect a circle over this interval as diameter. Then θ is the central angle of the point P' on the circumference of the circle which is projected down on the x-axis.

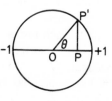

Sometimes, however, x ranges only between 0 and 1. In this case we put

$$x = \frac{1 - \cos \theta}{2} = \sin^2 \frac{\theta}{2} \tag{3-4.10}$$

and define the "shifted Chebyshev polynomials" $T_n^*(x)$ by the recurrence relation

$$T_{n+1}^*(x) = 2(1 - 2x)T_n^*(x) - T_{n-1}^*(x) \tag{3-4.11}$$

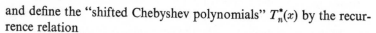

starting with

$$T_0^*(x) = 1, \quad T_1^*(x) = 1 - 2x$$

Similarly the "shifted Chebyshev polynomials of the second kind" are defined by the same recurrence relation

$$U_{n+1}^*(x) = 2(1 - 2x)U_n^*(x) - U_{n-1}^*(x) \qquad (3\text{-}4.12)$$

but starting with

$$U_0^*(x) = 1, \quad U_1^*(x) = 2(1 - 2x) \qquad (3\text{-}4.13)$$

5. Spectroscopic eigenvalue analysis. It is a frequent occurrence in vibration problems that the characteristic frequencies of an elastic structure are determined by the eigenvalues of a given differential operator. This operator is then approximated by a finite difference operator, and the problem becomes a matrix problem of finite order. Of particular physical interest are the *small* eigenvalues of this operator. In the realm of large eigenvalues (high frequencies) the error caused by the change from a differential to a difference operator becomes too pronounced. Moreover, the eigen-vectors associated with the large eigenvalues are usually excited with very small amplitudes and are thus of minor practical significance.

In the p,q algorithm considered in § 3 the eigenvalues appear *gradually*, from the top to the bottom. It is relatively easy to obtain the large eigenvalues of a matrix; the small eigenvalues, however, come only late in appearance, at the end of a careful orthogonaliza-tion process. And yet frequently our entire interest centers around the small eigenvalues and associated eigenvectors.

One possible way out of the difficulty is preliminary inversion of the matrix, which, however, in the case of large matrices is no easy task. Hence it is of interest that the Chebyshev polynomials provide a tool which opens a new perspective in the eigenvalue analysis of symmetric matrices or generally arbitrary matrices whose eigenvalues are real. We can search for the eigenvalues of a matrix in *any* range we like, somewhat as a spectroscope can scan the line-spectrum of a vibrating atom in an arbitrary prescribed frequency range. We resolve the eigenvalue spectrum into spectral lines which can be made practically independent of each other. The usual "drowning-out" of the small eigenvalues by the large ones is circumvented. The eigenvalues λ_i are projected up on a circle. On this circle the concept of "small" and "large" loses its significance. Any point of the circle is equivalent to any other point, since any point can be conceived as the beginning or end of the circle.

The Chebyshev polynomials are restricted to a definite range of the variable x. The transformation

$$x = \cos \theta \tag{3-5.1}$$

demands that x shall lie between ± 1. In order to insure this range, it will be necessary that the eigenvalues of A shall be properly normalized. This is made possible by a theorem of Gersgorin (1931) which establishes an upper bound for the absolutely largest eigenvalue $\lambda = \lambda_M$ of a matrix. This theorem is based on the following consideration. We start with the equations which define an arbitrary eigenvalue λ and its associated eigenvector:

$$a_{i1}x_1 + \cdots + a_{in}x_n = \lambda x_i \tag{3-5.2}$$

The particular solution which belongs to λ_M shall be denoted by

$$u_M = (\xi_1, \xi_2, \cdots, \xi_n)$$

Among these (generally complex) components we select the absolutely largest ξ_i; let it be a certain ξ_α. Then equation (2) which belongs to $i = \alpha$ gives

$$\lambda_M = a_{\alpha 1}\frac{\xi_1}{\xi_\alpha} + a_{\alpha 2}\frac{\xi_2}{\xi_\alpha} + \cdots + a_n\frac{\xi_n}{\xi_\alpha} \tag{3-5.3}$$

Now we know that the absolute value of a sum of complex numbers can never exceed the sum of the absolute values of these numbers.

$$|\lambda_M| \leq |a_{\alpha 1}|\left|\frac{\xi_1}{\xi_\alpha}\right| + |a_{\alpha 2}|\left|\frac{\xi_2}{\xi_\alpha}\right| + \cdots + |a_{\alpha n}|\left|\frac{\xi_n}{\xi_\alpha}\right| \tag{3-5.4}$$

But the second factors are necessarily numbers between 0 and 1, and thus

$$|\lambda_M| \leq |a_{\alpha 1}| + |a_{\alpha 2}| + \cdots + |a_{\alpha n}| \tag{3-5.5}$$

Let us now form the sum of the absolute values of the elements of each row.

$$|a_{i1}| + |a_{i2}| + \cdots + |a_{in}| = s_i \tag{3-5.6}$$

These are n positive numbers. One of them is the *largest*; let us denote this particular s_i by the symbol s. Then for all i, $s_i \leq s$, and hence according to (5),

$$|\lambda_M| \leq s_\alpha \leq s \qquad (3\text{-}5.7)$$

We have thus found a definite *upper bound* for the absolutely largest eigenvalue of an arbitrary matrix by forming the absolute sum of the elements of any row and then choosing the maximum of these n positive numbers. Since the transposed matrix has the same eigenvalues as the original one, we can perform the same operation with the *columns* instead of the rows. The *smaller* of the two maxima will give a better upper bound for $|\lambda_M|$.

The estimation of Gersgorin is always safe but frequently unrealistic; i.e., we may greatly *overrate* the largest eigenvalue by this estimate. A more realistic, but not necessarily safe, estimate is possible of the basis of the procedure of § 3 (cf. 3-3.9) which obtained the largest eigenvalue of an arbitrary matrix in approximation.

Now let us assume that we have a matrix A whose eigenvalues are known to be real, although they may be positive or negative. In this case we will operate with the following "scaled" matrix:

$$C = \frac{2}{s} A \qquad (3\text{-}5.8)$$

The eigenvalues of C will be bounded by ± 2. On the other hand, let it be known that all the eigenvalues of A are positive or zero. In this case we will put

$$C = 2I - \frac{4}{s} A \qquad (3\text{-}5.9)$$

Once more the eigenvalues of C will lie between ± 2.

We start with a "trial vector" b_0 composed of random numbers, and generate a sequence of vectors, characterized by the recurrence relation (cf. 4.3):

$$b_{k+1} = Cb_k - b_{k-1} \qquad (3\text{-}5.10)$$

and starting with

$$b_0 = b_0, \qquad b_1 = \tfrac{1}{2}Cb_0 \qquad (3\text{-}5.11)$$

We analyze the trial vector b_0 in the reference system of the eigenvectors of A.

$$b_0 = \beta_1 u_1 + \beta_2 u_2 + \cdots + \beta_n u_n \qquad (3\text{-}5.12)$$

An arbitrary b_k of the sequence generated on the basis of the recurrence relations (10), (11) becomes

$$b_k = \beta_1(\cos k\theta_1)u_1 + \beta_2(\cos k\theta_2)u_2 + \cdots + \beta_n(\cos k\theta_n)u_n \quad (3\text{-}5.13)$$

where the angles $\theta_1, \theta_2, \cdots, \theta_n$ are associated with the eigenvalues λ_i of the matrix C. In the case (8) the correlation becomes

$$\lambda_i = \cos \theta_i \quad (3\text{-}5.14)$$

while in the case (9) we have

$$\lambda_i = \frac{1 - \cos \theta_i}{2} = \sin^2 \frac{\theta_i}{2} \quad (3\text{-}5.15)$$

Now we can introduce the following function $f(t)$ of the continuous variable t:

$$f(t) = (\beta_1 \cos \theta_1 t)u_1 + (\beta_2 \cos \theta_2 t)u_2 + \cdots + (\beta_n \cos \theta_n t)u_n \quad (3\text{-}5.16)$$

Then the vectors b_k represent the values of $f(t)$ at integer points $t = k$:

$$b_k = f(k) \quad (3\text{-}5.17)$$

Our problem can thus be formulated as follows. "We have a function $f(t)$ composed of purely periodic components of variable amplitudes and variable frequencies. This function is observed at the equidistant time moments

$$t = 0, 1, 2, \cdots N$$

Find the unknown frequency θ_i of each one of the periodic components."

A problem of this kind is encountered in tide research, in meteorology, in astronomy, in economic research, and in other fields where the resultant interaction of certain periodic components is given, and our aim is to disentangle these components and obtain the amplitude and frequency of each of the constituent vibrations separately. The problem is often called "search for hid den periodicities" (cf. IV, 22). The solution is obtained with the help of the "Fourier transform" (cf. IV, 17). We transform the original $f(t)$ into a new function $F(p)$, with the help of the Fourier transform, which in our case assumes the form of a *sum* instead of an integral.

$$F(p) = \frac{1}{2} f(0) + f(1) \cos \frac{\pi}{N} p$$

$$+ f(2) \cos 2 \frac{\pi}{N} p + \cdots + \frac{1}{2} f(m) \cos N \frac{\pi}{N} p \quad (3\text{-}5.18)$$

This function reveals the existence of a periodic component $\cos \theta t$ by having a peak at the point

$$p = \frac{N}{\pi} \theta \qquad (3\text{-}5.19)$$

because if $f(t)$ is of the form $\cos \theta_i t$, the associated function $F(p)$ becomes

$$F(p) = K\left(\frac{N}{\pi} \theta_i + p\right) + K\left(\frac{N}{\pi} \theta_i - p\right) \qquad (3\text{-}5.20)$$

where

$$K(\xi) = \frac{1}{4} \sin \pi\xi \cot \frac{\pi\xi}{2N} \qquad (3\text{-}5.21)$$

The function $K(\xi)$ is essentially a function of ξ alone, since for small values of ξ we can write

$$K(\xi) = \frac{N}{2} \frac{\sin \pi\xi}{\pi\xi} \qquad (3\text{-}5.22)$$

The height of the peak increases uniformly with N but the shape of the function $F(p)$ in the neighborhood of the peak $\xi = 0$ does not change with N. Since θ ranges between 0 and π, the range of p extends from 0 to N; $F(p)$ is an even function of p and thus negative values of p need not be considered. The factor N on the right side of (19) shows that the "resolution power" of the method increases linearly with N, the number of iterations. While the shape of the mountain which characterizes a certain maximum of the function $F(p)$, does not change, the *sharpness* of the peak is proportional to N because with increasing N the peaks move further apart and eventually even very close peaks can be separated.

This method of finding the eigenvalues of a matrix by looking for the hidden frequencies of an assembly of periodic functions can be called a "spectroscopic method" since we imitate mathematically the operation of a spectroscope. A spectroscope detects the frequencies of which the light emission of an excited atom is composed. These frequencies can be evaluated from the position of the "spectral lines." The spectral lines are not lines in the mathematical sense but have a certain finite width. The distribution law of the amplitudes in the neighborhood of a peak is analogous to that given by $F(p)$. The only difference is that the excessive accuracy of spectroscopic measurements is caused by the high persistency of optical vibrations which corresponds to a very large value of N. We can cut down on the number of iterations and still maintain high accuracy, because of the large number of significant figures we have at our disposal. Although the peaks are now much closer together, we may correct our preliminary results by evaluating the interference from the neighboring peaks and subtracting their action. In this fashion the position of the maxima may be ascertained with a relative accuracy of 10^6 and more.

The successive vectors $b_0, b_1, \cdots, b_m$ need not be printed out in full. It suffices to print out *a single element* of each vector, as long as it is consistently the same element (e.g., the first, second, $\cdots$) of each vector. We thus obtain a one-dimensional sequence of numbers:

$$\gamma_0, \gamma_1, \gamma_2, \cdots, \gamma_N \qquad (3\text{-}5.23)$$

which can be subjected to a Fourier cosine analysis. Of particular interest are the *integer values* $p = k$ of p. The corresponding functional values $F(k)$ are denoted by y_k.

$$y_k = \frac{1}{2}\gamma_0 + \gamma_1 \cos\frac{\pi k}{N} + \gamma_2 \cos\frac{2\pi k}{N} + \cdots + \frac{1}{2}\gamma_N \cos\frac{N\pi k}{N} \quad (3\text{-}5.24)$$

If we evaluate these y_k systematically for $k = 0, 1, 2, \cdots, N$, we have transformed the original sequence (23) into a new sequence,

$$y_0, y_1, y_2, \cdots, y_N \qquad (3\text{-}5.25)$$

Let us denote

$$\omega_i = \frac{N}{\pi}\theta_i \qquad (3\text{-}5.26)$$

Then the relations (20) and (22) yield

$$K(\omega_i - p) = \frac{N}{2}\frac{\sin \pi(\omega_i - p)}{\pi(\omega_i - p)}$$

For integer values of p the numerator becomes $(-1)^k \sin \pi\omega_i$ and we see that, apart from the constantly alternating plus-minus signs, the interference of one peak on the other is given by the function,

$$\frac{\sin \pi\omega_i}{\pi(\omega_i - \omega_j)} \qquad (3\text{-}5.27)$$

If by accident a certain peak at $p = \omega_i$ happens to fall on an *integer value* $\omega_i = k$, we get a solitary peak without any "tails." One single maximum is then flanked by zeros on both sides. If, however, the maximum does not fall on an integer value, the maximum amplitude is flanked by smaller amplitudes on both sides. The slowest decrease occurs if the maximum is exactly halfway between two integer values of p. The amplitude pattern now becomes

$$\tfrac{1}{9}, -\tfrac{1}{7}, +\tfrac{1}{5}, -\tfrac{1}{3}, 1, 1, -\tfrac{1}{3}, +\tfrac{1}{5}, -\tfrac{1}{7}, \cdots$$

This slow decrease of the amplitudes can be considerably speeded up by operating with the *second differences* of the original amplitudes. In view of the alternating $\pm$ signs, the desired operation becomes

$$z_k = y_{k-1} + 2y_k + y_{k+1} \tag{3-5.28}$$

The previous pattern is now changed as follows.

$$-\tfrac{8}{693}, \tfrac{8}{315}, -\tfrac{8}{105}, \tfrac{8}{15}, \tfrac{8}{3}, \tfrac{8}{3}, \tfrac{8}{15}, -\tfrac{8}{105}, \tfrac{8}{315}, \cdots$$

The interference diminishes now with the *cube* of the distance between two peaks and will generally become negligible, except if two very near peaks have to be separated.

The exact position of a maximum, based on second differences (actually second *sums*, due to the $\pm$ pattern), can be obtained as follows. We examine the sequence y_k and pay attention particularly to the regular $\pm$ sequence of signs. At certain points we notice that this sequence is interrupted by a $++$, or $--$ sequence. We underline these irregular sequences and we know that a peak must occur between two such p values, $p = k$ and $p = k + 1$. We put

$$p = k + \varepsilon \tag{3-5.29}$$

We form the ratio of two successive z_p (cf. 28) values belonging to $p = k$ and $p = k + 1$.

$$q_k = \frac{z_k}{z_{k+1}} \tag{3-5.30}$$

Now z_k is proportional to

$$-\frac{1}{\varepsilon - 1} + \frac{2}{\varepsilon} - \frac{1}{\varepsilon + 1} = \frac{-2}{(\varepsilon + 1)\varepsilon(\varepsilon - 1)}$$

while z_{k+1} is proportional to

$$\frac{1}{\varepsilon} - \frac{2}{\varepsilon + 1} + \frac{1}{\varepsilon + 2} = \frac{2}{(\varepsilon + 2)(\varepsilon + 1)\varepsilon}$$

Forming the ratio, we get

$$q_k = \frac{z_k}{z_{k+1}} = -\frac{\varepsilon + 2}{\varepsilon - 1}$$

from which

$$\varepsilon = \frac{2 - q}{1 + q} \tag{3-5.31}$$

Finally

$$\theta = \frac{\pi}{N}(k + \varepsilon) \tag{3-5.32}$$

and

$$\lambda = \sin^2 \frac{\theta}{2} = \frac{1 - \cos \theta}{2} \qquad \text{[case (9)]} \tag{3-5.33}$$

$$\lambda = \cos \theta \qquad \text{[case (8)]}$$

and finally going back to the eigenvalues of the original unscaled matrix A,

$$\lambda_i' = s\lambda_i \tag{3-5.34}$$

The most surprising feature of this algorithm is that it is entirely free of a dangerous accumulation of rounding errors. While in the p,q algorithm of § 3 the rounding errors accumulate rapidly and have to be counteracted by a constant reorthogonalization process, the present algorithm allows a continuation to hundreds and perhaps even thousands of iterations, without undue danger or distortion from the part of rounding errors. Since the "signal" increases proportionally to the number of iterations, a not more than linear increase of the rounding errors would leave the "signal-to-noise ratio" unchanged. Experiments on small matrices could not detect any damage even after 2000 iterations. Experiments on large matrices are not yet available, nor is the general statistical behavior of the noise associated with this algorithm sufficiently investigated. It seems, however, that the great precision obtainable with this method in independent determination of eigenvalues and the high resolution power in the separation of close eigenvalues will not suffer by an increase of the size of the matrix to which it is applied.

It is convenient to choose N, the number of iterations, as some multiple of 180, since the trigonometric tables divide the half-circle into 180 parts. If, for example, we take $N = 720$, we scan the half-circle in units of $15'$. Since the second sum method gives good results if two neighboring peaks are at least 4 units apart, we can obtain with high precision the position of any eigenvalue on the circle which is separated from its neighbors by at least 1°.

6. Generation of the eigenvectors. In the previous section only a single component of the vectors b_k was employed. The method for isolating a definite frequency is a resonance method; we consistently increase the magnitude of a small periodic component whose frequency happens to agree with the impressed frequency. In the previous section we were interested in the exact *position* of the maximum only. But the *magnitude* of the peak is also of interest if our aim is to obtain the eigen*vector* which is associated with a certain eigenvalue. The "second sum" method (cf. 5.28) is a valuable tool again in isolating the peak from the contaminating influence of the neighboring peaks. If we find that a maximum is between $p = k$ and $p = k + 1$, with a larger amplitude at $p = k$ (or between $p = k$ and $k - 1$ with a larger amplitude at $p = k$), the quantity

$$z_k = y_{k-1} + 2y_k + y_{k+1} \qquad (3\text{-}6.1)$$

will be proportional to one of the components of the eigenvector u_j. In order to obtain the entire eigenvector u_j, we have to repeat the same calculation for every component.

This again can be done with maximum economy, without printing out the single vectors b_k. We have stored these vectors on tape. We now multiply each one of these successive vectors by a preassigned weight factor ρ_i and form the sum, thus obtaining

$$\bar{u}_j = b_0 + \sum_{\alpha=1}^{N-1} \rho_\alpha b_\alpha \qquad (3\text{-}6.2)$$

where the weights ρ_α are defined as

$$\rho_\alpha = \left(1 + \cos\frac{\pi}{N}\alpha\right)\cos\frac{\pi}{N}k\alpha \qquad (3\text{-}6.3)$$

Only the final sum $\bar{u}_j$ is printed out. The vector $\bar{u}_j$ (the bar refers to the fact that it is not the exact eigenvector u_j but only a close approximation of it) is not normalized in length. The contamination from the part of the other eigenvectors is negligible if the peak near $p = k$ and the next peak are at least 4 units apart.

By this method any particular eigenvector of A or all of them may be generated. If A is symmetric we can test the $\bar{u}_j$ thus obtained for orthogonality. If A is not symmetric, we have to operate simultaneously with A and $\tilde{A}$, repeating the entire process with another trial vector $\tilde{b}_0$ and replacing A by $\tilde{A}$. Again the $\tilde{b}_k$ are stored on tape, but the weighting (2) occurs once more in identical fashion. The resulting sum is again printed out and gives our $\bar{v}_j$. The vectors $\bar{u}_i$ and $\bar{v}_j$ must now show the biorthogonality relation (2-6.16).

7. Iterative solution of large-scale linear systems. Application of the Chebyshev polynomials to the eigenvalue problem of matrices with real eigenvalues leads to a method of solving large-scale linear systems by successive iterations, without actual matrix inversion. We encounter the preliminary difficulty that the Chebyshev polynomials operate properly only in the *real* range, while the eigenvalues of an arbitrary nonsymmetric matrix A are generally *complex* numbers. There are two ways in which to overcome this difficulty. One is that we reformulate the given equation

$$Ax = b \tag{3-7.1}$$

according to the method of least squares.

$$\tilde{A}^*Ax = \tilde{A}^*b \tag{3-7.2}$$

The new matrix $\tilde{A}^*A$ is "positive definite," i.e., its eigenvalues are all real and even positive numbers. If the original matrix A was properly scaled in rows and columns (cf. II, 26), the diagonal elements of $\tilde{A}^*A$ will be nearly 1, and the nondiagonal elements will lie between ± 1.

The symmetrization of A by premultiplication by $\tilde{A}^*$ is a simple but elaborate operation. It is equivalent to n iterations, and if n is high, this operation is by no means trivial. Moreover, if A itself has many zeros, this is no longer true of $\tilde{A}^*A$, and thus a great practical advantage of the original matrix A may be lost. Hence it is sometimes preferable not to generate the matrix $\tilde{A}^*A$ explicitly, but to obtain the operation $b_1 = \tilde{A}^*Ab_0$ in two steps, viz.,

$$b_1' = Ab_0, \qquad b_1 = \tilde{A}^*b_1' \tag{3-7.3}$$

However, there is an altogether different way for symmetrization of A which avoids premultiplication by $\tilde{A}^*$, at the cost of doubling the size of the matrix. Let us enlarge the given linear system as follows:

$$Ax = b, \qquad \tilde{A}^*y = 0 \tag{3-7.4}$$

considering

$$z = (y,x) \tag{3-7.5}$$

as a vector of $2n$ components. On the surface, addition of the second equation appears an unnecessary luxury, since it has the trivial solution $y = 0$. However, the extended matrix

$$B = \begin{bmatrix} 0 & A \\ \tilde{A}^* & 0 \end{bmatrix} \tag{3-7.6}$$

has the great advantage that it is always *Hermitian*.

$$\tilde{B}^* = B \tag{3-7.7}$$

Hence the eigenvalues of B are always *real*. The application of the Gersgorin method to the new matrix leads to interesting consequences. In § 5 we found that an upper bound on the eigenvalues of A could be found by considering the absolute sum of each *row* and selecting the maximum of these n numbers; or the absolute sum of each *column* and selecting the maximum of these n numbers. The *smaller* of these two values could be chosen as our s. Now the *larger* of these two numbers has to be chosen as the s of the enlarged matrix (6), and we will put

$$C = \frac{2}{s}B \tag{3-7.8}$$

The eigenvalues of C will again lie between the limits ± 2.

But this same s can serve still another purpose. The eigenvalue problem of the matrix B leads to the equations

$$Ax = \rho y, \qquad \tilde{A}^*y = \rho x \tag{3-7.9}$$

which gives

$$\tilde{A}^*Ax = \rho^2 x \tag{3-7.10}$$

This shows that the $2n$ eigenvalues of the matrix B are the *square roots* $\pm\sqrt{\overline{\lambda}}$ of the eigenvalues of $\tilde{A}^*A$. Since we have bounded the absolutely largest eigenvalue of B by s, we now find

$$\lambda_M \leq s^2 \tag{3-7.11}$$

as an estimated upper bound of $\tilde{A}^*A$, without actually generating the elements of this matrix. Moreover, the matrix

$$S_0 = \frac{1}{s^2}\,\tilde{A}^*A \qquad\qquad (3\text{-}7.12)$$

will have eigenvalues which lie between 0 and 1.

For our following discussions we will assume that we have symmetrized A by premultiplication, and not by the extension method (4), although the resulting algorithm can easily be reinterpreted for the case (4).

We will formulate our problem as an eigenvalue problem whose solution we have found in § 5. For this purpose we write the given equation (1) in the form

$$S_0 x + c = 0 \qquad\qquad (3\text{-}7.13)$$

with

$$c = -\frac{\tilde{A}^*b}{s^2} \qquad\qquad (3\text{-}7.14)$$

In many problems arising from self-adjoint differential equations we have a matrix which from the very beginning possesses only positive eigenvalues. In that case no symmetrization is needed and we can put directly

$$S_0 = \frac{1}{s}\,A, \qquad c = -\frac{1}{s}\,b \qquad\qquad (3\text{-}7.15)$$

Now our linear set of equations in n unknowns:

$$s_{11}x_1 + s_{12}x_2 + \cdots + s_{1n}x_n + c_1 = 0$$
$$\vdots$$
$$s_{n1}x_1 + s_{n2}x_2 + \cdots + s_{nn}x_n + c_n = 0$$

can be reformulated as a homogeneous set of equations in $n + 1$ unknowns:

$$s_{11}x_1 + s_{12}x_2 + \cdots + s_{1n}x_n + c_1 x_{n+1} = 0$$
$$\vdots$$
$$s_{n1}x_1 + s_{n2}x_2 + \cdots + s_{nn}x_n + c_n x_{n+1} = 0$$

This means that we add the right side c as an additional column to our matrix S_0. We thus get an extended $(n + 1)$ by $(n + 1)$ matrix S, and the vector x takes on one more element.

$$S = \begin{array}{|c|c|} \hline & \\ S_0 & c \\ & \\ \hline 0 & 0 \\ \hline \end{array} \qquad (3\text{-}7.16)$$

The equation (13) is now reformulated as the following homogeneous system of $(n + 1)$ equations in $(n + 1)$ unknowns:

$$S u_0 = 0 \qquad (3\text{-}7.17)$$

Since an arbitrary scale factor remains free in u_0, we may normalize this scale factor in such a way that x_{n+1} shall become 1. Then the $x_1, x_2, \cdots, x_n$ become identical with the desired solution x of the original set (13). However, we can also leave the scale factor arbitrary and agree that at the end we will form the ratios

$$\frac{x_1}{x_{n+1}}, \quad \frac{x_2}{x_{n+1}}, \cdots, \frac{x_n}{x_{n+1}}$$

These ratios will give the solution of the original inhomogeneous equations. The two systems (13) (17) are thus equivalent, and from now on we will operate with the homogeneous system (17).

The n eigenvalues and eigenvectors of the matrix S_0 hold also for the extended system, if the eigenvectors are complemented by the element $x_{n+1} = 0$. However, the extended matrix S has $n + 1$ eigenvalues and eigenvectors. The $(n + 1)$st eigenvalue of S is $\lambda = 0$, and the associated eigenvector is defined by the equation

$$S u_0 = \lambda u_0 = 0 \qquad (3\text{-}7.18)$$

The notation u_0 for the solution of the homogeneous set (17) was chosen because this solution has the significance of one of the *eigenvectors* or principal axes of the extended matrix S. The solution of a homogeneous system of equations can thus be reformulated as a special case of an eigenvalue problem. Let $\lambda = 0$ be known as one of the eigenvalues of a given matrix, and find the eigenvector which belongs to the eigenvalue $\lambda = 0$.

We will display the complete eigenvalue analysis of the matrix S and the transposed matrix $\tilde{S}$. In accordance with our usual procedure we put the components of the successive eigenvectors in successive columns of a matrix. In our problem the mutually orthogonal principal axes of S_0 are taken for granted. They are denoted by $u_1, u_2, \cdots, u_n$. The additional $(n+1)$st element is indicated separately.

	S					$\tilde{S}$			
$\lambda = 0$	λ_1	λ_2	$\cdots$	λ_n	$\lambda = 0$	λ_1	λ_2	$\cdots$	λ_n
x	u_1	u_2	$\cdots$	u_n	0	u_1	u_2	$\cdots$	u_n
1	0	0		0	1	$\dfrac{u_1 \cdot c}{\lambda_1}$	$\dfrac{u_2 \cdot c}{\lambda_2}$		$\dfrac{u_n \cdot c}{\lambda_n}$

$$(3\text{-}7.19)$$

We start with a trial vector b_0, chosen as follows.

$$b_0 = (0, 0, \cdots, 0, 1) \qquad (3\text{-}7.20)$$

We analyze it in the reference system of the principal axes of S.

$$b_0 = \beta_0 u_0 + \beta_1 u_1 + \cdots + \beta_n u_n \qquad (3\text{-}7.21)$$

The coefficients β_i are obtained by forming the scalar product of b_0 with the adjoint axes. Hence

$$\beta_0 = 1, \qquad \beta_i = \frac{u_i \cdot c}{\lambda_i} \qquad (3\text{-}7.22)$$

We let some properly chosen polynomial $P_m(S)$ operate on b_0. In view of (2.13), we obtain

$$P_m(S)b_0 = P_m(0)u_0 + P_m(\lambda_1)\beta_1 u_1 + \cdots + P_m(\lambda_n)\beta_n u_n \quad (3\text{-}7.23)$$

Our goal is to obtain the vector u_0 *alone* without any contamination from the other vectors u_i. Hence our aim will be to search for a polynomial $P_m(\lambda)$ which should take the value 1 for $\lambda = 0$ and should remain small for all other values of λ between 0 and 1. Although a polynomial cannot drop from 1 to 0 suddenly, we may succeed in constructing a polynomial which will decrease steeply from the initial value $y = 1$ at $\lambda = 0$ and then remain uniformly small. The

Chebyshev polynomials of the second kind will give an adequate solution of this problem.

We construct a sequence of vectors $b_1, b_2, \cdots, b_n$ by the following simple iterative routine. We form the matrix

$$C = 2I - 4S$$

and generate the b_{k+1} vectors by a uniform recurrence scheme which at each step involves two vectors, viz., the last vector b_k and the previous vector b_{k-1}.

$$b_{k+1} = Cb_k - b_{k-1} \tag{3-7.24}$$

The scheme starts with the two vectors

$$b_1 = b_0, \qquad b_2 = Cb_1$$

Then the normal routine follows:

$$b_3 = Cb_2 - b_1$$

and so on. Finally, after arriving at a certain b_m, we divide that vector by m. The polynomial operator thus generated can be written

$$P_m(\lambda) = \frac{\sin m\theta}{m \sin \theta} \quad \left(\lambda = \sin^2 \frac{\theta}{2}\right) \tag{3-7.25}$$

This function has for small θ the character of a universal function of ξ, if ξ is defined by

$$\xi = 4m^2\lambda \tag{3-7.26}$$

The polynomial $P_m(\lambda)$, considered as a function of ξ, is in good approximation equal to

$$\phi(\xi) = \frac{\sin \sqrt{\xi}}{\sqrt{\xi}} \tag{3-7.27}$$

This function begins with $y = 1$ at $\xi = 0$, decreases steeply with increasing ξ, but develops secondary maxima and minima. As m, i.e., the number of iterations, increases, a certain eigenvalue λ_i moves out along the curve with the speed m^2, according to the relation (26). If we increase the number of iterations, even a very small λ will eventually move out to large values of ξ at which $\phi(\xi)$ is already very small. But this process has very slow convergence if the given linear system is strongly skew-angular. In order to avoid an excessive number of iterations, it is preferable to terminate our recurrence scheme after m steps and then repeat the entire cycle a second time, and possibly a third and fourth time. This can be done with no difficulty by a slight modification of our routine.

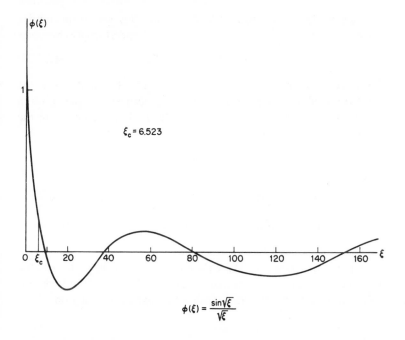

$$\phi(\xi) = \frac{\sin\sqrt{\xi}}{\sqrt{\xi}}$$

It is advisable to choose m, the length of each cycle, as some power of 2, e.g., $128 = 2^7$. Then the division by m is accomplished by a mere shift of 7 binary digits. We have thus obtained

$$b_1, b_2, \cdots, b_m, b'_m = b_m/m$$

and now we continue with our routine by choosing b'_m as the starting vector b_{m+1} of the second cycle. The interference with the regular routine occurs only in one step, viz., in the formation of b_{m+2}:

$$b_{m+2} = Cb_{m+1} \qquad (3\text{-}7.28)$$

where the subtraction of b_m is prevented. But immediately afterward we return to the normal routine:

$$b_{m+3} = Cb_{m+2} - b_{m+1}$$

and continue undisturbed, until b_{2m} and

$$b'_{2m} = b_{2m}/m$$

is reached. Then the third cycle starts and thus we go on, until ν cycles are finished. The resultant polynomial operator $Q(\lambda)$ becomes

$$Q_{\nu m}(\lambda) = [P_m(\lambda)]^\nu \qquad (3\text{-}7.29)$$

Let us see how the length m of each cycle and the number ν of the repetitions of the cycle are related to the accuracy of the solution obtained. For this purpose we substitute the last vector $b'_{\nu m}$ in the given equation (17) and form the residual. We denote

$$\nu m = N$$

Then

$$r = Sb_N = \beta_1\lambda_1 Q_N(\lambda_1)u_1 + \beta_2\lambda_2 Q_N(\lambda_2)u_2 + \cdots + \beta_n\lambda_n Q_N(\lambda_n)u_n$$

$$= (c\cdot u_1)Q_N(\lambda_1)u_1 + (c\cdot u_2)Q_N(\lambda_2)u_2 + \cdots + (c\cdot u_n)Q_N(\lambda_n)u_n \qquad (3\text{-}7.30)$$

Now $(c\cdot u_1), (c\cdot u_2), \cdots, (c\cdot u_n)$ are the components of the vector c in the reference system of the principal axes of S_0. If we think of our solution once more in terms of the inhomogeneous equation

$$S_0 x = -c$$

and put

$$x = b_N + y \qquad (3\text{-}7.31)$$

we obtain for the correction y the new equation

$$S_0 y = -r \qquad (3\text{-}7.32)$$

Even without evaluating y we can predict theoretically how accurate our solution b_N will be. The vectors $u_1, u_2, \cdots, u_n$ are orthogonal to each other. If we assume that the eigenvalues λ_i are arranged in increasing order, the smallest eigenvalue will be λ_1; the largest error will occur in the direction of the associated eigenvector u_1. The factor by which we have cut down the component of the vector c in this direction is $Q_N(\lambda_1)$. Now let us assume that the smallest eigenvalue λ_1 is beyond a certain critical value λ_c, defined by

$$\lambda_c = \frac{\xi_c}{4m^2} \tag{3-7.33}$$

where

$$\xi_c = (2.554)^2 = 6.523 \tag{3-7.34}$$

This particular value of ξ is chosen in view of the first secondary maximum of the function $(\sin x)/x$ which is at $x = 4.4934$. The value of this maximum is -0.2172. The same amplitude with a positive sign was reached earlier at the point $x = 2.5536$. Beyond this point the function $\phi(\xi)$ never rises above 0.22. Hence we can say that if the smallest eigenvalue λ_1 of our matrix S_0 does not drop below a certain critical value,

$$\lambda_1 \geq \left(\frac{1.277}{m}\right)^2 \tag{3-7.35}$$

the relative error of our solution will be

$$\eta \leq (0.22)^\nu \tag{3-7.36}$$

This "relative error" means the following. We cannot measure the accuracy of the solution of a linear system by comparing the calculated value of a certain component with its true value. The true value of a certain component x_i may accidentally be very small and even zero, while another component x_j may be very large. A fair measure of the accuracy is the length of the error vector divided by the length of the true vector, i.e., the square root of the sum of the squares of all the individual errors, divided by the square root of the sum of the squares of all the unknowns x_j:

$$\eta = \sqrt{\frac{(\delta x_1)^2 + (\delta x_2)^2 + \cdots + (\delta x_n)^2}{x_1^2 + x_2^2 + \cdots + x_n^2}} \tag{3-7.37}$$

Here δx_i denotes the difference between the true value x_i and the calculated value x_i'.

The relation (36) shows that the *number of repeated cycles decides the accuracy of our solution*. Two cycles will give us an accuracy of

4.72%, three cycles an accuracy of 1.02%, and four cycles an accuracy of 0.22%. These are estimated accuracies, and an error estimation has to be based on the worst possibility. The actual error may be much smaller, since the λ_i of our matrix may be nearer to the zeros than to the maxima of the previous graph.

The relation (35), on the other hand, shows that the *length of each cycle decides the separation power of the method*. In the case of a very small λ_1 the minimum number of iterations demanded for successful separation of the true solution from the contribution of the smallest eigenvector may become quite large. We have seen, however, that physical systems with a measured right side will hardly allow a λ_1 which drops below 10^{-4} (otherwise the noise of the measurements will drown out the mathematically correct solution). But then a cycle of $m = 128 = 2^7$ iterations will be sufficient to purify the solution from all contaminations caused by the eigenvectors of S_0. Moreover, the accuracy of the solution needs hardly surpass 5%, because of the inaccuracies of the physical observations. We can thus say that a double cycle of 128 iterations will cover practically all physical systems, irrespective of the size of the matrix.[1] For the sake of brevity we will call the scheme (24) of generating a set of vectors by the name "*C* iterations."

8. The residual test. If we have made a complete eigenvalue analysis of the matrix S_0, we can tell in advance from the length of the cycles of the previous algorithm and the number of cycles employed, what the minimum accuracy (maximum error) of our solution will be. The actual accuracy may be much more optimistic. Hence we would like to find out how close our solution comes to the mathematically perfect solution. The most natural method of finding out the accuracy of a solution would seem to be substitution of the alleged solution into the given equation. If we find that the given equation is satisfied, we know that our answer is correct.

---

[1] The polynomial (15) has a strong maximum not only at $\theta = 0$, but also at $\theta = \pi$. This means that the spotlight is put on the neighborhood of $\lambda = 0$ *and* $\lambda = 1$. Hence the eigenvalue spectrum must not extend completely up to $\lambda = 1$. The Gersgorin estimation usually overestimates the largest eigenvalue so strongly that the danger that S_0 has an eigenvalue at $\lambda = 1$ is very slight. However, in order to be on the safe side, we may define the matrix (3) as $\tilde{A}A$ divided by 1.05μ.

This method, however, has its great dangers. It is true that we can test in this way a solution of absolute accuracy. If we substitute in the expression $S_0 x + c$ and find that the residual vector

$$r = S_0 x + c \qquad (3\text{-}8.1)$$

comes out zero, we know that we have found the proper x. This will happen only very occasionally and usually only in particularly simple but not characteristic examples in which the elements of the matrix S_0 are chosen as simple integers. In most cases the limited accuracy of numerical calculations will prohibit a perfect answer. Hence we cannot expect that the residual vector r, obtained as a result of substituting the obtained solution into the given equation, will be truly zero, but only that it will be very small. Yet it is one of the paradoxes of linear systems that this simple test—we may call it the "residual test"—need not be a true measure of the accuracy of our solution. While a vanishing residual indicates a perfect answer, a small residual need *not* indicate a *close* answer.

The following example may illustrate the situation. Let us assume that the largest eigenvalue of S_0 is $\lambda_n = 1$, the smallest eigenvalue $\lambda_1 = 10^{-4}$. The true solution of our problem will be

$$x = u_1 + u_n$$

i.e., the smallest and the largest eigenvector participate in the solution with equal strength. However, we are not aware of the eigenvalues and eigenvectors of S_0. We have merely found by some numerical procedure a solution which happens to be

$$x' = u_1 + 1.01 u_n$$

On the other hand, by some other numerical procedure we have found another answer which happens to be

$$x'' = 4u_1 + u_2$$

In the first case the error vector is

$$x' - x = 0.01 u_n$$

In the second case the error vector is

$$x'' - x = 3u_1$$

Hence the relative error is in the first case,

$$\frac{0.01}{\sqrt{1+1}} = 0.0071$$

and in the second case, $\dfrac{3}{\sqrt{2}} = 2.13$

That is, the first solution is in error by less than 1%, the second by more than 200%. But making the residual test we find in the first case,

$$r' = 0.01 u_n$$

in the second case, $r'' = 0.0003 u_1$

If we go by the residual test, the odds are all against the first solution, since the error is 33 times larger. We would thus gladly accept the second solution as the superior one. And yet the first solution is a very satisfactory solution, the second solution a completely useless solution. Hence we see that the mere residual test, without any further investigation, is no measure of the accuracy or inaccuracy of a given solution, because the error in the direction of a small eigenvector is multiplied by a very small factor and thus practically obliterated, although that error is present in the solution itself, before multiplying by the matrix S_0.

The residual test can be made reliable, however, if we do not stop after obtaining the residual, but use this residual as a new right side and repeat the entire procedure a second time. Finally we form the residual once more. If the new residual is greatly diminished compared with the previous one, this is an indication that we did find a solution of the given problem. If, on the other hand, the residual remains essentially unchanged at the end of the new cycle, this indicates the presence of a very small eigenvalue with which our iteration process was unable to cope.

9. The smallest eigenvalue of a Hermitian matrix. The iterative procedure of § 7 gives a new attack on the problem of small eigenvalues. Ordinarily, if we want the smallest eigenvalues of a matrix, we first invert that matrix and then look for the largest eigenvalues. For this purpose we generate the vectors (cf. 33.8):

$$b_0, \quad A^{-1}b_0, \quad A^{-2}b_0, \quad \cdots \tag{3-9.1}$$

and the corresponding vectors of the adjoint system. But the equation $b_1 = A^{-1}b_0$ is equivalent to the equation

$$Ab_1 = b_0 \tag{3-9.2}$$

and since we can solve this equation by successive iterations, the inversion of the matrix A is no longer demanded. Then, after finding b_1, we replace b_0 by b_1 and repeat the procedure. We thus obtain the solution of the equation

$$Ab_2 = b_1 \tag{3-9.3}$$

It is true that we have obtained the iterative solution of linear equations only for the case of positive definite matrices. But the matrices appearing in vibration problems are usually of the positive definite type. Moreover, if A is an arbitrary generally complex matrix whose smallest eigenvalue is to be determined, this eigenvalue problem is closely related to an associated problem, involving the Hermitian matrix

$$S_0 = \tilde{A}^*A \tag{3-9.4}$$

Although A itself has complex elements, the matrix S_0 has always real eigenvalues, since the Hermitian condition

$$\tilde{S}_0{}^* = S_0 \tag{3-9.5}$$

is satisfied (cf. 2-8.29). Moreover, in view of the least-square character of the matrix (4), its eigenvalues are not only real, but even positive. The equation

$$S_0 b_1 = b_0 \tag{3-9.6}$$

is thus always solvable by the C iterations discussed in § 7.

In the realm of very small eigenvalues we did not succeed with the inversion problem. This, however, is no serious handicap. The principal aim of the inversion is to reverse the order of magnitude of the eigenvalues by making the small eigenvalues large and the large eigenvalues small. Our process achieved this aim. With two cycles of m iterations we have generated the following function of the eigenvalue λ:

$$4m^2 \, \frac{\xi - (\sin\sqrt{\xi})^2}{\xi^2} \qquad (\xi = 4m^2\lambda) \qquad (3\text{-}9.7)$$

This function differs from λ^{-1} only in the realm of small ξ, i.e., in the realm of very small λ. But the function (7) is a legitimate substitute for the function λ^{-1} even in this realm. We generate the scalars

$$c_0 = b_0 b_0^*, \quad c_1 = b_0 b_1^*, \quad c_2 = b_0 b_2^* = b_1 b_1^*, \quad c_3 = b_1 b_2^* \qquad (3\text{-}9.8)$$

These scalars are all real, although the vectors b_i are complex. Then the quadratic equation

$$\begin{vmatrix} 1 & \mu & \mu^2 \\ c_0 & c_1 & c_2 \\ c_1 & c_2 & c_3 \end{vmatrix} = 0 \qquad (3\text{-}9.9)$$

establishes the two largest eigenvalues of the "reversed" matrix. The *larger* of the two roots $\mu_1 > \mu_2$ is chosen, in order to obtain the smallest eigenvalue of A. The corresponding eigenvector u_1 is given by

$$u_1 = b_2 - \mu_2 b_1 \qquad (3\text{-}9.10)$$

This method operates even better if we start with a trial vector which has boosted up the smallest eigenvalue. We can do that by applying two or three double cycles of C iterations to the random vector r_0. Let the vector thus obtained be w_0 and let us apply one more double cycle of C iterations to it. This vector w_1 can be considered as a good approximation of the eigenvector u_1, and then the ratio

$$\mu = \frac{w_0 w_1^*}{w_0 w_0^*} \qquad (3\text{-}9.11)$$

gives the transformed value of the eigenvalue λ.

The transformation back from μ to λ can be accomplished as follows. We first evaluate

$$\frac{2m}{\pi\sqrt{\mu}} = v \qquad (3\text{-}9.12)$$

Then we obtain u by solving the equation

$$\frac{u}{\sqrt{1 - (\sin \pi u / \pi u)^2}} = v \tag{3-9.13}$$

This can be done by tabulation (cf. Table III), for $v \leq 5$. Beyond $v = 5$ we can put practically

$$u = v \sqrt{1 - \left(\frac{\sin \pi v}{\pi v}\right)^2} \tag{3-9.14}$$

After finding u, we obtain λ on the basis of the relation

$$\lambda = \left(\frac{\pi}{2m} u\right)^2 \tag{3-9.15}$$

10. The smallest eigenvalue of an arbitrary matrix. The problem of finding the smallest eigenvalue of an arbitrary matrix is of far-reaching significance. It may happen that we are interested in one particular eigenvalue of A which is neither the largest nor the smallest. We have, however, a fairly good approximation λ_0 of λ at our disposal. Hence we know that the desired λ differs from λ_0 by a small amount ε only. The question is how to obtain ε.

The boosting up of one particular eigenvalue is not an easy task, since in the customary iteration techniques the larger eigenvalues will inevitably overshadow the effect of the critical one. The following method is free of this difficulty.

Since

$$\lambda = \lambda_0 + \varepsilon \tag{3-10.1}$$

the matrix $A - \lambda_0 I - \varepsilon I$ has a zero eigenvalue. But this means that the matrix

$$A_0 = A - \lambda_0 I \tag{3-10.2}$$

has the small eigenvalue ε. This ε is generally complex even if λ_0 was given as real. It will thus be necessary to obtain a quadratic approximation of ε.

We will again think of the reciprocal matrix A_0^{-1} and try to obtain its largest eigenvalue by the method (3.9). For this purpose, however, we have to start with a trial vector b_0 which is strongly biased in favor of the largest eigenvalue. Now we argue as follows: if ε is neglected and A_0 has the eigenvalue zero, then the symmetrized

Hermitian matrix $S_0 = \tilde{A}_0^* A_0$ has also the eigenvalue zero. Moreover, the eigenvector associated with the zero eigenvalue is *common* to both matrices. But this means that even if ε is not zero but small, the eigenvector x_0 of S_0 associated with the smallest eigenvalue will have a strong component in the direction of the desired eigenvector of A_0.

Let us therefore obtain the smallest eigenvector of the Hermitian matrix S_0, employing the method of the previous section.

$$S_0 x_0 = \rho_0 x_0 \tag{3-10.3}$$

We either boost up an arbitrary random vector by several C iterations and consider the resulting vector directly as our x_0, or we make use of the more refined quadratic approximation (9.9). At all events we have now a starting vector $b_0 = x_0$. Now we need a similar starting vector for $\tilde{A}_0$. The symmetrized matrix here is another S_0, namely,

$$S_0' = A_0^* \tilde{A}_0 \tag{3-10.4}$$

But now the equation (3), if we premultiply by A_0, gives

$$A_0 \tilde{A}_0^* A_0 x_0 = \rho_0 A_0 x_0 \tag{3-10.5}$$

and changing i to $-i$,

$$(A_0^* \tilde{A}_0) A_0^* x_0^* = \rho_0 A_0^* x_0^* \tag{3-10.6}$$

This shows that

$$\tilde{x}_0 = A_0^* x_0^* \tag{3-10.7}$$

We now have a *pair* of starting vectors. If ρ_0 is very small, a linear approximation will suffice to obtain ε. In this case we get directly

$$\varepsilon = \frac{x_0 \cdot A \tilde{x}_0}{x_0 \cdot \tilde{x}_0} = \rho_0 \frac{x_0 \cdot x_0^*}{x_0 \cdot A^* x_0^*} \tag{3-10.8}$$

But if ρ_0 is not very small, we proceed to the construction of x_1 on the basis of $Ax_1 = x_0$, which means

$$S_0 x_1 = \tilde{A}_0^* x_0 \tag{3-10.9}$$

solvable by the previous iteration technique. For this purpose we go with N high enough to move ρ_0 in the safe zone: $\rho_0 \geq \lambda_c$ (cf. 7.33). Then we proceed as follows. We consider our trial vector $x_0, \tilde{x}_0$ not as $b_0, \tilde{b}_0$, but as the middle pair $b_1, \tilde{b}_1$. Then $b_2 = x_1$, while

$$b_0 = A_0 x_0 = \tilde{x}_0^*, \qquad \tilde{b}_0 = \tilde{A}_0 \tilde{x}_0 \tag{3-10.10}$$

[The last vector is replaceable by $\rho_0 x_0^*$, if (3) is exactly fulfilled. However we will not depend on the exact fulfillment of (3).] We now have the five vectors

$$(\tilde{x}_0^*, \tilde{b}_0) \qquad (x_0, \tilde{x}_0) \qquad x_1 \tag{3-10.11}$$

and we can form the four basic scalars,

$$c_0 = \tilde{x}_0^* \tilde{b}_0 \qquad\qquad \gamma_1 = \frac{\tilde{x}_0^* \tilde{b}_0}{\tilde{x}_0^* \tilde{x}_0}$$

$$c_1 = x_0 \tilde{b}_0 = \tilde{x}_0 \tilde{x}_0^*$$

$$c_2 = x_0 \tilde{x}_0 = \tilde{b}_0 x_1 \qquad \gamma_2 = \frac{x_0 \tilde{x}_0}{\tilde{x}_0^* \tilde{x}_0} \tag{3-10.12}$$

$$c_3 = x_1 \tilde{x}_0 \qquad\qquad \gamma_3 = \frac{x_1 \tilde{x}_0}{\tilde{x}_0^* \tilde{x}_0}$$

together with the quadratic equation

$$\begin{vmatrix} \varepsilon^2 & \varepsilon & 1 \\ \gamma_1 & 1 & \gamma_2 \\ 1 & \gamma_2 & \gamma_3 \end{vmatrix} = 0 \tag{3-10.13}$$

The terms with γ_1 are frequently negligibly small, in which case our equation is reduced to

$$\varepsilon^2(\gamma_3 - \gamma_2^2) + \varepsilon\gamma_2 - 1 = 0 \tag{3-10.14}$$

Here we have the correction in second approximation, which has to be added to the crude preliminary value λ_0 of the eigenvalue λ.[1]

Bibliographical References

[1] HORVAY, G., *Solution of Large Equation Systems and Eigenvalue Problems by Lanczos' Matrix Iteration Method* (General Electric Company, Report No. KAPL-1004, Technical Information Service, Oak Ridge, Tenn., 1953).

[2] HOUSEHOLDER, A. S., FORSYTHE, G. E., and GERMOND, H. H., *Monte Carlo Methods* (*Nat. Bur. Standards*, Washington, D.C., 1951, Applied Math. Series **12**).

[1] An interesting application of this method to the solution of algebraic equations had to be omitted, for the sake of economy.

Articles

[3] HESTENES, M. R., and KARUSH, W., "A Method of Gradients for the Calculation of the Characteristic Roots and Vectors of a Symmetric Matrix," *J. Research, Nat. Bur. Standards*, **47**, 45 (1951).

[4] HESTENES, M. R., and STIEFEL, E., "Method of Conjugate Gradients for Solving Linear Systems," *J. Research, Nat. Bur. Standards*, **49**, 409 (1952).

[5] LANCZOS, C., "An Iteration Method for the Solution of the Eigenvalue Problem of Linear Differential and Integral Operators," *J. Research, Nat. Bur. Standards*, **45**, 255 (1951).

[6] LANCZOS, C., "Solution of Systems of Linear Equations by Minimized Iterations," *J. Research, Nat. Bur. Standards*, **49**, 33 (1952).

IV

HARMONIC ANALYSIS

1. Historical notes. The discovery of the acoustical importance of a fundamental vibration and its overtones is attributed to the Greek sage Pythagoras (600 B.C.). The mathematical significance of analyzing a periodic function in terms of functions of the type $\sin kx$ and $\cos kx$, where k is an integer, was recognized in the eighteenth century by Euler and Lagrange. Since eighteenth century mathematics did not possess the exact notion of a limit, the true nature of an infinite series was not fully realized. Lagrange assumed that any superposition of analytical (i.e., infinitely differentiable) functions must again give an analytical function. Hence he restricted the possibility of a harmonic analysis to functions which were not only continuous but which could be differentiated any number of times. It was Fourier's brilliant discovery ["Théorie analytique de la chaleur" (1822)] that such a restriction is not demanded. The function may be entirely "unpredictable," i.e., composed of an arbitrary number of arcs which change their analytical law constantly. Nor is the continuity of the function necessary. Fourier gave examples for harmonic analysis of functions which had a finite number of discontinuities in the given fundamental interval. This fundamental interval, usually normalized to the range from $-\pi$ to $+\pi$, is all we need for the definition of the function $y = f(x)$. The periodicity of the function is accidental and enters the picture only if we leave the fundamental interval. The theory of harmonic analysis need not leave the given fundamental range. Another fundamental discovery of Fourier was the "Fourier integral" by which he generalized the methods of harmonic analysis to an infinite interval which ranges from $-\infty$ to $+\infty$, without requiring any periodicity for the function to be analyzed. Later Dirichlet investigated more specifically what conditions the "arbitrary function" of Fourier had to satisfy in order to be representable by a harmonic series. These conditions, called the "Dirichlet conditions," are sufficient for the convergence of the Fourier series; they are not always necessary, however. In recent times (1904) Fejér invented a new method of summing the Fourier series by which he greatly extended the validity of the series. Using the arithmetic means of the partial sums, instead of the partial sums themselves, he could sum series which were divergent in themselves. The only condition the function $f(x)$ still has to satisfy is the natural condition that the function shall be absolutely integrable.

2. Basic theorems. Let the function $y = f(x)$ satisfy the following conditions:

1. $f(x)$ is defined at every point of the interval

$$-\pi \leq x \leq +\pi$$

2. $f(x)$ is everywhere single valued, finite, and sectionally continuous. Hence $f(x)$ can have a finite number of discontinuities only, and two consecutive discontinuities must be separated by a finite interval.

3. $f(x)$ is of "bounded variation"; this means that $f(x)$ cannot have an infinite number of maxima and minima in the given interval.

These conditions imposed on $f(x)$ are called the "Dirichlet conditions." A function of this type can be expanded into a convergent infinite series of the following form.

$$f(x) = \tfrac{1}{2}a_0 + a_1 \cos x + a_2 \cos 2x + \cdots$$
$$+ b_1 \sin x + b_2 \sin 2x + \cdots \qquad (4\text{-}2.1)$$

where the coefficients a_k, b_k, called the "Fourier coefficients," are defined as follows.

$$a_k = \frac{1}{\pi} \int_{-\pi}^{+\pi} f(x) \cos kx \, dx \qquad (4\text{-}2.2)$$

$$b_k = \frac{1}{\pi} \int_{-\pi}^{+\pi} f(x) \sin kx \, dx$$

At a point of discontinuity $x = a$ we will define $f(a)$ as the arithmetic mean of the two limiting ordinates.

$$f(a) = \tfrac{1}{2}[f(a_+) + f(a_-)] \qquad (4\text{-}2.3)$$

This is the value to which the infinite Fourier series converges at the fixed point $x = a$.

A convenient alternate form of the Fourier series can be given in the following complex form:

$$f(x) = c_0 + c_1 e^{ix} + c_2 e^{2ix} + \cdots$$
$$+ c_{-1} e^{-ix} + c_{-2} e^{-2ix} + \cdots \qquad (4\text{-}2.4)$$
$$= \sum_{k=-\infty}^{+\infty} c_k e^{ikx}$$

where

$$c_k = \frac{1}{2\pi} \int_{-\pi}^{+\pi} f(x) e^{-ikx}\, dt \qquad (4\text{-}2.5)$$

Proof. If the existence and convergence of the series (1) is postulated at every point of the given range, we can immediately multiply on both sides by cos kx or sin kx and integrate from the lower limit $-\pi$ to the upper limit $+\pi$. Then the expressions (2) follow immediately for the expansion coefficients a_k, b_k because on the right side everything cancels out except the single term containing cos kx or sin kx. So far the results are simple, and were historically well established long before Fourier. But are we allowed to assume the validity of the expansion (2.1)? In order to answer this question, Dirichlet proceeded as follows. We start with the *finite* series

$$f_n(x) = \tfrac{1}{2}a_0 + a_1 \cos x + \cdots + a_n \cos nx$$
$$+ b_1 \sin x + \cdots + b_n \sin nx \qquad (4\text{-}2.6)$$

assuming for a_k, b_k their values (4-2.2), and investigate what happens if n increases to infinity. We obtain

$$f_n(x) = \int_{-\pi}^{+\pi} f(t) K_n(x - t)\, dt \qquad (4\text{-}2.7)$$

where $K_n(\xi)$ is the so-called "Dirichlet kernel":

$$K_n(\xi) = \frac{\sin (n + \tfrac{1}{2})\xi}{2\pi \sin (\tfrac{1}{2}\xi)} \qquad (4\text{-}2.8)$$

Our aim is to show that with n increasing to infinity, $f_n(x)$ approaches $f(x)$ with an error which can be made arbitrarily small. This requires a very strong *focusing power* of the function $K_n(\xi)$. It must put the spotlight on the immediate neighborhood of $\xi = 0$ and blot out everything else. In particular, the following two conditions should be expected as the exact mathematical expressions of the increasing focusing power of $K_n(\xi) = K_n(-\xi)$:

$$\lim_{n \to \infty} \int_{\varepsilon}^{\pi} \left| K_n(\xi) \right| d\xi = 0 \qquad (4\text{-}2.9)$$

$$\lim_{n \to \infty} \int_{-\varepsilon}^{+\varepsilon} K_n(\xi) \, d\xi = 1 \qquad (4\text{-}2.10)$$

The first condition guarantees that the kernel function blots out everything except the immediate neighborhood of the point $t = x$. The second condition guarantees that the point $t = x$ enters the integration with the proper weight; ε is assumed to be a prescribed arbitrarily small positive number, independent of n.

However, the Dirichlet kernel does *not* satisfy the first condition. The secondary maxima of the function (2.8) are not small enough to ensure predominance of the point $\xi = 0$. This is the reason that the expandibility of $f(x)$ into an infinite convergent Fourier series has to be restricted to a definite class of functions which are sufficiently smooth to counteract the insufficient focusing power of the Dirichlet kernel. The Dirichlet conditions 2 and 3 given above aim at establishing the desired smoothness of $f(x)$.[1]

Fejér's method. By an ingenious modification of the summation procedure, L. Fejér succeeded in increasing the focusing power of the Dirichlet kernel and thus extending the validity of the Fourier series to a much larger class of functions.[2] We will not restrict $f(x)$ by any conditions, except the single condition that $f(x)$ is "absolutely integrable," which means that the quantity

$$I_1 = \int_{-\pi}^{+\pi} \left| f(x) \right| \, dx \qquad (4\text{-}2.11)$$

exists ("absolutely integrable functions"). This condition is entirely

---

[1] The Dirichlet conditions are sufficient but unnecessarily stringent, although they comprise a very large class of functions. The minimum restrictions required for the convergence of the Fourier series are not yet known. Fejér's method obviates the problem by summing the series in a different way.

[2] Cf. {15}, p. 169.

natural, since otherwise the Fourier coefficients a_k and b_k, defined by (2), or c_k, defined by (5), would not exist either.

We can now construct the formal series (1) or (4). Generally, these series will not converge. This means that the sequence $f_n(x)$, defined by (6) taken at a definite x, will not approach a definite number as n increases to infinity. We call these $f_n(x)$ the "partial sums," since we have summed the series up to a certain point. We will change our notation and call them $s_n(x)$. We substitute a definite value for x and consider the sequence

$$s_0, s_1, s_2, s_3, \cdots$$

Out of this sequence we construct a new sequence by taking the arithmetic means of the previous sequence:

$$f_1 = s_0, \quad f_2 = \frac{s_0 + s_1}{2}, \quad \cdots, \quad f_n = \frac{s_0 + s_1 + \cdots + s_{n-1}}{n} \quad \text{(4-2.12)}$$

This new sequence has the remarkable property that it converges to a definite limit at all points at which the one-sided limits $f_1 = f(x_+)$ and $f_2 = f(x_-)$ exist, defined by

$$\lim_{\varepsilon \to 0} [f(x + \varepsilon) - f_1] = 0$$
$$\qquad\qquad\qquad\qquad (\varepsilon > 0) \quad \text{(4-2.13)}$$
$$\lim_{\varepsilon \to 0} [f(x - \varepsilon) - f_2] = 0$$

At all such points $f_n(x)$ converges to the arithmetic mean of the two limits (13). [In an ordinary point the two limits coincide, and the series simply converges to $f(x)$]. Hence $f_n(x)$ automatically converges to a reasonable value at all points at which a reasonable answer can be expected at all.

Fejér's results come about by the fact that his method is associated with the following kernel function.

$$K_n(\xi) = \frac{\sin^2(n\xi/2)}{2\pi n \sin^2(\xi/2)} \quad \text{(4-2.14)}$$

This kernel possesses the strong focusing properties expressed by the conditions (9) and (10).

3. Least-squares approximations. While the interest of pure analysis is essentially centered around the general convergence properties of the Fourier series, the interest of practical analysis is centered around the possibility of representing a fairly large but not too irregular class of functions effectively by a relatively *small* number of harmonic components. Here it is no longer a question of obtaining the given $y = f(x)$ with an arbitrarily small error but of approximating $f(x)$ with a finite error which, however, shall be made as small as possible. The Fourier series belongs to that class of expansions which approximate a given function $y = f(x)$ in a definite finite interval uniformly well, in the sense that we do not try to make the approximation very close in the neighborhood of a certain point at the cost of obtaining a large error away from that point, but we attach equal importance to *every* point of the region.

The principle involved in this kind of interpolation is given by the "method of least squares." We define the error $\eta(x)$ of a finite expansion by forming the difference between the given $y = f(x)$ and the finite sum

$$f_m(x) = c_1\varphi_1(x) + c_2\varphi_2(x) + \cdots + c_m\varphi_m(x) \qquad (4\text{-}3.1)$$

where $\varphi_1(x), \varphi_2(x), \cdots, \varphi_m(x)$ are prescribed functions, while the coefficients $c_1, c_2, \cdots, c_m$ are at our disposal. We call the difference $\eta_m(x)$.

$$\eta_m(x) = f(x) - f_m(x) \qquad (4\text{-}3.2)$$

Instead of trying to let $\eta_m(x)$ vanish at m more or less arbitrarily prescribed points, we characterize the "average error" by integrating over the finite range

$$a \leq x \leq b \qquad (4\text{-}3.3)$$

in which the function $f(x)$ shall be represented. We define this average error by the definite integral

$$\bar{\eta}^2 = \frac{1}{b-a} \int_a^b \eta_m^2(x) \, dx \qquad (4\text{-}3.4)$$

and we determine the coefficients c_i of the expansion (1) by the condition that $\bar{\eta}$ shall become as small as possible. This problem has always a definite solution, since an essentially positive quantity always assumes a definite absolute minimum for a certain choice of the variables.

Our problem is greatly facilitated if the given basic functions

$$\varphi_1(x), \varphi_2(x), \cdots, \varphi_m(x) \qquad (4\text{-}3.5)$$

satisfy the following "orthogonality condition"

$$\int_a^b \varphi_i(x)\varphi_k(x)\,dx = 0 \tag{4-3.6}$$

A set of functions which satisfies this condition is called an "orthogonal set." If in addition we apply for every $\varphi_k(x)$ such a factor of proportionality that we shall have

$$\int_a^b \varphi_i^2(x)\,dx = 1 \tag{4-3.7}$$

we speak of an "orthogonal and normalized" or briefly "orthonormal" set of functions.

Let us observe that for such a function system the quantity $f_m^2(x)$, integrated over the given range, appears in a particularly simple form.

$$\int_a^b f_m^2(x)\,dx = c_2^1 + c_2^2 + \cdots + c_m^2 \tag{4-3.8}$$

Furthermore, the integral of $\eta_m^2(x)$ becomes likewise greatly simplified.

$$\int_a^b \eta_m^2(x)\,dx = \int_a^b f^2(x)\,dx - 2(\gamma_1 c_1 + \gamma_2 c_2 + \cdots + \gamma_m c_m \tag{4-3.9}$$
$$+ c_1^2 + c_2^2 + \cdots + c_m^2$$

where

$$\gamma_k = \int_a^b f(x)\varphi_k(x)\,dx \tag{4-3.10}$$

Now the expression (9) is a simple algebraic function of the variables c_i, and the condition of minimum requires

$$c_i = \gamma_i \tag{4-3.11}$$

With this choice of the c_i the average error $\bar{\eta}_m$ of the approximation becomes

$$\bar{\eta}_m^2 = \frac{1}{b-a}\left[\int_a^b f^2(x)\,dx - (\gamma_1^2 + \gamma_2^2 + \cdots + \gamma_m^2)\right] \tag{4-3.12}$$

If the given $f(x)$ happens to be a linear combination of the m given $\varphi_k(x)$, the choice (11) of the c_i will give an expansion (1) which coincides with $f(x)$ identically. In that case $\bar{\eta}_m$ will turn out to be zero. For any other choice of $f(x)$, however, $\bar{\eta}_{lm}^2$ will remain a positive quantity which can be further reduced only by adding more functions $\varphi_{m+1}(x)$, $\cdots$ to the previous set.

The great advantage of an orthogonal set of functions is that addition of a new member to the set has no damaging influence on our previous calculations. We need not change anything on the previous coefficients, but merely determine the additional coefficient c_{m+1}, which depends solely on $\varphi_{m+1}(x)$. The average error $\bar{\eta}_{m+1}$ is now further diminished, since $-\gamma_{m+1}^2$ is added inside the bracket of the right side. If γ_{m+1} happens to come out as zero, this indicates that the addition of $\varphi_{m+1}(x)$ was not able to further decrease the average error; then $\varphi_{m+2}(x)$ has to be tried, and so on.

4. The orthogonality of the Fourier functions. The functions of the Fourier series give us an example for an orthogonal set. The range is from $-\pi$ to $+\pi$, or from 0 to 2π.

$$\int_{-\pi}^{+\pi} e^{imx}\, e^{inx}\, dx = \frac{1}{i(m+n)} \left[e^{i(m+n)x} \right]_{-\pi}^{+\pi} = 0 \qquad (m \neq -n)$$

$$(4\text{-}4.1)$$

However, for functions which assume complex values, the concept of orthogonality has to be slightly modified. Here we replace the quantity $\eta_m^2(x)$ by $\eta_m(x)\eta_m^*(x)$, where the * refers to the operation: "change i to $-i$." Accordingly, orthogonality shall mean

$$\int_a^b \varphi_i(x)\varphi_k^*(x)\, dx = 0 \qquad (i \neq k)$$

$$(4\text{-}4.2)$$

and normalization of the $\varphi_i(x)$ occurs by the condition

$$\int_a^b \varphi_k(x)\varphi_k^*(x)\, dx = 1$$

$$(4\text{-}4.3)$$

Furthermore, the γ_k are here defined by

$$\gamma_k = \int_a^b f(x)\varphi_k^*(x)\, dx$$

$$(4\text{-}4.4)$$

and these γ_k become the coefficients of the orthogonal expansion which minimizes the average error.

The ortho-normal functions of the complex Fourier series are the functions

$$\varphi_k = \frac{1}{\sqrt{2\pi}} \, e^{ikx} \qquad (4\text{-}4.5)$$

and we obtain

$$\gamma_k = \frac{1}{\sqrt{2\pi}} \int_{-\pi}^{+\pi} f(x) e^{-ikx} \, dx \qquad (4\text{-}4.6)$$

The resultant series is equivalent to the previous complex form (2.4), (2.5) of the Fourier series.

We observe that the infinite Fourier expansion (2.4) can be conceived as the process of constantly minimizing the average error, by taking in more and more orthogonal functions. One can show that with n growing to infinity the average error [cf. (3.4)] converges to zero for any "quadratically integrable function" of the range, i.e., for any function for which the integral

$$I_2 = \int_{-\pi}^{+\pi} |f(x)|^2 \, dx \qquad (4\text{-}4.7)$$

exists. And yet the peculiar phenomenon occurs that from this fact we can*not* draw conclusions as to the vanishing of the local error $\eta(x)$ itself. If the average error vanished *exactly*, this would necessarily mean the vanishing of $\eta(x)$ at every point, since the integral of an everywhere nonnegative function can vanish only if that function vanishes identically. This would prove that the Fourier series gives the correct functional value at every point of the range. What we can prove, however, is only that the average error can be made *as small as we wish*, by increasing n to a properly large value. This is not enough to prove that the local error itself converges to zero at every point of the range. For this purpose more has to be required of the function $f(x)$ than quadratic integrability (for example, the Dirichlet conditions, cf. § 2).

5. Separation of the sine and the cosine series. For many purposes of applied analysis it is convenient to separate the sine part and the cosine part of the Fourier series. This can be done by separating the function $f(x)$ into an even and an odd part. We put

$$g(x) = \tfrac{1}{2}[f(x) + f(-x)] \qquad (4\text{-}5.1)$$
$$h(x) = \tfrac{1}{2}[f(x) - f(-x)]$$

Then

$$f(x) = g(x) + h(x) \tag{4-5.2}$$

with

$$g(-x) = g(x) \tag{4-5.3}$$

$$h(-x) = -h(x) \tag{4-5.4}$$

A function with the symmetry property (3) is called an "even," one with the symmetry property (4) an "odd" function. In the first case the sine part of the Fourier series drops out, in the second case the cosine part. In both cases it suffices to give the function only in the range

$$0 \le x \le \pi \tag{4-5.5}$$

which is the half range of the regular Fourier series. In the other half the reflection properties already define the function.

We now obtain

$$g(x) = \tfrac{1}{2}a_0 + a_1 \cos x + a_2 \cos 2x + \cdots \tag{4-5.6}$$

with

$$a_k = \frac{2}{\pi}\int_0^\pi g(x) \cos kx \, dx \tag{4-5.7}$$

and

$$h(x) = b_1 \sin x + b_2 \sin 2x + \cdots \tag{4-5.8}$$

with

$$b_k = \frac{2}{\pi}\int_0^\pi f(x) \sin kx \, dx \tag{4-5.9}$$

Another point of departure is to assume that the function $f(x)$ is from the very beginning defined in the interval (5) only. In that case we have complete freedom to define the function for negative values of x either by the condition (3) or by the condition (4). Hence the very same function $f(x)$ of the interval $[0,\pi]$ can be expanded by either a cosine or a sine series. Both expansions are complete in themselves and converge to the given $f(x)$, provided that $f(x)$ belongs to the class of functions covered by the Dirichlet conditions. However, the convergence of these series may be vastly different. Very frequently we need the Fourier expansion of a function which is everywhere continuous and even differentiable in the given interval. In this case the convergence of the Fourier series will be determined by the properties of the function at the boundary points $x = 0$ and

$x = \pi$. If the value of $f(x)$ is not zero at the two boundary points, the reflection as an odd function (4) will cause a discontinuity at the two points $x = 0$ and $x = \pi$. The discontinuity at $x = 0$ follows directly from the law of reflection (4), the discontinuity at $x = \pi$ from the periodicity condition

$$f(x + 2\pi) = f(x) \tag{4-5.10}$$

which, if applied to the point $x = -\pi$ demands

$$f(\pi) = f(-\pi) \tag{4-5.11}$$

Both discontinuities are avoided if $f(x)$ is reflected as an *even* function. Then the discontinuity appears in the *first derivative* only but not in the function itself. Hence expanding a function which is not restricted by any particular boundary conditions, the cosine series (7) will have much better convergence properties than the sine series (9); the latter expansion will show the undesirable oscillations due to the "Gibbs' phenomenon" encountered in the neighborhood of a discontinuity; [cf. § 9].

On the other hand, let us assume that $f(x)$ *vanishes* at the two end points of the given range. Such a condition can always be met if it is permissible to subtract from $f(x)$ a linear trend of the form $\alpha + \beta x$, with suitably chosen α and β. The new function satisfies the boundary conditions

$$f(0) = 0, \quad f(\pi) = 0 \tag{4-5.12}$$

For a function of this type the sine series has much better convergence than the cosine series, since the reflection as an odd function now preserves continuity in function *and first derivative*, the discontinuity appearing only in the *second* derivative, at the points $x = 0$ and $x = \pi$.

We can estimate the order of magnitude of the Fourier coefficients. In an integral of the type

$$\int f(x) \cos kx \, dx \tag{4-5.13}$$

not much can be said as long as k is small. For large k, however, the first factor is smooth compared with the rapid oscillations of the second factor. Hence we can integrate by parts, obtaining

$$\int f(x) \cos kx \, dx = \frac{f(x) \sin kx}{k} - \frac{1}{k} \int f'(x) \sin kx \, dx \tag{4-5.14}$$

Repeating the process once more,

$$\int f'(x) \sin kx \, dx = -\frac{f'(x) \cos kx}{k} + \frac{1}{k} \int f''(x) \cos kx \, dx \quad (4\text{-}5.15)$$

Putting in the limits 0 and π, the dominant term for large k becomes

$$\int_0^\pi f(x) \cos kx \, dx = \frac{1}{k^2} \left| f'(x) \cos kx \right|_0^\pi = \frac{(-1)^k f'(\pi) - f'(0)}{k^2}$$

while $\quad (4\text{-}5.16)$

$$\int_0^\pi f(x) \sin kx \, dx = -\frac{1}{k} \left| f(x) \cos kx \right|_0^\pi = \frac{f(0) - (-1)^k f(\pi)}{k}$$

$$(4\text{-}5.17)$$

This shows that the coefficients of the cosine series decrease with the speed k^{-2}, the coefficients of the sine series only with the speed k^{-1}, if $f(x)$ does not satisfy any specific boundary conditions.

If, however, $f(x)$ satisfies the boundary conditions (12), the right side of (17) vanishes, and we obtain for the dominant term,

$$\int_0^\pi f(x) \sin kx \, dx = \frac{1}{k^3} \left| f''(x) \cos kx \right|_0^\pi = \frac{(-1)^k f''(\pi) - f''(0)}{k^3}$$

$$(4\text{-}5.18)$$

The speed with which the coefficients decrease is now k^{-3}, which will be satisfactory for many applications of the Fourier series.

We can transfer our results from the range $[0,\pi]$ to the range $[-\pi,\pi]$ by calling the previous variable x_1 and transforming it to the new variable

$$x = -\pi + 2x_1 \quad (4\text{-}5.19)$$

and finally returning to the notation x. We thus obtain for a function $f(x)$ which satisfies the boundary conditions

$$f(\pm\pi) = 0 \quad (4\text{-}5.20)$$

the expansion

$$f(x) = a_1 \cos \frac{x}{2} + a_2 \cos \frac{3x}{2} + \cdots + a_k \cos \frac{2k-1}{2} x + \cdots$$

$$+ b_1 \sin x f + b_2 \sin 2x + \cdots + a_k \sin kx + \cdots \quad (4\text{-}5.21)$$

with

$$a_k = \frac{1}{\pi} \int_{-\pi}^{+\pi} f(x) \cos \frac{2k-1}{2} x \, dx$$

$$b_k = \frac{1}{\pi} \int_{-\pi}^{+\pi} f(x) \sin kx \, dx$$

(4-5.22)

The series (21) has faster convergence than the original Fourier series (2.1) if the function $f(x)$ satisfies the boundary conditions (20); its coefficients decrease with the speed k^{-3}, instead of the speed k^{-2} which holds if the same function is analyzed in the original Fourier functions.

6. Differentiation of a Fourier series. Let us consider the finite expansion

$$f_m(x) = \sum_{k=-(m-1)}^{m-1} c_k e^{ikx}$$

(4-6.1)

together with the residual

$$\eta_m(x) = \sum_{k=m}^{\infty} (c_k e^{ikx} + c_{-k} e^{-ikx})$$

(4-6.2)

The residual may be written as follows.

$$\eta_m(x) = e^{imx} \sum_{k=0}^{\infty} c_{m+k} e^{ikx} + e^{-imx} \sum_{k=0}^{\infty} c_{-m-k} e^{-ikx}$$

(4-6.3)

It suffices to consider the first term of the right side only. We can put

$$\sum_{k=0}^{\infty} c_{m+k} e^{ikx} = \rho_m(x)$$

(4-6.4)

Then $\rho_m(x)$ is a generally smooth function which does not show any rapid oscillations. On the other hand, e^{imx} is a rapidly oscillating function. Hence the error term of the Fourier series has the character of a "modulated carrier wave" of high frequency. If we formally differentiate the truncated series (1) and compare it with the value of $f'(x)$, we obtain the large error,

$$\eta_m'(x) = ime^{imx}\rho_m(x) + e^{imx}\rho_m'(x) - ime^{-imx}\rho_{-m}(x) + e^{-imx}\rho_{-m}'(x)$$

(4-6.5)

The primed terms arise from the differentiation of the modulation and do not cause any serious difficulty. However, the other terms contain the large factor m. It is this fact which causes a serious loss of accuracy in formally differentiating the Fourier series. We have differentiated the modulation *and* the carrier wave. We should differentiate the modulation only, *without* the carrier wave.

This aim can be achieved by the following procedure. Let us replace the process of differentiation by the following differencing process:

$$\mathscr{D}_m y = \frac{f(x + \pi/m) - f(x - \pi/m)}{2\pi/m} \tag{4-6.6}$$

The operator $\mathscr{D}$ does not coincide with the operator D of ordinary differentiation but the difference gets smaller and smaller as m increases to infinity. In the limit

$$D = \frac{d}{dx} = \lim_{m \to \infty} \mathscr{D}_m \tag{4-6.7}$$

Now the operator $\mathscr{D}_m$, if applied to $\eta_m(x)$, picks out two points on the carrier wave which are exactly in the same phase, viz., $\pm 180°$ away from the phase at the point x. For this reason the operator $\mathscr{D}_m$ applies merely the factor -1 to the carrier wave, without differentiating it. We obtain

$$\mathscr{D}_m \eta_m(x) = - e^{imx} \mathscr{D}_m \rho_m(x) - e^{-imx} \mathscr{D}_m \rho_{-m}(x) \tag{4-6.8}$$

Now $\rho_m(x), \rho_{-m}(x)$ are smooth functions whose differentiation does not cause any increase in the order of magnitude of the error. Consequently the convergence of the Fourier series was not damaged by this operation. In the limit, as m grows to infinity, the derivative of $f(x)$ is obtained at all points where this derivative exists.

This method of differentiating the Fourier series is equivalent to multiplying the coefficients of the formally differentiated series by certain preassigned weight factors σ_k which depend on m, the order of the partial sums. Consider the term associated with e^{ikx}.

$$\mathscr{D}_m e^{ikx} = i \frac{\sin \pi k/m}{\pi/m} e^{ikx} = \frac{\sin \pi k/m}{\pi/m} ike^{ikx} \tag{4-6.9}$$

The correction factor is thus

$$\sigma_k = \frac{\sin \pi k/m}{\pi/m} \qquad (4\text{-}6.10)$$

Since the same σ_k holds for both $\pm k$, the same factor is applied to both c_k and c_{-k}. Moreover, the a_k, b_k coefficients of the real form of the Fourier series are linear combinations of c_k and c_{-k}.

$$a_k = c_k + c_{-k}, \qquad b_k = i(c_k - c_{-k}) \qquad (4\text{-}6.11)$$

Hence the coefficients a_k and b_k receive likewise the same "attenuation factors" σ_k. The factor σ_k is 1 only for $k = 0$; for increasing k the σ_k decrease monotonously and become almost zero for the highest subscript $k = m - 1$.

The strong attenuation of the Fourier coefficients of high order successfully counteracts their tendency to make the series divergent. Hence a series which would otherwise be completely divergent can be made convergent by application of the σ_k-factors.

7. Trigonometric expansion of the delta function. A good example is provided by the "square wave," defined by the conditions

$$f(-x) = -f(x)$$
$$f(x) = \tfrac{1}{2} \qquad (0 < x < \pi) \qquad (4\text{-}7.1)$$
$$f(0) = f(\pi) = 0$$

The corresponding Fourier series,

$$y(x) = \frac{2}{\pi}\left(\sin x + \frac{\sin 3x}{3} + \frac{\sin 5x}{5} + \cdots\right) \qquad (4\text{-}7.2)$$

converges at every point of the interval $[-\pi, +\pi]$ to $f(x)$, but the convergence is slow. If we differentiate the given $f(x)$, we obtain

$$f'(x) = 0 \qquad (4\text{-}7.3)$$

at every point, except at $x = 0$ where the derivative does not exist. The formal differentiation of the Fourier series, on the other hand, yields

$$y'(x) = \frac{2}{\pi}(\cos x + \cos 3x + \cos 5x + \cdots) \qquad (4\text{-}7.4)$$

This series does not converge to a definite limit at *any* point x (except at $x = \pm\pi/2$). Hence the strong singularity at the point $x = 0$ irradiates its effect to every point of the region and destroys the convergence of the Fourier series everywhere, although $f'(x)$ is entirely regular at every point, except at the origin $x = 0$.

If we now replace $y'(x)$ by the series

$$y'_{2m}(x) = \frac{2}{\pi}\left[\frac{\sin(\pi/2m)}{\pi/2m}\cos x + \frac{\sin(3\pi/2m)}{3\pi/2m}\cos 3x\right.$$

$$\left. + \cdots + \frac{\sin(2m-1)\pi/2m}{(2m-1)\pi/2m}\cos(2m-1)x\right] \quad (4\text{-}7.5)$$

we get an entirely different behavior. As m grows to infinity, $y'_{2m}(x)$ converges to zero at every fixed point x, except the point $x = 0$, where the function grows beyond all bounds. The function grows to infinity in such a way that the area under the curve, taken between the points $x = \pm\varepsilon/2$, where ε is an arbitrarily small but fixed quantity, converges to 1.

The derivative of the function (1) is frequently called the "delta function," and is denoted by $\delta(x)$. It is not a legitimate function, since the original function cannot be differentiated at the point $x = 0$. It is, however, the limit of a legitimate function and useful for many analytical purposes. We define this function as zero everywhere, except between the limits $\pm\varepsilon/2$, where ε tends toward zero. At $x = 0$ the function goes to infinity in such a way that the area under the function shall be 1.

In physical interpretation the delta function represents a pulse of the intensity 1, applied during a time interval ε around the point $x = 0$, letting ε converge to zero. The formal definition of the delta function leads to the infinite series

$$y = \frac{1}{\pi}(\tfrac{1}{2} + \cos x + \cos 2x + \cdots) \quad (4\text{-}7.6)$$

which does not converge at any point. If, however, the σ-factors are applied and we consider the expansion

$$y_m = \frac{1}{\pi}(\tfrac{1}{2} + \sigma_1\cos x + \sigma_2\cos 2x + \cdots + \sigma_{m-1}\cos(m-1)x]$$

$$(4\text{-}7.7)$$

where the σ_k are defined by (6.10), we obtain an expansion which can be considered the trigonometric representation of the delta function. The following figure plots the course of the function thus obtained.

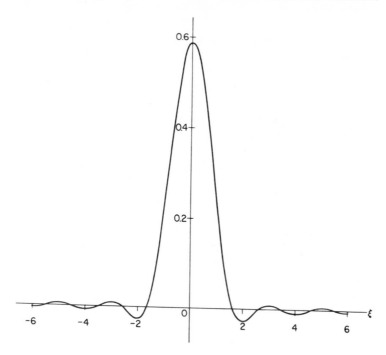

In this figure the variable ξ represents the product

$$\xi = \frac{m}{\pi} x \qquad (4\text{-}7.8)$$

Hence the points ± 1 of the figure are the points $\pm \pi/m$ in terms of x. Moreover the amplitudes of $y(\xi)$ are in the following relation to $y_m(x)$.

$$y_m(x) = my(\xi) = my\left(\frac{m}{\pi} x\right) \qquad (4\text{-}7.9)$$

As m increases, the figure is reduced to smaller and smaller portions of the x axis, while the maximum amplitude gets larger and larger. The figure shrinks by the factor m in the x direction, while it grows by the factor m in the y direction. As m grows to infinity, y_m converges to 0 at every point x, except at $x = 0$. At the same time the area under the curve, evaluated with the help of the series, remains constantly 1.

8. Extension of the trigonometric series to nonintegrable functions. We have seen that the absolute integrability of $f(x)$ had to be demanded in order to expand $y = f(x)$ into a Fourier series. Hence $y = \log x$ is a function which between 0 and π can be expanded into a cosine (or a sine) series because the integral of the function is bounded, although the function itself becomes infinite at $x = 0$. On the other hand, the function $y = x^{-1}$ cannot be expanded, since now the function goes so strongly to infinity that even the area under the curve becomes infinite. The Fourier coefficients a_k, b_k go out of bound for such a function.

However, x^{-1} can be considered as the derivative of $\log x$. If we have a procedure by which we can differentiate a convergent Fourier series without losing its convergence, then we can first expand $y = \log x$ into a Fourier series and then by differentiation obtain x^{-1}. This can be done by application of the σ_k factors. The formal derivative of the series for $\log x$ does not converge at any point. But if the formal coefficients are multiplied by $\sigma_k(m)$, then with increasing m the modified series converges to the function x^{-1}, and the error can be made arbitrarily small, with only the exception of the point $x = 0$, where the derivative does not exist and the function grows to infinity. The cosine series likewise grows to infinity at this point, while the sine series remains constantly zero. And yet we can choose an arbitrarily small ε, and the sine series will converge to the very large value $1/\varepsilon$, in spite of the fact that it has to start from the value zero at the point $x = 0$.

If the same process is applied several times, we have a simple method of obtaining trigonometric expansions for functions which are not integrable in themselves but which are the μth derivatives of an absolutely integrable function. In this case we start with the generating function, obtain its Fourier series, differentiate it μ times formally, and then apply the factors σ_k^μ to the formal coefficients of the differentiated function, The modified series will converge to the μth derivative of the original function at every point where this derivative exists. By this method even a function of the type $x^{-\mu}$

with arbitrarily large positive μ can be expanded into a trigonometric series and the error can be made arbitrarily small at any point excluding the origin, although the classical Fourier series does not exist for these functions.

9. Smoothing of the Gibbs oscillations by the σ factors. Application of the σ factors can serve not only to transform a divergent Fourier series into a convergent series, but also to increase the slow convergence of a Fourier series. Since any function can be conceived as the derivative of its integral, the reduction of the error in the derivative by the factor m must show its beneficial effect on the convergence of any given Fourier series.

The slow convergence of a Fourier series is particularly undesirable if a point of discontinuity is involved. The effect of this discontinuity is that the series oscillates around the true function with amplitudes which decrease only very slowly with increasing number of terms. The high-frequency oscillations of the truncated series around the true function are always present, as the error term (6.3) has shown. Ordinarily, the amplitudes of these oscillations are sufficiently small to be of no major consequence. In the case of a discontinuity, however, they are very noticeable and interfere with an efficient harmonic synthesis of the square wave.[1]

Fejér's arithmetic mean method completely eliminates the Gibbs oscillations. The approach to the square wave now occurs in the form of completely smooth, monotonous functions which remain constantly below the curve. The method of the σ factors does not eliminate the oscillations of the Fourier series but cuts down their amplitudes. The former amplitude of 0.08949 of the jump in the first oscillation is reduced to 0.01187; the next minimum of 0.04859 is reduced to only 0.00473, and so on.

The accompanying figure plots the course of the original truncated Fourier sum with the Gibbs oscillations, and likewise the course of the arithmetic mean and the result of the σ smoothing. We notice that the method of the arithmetic mean, while avoiding the undesirable oscillations of the Fourier series, has the disadvantage that it approaches the asymptotic value very slowly and that it starts with a relatively small slope. The method of the σ factors yields a faster increase of the approximating function, together with a sharp turn

---

[1] The "Gibbs phenomenon"; cf. [2], Chapter IX; see also {1}, p. 105.

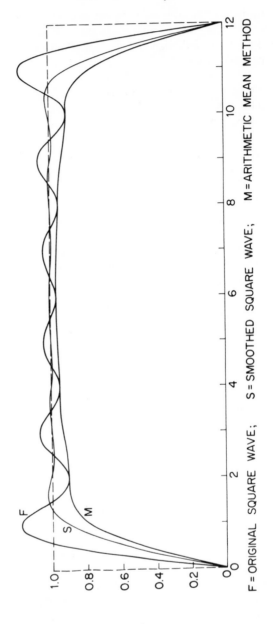

F = ORIGINAL SQUARE WAVE;   S = SMOOTHED SQUARE WAVE;   M = ARITHMETIC MEAN METHOD

after the maximum level has been reached. It is true that the oscillations still exist, but their amplitude is strongly attenuated and quickly damped out. The fidelity of the approximation is thus markedly better than that of the arithmetic mean method, while in comparison with the original series we can state that we have practically eliminated the cumbersome oscillations of the Gibbs phenomenon, at the cost of a somewhat less steep ascent at the beginning of the curve.

10. General character of the σ smoothing. Application of the σ_k factors to the classical Fourier coefficients is a somewhat simpler process than taking the arithmetic mean of the partial sums, and is less damaging from the fidelity point of view. The power of this smoothing process is equivalent to that of the arithmetic mean process. One can show that the outstanding properties of the Fejér's kernel (2.14) hold also for the kernel associated with the σ process. The multiplication of the classical Fourier coefficients by the factors σ_k has thus the same effect on the convergence of the series as the taking of the partial sums; convergence is obtained at all points where a definite limit can reasonably be expected.

In contrast to the summing by partial sums, the method of the σ factors is not merely a technical device for summing an infinite series in a definite manner but has an invariant significance in relation to the given function $f(x)$. Replacement of the D-process by the $\mathscr{D}_m$ process of § 6 can be approached from an altogether different viewpoint. Instead of changing the process of differentiation we will now change the function involved in the operation but leave the operation itself unchanged. We can write

$$\mathscr{D}_m f(x) = D\bar{f}(x) \tag{4-10.1}$$

where
$$\bar{f}(x) = \frac{m}{2\pi} \int_{-\pi/m}^{+\pi/m} f(x + t)\, dt \tag{4-10.2}$$

This means that the function $f(x)$ is replaced by a new function $\bar{f}(x)$ which smooths the original function by taking at every point the arithmetic mean of $f(x)$ around the point x, between the limits $\pm\pi/m$. Instead of saying that we multiply the coefficients of the truncated Fourier series by σ_k, we can also say that we operate with the truncated

Fourier series of a modified function $\bar{f}(x)$ which is associated with $f(x)$ in a definite way, viz., on the basis of local averaging. If m is sufficiently large, the local averaging extends to a very small region around the point x only, and thus the distorting effect of this operation is negligibly small, except in the neighborhood of a singularity, where the smoothing of the original roughness of the function is actually desirable.

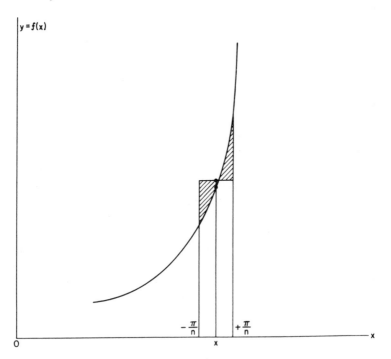

We thus see that at the cost of a relatively small distortion of the function the convergence of the Fourier series can be considerably improved in all instances in which the convergence of the original series is not satisfactory. At a regular point of the function the distortion thus introduced can be characterized in terms of the second derivative. By the Taylor series we have

$$f(x + h) = f(x) + hf'(x) + \frac{h^2}{2}f''(x) + \cdots \qquad (4\text{-}10.3)$$

Considering h as the variable and integrating between the limits $\pm \pi/m$, we obtain

$$\bar{f}(x) \doteq f(x) + \frac{\pi^2}{6m^2} f''(x) \tag{4-10.4}$$

11. The method of trigonometric interpolation. The harmonic analysis of functions is frequently handicapped by the difficulty of obtaining the integrals (2.2). Even for functions of relatively simple structure, actual evaluation of the integrals (2.2) may meet insuperable difficulties; the integral of the product of a given function with a trigonometric function of the type $\sin nx$ and $\cos nx$ is seldom expressible in terms of simple functions. Hence the question has to be raised as to what steps shall be taken for practical evaluation of the Fourier coefficients a_k and b_k.

A similar problem arises in connection with empirically observed functions. The data are available in a finite set of points only, namely at the points of observation. In most practical cases the data points x_α are equidistantly spaced. The same can be said of tabulated functions. Here, too, the tabulation usually occurs in equidistant steps of the independent variable x.

It is true that for intermediate points we can frequently interpolate the given fundamental data. This is particularly true for tabulated functions which can be given with sufficient accuracy to allow an effective interpolation at any point x of the given range. We are in a less favorable position, however, in relation to empirical functions, since here the mathematical law of the function is frequently not known; moreover, the "noise" superimposed on the true course of the function makes preliminary smoothing of the data necessary, which cannot be done with any degree of certainty.

Under these circumstances we save a great deal in difficulties if we can eliminate the interpolation problem altogether by operating directly with the given discrete set of data. Very fortunately, harmonic analysis is exceedingly well suited to the nature of equidistant data. The numerical procedure for obtaining the harmonic components of an unknown function given in equidistant intervals is simple and straightforward, and at the same time well convergent. The fundamental tool of the Fourier series is thus greatly extended in its usefulness since it is applicable to a discrete set of equidistant data no less than to a continuous set of data.

The general problem of orthogonal expansions with respect to discrete data may be formulated as follows. Let a function $y = f(x)$ be given at the discrete points $x = x_\alpha$ and let the functional values at these points be denoted by

$$y_\alpha = f(x_\alpha) \tag{4-11.1}$$

Let us approximate $y = f(x)$ by a linear superposition of given functions $\varphi_k(x)$:

$$\bar{y} = \sum_{k=1}^{m} c_k \varphi_k(x) \tag{4-11.2}$$

We assume that m, the number of given functions, is generally less than n, the number of data, but the limiting case $m = n$ shall not be excluded. We also assume that the functions $\varphi_k(x)$ are linearly independent.

Generally the approximation $\bar{y}$ at the data points x_α will not coincide with the given functional values y_α. We characterize the square of the average error of our approximation by forming the sum of the squares of the residuals at the data points $x = x_\alpha$.

$$\eta^2 = \sum_{\alpha=1}^{n} (y_\alpha - \bar{y}_\alpha)^2 = \sum_{\alpha=1}^{n} \left[y_\alpha - \sum_{k=1}^{m} c_k \varphi_k(x_\alpha) \right]^2 \tag{4-11.3}$$

We determine the "best" approximation by minimizing η^2 with respect to the c_k. This demands that the partial derivative of η^2 with respect to c_k shall become zero.

$$\sum_{\alpha=1}^{n} \left[y_\alpha - \sum_{\alpha=1}^{k} c_k \varphi_k(x_\alpha) \right] \varphi_i(x_\alpha) = 0 \tag{4-11.4}$$

This leads to a simultaneous system of linear equations for determination of the c_k. The resulting system becomes greatly simplified if the given functions $\varphi_k(x)$ satisfy the following orthogonality conditions.

$$\sum_{\alpha=1}^{n} \varphi_i(x_\alpha) \varphi_k(x_\alpha) = 0 \qquad (i \neq k) \tag{4-11.5}$$

In these conditions we recognize the earlier orthogonality conditions (3.6) but now translated from the process of integration to the process of summation. The geometrical term "orthogonality" refers to the following concept of analytical geometry. Let us plot in an imaginary Euclidean space of n dimensions the functional values $\varphi(x_1)$, $\varphi(x_2)$, $\cdots$, $\varphi(x_n)$ of an arbitrary function $y = \varphi(x)$, as rectangular components of a vector. The function $\varphi(x)$, taken at the data points, is thus represented by some vector of an n-dimensional space. The given function $y = f(x)$ represents *one* such vector. The m functions $\varphi_1(x)$, $\varphi_2(x)$, $\cdots$, $\varphi_m(x)$ represent m *other* vectors, which in view of their linear independence include a linear subspace of m dimensions. The approximation (2) has the significance of an arbitrary vector of this linear subspace. The method of minimizing the average error in the sense of minimizing the sum (3) has geo-metrically the significance of minimizing the distance of the given vector y from the subspace of the base vectors φ_1, φ_2, $\cdots$, φ_m. This means that we *project* y on the subspace of the φ_i; (cf. also V, 16).

Now the process of projection is greatly simplified if the base vectors φ_i are mutually perpendicular to each other. This means that the "dot product" of any two vectors φ_i and φ_k vanishes.

$$\varphi_i \cdot \varphi_k = \sum_{\alpha=1}^{n} \varphi_i(x_\alpha)\varphi_k(x_\alpha) = 0 \qquad (i \neq k) \qquad (4\text{-}11.6)$$

Hence we are back at our previous orthogonality condition (5) but now interpreted in the language of geometry.

As a consequence of this orthogonality condition the equation (4) is simplified to

$$c_i \sum_{\alpha=1}^{n} \varphi_i^2(x_\alpha) = \sum_{\alpha=1}^{n} y_\alpha \varphi_i(x_\alpha) \qquad (4\text{-}11.7)$$

Hence we have obtained an explicit solution of the given least square problem (3) in the form

$$c_i = \frac{y \cdot \varphi_i}{\varphi_i \cdot \varphi_i} = \frac{\displaystyle\sum_{\alpha=1}^{n} y_\alpha \varphi_i(x_\alpha)}{N_i} \qquad (4\text{-}11.8)$$

if we introduce the "norms" of the given vectors φ_i by putting

$$N_i = \sum_{\alpha=1}^{n} \varphi_i^2(x_\alpha) \qquad (4\text{-}11.9)$$

Once more we can call our function system φ_k "ortho-normal" if the norm of each function $\varphi(x)$—i.e., the length square of each base vector φ_k—is normalized to 1. We then have an orthogonal set of base vectors of length 1 which are particularly well suited for analytical operations.

Assuming such an ortho-normal set, the earlier equation (3.8) appears now in the following form.

$$\sum_{\alpha=1}^{n} \bar{y}_\alpha^2 = c_1^2 + c_2^2 + \cdots + c_m^2 \qquad (4\text{-}11.10)$$

Moreover, the sum of the squares of the residuals, defined by (11.3), now becomes

$$\sum_{\alpha=1}^{n} (y_\alpha - \bar{y}_\alpha)^2 = \sum_{\alpha=1}^{n} y_\alpha^2 - (c_1^2 + c_2^2 + \cdots + c_m^2) \qquad (4\text{-}11.11)$$

For not-normalized but orthogonal systems the last two equations have to be generalized as follows.

$$\sum_{\alpha=1}^{n} \bar{y}_\alpha^2 = N_1 c_1^2 + N_2 c_2^2 + \cdots + N_m c_m^2 \qquad (4\text{-}11.12)$$

$$\sum_{\alpha=1}^{n} (y_\alpha - \bar{y}_\alpha)^2 = \sum_{\alpha=1}^{n} y_\alpha^2 - (N_1 c_1^2 + \cdots + N_m c_m^2) \qquad (4\text{-}11.13)$$

In the limiting case $m = n$ our approximation becomes exact in the data points, since now the m-dimensional subspace is extended to the entire space. Then the two vectors y and $\bar{y}$ coincide and we have

$$y^2 = N_1 c_1^2 + N_2 c_2^2 + \cdots + N_n c_n^2 \qquad (4\text{-}11.14)$$

If the given functions $\varphi_k(x)$ have complex values, the previous operations have to be modified in the sense that the square φ^2 is to be replaced by the product $\varphi\varphi^*$ (cf. § 4). Correspondingly in the last three equations, c_i^2 is to be replaced by $c_i c_i^*$. The orthogonality condition (6) now becomes $\varphi_i \varphi_k^* = 0$ and the formula (8) for the expansion coefficient c_i becomes

$$c_i = \frac{\displaystyle\sum_{\alpha=1}^{n} y_\alpha \varphi_i^*(x_\alpha)}{N_i} \qquad (4\text{-}11.15)$$

$$N_i = \sum_{\alpha=1}^{n} \varphi_i(x_\alpha) \varphi_i^*(x_\alpha) \qquad (4\text{-}11.16)$$

We now consider the following fundamental trigonometric identity.

$$\tfrac{1}{2}e^{-ni\theta} + e^{-(n-1)i\theta} + \cdots + e^{(n-1)i\theta} + \tfrac{1}{2}e^{ni\theta} = \sin n\theta \cotan \theta/2$$

$$(4\text{-}11.17)$$

and define our functions $\varphi_k(x)$ as the trigonometric powers

$$\varphi_k(x) = e^{ikx} \qquad (4\text{-}11.18)$$

while the position of the data points x_α shall be chosen as follows:

$$x_\alpha = \alpha \frac{\pi}{n} \qquad [\alpha = -n, -(n-1), \cdots, (n-1), n] \quad (4\text{-}11.19)$$

The range of x is thus normalized to $[-\pi, +\pi]$ and divided into $2n$ equal intervals. Hence the number of data points is $2n + 1$. In our problem,

$$\varphi_j(x_\alpha)\varphi_k^*(x_\alpha) = e^{i(j-k)\alpha\pi/n} \qquad (4\text{-}11.20)$$

If now we sum over α with the understanding that the two limiting terms $\alpha = \pm n$ are taken with half weight, we obtain the left side of the identity (17), with

$$\theta = (j - k)\frac{\pi}{n} \qquad (4\text{-}11.21)$$

Since $j - k$ is an integer $\neq 0$ for $j \neq k$, the right side of (17) vanishes and we obtain

$$\varphi_j \cdot \varphi_k^* = \sum_{\alpha=-n}^{+n}{}' \varphi_j(x_\alpha)\varphi_k^*(x_\alpha) = 0 \qquad (4\text{-}11.22)$$

The prime in Σ' refers to the fact that the two limiting terms of the sum are taken with half weight.

The normalization factors of the functions $\varphi_k(x)$ are likewise simple. Putting $j = k$, we obtain

$$\varphi_k \cdot \varphi_k^* = \sum_{\alpha=-n}^{+n}{}' \varphi_k(x_\alpha)\varphi_k^*(x_\alpha) = 2n \qquad (4\text{-}11.23)$$

In conclusion we can say that the least-square solution of the problem of fitting the given data by an expansion of the form

$$\bar{y} = \sum_{k=-m}^{m} c_k e^{ikx} \qquad (m \leq n) \qquad (4\text{-}11.24)$$

is given explicitly by the formula

$$c_k = \frac{1}{2n} \sum_{\alpha=-n}^{n}{}' y_\alpha e^{-ikx\alpha} \qquad (4\text{-}11.25)$$

The same expansion can be given more conveniently in real form, separating the sine and the cosine terms. Every function can be written as the sum of an even and an odd function.

$$f(x) = \tfrac{1}{2}[f(x) + f(-x)] + \tfrac{1}{2}[f(x) - f(-x)] \qquad (4\text{-}11.26)$$

Hence it suffices to consider a function $g(x)$ with the symmetry property

$$g(-x) = g(x) \qquad (4\text{-}11.27)$$

and a function $h(x)$ with the symmetry property

$$h(-x) = -h(x) \qquad (4\text{-}11.28)$$

In the first case $c_{-k} = c_k$, while in the second case $c_{-k} = -c_k$. Hence in the first case we obtain

$$g(x) = \tfrac{1}{2}a_0 + a_1 \cos x + \cdots + a_m \cos mx \qquad (4\text{-}11.29)$$

(last term gets weight $\tfrac{1}{2}$ if $m = n$) with

$$a_k = \frac{2}{n} \sum_{\alpha=0}^{n}{}' g_\alpha \cos k\alpha \frac{\pi}{n} \qquad (4\text{-}11.30)$$

In the second case we obtain

$$h(x) = b_1 \sin x + \cdots + b_m \sin mx \qquad (4\text{-}11.31)$$

with

$$b_k = \frac{2}{n} \sum_{\alpha=1}^{n-1} h_\alpha \sin k\alpha \frac{\pi}{n} \qquad (4\text{-}11.32)$$

In the limiting case $m = n$ we have enough functions to fit the given data *exactly*. In this case $\bar{y}(x)$ is no longer an approximation but an *interpolation* of the given data. We obtain an analytical expression in the form of a trigonometric polynomial of lowest order which fits the given data exactly and which fits the functional values between with a certain accuracy. How great this accuracy is depends on the given function. The power of trigonometric interpolation lies in the fact that with increasing n the approximation $\bar{y}(x)$ approximates $y(x)$ with ever-diminishing oscillations. For every function of bounded variation, the trigonometric interpolation converges unlimitedly to the given $y = f(x)$ at every point of the given range, as the number of data points increases to infinity.

This behavior of the trigonometric interpolation is in marked contrast to the interpolation of equidistant data by powers. While we can always find a polynomial of $2n$th order which will fit $2n + 1$ equidistant data of a given finite range exactly, the error oscillations between the data points need not have the tendency to diminish in amplitudes, as n increases. Around the end of the range, the error oscillations may increase to infinity, thus giving an arbitrarily large error everywhere except in the data points. This happens with such a simple and regular function as

$$y = \frac{1}{1 + 25x^2} \qquad (4\text{-}11.33)$$

in the range between $[-1,+1]$, as shown by O. Runge.[1]

The trigonometric kind of interpolation is entirely free of this peculiar difficulty. The error oscillations do not have the tendency to increase toward the end points of the range but continue to maintain the same order of magnitude throughout the range. The trigonometric kind of interpolation is thus both analytically and practically vastly superior to the ordinary polynomial interpolation for data which are given equidistantly.

12. Interpolation by sine functions. Let $f(x)$ be an *odd* function. The formula (11.31) comes into operation, with $m = n - 1$.

$$h(x) = b_1 \sin x + b_2 \sin 2x + \cdots + b_{n-1} \sin (n - 1)x \qquad (4\text{-}12.1)$$

with

$$b_k = \frac{2}{n} \sum_{\alpha=1}^{n-1} y_\alpha \sin k_\alpha \frac{\pi}{n} \qquad (4\text{-}12.2)$$

The term $b_n \sin nx$ cannot be added, since the function $\sin nx$ has zeros at all the data points, which leaves b_n undetermined. Actually the number of data points is in effect not more than $n - 1$, since $h(0) = 0$ because of the odd character of the function, while the value $h(\pi)$ cannot come into evidence in view of the fact that all functions $\sin kx$ vanish at the point $x = \pi$. The expansion (11.31) will have

---

[1] Cf. the discussion of the Runge phenomenon in V, 15.

very slow convergence if $h(x)$ does not satisfy the boundary condition

$$h(\pi) = 0 \qquad (4\text{-}12.3)$$

It is thus advisable to apply the sine analysis to a given function only if that function vanishes at the points $x = 0$ and $x = \pi$. If these conditions are not satisfied, we can subtract a linear trend from the given function $f(x)$ by putting

$$h(x) = f(x) - (\alpha + \beta x) \qquad (4\text{-}12.4)$$

We determine α and β from the conditions $h(0) = h(\pi) = 0$, obtaining

$$\alpha = f(0) \qquad (4\text{-}12.5)$$

$$\beta = \frac{f(\pi) - f(0)}{\pi}$$

The sine analysis of $h(x)$ has now satisfactory convergence.

If it is unavoidable that the sine analysis of the original $f(x)$ shall be found, it is still advisable to proceed in the given manner and then add the theoretically known sine expansion of the function $\alpha + \beta x$.

Computationally the coefficients b_k of a sine analysis can be found as follows. We construct a matrix with the elements

$$b_{\alpha\beta} = \sin \alpha\beta(\pi/n) \qquad (4\text{-}12.6)$$

We write the given data y_α in a row.

$$(0), y_1, y_2, \cdots, y_{n-1}, (0)$$

and multiply this row with the successive rows of the $b_{\alpha\beta}$ matrix. The coefficients thus obtained are then multiplied by the constant $2/n$. This gives the successive amplitudes

$$b_1, b_2, \cdots, b_{n-1}$$

of the sine analysis (1).

The construction of the matrix (6) is simplified if we first set it up in *coded form*. We use the code numbers $1, 2, \cdots, n - 1$, and start out with the "guiding line"

$$0, 1, 2, \cdots, n - 1, -0, -1, -2, \cdots, (n-1), 0 \qquad (4\text{-}12.7)$$

We imagine that these elements are written along the periphery of a circle, the last zero coinciding with the first zero. We have thus a complete cycle which has no beginning and no end.

We now pick out every *first*, every *second*, every *third*, $\cdots$ element of the guiding line and write them as the successive rows of a matrix, omitting the element 0 at the beginning and at the end. Every row has thus $n - 1$ elements. For example, the case $n = 6$ leads to the following construction.

Guiding line:

$$0, 1, 2, 3, 4, 5, -0, -1, -2, -3, -4, -5, 0$$

Coded matrix:

$$
B_6 =
\begin{array}{c|ccccc}
 & y_1 & y_2 & y_3 & y_4 & y_5 \\
\hline
 & 1 & 2 & 3 & 4 & 5 \\
 & 2 & 4 & -0 & -2 & -4 \\
 & 3 & -0 & -3 & 0 & 3 \\
 & 4 & -2 & 0 & 4 & -2 \\
 & 5 & -4 & 3 & -2 & 1 \\
\end{array}
\qquad (4\text{-}12.8)
$$

The significance of these code numbers is as follows: We replace the code number k by $\sin(k\pi/n)$. Hence in our case ($n = 6$) the code numbers have to be replaced by the following actual numbers:

$$
\begin{aligned}
0 &= 0 \\
1 &= \sin\ 30° = 0.5 \\
2 &= \sin\ 60° = 0.86603 \\
3 &= \sin\ 90° = 1 \\
4 &= \sin 120° = 0.86603 \\
5 &= \sin 150° = 0.5
\end{aligned}
$$

The multiplication matrix is now constructed. If the data row $y_1, \cdots, y_5$ is multiplied by the successive rows of the matrix B, we obtain 5 quantities $b_1', \cdots, b_5'$. These b_i' are now multiplied by $2/n = 1/3$, thus obtaining the final Fourier coefficients

$$b_i = \frac{2}{n} b_i' \qquad (4\text{-}12.9)$$

13. Interpolation by cosine functions. Now let $f(x)$ be an *even* function. Then formula (11.29) comes into operation, with the following slight modification: the expansion now extends up to $m = n$, but the last term receives the factor $\frac{1}{2}$.

$$g(x) = \tfrac{1}{2}a_0 + a_1 \cos x + \cdots + a_{n-1} \cos(n-1)x + \tfrac{1}{2}a_n \cos nx \qquad (4\text{-}13.1)$$

with

$$a_k = \frac{2}{n} \sum_{\alpha=0}^{n}{}' y_\alpha \cos k\alpha \frac{\pi}{n} \qquad (4\text{-}13.2)$$

The multiplication matrix is now composed of the elements

$$a_{\alpha\beta} = \cos \alpha\beta \frac{\pi}{n} \qquad (4\text{-}13.3)$$

The coded matrix is identical with the previous matrix (12.8) with the only difference that the columns 0 and n cannot be omitted. The zero column—which belongs to the ordinate $y_0 = g(0)$—is composed of all zeros, while the column n—which belongs to the ordinate $y_n = g(\pi)$—is composed of the elements $0, -0, 0, -0, \cdots$. Moreover, there is a zero row, composed of all zeros, corresponding to the coefficient $\frac{1}{2}a_0$, which did not exist in the case of the sine analysis; similarly there is an nth row, composed of alternate zeros, $0, -0, 0, \cdots$, corresponding to the last coefficient $\frac{1}{2}a_n$. The complete multiplication matrix A thus becomes (for the example $n = 6$)

	$\frac{1}{2}y_0$	y_1	y_2	y_3	y_4	y_5	$\frac{1}{2}y_6$
	0	0	0	0	0	0	0
	0	1	2	3	4	5	-0
	0	2	4	-0	-2	-4	0
$A_6 =$	0	3	-0	-3	0	3	-0
	0	4	-2	0	4	-2	0
	0	5	-4	3	-2	1	-0
	0	-0	0	-0	0	-0	0

The code numbers have now the following significance: the code number k has to be replaced by $\cos(k\pi/n)$. Hence

$$
\begin{aligned}
0 &= \cos \quad 0° = 1 \\
1 &= \cos \quad 30° = 0.86603 \\
2 &= \cos \quad 60° = 0.5 \\
3 &= \cos \quad 90° = 0 \\
4 &= \cos 120° = 0.5 \\
5 &= \cos 150° = 0.86603
\end{aligned}
$$

After obtaining the coefficients a_k' we multiply by $2/n$ in order to obtain the final coefficients a_k.

$$a_k = \frac{2}{n} a_k'$$

In the chosen example ($n = 6$) all the a'_k are divided by 3.

The weight factor $\frac{1}{2}$. The cosine analysis shows a peculiarity which was not encountered in the sine analysis; this is the weight factor $\frac{1}{2}$ at the two ends. We must remember that the two limiting ordinates y_0 and y_n enter all calculations with the weight $\frac{1}{2}$. Moreover, the two limiting terms of the cosine expansion are likewise multiplied by the factor $\frac{1}{2}$. These irregularities do not occur in the sine expansion since there all these terms are zero.

Symmetry properties of the multiplication matrices A and B. Both matrices A and B, defined by (13.3) and (12.6), are symmetric, because of the interchangeability of α and β. They have, however, some valuable additional symmetry properties, caused by the symmetry properties of the sine and cosine functions in the four quadrants of the circle. We can benefit from these symmetry properties in reducing the number of multiplications by the factor 2 and even the factor 4. We notice from the interpretation of the code numbers that the elements k and $n - k$ are interrelated. The number of independent elements is only $n/2$, instead of n. Moreover, the columns k and $n - k$, and likewise the rows k and $n - k$ are interrelated, differing from each other only in sign. If we consider all the even columns separately and all the odd columns separately, and do the same with the even and odd rows, the entire matrix can be reduced to one-fourth of its previous size. We break the matrix into four smaller matrices of $n/2$ rows and columns, operating with the sums and differences of the ordinates y_k and y_{n-k}. By this procedure we gain by the factor 4 in the number of multiplications but we lose by the necessity of writing down a larger number of partial results. We separate the coefficients of even and odd order, obtaining each coefficient as the sum or difference of two partial results.[1]

Numerical checks. Numerical checks of the computations are provided by the fact that the Fourier synthesis of the obtained coefficients has to restore the original data. The orthogonality of the matrices A and B has the consequence that while the product of the y_i with the matrices A and B gives the Fourier coefficients a_k and b_k (except for the factor $2/n$), the product of the Fourier coefficients a_k with the matrix A, or the coefficients b_k with the matrix B, restores the original ordinates y_i.

---

[1] For further numerical details and practical examples cf. [10].

14. Harmonic analysis of equidistant data. Let a periodic function of the period $2l$ be given in the range $[-l,+l]$ by observing it at the $2n + 1$ equidistant points

$$x_\alpha = \alpha \frac{l}{n} \qquad (\alpha = -n, \cdots, +n) \qquad (4\text{-}14.1)$$

This function can be analyzed by laying a trigonometric polynomial of lowest order through the given ordinates. The functions of this expansion are

$$\cos k \frac{\pi}{l} x, \qquad \sin k \frac{\pi}{l} x \qquad (4\text{-}14.2)$$

They take the place of the previous functions $\cos kx$, $\sin kx$, which are adjusted to the range $[-\pi,+\pi]$. The complete analysis of the a_k and b_k coefficients occurs exactly according to the previous algorithm, using the ordinates $\frac{1}{2}[f(x_\alpha) + f(-x_\alpha)]$ for the cosine analysis, and the ordinates $\frac{1}{2}[f(x_\alpha) - f(-x_\alpha)]$ for the sine analysis.

In many problems of applied analysis the basic function is not periodic in itself but defined in a given finite interval of x, let us say between $x = 0$ and $x = l$. The method of trigonometric expansion may be employed as a tool of interpolating the given data by giving an analytical expression for the entire function $f(x)$, observed or tabulated in a discrete set of equidistant points only. Such an analytical expression may be of importance for evaluating $f(x)$ at points which lie between the data points. But even more frequently the value of such an analytical expression may lie in its *operational* advantages. Many operations of advanced analysis can be performed with exponential functions relatively easily, while the same operation with a given algebraic or transcendental function can only formally be indicated; the actual numerical answer would require a prohibitive amount of work. In such cases it is of inestimable value if the given function is first replaced parexically by a trigonometric expansion which gives a close approximation of that function. For this purpose the method of trigonometric interpolation can frequently be employed.

In such problems it is of decisive importance that the approximating trigonometric series shall have good convergence. This requires that function and first derivative shall return to the same values at

the beginning and the end of the chosen period. Since this condition is usually not satisfied at the two end points of the given range,
we cannot choose the given interval as the full period of a harmonic
analysis. If we choose the interval $[0,l]$ as the half period of a harmonic
analysis, defining $f(x)$ in the negative half as an even function
$f(-x) = f(x)$, we avoid the discontinuity of the function at the two
end points $\pm l$ of the range, since now $f(-l) = f(l)$. However,
$f'(-l) = -f'(l)$, and thus we have generally a discontinuity in the
first derivative. We obtain better convergence by subtracting a linear
function $\alpha + \beta x$ from the given $f(x)$, thus operating with a function

$$h(x) = f(x) - (\alpha + \beta x) \qquad (4\text{-}14.3)$$

which vanishes at the two end points $x = 0$ and $x = l$ [cf. (12.5),
replacing π by l]. If we now reflect this $h(x)$ as an *odd* function
$h(-x) = -h(x)$ and consider $2l$ as the full period of the harmonic
analysis, we have continuity of function *and first derivative* at the two
end points of the period, since

$$h(-l) = h(l) = 0, \qquad h'(-l) = h'(l) \qquad (4\text{-}14.4)$$

The expansion of $h(x)$ into a pure sine series

$$h(x) = \sum_{k=1}^{n-1} b_k \sin k \frac{\pi}{l} x \qquad (4\text{-}14.5)$$

where the b_k are obtained according to § 12, will now give satisfactory convergence, since the coefficients b_k decrease with the *third
power* of k (cf. § 5).

The importance of the method of trigonometric interpolation of
equidistant data can hardly be overestimated. It makes the mighty
tool of Fourier analysis accessible to difficult functions whose
Fourier coefficients, based on the original definition as definite
integrals [cf. (2.2)] are not calculable. The method of trigonometric
interpolation replaces these integrals by a simple summation process,
carried out over a relatively small number of equidistantly spaced
ordinates; (cf. also, VI, 18).

15. The error of trigonometric interpolation. It is not always of
advantage to use the classical Fourier coefficients as a unique frame
of reference in discussing the nature of trigonometric expansions.
It is true that the truncated Fourier series (3.1) is at every value of m

a best approximation in the sense of minimizing the average error by the method of least squares. However, the average error is not always the best gauge of the error of a certain approximation. Moreover, if the coefficients of this best approximation are not accessible because of the excessive labor involved in their evaluation, we pay a relatively small price if we replace the classical series by a modified series which is easily calculated, and whose error is nevertheless not essentially worse than the previous error. It will thus be of interest to compare the accuracy of the trigonometric series obtained by interpolation with the accuracy of the truncated Fourier series of the same number of terms.

Let us first assume that the given $f(x)$ does not contain more harmonic components than $n + 1$ cosine terms and $n - 1$ sine terms. Then the given $2n + 1$ equidistant ordinates—their actual number is $2n$ because of the boundary condition $f(x_{-n}) = f(x_n)$—determines $f(x)$ *exactly*, and the series obtained by trigonometric interpolation coincides with the truncated Fourier series of $2n$ terms, both of these series giving $f(x)$ without any error.

Let us now assume that the given $f(x)$ contains further harmonic components, the frequency spectrum going up to $2n$ instead of n. Hence $f(x)$ contains terms proportional to

$$\sin (n + k)x, \qquad \cos (n + k)x \qquad (k = 1, 2, \cdots, n) \qquad (4\text{-}15.1)$$

Now the Fourier coefficients of the truncated Fourier series take no account of the presence of these higher harmonics. They strictly separate the contribution of the various frequencies, thus giving the exact harmonic analysis of the given function.

The series obtained by trigonometric interpolation behaves differently. It is sensitive to the presence of these higher harmonics by converting them into lower frequencies. The error of the interpolation process must be of the form

$$\eta_n(x) = (\sin nx)u(x) \qquad (4\text{-}15.2)$$

since the error is zero at the data points, i.e., at the zeros of $\sin nx$. Now the trigonometric identities

$$\sin (n + k)x + \sin (n - k)x = 2 \cos kx \sin nx$$
$$\cos (n + k)x - \cos (n - k)x = -2 \sin kx \sin nx \qquad (4\text{-}15.3)$$

show that the frequency $n + k$ is equivalent to the frequency $n - k$

in all the data points. The trigonometric interpolation thus converts the frequency $n + k$ into the frequency $n - k$ and records an amplitude of this frequency with full strength and no phase shift in the cosine case, with full strength and phase shift of 180° in the sine case.

The same conversion phenomenon recurs between the frequencies $2n$ and $3n$, since the frequency $2n + k$ is equivalent to a frequency $2n - k = n + (n - k)$ and thus to the frequency $n - (n - k) = k$ and so on.

This "contamination" from the part of the higher frequencies shows that the classical Fourier coefficients are superior to the coefficients obtained by interpolation, if we are interested in the true harmonic spectrum of the given function. If, however, our aim is merely to *approximate* the given $f(x)$ by a harmonic series, we cannot say a priori that the error of the truncated series will necessarily be smaller than the error of the interpolated series. Both errors are of the same order of magnitude, since the error function $\eta_n(x)$ of the interpolated series can be conceived as a spectrally garbled version of the error of the truncated series. The *estimated* maximum error is the same in both cases. The *actual* maximum error may be in favor of the one or the other series.

Comparison of the Fourier series with the series obtained by trigonometric interpolation deserves more than passing attention. Derivation of the Fourier series by elementary means seems to indicate that the coefficients a_k, b_k of the infinite trigonometric series representing $f(x)$ cannot be anything else but the customary definite integrals (2.2). And thus the erroneous notion prevails that only the Fourier series gives "exact" results, while other series whose coefficients have been obtained by different means, are only "approximate" in nature. The mistake can be traced to the historical fact that a certain type of infinite limit process received overwhelming emphasis in the evolution of mathematics and thus acquired a feature of uniqueness which leads to wrong interpretations.

If we write down an infinite ortho-normal series of the form

$$f(x) = c_1 \varphi_1(x) + c_2 \varphi_2(x) + \cdots \qquad (4\text{-}15.4)$$

and then multiplying by $\varphi_k(x)$ and integrating term by term obtain

$$c_k = \int_a^b f(x) \varphi_k(x)\, dx \qquad (4\text{-}15.5)$$

the process seems unique. It is important, however, to keep certain *tacit assumptions* in mind which were not clear in Fourier's time which, however, became completely clarified when the exact limit concept emerged during the nineteenth century. Two remarks are of particular interest.

In the first place we have to realize that the "equal" sign in equation (4) is *not* used in the legitimate sense and is meaningful only if we know its significance. Lagrange objected to the thesis that a nonanalytical function can be represented by an infinite superposition of sine and cosine functions, arguing that these functions are analytical functions, and any number of them remains still analytical. The proper answer to Lagrange's objection is that the equal sign in an infinite expansion of the type (4) does not mean real equality. It means only that the right side, as we add more and more terms, comes nearer and nearer to the left side, and by adding up enough terms we can make the error in absolute value *as small as we wish, but not zero.* Hence the Fourier series, like all infinite expansions, does not give exact results, but only *arbitrarily close* results. It represents a never ending *approximation process.*

The second remark is concerned with the special type of infinite approximation which is achieved by an equation of the form (4), that is, an equation of the "dot-dot-dot type." We add up more and more terms which means in exact analysis that the step from n to $n + 1$ in this process occurs as follows. We add one more term, *without changing anything in the previous terms.* This, however, is by no means necessary. We could add one more term *and* at the same time change the coefficients of the previous terms. If we do so, we lose in simplicity, since now we have to give $n + 1$ new coefficients instead of *one* new coefficient. A one-dimensional sequence of terms changes to a triangular matrix. But this sacrifice in simplicity may lead to simplicity in another direction. For example, the usually inaccessible definite integrals which are needed for the evaluation of the Fourier coefficients may be replaced by the easily calculated sums which are demanded in the process of trigonometric interpolation. Moreover, the validity of the trigonometric expansion may be extended to a class of functions which go far beyond the class of absolutely integrable functions; (cf. § 8). These series differ from the Fourier series in the *manner of approximation* but *not* in the degree of accuracy which is of the "not absolute but arbitrarily close" type in all cases in which the series converges at all (cf. also VII, 5).

16. Interpolation by Chebyshev polynomials. The disadvantage of trigonometric interpolation of equidistant data is usually the fact that $f(x)$ is basically *not* a periodic function of x, and is made periodic only by definition. The best we can hope for is continuity of function and first derivative at the beginning and end of the period. The discontinuities in the higher derivatives make the resulting series relatively slowly convergent. This slow convergence is avoided by a modified form of the Fourier series which is frequently of eminent importance. Let the range of $f(x)$ be normalized to $[-1, +1]$. Let $f(x)$ be analytic inside this range and on the boundary, but without any further boundary conditions. We now make the transformation from x to a new variable θ, by the definition

$$x = \cos \theta \qquad (4\text{-}16.1)$$

The given $f(x)$ now becomes $f(\cos \theta) = \phi(\theta)$ and is thus transformed into a *genuinely* periodic function of the new variable θ.

If x changes between -1 and $+1$, the angle variable θ changes between 0 and π. Since, however, a change of θ to $-\theta$ leaves x unchanged, we possess the function $\phi(\theta)$ at any value of θ. It is an even function of θ.

$$\phi(-\theta) = \phi(\theta) \qquad (4\text{-}16.2)$$

Moreover, if $f(x)$ can be differentiated any number of times with respect to x, the transformed function $\phi(\theta)$ can be differentiated any number of times with respect to θ, at any point of the range, inclusive of the boundaries.

Under these circumstances the Fourier expansion of $\phi(\theta)$ will have much faster convergence than if the original $f(x)$ had been expanded into a Fourier series. This Fourier expansion has only cosine terms since $\phi(\theta)$ is an even function of θ.

$$\phi(\theta) = \frac{1}{2}\gamma_0 + \sum_{k=1}^{\infty} \gamma_k \cos k\theta \qquad (4\text{-}16.3)$$

Let us interpret this series in terms of the original variable x.

The functions $\cos k\theta$, if written in the variable x, become polynomials of x, called the Chebyshev polynomials $T_n(x)$; (cf. V, 20). They are alternatingly even and odd functions of x which can be generated on the basis of the recurrence relation

$$T_{n+1}(x) = 2xT_n(x) - T_{n-1}(x) \qquad (4\text{-}16.4)$$

starting with $\qquad T_0(x) = 1, \qquad T_1(x) = x$

The higher polynomials are tabulated in Table V.

The expansion (3), if written in the original variable x, becomes

$$f(x) = \sum_{k=0}^{\infty} \gamma_k T_k(x) \qquad (4\text{-}16.5)$$

The coefficients of this expansion are determined by the customary definite integrals.

$$a_k = \frac{2}{\pi} \int_0^\pi \phi(\theta) \cos k\theta \, d\theta \qquad (4\text{-}16.6)$$

Translated into the variable x we obtain

$$a_k = \frac{2}{\pi} \int_{-1}^{+1} f(x) T_k(x) \frac{dx}{\sqrt{1 - x^2}} \qquad (4\text{-}16.7)$$

Once more we can raise the objection that evaluation of these definite integrals will be frequently beyond our capacities. Hence we will change over to the method of trigonometric interpolation, described before in § 13. For this purpose the basic functional values y_α have to be given at the points

$$\theta_\alpha = \alpha \frac{\pi}{n} \qquad (\alpha = 0, 1, 2, \cdots, n) \qquad (4\text{-}16.8)$$

This means in the variable x that the data points are placed according to the law

$$x_\alpha = \cos \alpha \frac{\pi}{n} \qquad (4\text{-}16.9)$$

This is a strongly nonuniform distribution of the points of interpolation, which crowds the points near to the two end points $x = \pm 1$ of the range. The geometrical interpretation of the law (16.9) is that the semicircle of the radius 1 is divided into n equal parts and the points projected down on the x axis. The equidistant distribution in the angle variable θ causes a strongly nonequidistant distribution of the points in the projection.

The functional values

$$y_\alpha = f(x_\alpha) = f(\cos \alpha \frac{\pi}{n}) \qquad (4\text{-}16.10)$$

have to be given in these $n + 1$ generally irrational points. Apart from this inconvenience, the interpolation itself is an easily performed

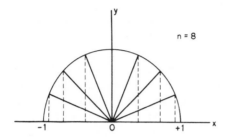

routine process, since we evaluate the coefficients with the help of the matrix (13.3), discussed before in § 13. We thus obtain the expansion

$$\bar{f}(x) = \tfrac{1}{2}a_0 + a_1 T_1(x) + \cdots + \tfrac{1}{2}a_n T_n(x) \qquad (4\text{-}16.11)$$

which can now be rearranged into an ordinary power series. The convergence of this series is far superior to the convergence of the ordinary Taylor series. In fact, the Taylor expansion may diverge completely, while the convergence of the expansion (11) is guaranteed by the fact that any continuous function of bounded variation can be expanded into a uniformly convergent Fourier series.

The nonequidistant distribution of the data points has a highly beneficial effect on the convergence of the resulting interpolation. While equidistant polynomial interpolation gives error oscillations which are strongly increased around the two ends of the range, proper crowding of the data points toward the two end points prevents the oscillations from becoming damaging. The error now oscillates with the same order of magnitude throughout the range. The resulting power series is thus distinguished by the fact that it approximates the given function with an absolutely smaller maximum error than approximations obtained by other polynomials. Moreover, the process of interpolation avoids evaluation of definite integrals, and is numerically simple and straightforward. Thus the process discussed here has many practical applications. It translates the outstanding analytical properties of the trigonometric type of interpolation into the realm of powers.

17. The Fourier integral. While the series named after Fourier was already well established at the time of Fourier—based on the pioneering work of J. Bernoulli, Euler, and Lagrange—the Fourier integral is the undisputed discovery of Fourier. It has become one of the most powerful tools of mathematical analysis, and is particularly fundamental in all problems pertaining to the input-output relation of electric networks. Fourier found that decomposition of arbitrary functions into harmonic components remains possible even if the realm of the function $f(x)$ extends on both sides to infinity. In this case the fundamental frequency converges to zero, and thus the process of summation changes into one of integration. Moreover, the limits of the integrals which define the Fourier coefficients are no longer $\pm\pi$ but $\pm\infty$.

The Fourier series analyzes a function of a definite finite range in terms of sine and cosine functions of given frequencies. If we extend the given function beyond its range of definition, one of two things can happen. The function may be a truly periodic function, or may exist in a finite interval $[-l,+l]$ only and we may force periodicity on $f(x)$ in order to make the Fourier series applicable for its representation. In the latter case the periodicity of the function is not given by nature, but is employed as a mathematical artifice only. If we use a physical instrument such as a wave analyzer for determination of the harmonic components of the function, we will in the first case actually obtain the coefficients of the Fourier series as measurable physical quantities. In the second case the situation is quite different. The wave analyzer does not recognize the given finite range of $f(x)$. The variable x represents now the time t, and the wave analyzer takes account of the given function not only during the finite time interval $2l$ but during all times. And thus the question arises, What happens to the harmonic analysis if the fundamental period of the analysis is not specified to a definite quantity but can become arbitrarily large?

We start with the function $y = f(x)$, defined in the range between $-l$ and $+l$, satisfying the Dirichlet conditions. This range can be transformed to the normal range $\pm\pi$ by the scale transformation $\xi = \pi x/l$. Going back to the original variable x, we can write the

Fourier series, now formulated for arbitrary limits $\pm l$. We will use the mathematically most convenient complex form of the Fourier series (cf. 2.4).

$$f(x) = \sum_{k=-\infty}^{+\infty} c_k e^{ik\pi x/l} \tag{4-17.1}$$

with

$$c_k = \frac{1}{2l} \int_{-l}^{+l} f(x) e^{-ik\pi x/l}\, dx \tag{4-17.2}$$

The customary "real" form of the series arises if we put

$$c_k = \tfrac{1}{2}(a_k - ib_k), \qquad c_{-k} = \tfrac{1}{2}(a_k + ib_k) \tag{4-17.3}$$

and combine the terms of the subscripts k and $-k$.

Although the function $f(x)$ was originally given in the realm $\pm l$ only, there is no reason why we should not enlarge the range to a more extended interval $\pm L$. Outside the original range we will define $f(x)$ as zero.

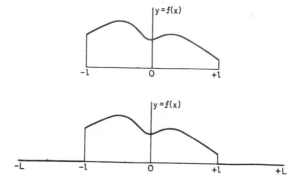

The new function $y = f(x)$, defined in the enlarged range, can again be expanded into a Fourier series. The formulas (1) and (2) hold again, replacing l by L.

$$f(x) = \sum_{k=-\infty}^{+\infty} c_k e^{ik\pi x/L} \tag{4-17.4}$$

$$c_k = \frac{1}{2L} \int_{-l}^{+l} f(x) e^{-ik\pi x/L}\, dx \tag{4-17.5}$$

Although the new series (4) analyzes the function $f(x)$ in entirely new frequencies and entirely new coefficients, it approaches nevertheless the same $f(x)$ for any x value between $\pm l$, while outside of that range (up to the bounds $\pm L$), the new series approaches the value zero.

Let us analyze the harmonic contents of the function $f(x)$. If a sine or a cosine function is written in the form $\sin 2\pi\nu t$, $\cos 2\pi\nu t$, we call ν—the number of vibrations per second— the "frequency" of the harmonic vibrations. The harmonic functions of the first series (1) can be written in the form

$$\cos k \frac{2\pi x}{2l} + i \sin k \frac{2\pi x}{2l}$$

Hence the frequencies present in our first analysis are

$$\nu_1 = \frac{1}{2l}, \qquad \nu_2 = \frac{2}{2l}, \qquad \cdots, \qquad \nu_k = \frac{k}{2l}, \qquad \cdots$$

The corresponding frequencies present in our second analysis become

$$\nu_1 = \frac{1}{2L}, \qquad \nu_2 = \frac{2}{2L}, \qquad \cdots, \qquad \nu_k = \frac{k}{2L}, \qquad \cdots$$

If L is twice as large as l, the fundamental frequency becomes *one-half* of the previous fundamental frequency. Hence the harmonics $1, 2, 3, 4$ of the previous analysis become now the harmonics $2, 4, 6, 8, \cdots$, and our complete analysis contains twice as many frequencies as before.

In order to study the distribution of the harmonic components, we will plot them in a graph. As abscissa we choose the frequency to which they belong. Hence the chart of a harmonic analysis looks as follows, separating real and imaginary parts of c_k.

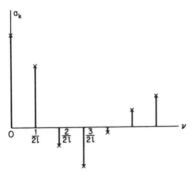
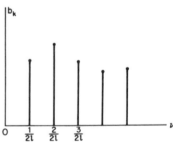

If the same analysis is performed in the second case, we get more lines, since more frequencies are present. We will agree to omit the constant factor $1/2L$ in (5), and merely plot the quantities

$$\gamma_k = \int_{-l}^{+l} f(x)e^{-2\pi i k x/2L} \qquad (4\text{-}17.6)$$

Then

$$c_k = \frac{\gamma_k}{2L} \qquad (4\text{-}17.7)$$

The amplitudes γ_k have the advantage that we need not change them as we make L larger and larger; we merely have to fill in newer and newer lines. In fact we notice that our graph merely picks out certain *definite ordinates of a universal function*. Let us define the following function of the continuous variable ν.

$$F(\nu) = \int_{-l}^{+l} f(x)e^{-2\pi i \nu x} \, dx \qquad (4\text{-}17.8)$$

This function contains all possible harmonic components of $f(x)$, no matter how small or large L may be.

$$\gamma_k = F\left(\frac{k}{2L}\right) \qquad (4\text{-}17.9)$$

Our previous graph is now replaced by the accompanying picture.

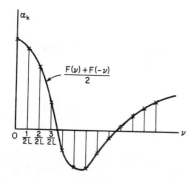
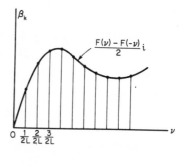

It is a remarkable fact that, no matter how irregular $f(x)$ was, the new function $F(\nu)$ is always a *continuous* and even *analytical* function (i.e., differentiable to any degree) of ν. The original $f(x)$ may have any number of discontinuities or other irregularities, but the new

$F(\nu)$ is nevertheless analytical for all (real or complex) values of ν. It is called the "Fourier transform" of the original function $f(x)$. It does not resemble the original function at all; it is merely *associated* with it, somewhat as the logarithm of a number is associated with the original number.

Let us now see what happens if we go with L to larger and larger values, increasing the period of our analysis gradually to infinity. Then the lines we have to fill in get constantly denser and denser, and in the limit, as L grows to infinity, there is no longer discrimination in favor of any specific frequencies, but *all frequencies are equally represented*. The previous line spectrum changes in the limit to a *continuous spectrum*. The function gives now *in its totality* the distribution of the harmonic amplitudes of $f(x)$.

What can we say about the Fourier *synthesis*, expressed in the form of the Fourier series? This series is in itself in very close relation to the transform function $F(\nu)$. Equations (4) and (5) can be combined in terms of the Fourier transform $F(\nu)$.

$$f(x) = \frac{1}{2L} \sum_{k=-\infty}^{+\infty} F\left(\frac{k}{2L}\right) e^{2\pi i k x / 2L} \qquad (4\text{-}17.10)$$

We define the following function of the frequency ν:

$$G(\nu) = F(\nu) e^{2\pi i \nu x} \qquad (4\text{-}17.11)$$

Then

$$f(x) = \frac{1}{2L} \sum_{k=-\infty}^{+\infty} G\left(\frac{k}{2L}\right) \qquad (4\text{-}17.12)$$

or, putting

$$\frac{1}{2L} = \varepsilon \qquad (4\text{-}17.13)$$

we can write

$$f(x) = \varepsilon \sum_{k=-\infty}^{+\infty} G(k\varepsilon) \qquad (4\text{-}17.14)$$

(The figure assumes that x is a given constant. The fact that the values of $G(\nu)$ are complex is discarded for the purpose of the illustration.) The larger the fundamental interval $2L$, the denser become the ordinates of $G(v)$ which participate in the Fourier sum. Let us now increase L to infinity. This means that ε tends to zero. By the fundamental theorem of integral calculus the sum (14)

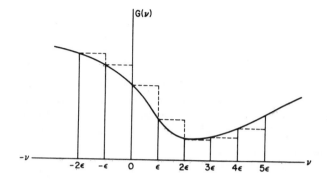

approaches a definite limit as ε decreases to zero. This limit is the *total area* under the function $y = G(v)$.

$$f(x) = \int_{-\infty}^{+\infty} G(v) \, dv = \int_{-\infty}^{+\infty} F(v)e^{2\pi i v x} \, dv \qquad (4\text{-}17.15)$$

The Fourier *sum* approaches more and more the Fourier *integral*. While the Fourier transform $F(v)$ resolves the given $f(x)$ into its harmonic components, the Fourier integral *synthesizes* these harmonic components to the original function. Observe the remarkable reciprocity of the two equations:

$$F(v) = \int_{-l}^{+l} f(x)e^{-2\pi i v x} \, dx$$

$$f(x) = \int_{-\infty}^{+\infty} F(v)e^{2\pi i x v} \, dv$$

The only discrepancy is that in the first integral the limits are $\pm l$, in the second $\pm\infty$. But this happened only because we have assumed that $f(x)$ was given only between $\pm l$ and was zero outside of that interval. We can stretch the realm of defining $f(x)$ to an arbitrarily large interval, and we can gradually approach infinity without invalidating our results. In the limit we obtain

$$F(\nu) = \int_{-\infty}^{+\infty} f(x)e^{-2\pi i \nu x}\, dx \qquad (4\text{-}17.16)$$

$$f(x) = \int_{-\infty}^{+\infty} F(\nu)e^{2\pi i x \nu}\, d\nu \qquad (4\text{-}17.17)$$

The reciprocity is now complete. The only thing we lose by the infinite limits of the first equation is that $F(\nu)$ is no longer an analytical or even necessarily continuous function of ν. Nor is it permissible to substitute anything but *real* values for ν. But $F(\nu)$ still approaches a definite limit, provided that $f(x)$ is absolutely integrable, which means the existence of

$$\int_{-\infty}^{+\infty} |f(x)|\, dx \qquad (4\text{-}17.18)$$

and that $f(x)$ is of bounded variation in every finite interval.

In order to display the characteristic similarities and dissimilarities of the Fourier series and the Fourier integral, we display once more the fundamental formulas in juxtaposition:

Fourier *series*; *discrete* spectrum (frequencies $1/2l, 2/2l, \cdots, k/2l$):

$$f(x) = \sum_{k=-\infty}^{+\infty} c_k e^{2\pi i k(x/2l)}, \qquad c_k = \frac{1}{2l}\int_{-l}^{+l} f(x)e^{-2\pi i k(x/2l)}\, dx$$

Fourier *integral*; *continuous* spectrum (all frequencies between $-\infty$ and $+\infty$):

$$f(x) = \int_{-\infty}^{+\infty} F(\nu)e^{2\pi i \nu x}\, d\nu, \qquad F(\nu) = \int_{-\infty}^{+\infty} f(x)e^{-2\pi i \nu x}\, dx$$

Question: What is a negative frequency?

Answer: If we operate with positive frequencies only, every frequency is associated with *two* functions, viz., $\sin 2\pi\nu x$ and $\cos 2\pi\nu x$. This duplicity can be avoided by introducing positive and negative frequencies and associating with ν the single function $e^{2\pi i \nu x}$. An arbitrary real vibration is then always an interaction of the two frequencies $+\nu$ and $-\nu$.

In practical applications of the Fourier transform it is frequently of greater advantage to introduce the "angular frequency" $\omega = 2\pi\nu$ instead of the ordinary frequency ν, and to write the formulas of harmonic analysis and synthesis in the following form.[1]

$$F(\omega) = \int_{+\infty}^{+\infty} f(x)e^{-i\omega x}\,dx \qquad (4\text{-}17.19)$$

$$f(x) = \frac{1}{2\pi} \int_{-\infty}^{+\infty} F(\omega)e^{i\omega x}\,d\omega$$

18. The input-output relation of electric networks. The Fourier transform is one of the most important tools of applied analysis. It plays a fundamental role in all electric network problems, but its applicability reaches over to a much wider field because the conditions which prevail in electric networks occur equally in many other problems of physics and engineering.

The basic situation can be described as follows: We have a measuring device of the galvanometer type which records a certain given function of the time t. We have thus two functions, viz., the "input function" or "signal," $f(t)$, and the "output function" or "response," $g(t)$. The measuring device may take the form of a telephone or loud-speaker or other communication device. It may equally take the form of a servo-mechanism which responds to a given "command." Or it may take the form of a boundary value problem in which the right side of the given partial differential equation is the input, and the solution is the output. In all these cases the general pattern of action can be characterized by a number of features which are common to the entire group of problems.

We have two functions $f(t)$ and $g(t)$, the input and the output. The given physical mechanism determines the relation between these two functions. We can conceive $g(t)$ as a certain "mapping" if the function $f(t)$ and we will say that this mapping is "of the C type" (C

---

[1] For historical reasons almost the entire mathematical literature discusses the theory of the Fourier integral on the basis of the so-called "Fourier double integral" which combines harmonic analysis and synthesis into a single operation, thus obscuring the issue to such an extent that the theory of the Fourier integral—in spite of its fundamental importance in extended fields of physics and engineering—becomes frequently one of the most elusive and least understood chapters of advanced analysis.

referring to communication) if the following general conditions are realized:

1. The mapping is *linear*. This means that if $f(t)$ is changed into $\alpha f(t)$, also $g(t)$ is changed into $\alpha g(t)$. Moreover, if $f(t)$ is a linear superposition of any number of functions:

$$f(t) = \alpha_1 f_1(t) + \alpha_2 f_2(t) + \cdots + \alpha_n f_n(t) \qquad (4\text{-}18.1)$$

then $g(t)$ is the same superposition of the corresponding mapped functions.

$$g(t) = \alpha_1 g_1(t) + \alpha_2 g_2(t) + \cdots + \alpha_n g_n(t) \qquad (4\text{-}18.2)$$

2. If $f(t)$ is a *periodic* function, written in the complex form:

$$f(t) = e^{i\omega t} \qquad (4\text{-}18.3)$$

then $g(t)$ reproduces this function with a mere factor of proportionality.

$$g(t) = \phi(\omega) e^{i\omega t} \qquad (4\text{-}18.4)$$

The factor $\phi(\omega)$ is generally *complex* and can thus be split into a real and imaginary part.

$$\phi(\omega) = A(\omega) + B(\omega)i \qquad (4\text{-}18.5)$$

The factor $\phi(\omega)$ is called the "transfer function." The physical significance of this function is that if the input is a strictly periodic function of definite frequency and constant amplitude, the output is likewise a periodic function of the same frequency but modified amplitude and modified phase. The modification of amplitude and phase is a function of the frequency and is characterized by $\phi(\omega)$. The response $g(t)$ to a strictly sinusoidal input function is called the "steady-state response" or "frequency response."

Now the Fourier integral resolves an arbitrary $f(t)$ into its harmonic components (cf. 17.19).

$$f(t) = \frac{1}{2\pi} \int_{-\infty}^{+\infty} F(\omega) e^{i\omega t} \, d\omega \qquad (4\text{-}18.6)$$

In view of the superposition principle, the communication device takes each one of these harmonic components and applies to it its own transfer function. Then the harmonic components are synthesized again. The result of this operation is

$$g(t) = \frac{1}{2\pi} \int_{-\infty}^{+\infty} F(\omega) \phi(\omega) e^{i\omega t} \, d\omega \qquad (4\text{-}18.7)$$

Hence we see that we can find to any given $f(t)$ the corresponding $g(t)$ if the transfer function $\phi(\omega)$ is given.

However, the relation between $f(t)$ and $g(t)$ can still be expressed in a totally different form, introducing a second fundamental function by which the action of the network may be characterized. We have seen (cf. 17.19) that by the Fourier's reciprocity theorem the relation (18.6) is reversible.

$$F(\omega) = \int_{-\infty}^{+\infty} f(t)e^{-i\omega t}\, dt \tag{4-18.8}$$

We can introduce this $F(\omega)$ in the equation (7) and write the result as follows.[1]

$$g(t) = \int_{-\infty}^{+\infty} f(\tau)K(t-\tau)\, d\tau \tag{4-18.9}$$

where

$$K(\xi) = \frac{1}{2\pi}\int_{-\infty}^{+\infty} \phi(\omega)e^{i\omega\xi}\, d\omega \tag{4-18.10}$$

Now we know in advance from the nature of a "response" that a response cannot anticipate but must *follow* the signal. The future can have no influence on the past. Hence $g(t)$ cannot depend on values of $f(t)$ which lie beyond the time moment $t = \tau$. For this reason we must demand for all "stable" mappings of the C type that the following additional condition shall hold.

$$K(t-\tau) = 0 \qquad \text{for } \tau > t$$

which means

$$K(\xi) = \frac{1}{2\pi}\int_{-\infty}^{+\infty} \phi(\omega)e^{i\omega\xi}\, d\omega = 0 \qquad (\xi < 0) \tag{4-18.11}$$

In this case the upper limit of the integral (9) becomes t instead of infinity.

$$g(t) = \int_{-\infty}^{t} f(\tau)K(t-\tau)\, d\tau \tag{4-18.12}$$

A physical signal $f(t)$ usually does not start at $t = -\infty$ but at a definite time moment which may be normalized to $t = 0$. In this case the output $g(t)$ becomes

$$g(t) = \int_{0}^{t} f(\tau)K(t-\tau)\, d\tau \tag{4-18.13}$$

---

[1] This move is not self-evident since a double limit process is involved. But the justification can be given for all functions which are of bounded variation.

In the new representation the fundamental quantity which characterizes the input-output relation of the given mechanism is the function $K(\xi)$. This function has a very definite physical significance. Let us employ a signal $f(t)$ which shall last only during the small time interval between $t = 0$ and $t = \varepsilon$. During this time interval $f(t)$ shall jump from 0 to the large constant value $1/\varepsilon$ and then fall back again to the value 0. We now make ε arbitrarily small. A signal of this type (comparable to a hammer blow) is called a "unit pulse"; the corresponding response is called the "pulse response."

Since now $f(t)$ lasts only for an infinitesimal time, during which the function $K(t - \tau)$ is practically a constant, we can write equation (13) in the form

$$g(t) = K(t) \int_0^\varepsilon f(\tau)\, d\tau = K(t) \qquad (4\text{-}18.14)$$

This shows that the function $K(t)$ has the significance of the *pulse response* of the mechanism.[1]

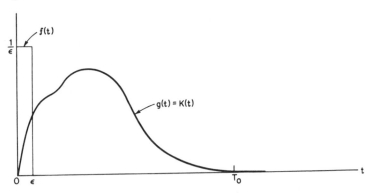

According to (10) the pulse response and the steady-state response are in a definite relation to each other. One is the Fourier transform of the other. If the transfer function $\phi(\omega)$ is given, we can obtain $K(t)$ on the basis of the Fourier integral (10). But then, by Fourier's reciprocity theorem we can also put

$$\phi(\omega) = \int_0^\infty K(t) e^{-iwt}\, dt \qquad (4\text{-}18.15)$$

---

[1] For historical reasons electrical engineers frequently prefer to characterize the transient behavior of a network by the "unit step function response." The two responses are in a simple relation to each other inasmuch as the pulse response is the time derivative of the step function response.

(the lower limit being zero, since $K(t)$ vanishes for negative t). Splitting real and imaginary parts we obtain according to (5),

$$A(\omega) = \int_0^\infty K(t) \cos \omega t \, dt$$

(4-18.16)

$$B(\omega) = -\int_0^\infty K(t) \sin \omega t \, dt$$

This shows that $A(\omega)$ is always an *even*, $B(\omega)$ always an *odd* function of ω.

$$A(-\omega) = A(\omega), \qquad B(-\omega) = -B(\omega) \qquad (4\text{-}18.17)$$

Going back to equation (10) and writing it in real form we obtain

$$K(t) = \frac{1}{\pi} \left[\int_0^\infty A(\omega) \cos \omega t \, d\omega - \int_0^\infty B(\omega) \sin \omega t \, d\omega \right] \quad (4\text{-}18.18)$$

However, the stability condition (11) establishes a further relation between $A(\omega)$ and $B(\omega)$, viz., that for all positive t we must have

$$\int_0^\infty A(\omega) \cos \omega t \, d\omega = -\int_0^\infty B(\omega) \sin \omega t \, d\omega \quad (4\text{-}18.19)$$

Hence it suffices to know *either* the real *or* the imaginary part of the transfer function.

$$K(t) = \frac{2}{\pi} \int_0^\infty A(\omega) \cos \omega t \, d\omega = -\frac{2}{\pi} \int_0^\infty B(\omega) \sin \omega t \, d\omega \quad (4\text{-}18.20)$$

19. Empirical determination of the input-output relation. It is frequently possible to determine the frequency response of a mechanism by direct observations. We bring the mechanism in forced vibrations by using a sinusoidal input function and waiting until the transient response has died away. If we then measure the amplitude of the output and the phase shift between the two functions, we have obtained $\phi(\omega)$ for a given frequency ω. The measurements have to be repeated for a sufficiently wide range of frequencies.

In many cases, application of a pulse as an input is more convenient. By observing the output we are in possession of the function $K(t)$ whose Fourier transform gives $\phi(\omega)$. The difficulty is only that the physical realization of a very sudden "hammer blow" is frequently not possible without doing serious damage to the given mechanism.

We have to be satisfied with a more general situation; we use some observed function $f(t)$ as input, and obtain some observed function $g(t)$ as output. From these data we have to obtain the pulse response $K(t)$ or the frequency response $\phi(\omega)$ by calculation. Although we will leave the shape of $f(t)$ arbitrary, we still assume one general feature in which it resembles a pulse: it shall start from zero at $t = 0$ and it shall come down to zero again, after a certain finite time t_0.

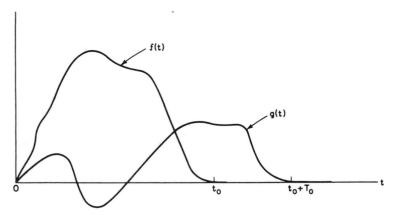

There is one further feature of the pulse response which was not included in our general discussions. All the devices used in communications and servo-mechanisms have the property that they dissipate the energy which was transferred to them by the input pulse. Hence $K(t)$ does not extend to infinity but becomes practically zero after a certain time T_0, which we want to call the "memory time" of the device.[1] Since we have assumed that the function $f(t)$ vanishes after the time interval t_0, we can now add that the output $g(t)$ will practically vanish after the time interval

$$T_1 = t_0 + T_0 \qquad (4\text{-}19.1)$$

Let us now imagine that the same $f(t)$ was applied again and again, before and after $t = 0$, in intervals of T_1, i.e., with the starting times $t = 0, \pm T_1, \pm 2T_1, \cdots$. Then $f(t)$ becomes *periodic* with the period

[1] This memory time is not a sharply defined quantity since $K(t)$ becomes zero only *asymptotically*. But for any given accuracy a certain T_0 exists beyond which $K(t)$ becomes negligible. The greater the demanded accuracy, the greater is the memory time T_0.

T_1 and the same is true of $g(t)$. Yet within the time interval T_1 no change occurred in either of the functions, since we have waited until $g(t)$ became zero. But now we can analyze both $f(t)$ and $g(t)$ in terms of harmonic functions by expanding both functions in a Fourier series of the base T_1:

$$f(t) = \sum_{k=-\infty}^{+\infty} c_k e^{ik2\pi t/T_1} \qquad g(t) = \sum_{k=-\infty}^{+\infty} \bar{c}_k e^{ik2\pi t/T_1} \qquad (4\text{-}19.2)$$

Practically only a finite number of terms will be present in both expansions, and the coefficients c_k and $\bar{c}_k$ will be determined by trigonometric interpolation rather than by integration (cf. § 11 to § 13). Our input function is now a superposition of strictly sinusoidal functions, and the same can be said of the output. By the definition of the transfer function $\phi(\nu)$ we obtain

$$\phi\left(\frac{k}{T_1}\right) = \frac{\bar{c}_k}{c_k} \qquad (k = 0, \pm 1, \pm 2, \cdots) \qquad (4\text{-}19.3)$$

Although $\phi(\nu)$ is thus obtained only at a discrete set of equidistant points, we can interpolate between by using linear (or quadratic) interpolation.

Even so, the difficulty remains that the given $f(t)$ may have been too smooth for our purpose. If $f(t)$ does not contain enough harmonic overtones, determination of the ratio (3) may be rendered impossible beyond a certain k, because of the smallness of the denominator. Then we have to stop with a $\phi(\nu)$ which is still far from being negligibly small. However, it suffices that we shall be able to establish the *asymptotic law* of $\phi(\nu)$. The speed with which $\phi(\nu)$ decreases to zero is at least ν^{-1}. If the last few observable $\phi(\nu)$ fit the pattern

$$\phi(\nu) = \frac{ia_1}{\nu} \qquad (4\text{-}19.4)$$

or more accurately the pattern

$$\phi(\nu) = \frac{ia_1}{\nu} + \frac{a_2}{\nu^2} \qquad (4\text{-}19.5)$$

we can be satisfied, since for all frequencies beyond the range of our measurements the value of $\phi(\nu)$ can be obtained by extrapolation.

If the frequency response was evaluated in this fashion, the pulse response too can be found, on the basis of the same data. We use the device that we repeat the input pulse in regular intervals T_0. The Fourier series of this pulse ("delta function") is the divergent series

$$f(t) = \frac{1}{T_0} \sum_{k=-\infty}^{+\infty} e^{2\pi ikt/T_0} \qquad (4\text{-}19.6)$$

This series, although useless in itself, becomes convergent by applying the transfer function to each of the frequencies present.

$$K(t) = \frac{1}{T_0} \sum_{k=-\infty}^{+\infty} \phi\left(\frac{k}{T_0}\right) e^{2\pi ikt/T_0} \qquad (4\text{-}19.7)$$

While this infinite series converges, the convergence may be too slow for practical purposes, particularly if the asymptotic pattern of $\phi(\nu)$ is of the type (4). We speed up the convergence by putting

$$K(t) = A_1 e^{-\alpha t} + A_2 t e^{-\alpha t} + K_1(t) \qquad (4\text{-}19.8)$$

The transfer function of $K(t)$ becomes

$$\phi(\nu) = \frac{A_1}{\alpha + 2\pi i\nu} + \frac{A_2}{(\alpha + 2\pi i\nu)^2} + \phi_1(\nu) \qquad (4\text{-}19.9)$$

Hence for large ν,

$$\phi_1(\nu) = \phi(\nu) + \frac{A_1 i}{2\pi\nu} + \frac{A_2 + \alpha A_1}{(2\pi\nu)^2} \qquad (4\text{-}19.10)$$

Let us now choose

$$\begin{aligned} A_1 &= -2\pi a_1 \\ A_2 &= -4\pi^2 a_2 - \alpha A_1 \end{aligned} \qquad (4\text{-}19.11)$$

Then the asymptotic law (5) is absorbed by the correction terms, and $\phi_1(\nu)$ decreases to zero with the *third power* of ν^{-1}. The expansion (7) of the function $K_1(t)$ has now sufficiently quick convergence. The exponent α may be chosen as

$$\alpha = 2\pi/T_0 \qquad (4\text{-}19.12)$$

because $e^{-2\pi}$ is sufficiently small to be considered negligible.

20. Interpolation of the Fourier transform. The infinite limits of the Fourier transform can often be changed into finite limits, which is sometimes of great operational advantage. Even if $f(t)$ extends to infinity, we may subtract from it some suitably chosen function $f_0(t)$ which imitates the asymptotic behavior of $f(t)$ for large values of t. We then replace the original function $f(t)$ by the new function $f_1(t) = f(t) - f_0(t)$ which is practically zero beyond a certain $t = \pm l$. Thus we obtain

$$F(\nu) = F_1(\nu) + F_0(\nu) \tag{4-20.1}$$

where $F_0(\nu)$ is the Fourier transform of the suitably chosen function $f_0(t)$, while

$$F_1(\nu) = \int_{-l}^{+l} f_1(t) e^{-2\pi i \nu t} \, dt \tag{4-20.2}$$

This new function has the remarkable property that it suffices to give a certain set of fundamental data as "key values" from which all other values of $F_1(\nu)$ are obtainable by interpolation.

We assume that $f_1(t)$ is a function of bounded variation defined in the range $\pm l$. Hence $f_1(t)$ can be expanded into a convergent Fourier series.

$$f_1(t) = \sum_{k=-\infty}^{+\infty} c_k e^{2\pi i k t / 2l} \tag{4-20.3}$$

where

$$c_k = \frac{1}{2l} \int_{-l}^{+l} f_1(t) e^{-\pi i k t / l} \, dt = \frac{1}{2l} F_1 \left(\frac{k}{2l} \right) \tag{4-20.4}$$

If the infinite expansion (3) is introduced in (2) and we integrate term by term, we obtain

$$F_1(\nu) = \frac{\sin 2\pi \nu l}{\pi} \sum_{k=-\infty}^{+\infty} F_1 \left(\frac{k}{2l} \right) \frac{(-1)^k}{2l\nu - k} \tag{4-20.5}$$

This formula is an interpolation (or extrapolation) formula. It gives $F_1(\nu)$ for all (real or complex) values of ν in the form of a convergent infinite expansion, expressed in terms of the basic ordinates

$$y_k = F_1(k/2l)$$

If ν is of the form

$$\nu = \frac{k}{2l} + \varepsilon$$

and ε is small, $F_1(\nu)$ will be determined predominantly by y_k and its immediate neighbors. But if ε is around the halfway between two key points, the convergence of the series (5) is very slow. We can speed up the convergence by the following device. We define a set of ordinates u_k by the following recurrences:

$$
\begin{aligned}
u_0 &= y_0 \\
u_1 &= y_1 - u_0 & u_{-1} &= y_{-1} - u_0 \\
u_2 &= y_2 - u_1 & u_{-2} &= y_{-2} - u_{-1} \\
&\;\vdots & &\;\vdots
\end{aligned}
\tag{4-20.6}
$$

Then the sum (5) can be replaced by the faster converging sum

$$
F_1(\nu) = \frac{\sin 2\pi\nu l}{\pi} \left[-\frac{u_0}{2l\nu + 1} - \sum_{k=0}^{\infty} \frac{(-1)^k u_k}{(k - 2l\nu)(k - 2l\nu + 1)} \right.
$$

$$
\left. + \sum_{k=1}^{\infty} \frac{(-1)^k u_{-k}}{(k + 2l\nu)(k + 2l\nu + 1)} \right]
\tag{4-20.7}
$$

Here the weights of the ordinates away from the central ordinate fall off *quadratically* with the distance and the number of terms needed for satisfactory convergence becomes reasonably small.[1]

[1] Cf. [10], p. 441.

21. Interpolatory filter analysis. We have seen in §18 that the frequency response of a communication device is the Fourier transform of its pulse response. In view of the fact that the pulse response $K(t)$ is not an arbitrary function of t but a function which vanishes for all negative values of t, the frequency response is by no means a freely choosable function but a function which has to satisfy some very definite analytical conditions, expressed in the symmetry relations (18.17) and in the integral condition (18.19). Now in the construction of electric filters we would like to obtain a $\phi(\nu)$ of prescribed properties, and the question arises to what extent we can satisfy certain given filter properties, without violating the analytical conditions which are demanded by the general nature of the C type kind of mapping.

The interpolation formula of the previous section is of great value in the discussion of this problem. We have seen that the pulse response of any stable network dissipates the input energy and thus comes to practically zero after a certain time T_0, the "memory time" of the network. Consequently the function $\phi(v)$ is of a very definite character. Not only is the lower limit of the integration zero, but the upper limit is a finite time T_0.

$$\phi(v) = \int_0^{T_0} K(t)e^{-2\pi i v t}\, dt \qquad (4\text{-}21.1)$$

If we now transform t to a new variable t_1 by the transformation

$$t = t_1 + \tfrac{1}{2}T_0 \qquad (4\text{-}21.2)$$

and put

$$K(t_1 + \tfrac{1}{2}T_0) = K_1(t_1) \qquad (4\text{-}21.3)$$

we obtain

$$\phi(v) = e^{-\pi i v T_0}\phi_1(v) \qquad (4\text{-}21.4)$$

where

$$\phi_1(v) = \int_{-T_0/2}^{T_0/2} K_1(t_1)e^{-2\pi i v t_1}\, dt_1 \qquad (4\text{-}21.5)$$

This expression is of the form (20.2) studied before, the limit l being replaced by $T_0/2$. Hence the interpolation formula (20.5) becomes valid again. The key values are

$$y_k = \phi_1(k/T_0) \qquad (4\text{-}21.6)$$

They can be prescribed with considerable freedom, except for the reality condition

$$y_{-k} = y_k^* \qquad (4\text{-}21.7)$$

and the convergence condition

$$\sum_{k=0}^{\infty} |y_k| = \text{finite} \qquad (4\text{-}21.8)$$

It is of interest to observe that while not even an arbitrarily small *continuous* portion of $\phi(v)$ can be freely prescribed, yet it is permissible to prescribe $\phi_1(v)$ practically freely in an infinity of equidistant points, belonging to the frequencies

$$v = 0, \qquad v_0 = 1/T_0, \qquad 2v_0, \qquad 3v_0, \qquad \cdots \qquad (4\text{-}21.9)$$

The question remains whether the interpolation of $\phi_1(\nu)$ *between* these points will be sufficiently smooth. Now the function generated by one single ordinate $y_k = 1$ is given by

$$\varphi_k(\nu) = \frac{\sin \pi(T_0\nu - k)}{\pi(T_0\nu - k)} \qquad (4\text{-}21.10)$$

It is of the character of the Dirichlet kernel (2.8). In view of the secondary maxima of this function the concentration to the immediate neighborhood of $\nu = k\nu_0$ is not very strong. The consequence is that the interpolation is smooth only if the y_k follow a smooth pattern. Near a discontinuity, however, the cumbersome Gibbs oscillations will come into play (cf. § 9). But we have seen how beneficially the large amplitudes of these oscillations can be cut down by the method of σ smoothing. This smoothing can be applied to our present problem. It consists in taking the local arithmetic average of $\phi(\nu)$ between $\nu + 1/T_0$ and $\nu - 1/T_0$. The result of this operation is that the function (10) is replaced by the function

$$\overline{\varphi}_k(\nu) = \frac{1}{2\pi} \left[Si\,\pi(T_0\nu - k + 1) - Si\,\pi(T_0\nu - k - 1) \right] \quad (4\text{-}21.11)$$

where

$$Si(\xi) = \int_0^\xi \frac{\sin x}{x}\, dx \qquad (4\text{-}21.12)$$

This function is called the "sine integral"; it is a well-investigated and well-tabulated function.[1] The new function is satisfactorily concentrated to the immediate neighborhood of the central maximum.

The effect of this smoothing operation, if applied to the equation (5), is that $K_1(t_1)$ is to be replaced by

$$\bar{K}_1(t_1) = K_1(t_1) \frac{\sin (2\pi t_1/T_0)}{2\pi t_1} T_0 \qquad (4\text{-}21.13)$$

The function $K_1(t_1)$ itself is expressible in terms of the prescribed ordinates y_k, according to (20.3).

$$K_1(t_1) = \frac{1}{T_0} \sum_{k=-\infty}^{+\infty} y_k e^{-2\pi ikt/T_0} \qquad (4\text{-}21.14)$$

[1] *Tables of the Sine, Cosine and Exponential Integrals*, Nat. Bur. Standards, Vols. I, II.

One of the classical problems of filter analysis is the construction of an "ideal low-pass filter." This filter would absorb all frequencies beyond a certain cutoff frequency ν_c but would let through all frequencies below ν_c without any change of amplitude or phase. The nearest solution of this problem can be given by the interpolatory treatment and subsequent elimination of the Gibbs oscillations by σ smoothing. The pulse response associated with this filter comes out as follows

$$
\begin{aligned}
\bar{K}(_1 t_1) &= \frac{\sin 2\pi\nu_c t_1 \cos (\pi t_1/T_0)}{2\pi t_1} \qquad (|t_1| \le T_0/2) \\
&= 0 \qquad\qquad\qquad\qquad\quad (|t_1| \ge T_0/2)
\end{aligned}
\qquad (4\text{-}21.15)
$$

In all our discussions we were concerned with the function $\phi_1(\nu)$ and not with the original function $\phi(\nu)$. The relation between these two functions is given by (4). By definition the significance of $\phi(\nu)$ is given as follows:

Input:
$$f(t) = e^{2\pi i\nu t}$$

Output:
$$
\begin{aligned}
g(t) &= \phi(\nu)e^{2\pi i\nu t} \\
&= \phi_1(\nu)e^{2\pi i\nu(t - T_0/2)}
\end{aligned}
$$

We see that the output constructed on the basis of $\phi_1(\nu)$ *lags behind* the input by the constant time delay $T_0/2$. Hence the freedom we have in constructing filters of prescribed characteristics is restricted by the fact that there is an inevitable *constant time lag*, equal to one-half of the memory time, associated with the use of the filter. In many communication problems this time lag is of no further consequence. But if we cannot permit an adequate time lag and have to cut down T_0 to a too insignificant amount, we gradually lose accuracy in realizing the filter characteristics, because the points at which $\phi(\nu)$ can be prescribed grow too far apart—on account of the largeness of the basic frequency $\nu_0 = 1/T_0$—and we miss important details of the desired curve.

22. Search for hidden periodicities. In certain meteorological and astronomical problems, in analysis of tides, and in all situations in which hidden periodicities are suspected, the following mathematical problem is encountered. We know that a certain function $f(t)$ is resolvable into components which are strictly periodic, although the periodicities are generally not in a harmonic ratio to each other. We want to obtain the unknown frequencies and amplitudes of each of the components. What we have at our disposal are a large number of observations, taken at equidistant time intervals. Our conclusions have to be drawn from the information conveyed to us by these data.

We will assume that the total number of observational data is the odd number $2N + 1$. Moreover, we will put the time moment $t = 0$ in the mid-point of our data. Finally, if our observations were made in intervals of τ, we will rescale the original variable t_1 to a new variable

$$t = \frac{1}{\tau} t_1 \qquad (4\text{-}22.1)$$

The observations now belong to the time moments

$$t_k = 0, \quad \pm 1, \quad \pm 2, \quad \cdots, \quad \pm N \qquad (4\text{-}22.2)$$

The readings shall be denoted by

$$f_k = f(t_k) \qquad (4\text{-}22.3)$$

The function $f(t)$ is generally of the following form:

$$f(t) = \sum_{\alpha=1}^{j} (A_\alpha \cos \theta_\alpha t + B_\alpha \sin \theta_\alpha t) \qquad (4\text{-}22.4)$$

The number of periodic components, denoted by j, is usually not known in advance. Nor can we say anything about the range of the angular frequencies $\omega_\alpha = \theta_\alpha / \tau$. However, we can say in advance that in the new variable t the angular frequencies θ_α have to be restricted by the inequality

$$\theta_\alpha < \pi \qquad (4\text{-}22.5)$$

because, if θ_α surpasses this limit, two frequencies $\pi + \beta$ and $\pi - \beta$ cannot be distinguished. We will put

$$\theta_\alpha = \frac{\pi}{N} p_\alpha \qquad (4\text{-}22.6)$$

and let p_α range between 0 and N.

We separate the sine and the cosine functions by forming the sums and differences

$$f(t) + f(-t) = 2 \sum_{\alpha=1}^{j} A_\alpha \cos \theta_\alpha t$$

$$f(t) - f(-t) = 2 \sum_{\alpha=1}^{j} B_\alpha \sin \theta_\alpha t$$

(4-22.7)

Correspondingly our ordinates are separated into two groups:

$$u_k = f_k + f_{-k}$$
$$v_k = f_k - f_{-k}$$

$(k = 0, 1, 2, \cdots, N)$ (4-22.8)

We will employ the method of the Fourier transform (cf. 17.8) but adapted to *summation* instead of integration. We transform the original set of u_k data into a new set of $N + 1$ cosine amplitudes a_k and the v_k data into a set of $N - 1$ sine amplitudes b_k. These amplitudes are obtained by multiplying the basic data by a prescribed matrix, containing the cosines and sines of the multiples of π/N.

$$a_k = \sum_{\alpha=0}^{N}{}' u_\alpha \cos \frac{\pi}{N} \alpha k$$

$$b_k = \sum_{\alpha=1}^{N-1} v_\alpha \sin \frac{\pi}{N} \alpha k$$

(4-22.9)

The symbol Σ' refers to the fact that the first and the last functional data u_0 and u_N enter the sum with *half weight* only. These a_α and b_α can be plotted as a "line spectrum," belonging to integer values $p = \alpha$ of the continuous parameter p. The entire further analysis will be based on these two new sequences.

Let us first consider the case that the given frequencies θ_α are such that all the p_α of the relation (6) become *integers*. Then the amplitudes (9) give us directly the solution of our problem. Most of the a_k and b_k will be zero. If a certain a_k or b_k is not zero, this indicates that the frequency

$$\theta = \frac{\pi}{N} k$$

(4-22.10)

is present in our data. The associated cosine amplitude becomes

$$A = \frac{1}{N} a_k \qquad (4\text{-}22.11)$$

while the associated sine amplitude becomes

$$B = \frac{1}{N} b_k \qquad (4\text{-}22.12)$$

In the general case the exact p value associated with a certain frequency θ_α will lie between two integers m and $m + 1$, recognizable by the fact that the regular $\pm$ pattern of the amplitudes a_k and b_k is occasionally interrupted by a $++$ or $--$ pattern, We now put

$$p_\alpha = m + \varepsilon \qquad (4\text{-}22.13)$$

and obtain ε by *interpolation*. For this purpose we use the "second sum" method (cf. 3-5.28), described in III, 5. By this method the mutual interference of peaks is greatly diminished. The ratio q of two neighboring second sums is formed (cf. 3-5.30) and ε is obtained on the basis of (3-5.31). An excellent check on the accuracy of our observations is provided by the fact that the position of the peaks of both the a_k and the b_k amplitudes must be the same. Hence we have two independent determinations of θ_α, once using the a_k and once the b_k.

After obtaining ε we now possess the frequency θ_α due to the relation

$$\theta_\alpha = \frac{\pi}{N} (m + \varepsilon) \qquad (4\text{-}22.14)$$

Moreover, the amplitudes A_k, B_k associated with this frequency can be calculated on the basis of the Fourier amplitudes a_m and b_m.

$$A_\alpha = \frac{a_m}{N} \frac{\pi\varepsilon}{\sin \pi\varepsilon}, \qquad B_\alpha = \frac{b_m}{N} \frac{\pi\varepsilon}{\sin \pi\varepsilon} \qquad (4\text{-}22.15)$$

This method of isolating periodic components operates very satisfactorily if the total number of observations $2N + 1$ is sufficiently large. It is necessary that two p_α values of the relation (6) shall be separated by at least 4 units in order to isolate two neighboring peaks. If they move nearer together, their mutual interference increases and it becomes gradually more difficult to separate them properly.

If we cannot count on a sufficiently large value of N and the peaks move closer together, an altogether different approach is advocated. As we have seen earlier (cf. § 21), the focusing power of the function (2.8) can be greatly increased by the method of the σ smoothing. In our present problem this method amounts to a modification of the basic data u_k and v_k which appear in the Fourier sums (9). Before forming the Fourier sums we modify the given data by applying to them a properly chosen weight factor, according to the following definitions:

$$\bar{u}_k = u_k \sigma_k, \qquad \bar{v}_k = v_k \sigma_k \qquad (4\text{-}22.16)$$

where

$$\sigma_k = \frac{\sin (k\pi/N)}{k\pi/N} \qquad (4\text{-}22.17)$$

The a_k, b_k of (9) are now calculated with the help of these new u_α, v_α.

The result of this modification is that the function $(\sin \pi x)/\pi x$ is replaced by the function

$$S(x) = \frac{1}{2\pi} [Si(x + \pi) - Si(x - \pi)] \qquad (4\text{-}22.18)$$

The secondary peaks of this function are 4.8, 2.0, 1.1%, $\cdots$ of the central peak compared with 21.7, 12.8, 9.1%, $\cdots$ of the previous function (cf. Figure in § 7). This demonstrates the stronger focusing power of the new function, and the practical independence of the new maxima. At the same time the peaks are now somewhat broader than they were before. But this broadening of the lines has the beneficial effect that a parabola of second order can be laid through the maximum amplitude and two of its neighbors, obtaining the position and magnitude of the true maximum by this parabolic interpolation. We operate with the maximum amplitude a_m and its left and right neighbor a_{m-1}, a_{m+1}. Then the ε of the formulas (13) and (14) becomes

$$\varepsilon = \frac{1}{2} \frac{a_{m+1} - a_{m-1}}{2a_m - (a_{m+1} + a_{m-1})} \qquad (4\text{-}22.19)$$

while the maximum ordinate a_μ becomes

$$a_\mu = a_m + \frac{\varepsilon}{4} (a_{m+1} - a_{m-1}) \qquad (4\text{-}22.20)$$

Then

$$A_\alpha = \frac{a_\mu}{N}\, 1.6963 \qquad (4\text{-}22.21)$$

(The numerical factor represents the reciprocal of $S(0)$). The same calculation holds for the sine amplitude B_α, replacing the a_k by the b_k.[1]

23. Separation of exponentials. In the previous section the question was discussed of analyzing a function which is composed of periodic components. In radioactive decay measurements a similar problem arises, but here the periodic functions are replaced by exponential functions. Given is a function of the following form.

$$f(x) = A_1 e^{-\lambda_1 x} + A_2 e^{-\lambda_2 x} + \cdots + A_m e^{-\lambda_m x} \qquad (4\text{-}23.1)$$

Our aim is to find the "decay constants" λ_i and the amplitudes A_i. On the surface the problem is very similar to the previous one: the frequency ω_i has changed to an imaginary frequency $i\omega_i$. But in actual fact the two problems are far apart because the exponential functions in no way display the remarkable orthogonality properties of periodic functions. For this reason it is not enough to discuss the purely mathematical solution of the problem but we have to go into its numerical aspects as well. We will first deal with the theoretical side of the problem and then come to its numerical implications.

We know that an ordinary differential equation with constant coefficients has as its solution a linear combination of exponentials. The solution is thus of the form (1). The same is true of a difference equation with constant coefficients in which the operation d/dx is replaced by the operation $\Delta/\Delta x$. Generally, let us assume that the function $y = f(x)$ has the property that there exists a definite linear relation between $m + 1$ equidistant ordinates:

$$c_0 f(x) + c_1 f(x+h) + c_2 f(x+2h) + \cdots + c_m f(x+mh) = 0 \qquad (4\text{-}23.2)$$

The solution of this functional equation is

$$f(x) = A_1 e^{-\lambda_1 x} + A_2 e^{-\lambda_2 x} + \cdots + A_m e^{-\lambda_m x} \qquad (4\text{-}23.3)$$

---

[1] The numerical schemes discussed in this section are much more accurate than the traditional "periodogram" methods which fail to recognize the fundamental connection between the problem of hidden periodicities and the theory of the Fourier transform.

Substitution in (2) shows that the exponents λ_i are obtainable by the following method. We put

$$e^{-\lambda_i h} = \xi_i \tag{4-23.4}$$

and obtain an algebraic equation for the determination of the ξ_i:

$$c_0 + c_1 \xi + c_2 \xi^2 + \cdots + c_m \xi^m = 0 \tag{4-23.5}$$

The m roots of this equation are $\xi = \xi_1, \xi_2, \cdots, \xi_m$. Then, taking the natural logarithms,

$$\lambda_i = -\frac{1}{h} \log \xi_i \tag{4-23.6}$$

The smallest number of ordinates we must have at our disposal is $2m$ since both the exponents λ_i and the amplitudes A_i are unknown. Let these ordinates be denoted by $y_1, y_2, \cdots, y_{2m}$. We now obtain the c_i by solving the following set of linear equations (we normalize the highest coefficient c_m to 1):

$$
\begin{aligned}
y_1 c_0 + y_2 c_1 + \cdots + y_m c_{m-1} + y_{m+1} &= 0 \\
y_2 c_0 + y_3 c_1 + \cdots + y_{m+1} c_{m-1} + y_{m+2} &= 0 \\
&\vdots \\
y_m c_0 + y_{m+1} c_1 + \cdots + y_{2m-1} c_{m-1} + y_{2m} &= 0
\end{aligned}
\tag{4-23.7}
$$

In actual fact we will not choose neighboring data for $y_1, y_2, \cdots$, but will try to make h as large as possible in order to reduce the effect of observational errors. If for example our task is to separate $m = 4$ exponentials and we have 40 equidistant data at our disposal, we will divide these data into $2m = 8$ groups of 5 consecutive observations each. The sum of the 5 ordinates in each group provides us with 8 new ordinates $\bar{y}_\alpha$ which will be used as the y_α of the system (7). Then the h of the equations (2) and (6) is not the time interval Δx of two neighboring measurements, but 5 times that interval.

Let us assume that we have found the solution of the system (7), then formed the algebraic equation (5) and found its roots. Finally we have obtained the λ_i according to (6). We now come to the *second* half of our problem, viz., the determination of the amplitudes A_i. For this purpose only m data are needed. Let us assume that we use the data belonging to the time moments

$$x = 0, \quad h', \quad 2h', \quad \cdots, \quad (m-1)h'$$

These data shall be denoted by $y_0, y_1, \cdots, y_{m-1}$. The equations to be solved are now

$$\begin{aligned}
A_1 &+ A_2 &+ \cdots + A_m &= y_0 \\
A_1 p_1 &+ A_2 p_2 &+ \cdots + A_m p_m &= y_1 \\
A_1 p_1^2 &+ A_2 p_2^2 &+ \cdots + A_m p_m^2 &= y_2 \\
&\vdots & &\vdots \\
A_1 p_1^{m-1} &+ A_2 p_2^{m-1} &+ \cdots + A_m p_m^{m-1} &= y_{m-1}
\end{aligned} \qquad (4\text{-}23.8)$$

if we put

$$p_1 = e^{-\lambda_1 h'}, \quad p_2 = e^{-\lambda_2 h'}, \quad \cdots, \quad p_m = e^{-\lambda_m h'} \qquad (4\text{-}23.9)$$

The problem (8) is known as the "problem of weighted moments." It is a fundamental problem of applied analysis which appears in numerous combinations (cf. e.g., VI, 13). It is solvable by a particularly simple and elegant numerical algorithm. We first construct the "fundamental polynomial"

$$\begin{aligned}
F_m(p) &= (p - p_1)(p - p_2) \cdots (p - p_m) \\
&= f_0 + f_1 p + \cdots + f_{m-1} p^{m-1} + p^m
\end{aligned} \qquad (4\text{-}23.10)$$

Then a second polynomial $G_{m-1}(p)$ is constructed by multiplying $F_m(p)$ by a polynomial which proceeds in *reciprocal* powers of p:

$$(f_0 + f_1 p + \cdots + p^m)(y_0 p^{-1} + y_1 p^{-2} + \cdots + y_{m-1} p^{-m})$$

The numerical scheme is similar to our ordinary longhand multiplication scheme of two decimal numbers. We omit the powers of p and write down only the coefficients $f_1, f_2, f_3, \cdots$, (they correspond to the digits of a decimal number). They are in succession multiplied by y_0. Then we multiply similarly by y_1 but we *indent* one place to the left. Then we multiply equally by y_2, again indenting one place to the left. Thus we continue until y_{m-1} is reached. All elements corresponding to negative powers of p are omitted. For this reason the scheme does not extend to the left beyond the first column. Hence the second row contains only $m - 1$ elements, the third row only $m - 2$ elements, $\cdots$, the last row only 1 element. Finally (just as in ordinary multiplication), the sum of each column is formed:

$$\begin{array}{llll}
f_1 y_0, & f_2 y_0, & \cdots, \ f_{m-1} y_0, & y_0 \\
f_2 y_1, & f_3 y_1, & \cdots, \ y_1 & \\
\vdots & & & \\
f_{m-1} y_{m-2}, & y_{m-2} & & \\
y_{m-1} & & & \\
\hline
g_0, & g_1, & \cdots, \ g_{m-2} & g_{m-1}
\end{array} \qquad (4\text{-}23.11)$$

Sum:

This yields the polynomial

$$G_{m-1}(p) = g_0 + g_1 p + g_2 p^2 + \cdots + g_{m-1} p^{m-1} \quad (4\text{-}23.12)$$

Now the unknowns A_i of our problem (8) are obtained by a simple substitution scheme. We substitute $p = p_i$ in $G_{m-1}(p)$ and we do likewise in the derivative of $F_m(p)$. The ratio of these two numbers gives A_i:

$$A_i = \frac{G_{m-1}(p_i)}{F'_m(p_i)} \quad (4\text{-}23.13)$$

Another method of obtaining the solution of (8) is to construct the inverse of the matrix of the linear system (8). If the elements of the inverse are denoted by q_{ik}, we obtain

$$q_{ik} = \frac{F_{k-1}(p_i)}{F'_m(p_i)} \quad (i, k = 1, 2, \cdots, m) \quad (4\text{-}23.14)$$

where the polynomial $F_{k-1}(p)$ is formed with the help of the *last* elements of the fundamental polynomial $F_m(p)$, e.g.,

$$F_2(p) = f_{m-2} + f_{m-1} p + p^2$$

The successive numerators of q_{ik} (for fixed i) can be obtained also by synthetic division (cf. I, 8). If $F_m(p)$ is divided by the root factor $p - p_i$, the successive coefficients of the division scheme yield $F_0(p_i), F_1(p_i), \cdots, F_{m-1}(p_i)$.

This simple and straightforward mathematical solution of the separation problem would hardly indicate what enormous practical difficulties arise if we try to apply it to physical problems. The difficulty is caused by the fact that the solution of the equations (7) for the coefficients c_i succeeds only if the data are given with *excessive accuracy*. If the separation of four or five exponentials is demanded, the associated linear system (7) becomes so strongly skew-angular that an accuracy of 6 to 8 significant figures would be needed in the y_α for their successful solution. Such an accuracy is completely unrealistic if compared with the actual accuracy of decay measurements. Even the separation of three exponentials might encounter already unsurmountable difficulties. The following example is well suited to demonstrate the surprising numerical snags which may develop on account of the exceedingly nonorthogonal behavior of exponential functions.

The following set of 24 decay observations were obtained in time intervals of 3 minutes, i.e., 0.05 hour, if the hour is accepted as the unit of time in our problem; hence $\Delta x = 0.05$, starting with the time moment $x = 0$:

k	y_k	k	y_k	k	y_k	k	y_k
1	2.51	7	0.77	13	0.27	19	0.11
2	2.04	8	0.64	14	0.23	20	0.10
3	1.67	9	0.53	15	0.20	21	0.09
4	1.37	10	0.45	16	0.17	22	0.08
5	1.12	11	0.38	17	0.15	23	0.07
6	0.93	12	0.32	18	0.13	24	0.06

The observations are considered as accurate to $\frac{1}{2}$ unit of the second decimal.

We do not know how many exponentials are present in our data. We first try a separation in *three* exponentials. Hence we divide our data into 6 groups, taking in each group the sum of four consecutive ordinates: $k = 1$ to 4, 5 to 8, etc. The new data become (omitting the decimal point which for our present purposes is irrelevant):

$$759, \quad 346, \quad 168, \quad 87, \quad 49, \quad 30$$

Hence the equations for the determination of the c_i become

$$759c_0 + 346c_1 + 168c_2 + 87 = 0$$
$$346c_0 + 168c_1 + 87c_2 + 49 = 0$$
$$168c_0 + 87c_1 + 49c_2 + 30 = 0$$

Dividing by the factor of c_0 we get

$$C_0 + 0.4559c_1 + 0.2213c_2 + 0.1146 = 0$$
$$C_0 + 0.4855c_1 + 0.2514c_2 + 0.1416 = 0$$
$$C_0 + 0.5179c_1 + 0.2917c_2 + 0.1786 = 0$$

By subtraction the following two equations result for the determination of c_1 and c_2:

$$0.0296c_1 + 0.0301c_2 + 0.0270 = 0$$
$$0.0324c_1 + 0.0402c_2 + 0.0370 = 0 \qquad (4\text{-}23.15)$$

If the second equation is multiplied by 0.75, it becomes

$$0.0242c_1 + 0.0302c_2 + 0.0277 = 0$$

Now the accuracy of our observations is such that we cannot guarantee the accuracy of these coefficients with more than 1 unit in the second decimal. But this shows that the two equations (15) are redundant within the errors of our observations. Under no circumstances can they serve for an independent determination of c_1 and c_2. This shows that our measurements will be describable by only *two* exponentials. Hence we divide our data now into four groups of 6 ordinates each and form the sum within each group. This gives the new data

$$964, \quad 309, \quad 115, \quad 51$$

and we obtain the two equations

$$964c_0 + 309c_1 + 115 = 0$$
$$309c_0 + 115c_1 + 51 = 0$$

with the solution

$$c_0 = 0.1640, \qquad c_1 = -0.8839$$

This yields the quadratic equation

$$\xi^2 - 0.8839\xi + 0.1640 = 0$$

which has the two roots

$$\xi_1 = 0.619, \qquad \xi_2 = 0.265$$

The h of the formula (6) is now $6 \cdot 0.05 = 0.3$. The two exponents λ_1 and λ_2 thus become

$$\lambda_1 = 4.45, \qquad \lambda_2 = 1.58$$

We now come to the determination of the two amplitudes A_1 and A_2. In principle two observations are sufficient for this purpose. We will use, however, the abundance of our data for checking up on the constancy of A_i. We can conceive the solution of the linear set (8) in the following light. We choose a linear combination of m equidistant data which reduces the weight factors of all the A_α to zero, with the exception of the single weight A_i. Now if this linear combination is systematically applied in succession to the entire table of data, we succeed in isolating the single exponential function $A_i e^{-\lambda_i x}$. For example in our problem we may combine the data $k = 1$ and 5 for the annihilation of A_2. But then we can take the same linear combination of the data $k = 2$ to 6, then 3 and 7, and so on. We thus obtain the successive values of the function $A_1 e^{-\lambda_1 x}$. If now we multiply by $e^{\lambda_1 x}$, we should get a constant, except for the inevitable scatter caused by the inaccuracy of the data. If we can detect a linear trend in this series, we will lay a least-square straight line through the set, of the form

$$A + Bx = A(1 + \beta x), \qquad \beta = \frac{B}{A}$$

which in view of the smallness of β may be written in the form $Ae^{\beta x}$ and used as a correction of the previous value of λ_1:

$$Ae^{\beta x}e^{-\lambda_1 x} = Ae^{-(\lambda_1 - \beta)x}$$

Carrying through this process in our example, we find that no linear trend can be detected beyond the natural scatter of the results. For example the first 12 consecutive values for the determination of A_1, apart from a constant factor which can be applied later, become:

887, 870, 872, 842, 872, 867, 861, 869, 862, 908, 890, 914

(Here we stopped, since the magnification factor due to multiplication by $e^{\lambda_1 x}$ is here already 14.4). By taking the arithmetic mean of these values we obtain 876.1 which leads to

$$A_1 = 2.202$$

Similar is the situation concerning A_2. Here too a consecutive set of 16 values for A_2 failed to show a linear trend. The arithmetic mean gave

$$A_2 = 0.305$$

and thus the final result becomes

$$f(x) = 2.202e^{-4.45x} + 0.305e^{-1.58x} \qquad (4\text{-}23.16)$$

Here we have the mathematical representation of our data with the help of two exponentials. As a check we now generate this $f(x)$ at all the 24 data points $x = 0, 0.05, 0.1, 0.15, \cdots, 1.15$, obtaining the following table:

k	y_k	k	y_k	k	y_k	k	y_k
1	2.507	7	0.769	13	0.270	19	0.114
2	2.044	8	0.639	14	0.230	20	0.100
3	1.672	9	0.533	15	0.197	21	0.088
4	1.370	10	0.447	16	0.173	22	0.079
5	1.126	11	0.376	17	0.148	23	0.070
6	0.929	12	0.318	18	0.130	24	0.063

If we compare this table with the original table of our data, we notice that the deviation is never larger than 0.005, except in the single instance of $k = 5$, where the error reaches the magnitude 0.006. We can characterize the "average deviation" by taking the square root of the following quantity: the sum of the squares of the individual deviations, divided by 24. What we get is only 0.0026, which is well within the error limits of our data. Hence we can consider the law (16) as a perfectly satisfactory representation of our data.

In actual fact we have failed dismally in our task, since the given data were constructed on the basis of the mathematical law

$$f(x) = 0.0951e^{-x} + 0.8607e^{-3x} + 1.5576e^{-5x} \qquad (4\text{-}23.17)$$

Not only did we lose the relatively weakly represented exponent, but the presence of this component has a contaminating influence on the other two exponents by reducing the exponent 3 to 1.58 and the exponent 5 to 4.45. Moreover, the approximate ratio 1 : 2 of the amplitudes was distorted to 1 : 7. Our result is thus completely unsatisfactory. And yet our solution is "numerically equivalent" to the true solution since it gives a perfectly satisfactory fit of the data within the experimental errors. It would be idle to hope that some other modified mathematical procedure could give better results, since the difficulty lies not with the manner of evaluation but with the extraordinary sensitivity of the exponents and amplitudes to very small changes of the data, which no amount of least-square or other form of statistics could remedy. The only remedy would be an increase of accuracy to limits which are far beyond the possibilities of our present measuring devices.

Under these circumstances we will aim at a more modest task. Let us assume that we have some preliminary information about the λ_i, because of which we know the number of exponential components and an approximate value of each exponent. Our problem is now reduced to the determination of the A_i, together with a correction of the given λ_i.

Again we will proceed in a manner similar to the last phase of our previous problem. On the basis of the given λ_i we construct such a linear combination of the data which isolates the single exponential function $A_i e^{-\lambda_i x}$. By applying the same linear combination to a homologous set of data we will obtain a series of let us say $2n$

equidistant values of this function. We divide the sum of the first n values by the sum of the second n values, which gives

$$e^{\lambda_i n \Delta x}$$

from which λ_i can be determined. We now have one corrected λ_i, and we use this new value of λ plus the previously given λ to carry out the correction scheme for another of the λ_i. Thus we continue, taking into account at every new step the corrected values of the exponents. Finally we have corrected all the λ_i and then we come to the determination of the A_i in a similar manner as we have done before.

Let us apply this procedure to our previous example. The given values of the exponents shall be

$$\lambda_1 = 1.2 \text{ (instead of 1)}, \quad \lambda_2 = 2.7 \text{ (instead of 3)}, \quad \lambda_3 = 6 \text{ (instead of 5)}$$

The result of our calculations is as follows. We obtain

$$f(x) = 0.041e^{-0.50x} + 0.79e^{-2.73x} + 1.68e^{-4.96x} \qquad (4\text{-}23.18)$$

Compared with the correct expression (17) the result is again disappointing. But again we get a perfectly satisfactory fit of the given data within the accuracy of our measurements. This example shows that we cannot expect satisfactory results in the separation problem of exponentials even if we know in advance the approximate values of the decay constants and we have merely to refine these values. On the other hand, the picture would have been quite different if our data had had a ten times greater accuracy.

24. The Laplace transform. Closely related to the theory of the Fourier transform is a somewhat different transform which plays a superior role in many problems of mathematical physics and is of particular usefulness in problems of network analysis. This is the so-called "Laplace transform," defined by

$$\mathcal{L}(z) = \int_0^\infty e^{-xz} f(x)\, dx \qquad (4\text{-}24.1)$$

If imaginary values are assigned to z, then (1) assumes the form (17.16) which defined the Fourier transform of a function $f(x)$. This function is now specified to exist only between 0 and ∞, while for negative values we put $f(x) = 0$. However, the new lower limit 0 instead of $-\infty$ has a profound effect on the analytical nature of

$\mathcal{L}(z)$. While the previous $F(\nu)$ (cf. 17.16) could generally be defined only for *real* values of ν, the new transform is an analytical function of the complex variable z throughout the complex half plane $\mathscr{R}(z) > 0$, provided that $f(x)$ is an absolutely integrable function of x in the infinite interval $(0,\infty)$. Even if $f(x)$ does not decrease to infinity fast enough to be absolutely integrable, it is possible that $f(x)e^{-\varepsilon x}$ is already integrable, with an arbitrarily small ε. Then we can put

$$z = \varepsilon + z_1, \qquad f_1(x) = f(x)e^{-\varepsilon x} \qquad (4\text{-}24.2)$$

and consider the Laplace transform of $f_1(x)$. We then lose for our original variable z a small strip of the width ε next to the imaginary axis. Since, however, ε can be made arbitrarily small, the analytical nature of $\mathcal{L}(z)$ everywhere to the right of the imaginary axis is still guaranteed. For example, for $f(x) = 1$ we obtain

$$\mathcal{L}(z) = \int_0^\infty e^{-zx}\,dx = \frac{1}{z} \qquad (4\text{-}24.3)$$

This function of z, originally defined only for $\mathscr{R}(z) > 0$, can now by analytical continuation be extended to the entire complex plane, with the only exception of the singular point $z = 0$.

By a simultaneous change of the signs of x and z we can define a Laplace transform $\mathcal{L}_1(z)$ for the *negative* half plane

$$\mathcal{L}_1(z) = \int_{-\infty}^0 e^{-xz}f_1(x)\,dx \qquad (4\text{-}24.4)$$

and then combine $\mathcal{L}(z) + \mathcal{L}_1(z)$ to one single function. The common boundary of the two definitions is the imaginary axis, where we now get

$$\mathcal{L}(i\omega) + \mathcal{L}_1(i\omega) = \int_{-\infty}^{+\infty} e^{-i\omega x}f(x)\,dx \qquad (4\text{-}24.5)$$

This relation shows that the Fourier transform of an absolutely integrable function of the range $[-\infty, +\infty]$ can be conceived as the sum of two Laplace transforms, taken along the imaginary axis, one analytical in the right, the other analytical in the left half plane.

There is no point-to-point relation between the function $f(x)$ and the Laplace transform $\mathcal{L}(z)$. The value of $\mathcal{L}(z)$ at any point depends on the totality of values $f(x)$. There is one exception, however. Let us assume that $f(x)$ is analytical around the origin $x = 0$. Furthermore, let us assign a large positive real value to z. Then the smallness of e^{-zx} for any x which is appreciably different from zero reduces the realm of integration to a very small neighborhood of $x = 0$. Under these circumstances we obtain

$$\int_0^\infty f(x)e^{-xz}\,dx = f(0)\int_0^\infty e^{-xz}\,dx = \frac{f(0)}{z} \qquad (4\text{-}24.6)$$

More generally, expanding $f(x)$ around $x = 0$ into a convergent Taylor series:

$$f(x) = c_0 + c_1 x + c_2 x^2 + \cdots \qquad (4\text{-}24.7)$$

and integrating term by term we obtain for the corresponding Laplace transform, for large values of z (assuming that the real part of z is positive):

$$\mathcal{L}(z) = \frac{c_0}{z} + \frac{c_1}{z^2} + \frac{2!c_2}{z^3} + \frac{3!c_3}{z^3} + \cdots \qquad (4\text{-}24.8)$$

Hence there is a point-to-point correlation between the infinitesimal neighborhood of the point $x = 0$ of $f(x)$ and the infinitesimal neighborhood of the point $z = \infty$ of the Laplace transform. However, the series (8) need not converge for any value of z.

25. Network analysis and Laplace transform. Application of the Laplace transform to the equations of electric networks opened a new perspective which was anticipated by Heaviside's operational methods. The justification of Heaviside's ingenious intuitions was given when the discovery was made that the Laplace transform of the input and output functions of electric networks automatically satisfied the algebraic equations which Heaviside postulated in a less rigorous manner. A simultaneous set of ordinary differential equations with constant coefficients is transformed into a set of simple linear algebraic equations which connects the Laplace transforms of the original unknowns. If the block diagram of the electric network is given, we can solve the input-output relation of electric networks by solving a set of simultaneous linear equations. The solution gives the input-output relation in the form of a ratio of two algebraic polynomials $p(z)$ and $q(z)$. The fundamental quantity is here the pulse response of the network. The Laplace transform of the unit pulse becomes 1. The Laplace transform of the output function—i.e., of the pulse response—becomes

$$\mathcal{L}(z) = \int_0^\infty e^{-zt}\, K(t)\, dt \tag{4-25.1}$$

It is this $\mathcal{L}(z)$ which is directly derivable from the given block diagram of the network:

$$\mathcal{L}(z) = \frac{p(z)}{q(z)} \tag{4-25.2}$$

where the order of $q(z)$ is at least one higher than the order of $p(z)$.

The first physical quantity which is directly at our disposal is the frequency response of the network. The frequency response $\phi(\omega)$, expressed in terms of the angular frequency (cf. 17.19), is obtained by replacing z by $i\omega$.

$$\phi(\omega) = \mathcal{L}(i\omega) \tag{4-25.3}$$

The frequency response is thus available without any integration, purely on the basis of knowing the elements of the network and their interconnection. The pulse response, on the other hand, is the result of a Fourier transform (cf. 18.10):

$$K(t) = \frac{1}{2\pi} \int_{-\infty}^{+\infty} \mathcal{L}(i\omega)e^{i\omega t}\, d\omega \tag{4-25.4}$$

However, this integration—as it was observed by Heaviside—need not be carried out because we can resolve the ratio (2) into partial fractions.

$$\frac{p(z)}{q(z)} = \sum_{i=1}^{n} \frac{a_i}{z-\lambda_i} \tag{4-25.5}$$

Here λ_i are the roots of the denominator, while

$$a_i = \frac{p(\lambda_i)}{q'(\lambda_i)}$$

Since we know by elementary integration that

$$\int_0^\infty e^{\lambda_i t - zt}\, dt = \frac{1}{z-\lambda_i} \tag{4-25.6}$$

we find

$$K(t) = \sum_{i=1}^{n} a_i e^{\lambda_i t}$$ (4-25.7)

This is Heaviside's celebrated "expansion theorem" (in slight modification, since Heaviside was interested in the step function response which is the integral of the pulse response).

26. Inversion of the Laplace transform. Although Heaviside's expansion theorem solves the problem of the transient response of electric networks, yet in actual practice we frequently prefer other solutions which circumvent finding the generally complex roots of an algebraic equation. In fact, in the case of complicated block diagrams encountered in servo-problems of the guided missile type, even the construction of the polynomials $p(z)$, $q(z)$ as actual algebraic operators may encounter insuperable practical difficulties, although it may not be too difficult to obtain the numerical value of $\mathcal{L}(z)$ for any prescribed value of z. Hence we can ask the following question, and it is a question which is encountered in many other physical situations beside the case of electric networks: Given the Laplace transform of a function $f(t)$ (the "indicial function"), at certain suitably chosen points of the complex plane, find the original (indicial) function $f(t)$.

Since the range of t is infinite, the scale in which t is measured is in principle arbitrary. In actual fact the proper scaling of t is of utmost importance. The original scale in which t is measured may be entirely inadequate to our problem. In network problems $K(t)$ has the significance of the pulse response of the network. In view of the practically finite "memory time" T_0 of the network, $K(t)$ is of interest only up to a certain T_0 but generally this may not be known in advance. If the time unit is chosen as too large, the essential portion of $K(t)$ is crowded into a very small interval around $t = 0$. If, on the other hand, the time unit is chosen as too small, the essential portion of $K(t)$ spreads out too far. In both cases the convergence of the expansions to be studied in the next sections will suffer.

The memory time of the network could well be chosen as a natural unit of time. But in the absence of this knowledge we can still introduce a reasonable, although not necessarily safe normalization of the time scale. Even without detailed investigation of the distribution of the roots of the denominator $q(x)$ we can say something about their general location. Let the factor of z^n of $q(z)$ be normalized to 1. Then the factor q_{n-1} of the power z^{n-1} has the following significance. It gives the *negative sum* of all the roots. Hence

$$\frac{-q_{n-1}}{n} = \frac{\lambda_1 + \lambda_2 + \cdots + \lambda_n}{n} \tag{4-26.1}$$

gives the center of mass of all the generally complex roots of the denominator (i.e., of the characteristic frequencies of the network). Since the λ_i must all have a negative real part in order to make the network stable, the coefficient q_{n-1} must be *positive*. We can assign any value to q_{n-1} by the scale transformation

$$z_1 = \alpha z \tag{4-26.2}$$

where we have denoted with z_1 the original inadequately normalized variable of the Laplace transform, while z denotes the final variable with which we want to operate. We will choose the scale factor α by the condition that the new q_{n-1} shall become n. This means that the new time unit is chosen in such a way that the center of mass of the characteristic vibrations becomes -1. The transformation (2) of z is parallelled by the transformation of the time scale.[1]

$$t_1 = t/\alpha \tag{4-26.3}$$

We will consider four different solutions of the inversion problem. According to the given circumstances we may prefer, for numerical purposes, one or the other of these solutions. But each of these solutions has its own mathematical merits.

27. Inversion by Legendre polynomials. Our first solution utilizes equidistant *real* values of the variable z. We make the transformation

$$e^{-t} = \xi \tag{4-27.1}$$

The merit of this transformation is that it transforms the infinite range $[0, \infty]$ of t into the finite range $[0, 1]$ of ξ. We consider now

---

[1] A mere substitution of the form (2) in $\mathcal{L}(z_1)$ assumes that the Laplace transform is an invariant of a linear transformation. Actually the transformation of dt_1 requires that $\mathcal{L}(z)$ should be divided by α. Since, however, in network problems $f(t)$ has the significance of the unit pulse, and this pulse changes by the same factor because of the scale transformation, we are justified in the assumption that the Laplace transform is an invariant of the scale transformation (2).

$f(t)$ as a function of the new variable ξ—we indicate that by writing $f(\xi)$—and obtain the Laplace transform in the new form

$$\mathcal{L}(z) = \int_0^1 f(\xi)\xi^{z-1} \, d\xi \qquad (4\text{-}27.2)$$

We will consider the equidistant set of points

$$z = 1, 2, 3, \cdots \qquad (4\text{-}27.3)$$

Then we possess the "moments" of $f(\xi)$.

$$y_k = \mathcal{L}(k+1) = \int_0^1 f(\xi)\xi^k d\xi \qquad (4\text{-}27.4)$$

Moreover, if $p_n(\xi)$ is any polynomial of ξ,

$$p_n(\xi) = p_n^0 + p_n^1\xi + \cdots + p_n^n\xi^n \qquad (4\text{-}27.5)$$

and we weight the y_k by the coefficients of this polynomial,

$$c_n = \sum_{\alpha=0}^{n} y_\alpha p_n^\alpha \qquad (4\text{-}27.6)$$

we obtain the definite integral

$$c_n = \int_0^1 f(\xi)p_n(\xi) \, d\xi \qquad (4\text{-}27.7)$$

We will use the operational notation $p_n(y)$ in the following sense. We write out the polynomial $p_n(y)$ and replace y^k by y_k. With this convention, formula (7) may be written in the operational form,

$$c_n = p_n(y) \qquad (4\text{-}27.8)$$

We will now introduce the Legendre polynomials $P_n(x)$, [cf. (5-20.11)], but normalizing their range of orthogonality to the range [0,1], instead of the customary range [−1,1]. These polynomials are now directly expressible in terms of the hypergeometric function.

$$P_k^*(x) = F(-k, k+1, 1; \, x) \qquad (4\text{-}27.9)$$

Their "norm" is

$$N_k = \int_0^1 P_k^{*2}(w)\, dx = \frac{1}{2k+1} \qquad (4\text{-}27.10)$$

We can expand our $f(\xi)$ in these polynomials [cf. (5-16.9)]

$$f(\xi) = \sum_{k=0}^{\infty} (2k+1)c_k P_k^*(\xi) \qquad (4\text{-}27.11)$$

with

$$c_k = \int_0^1 f(\xi)P_k^*(\xi)\, d\xi = P_k^*(y) \qquad (4\text{-}27.12)$$

Hence we have obtained an explicit representation of $f(\xi)$ in the form of an infinite convergent expansion, whose coefficients are calculable in terms of y_k, i.e., in terms of equidistant values of the Laplace transform along the real axis.

The coefficients of $P_n^*(\xi)$ possess the numerically valuable property that they are all *integers*; they can be pretabulated (cf. Table IV), in the form of a triangular matrix. Multiplication of the y_k with this matrix provides us with the coefficients c_k, which, after multiplication by $(2k+1)$, give the coefficients of the expansion (12). Unfortunately, the elements of this matrix increase very rapidly. This has the consequence that the values y_k have to be given with *excessive accuracy*, in order to evaluate the c_k with even moderate accuracy. At $k = 10$ the matrix elements increase up to $2 \cdot 10^6$. Hence it is unsafe to go beyond $k = 10$, even if the ten-digit accuracy of the customary desk-machines is fully utilized. The accuracy with which the y_k have to be given is here already 10 significant figures. Mere measurements can never cope with such an accuracy, and thus it is demonstrated that physical *observations* of the Laplace transform can never solve the problem of restoring the indicial function with any degree of accuracy. The relation between the data and the original function is excessively nonorthogonal, and the slightest "noise" of the data destroys all hope of inverting an empirically given Laplace transform.

Here we observe a repetition of the problem encountered in the resolution of a function into a set of exponentials (cf. § 23), and the example is more than accidental, since a function of this type can actually be conceived as a Laplace transform.

The situation is quite different, however, if $\mathcal{L}(z)$ is a *theoretically given quantity*, such as the ratio of two polynomials which characterizes electric network problems:

$$\mathcal{L}(z) = p(z)/q(z)$$

In order to obtain the y_k we have to substitute the successive integers 1, 2, 3, $\cdots$ in two polynomials, which is a quick and simple process. The coefficients of these polynomials should not be given with more than a minimum number of significant figures (2 or 3). But in all subsequent calculations, full ten-place accuracy is required. In fact, if we want to go beyond 11 terms of the expansion (beyond $n = 10$), it is necessary that the basic data shall be given with 15 decimal places. This is by no means difficult since $p(k)$, $q(k)$, substituting the subsequent integers, will seldom go beyond 10 significant figures, even if absolute accuracy is used. Hence is is necessary only to perform the division with 15 significant figures, which requires that we shall put the remainder of the first division process back in the machine and repeat the division with this new numerator, with no change of the denominator. Moreover, the coefficients of $P_n^*(x)$ do not grow beyond 10 decimal places up to $n = 14$, and thus require no double precision. An expansion of more than 15 terms will seldom be demanded if the scale factor α was chosen properly.

28. Inversion by Chebyshev polynomials. Although the Legendre polynomials $P_k(x)$ are tabulated,[1] a more convenient expansion is obtainable in the form of a *Fourier series*. We introduce the angle variable θ by putting

$$e^{-t} = \xi = \frac{1 + \cos\theta}{2} = \cos^2\frac{\theta}{2} \qquad (4\text{-}28.1)$$

We will consider $f(\xi)$ a function of the angle variable θ—range $[0,\pi]$—and expand it in a Fourier sine series.

$$f(\theta) = \frac{4}{\pi}(b_1\sin\theta + b_2\sin 2\theta + \cdots) \qquad (4\text{-}28.2)$$

For this purpose the boundary conditions $f(0) = f(\pi) = 0$ have to be satisfied. Now the point $\theta = \pi$ corresponds to $t = \infty$, and there the pulse response is automatically zero. On the other hand, at the

[1] Cf. *Tables of the Legendre Polynomials* (Macmillan, New York, 1946).

point $\theta = 0$, i.e., $t = 0$, the function is generally not zero. We can make it zero, however, by a simple artifice. If $\mathcal{L}(z)$ is an infinity of the form p_1/z (cf. 24.8), we will put

$$f_1(t) = f(t) - p_1 e^{-t} \tag{4-28.3}$$

and

$$\mathcal{L}_1(z) = \mathcal{L}(z) - \frac{p_1}{z + 1} \tag{4-28.4}$$

Then $\mathcal{L}_1(z)$ decreases to infinity with z^{-2}, and $f_1(t)$ is zero at $t = 0$. The natural boundary conditions of the expansion (16) are now satisfied, and the convergence of the series will be satisfactory.

The coefficients b_k of the series (6) are determined in the standard manner.

$$b_k = \frac{1}{2} \int_0^\pi f_1(\theta) \sin k\theta \; d\theta = -\int_0^1 f_1(\xi) \frac{\sin k\theta}{\sin \theta} \; d\xi \tag{4-28.5}$$

Now the function $(\sin k\theta)/\sin \theta$ conceived as a function of ξ is a polynomial, called "Chebyshev polynomial of the second kind" (cf. III, 4 and V, 20).

$$\frac{\sin k\theta}{\sin \theta} = U_{k-1}^*(\xi) = \frac{1}{2k} T_k^{*\prime}(x) \tag{4-28.6}$$

The coefficients of these polynomials are once more integers and once more a direct relation to the hypergeometric series can be established.

$$U_{k-1}^*(\xi) = (-1)^{k-1} k F(k + 1, -k + 1, \tfrac{3}{2}; \; \xi) \tag{4-28.7}$$

Application of (27.7–8) to our case gives

$$b_k = U_{k-1}^*(y) \tag{4-28.8}$$

The coefficients of the polynomials can be pretabulated (cf. Table IX). Again they form a triangular matrix. Multiplication of the basic values y_k with this matrix again generates the coefficients of an orthogonal expansion, in analogy to the previous solution, but with the added advantage that the resulting series can be written as a Fourier sine series in the angle variable θ. Once more extreme numerical accuracy has to be maintained and we cannot come beyond $n = 10$ if the y_k are not given with more than 10-place accuracy. The 10-place stage for the coefficients of $U_{k-1}(\xi)$ is reached at $k = 14$.

29. Inversion by Fourier series. The strongly nonorthogonal behavior of equidistant points on the *real* axis of $\mathcal{L}(z)$ changes to an entirely different behavior, if equidistant points of the *imaginary axis* are considered. We will make use of the fact that in communication problems the indicial function (the pulse response) is practically limited to a finite interval, since beyond the memory time T_0 the pulse response becomes negligibly small. This fact permits us to develop a solution of the inversion problem which utilizes only equidistant values of $\mathcal{L}(z)$ along the imaginary axis.

We start our discussion with the following imaginary experiment. Instead of giving the unit pulse only once, we will repeat it rhythmically in intervals of T_0 or in intervals larger than T_0. Then the pulse response is also repeated rhythmically, but the successive responses do not overlap, since the time which separates them is enough to extinguish the response of the previous period. We now have a situation in which both $f(t)$ and $g(t)$ are periodic functions of the period

$$T \geq T_0$$

We will put

$$\omega_0 = 2\pi/T_0 \tag{4-29.1}$$

The input function, if resolved into its harmonic components, can be written as follows (this series does not converge):

$$f(t) = \frac{2}{T}\left[\frac{1}{2} + \cos\frac{2\pi}{T}t + \cos 2\frac{2\pi}{T}t + \cdots\right]$$

$$= \frac{\omega_0}{\pi}\left[\frac{1}{2} + \cos\omega_0 t + \cos 2\omega_0 t + \cdots\right] \tag{4-29.2}$$

The corresponding output $g(t)$ becomes, if we remember the definition (18.4) of the frequency response $\phi(\omega) = \mathcal{L}(i\omega)$:

$$g(t) = K(t)$$

$$= \frac{\omega_0}{2\pi}\sum_{k=0}^{\infty}{}'\left[\mathcal{L}(ik\omega_0)e^{ik\omega_0 t} + \mathcal{L}(-ik\omega_0)e^{-ikw_0 t}\right] \tag{4-29.3}$$

or, separating real and imaginary parts of $\mathcal{L}(i\omega)$,

$$\mathcal{L}(i\omega) = A(\omega) + iB(\omega)$$

$$K(t) = \frac{\omega_0}{\pi}\left[\frac{A(0)}{2} + \sum_{k=1}^{\infty} A(k\omega_0)\cos k\omega_0 t\right.$$

$$\left. - \sum_{k-1}^{\infty} B(k\omega_0)\sin k\omega_0 t\right] \qquad (4\text{-}29.4)$$

Here we have a resolution of $K(t)$ into a Fourier series, based on equidistant values of the Laplace transform along the imaginary axis. If we introduce the new variable z_1 by putting

$$z = i\omega z_1$$

and rewrite the polynomials $p(z)$ and $q(z)$ in the new variable, we have again the simple task of substituting the integers 0, 1, 2, $\cdots$ in two polynomials and forming their ratio. The only inconvenience is that the coefficients of these polynomials are now alternatingly real and imaginary numbers. The result of the substitution will be a complex number, and in the end we have to divide one complex number by another. The resultant complex number gives in its real part the coefficients of the cosine series, in its imaginary part the negative coefficients of the sine series.

Two difficulties still remain. The convergence of the resulting series may be too slow. Moreover, we may not know in advance the value of the memory time T_0. However, these difficulties can be surmounted. The convergence of the series can be speeded up by modifying the function $K(t)$ properly. For this purpose we first expand $\mathcal{L}(z)$ around the point $z = \infty$. We know in advance that for large values of z we will have

$$\frac{p_1}{z} + \frac{p_2}{z} + \frac{p_3}{z^3} \qquad (4\text{-}29.5)$$

This means that the terms of the Fourier expansion (4) decrease too slowly, namely with k^{-1}. But let us correct $\mathcal{L}(z)$ by putting

$$\mathcal{L}_1(z) = \mathcal{L}(z) - \frac{p_1}{z+1} - \frac{p_2+p_1}{(z+1)^2} \qquad (4\text{-}29.6)$$

Then for large values of z the decrease occurs according to the law z^{-3}, and the convergence of the series for $K(t)$ becomes satisfactory. We can terminate the series at a properly chosen k by scanning the imaginary axis for the maximum of $|\mathcal{L}_1(i\omega)|$ and then neglecting the terms from a point ω_l on where the associated $|\mathcal{L}_1(i\omega_l)|$ has dropped to a few per cent of the maximum. We now have to decide about the choice of the frequency ω_0, defined by (1). It is important only that we shall not overstep the upper limit $2\pi/T_0$, while smaller values are not harmful. If we adopt for ω_0 the value

$$\omega_0 = \frac{1}{12}\,\omega_l \qquad\qquad (4\text{-}29.7)$$

we have a fairly reliable guarantee that we did not overestimate ω_0. The division of ω_l into 12 parts means that we assume that 12 sine terms and 13 cosine terms will be sufficient to represent $K(t)$ with an accuracy of a few per cent. We will seldom go wrong with this assumption, since a Fourier series of 25 terms has great flexibility in representing even rather rugged functions. Afterwards we can check up on the correctness of our hypothesis. The function obtained by the series is not the original $K(t)$ but the following modification of it:

$$K_1(t) = K(t) - p_1 e^{-t} - (p_2 + p_1)t e^{-t} \qquad (4\text{-}29.8)$$

This function starts from zero and has a horizontal tangent at zero.

$$K_1(0) = 0, \qquad K_1'(0) = 0 \qquad\qquad (4\text{-}29.9)$$

These boundary conditions are responsible for the good convergence of the Fourier series. If we plot the course of $K_1(t)$, evaluated on the basis of the series (9), we can verify the adequacy of our choice (7) by finding that $K_1(t)$ flattens out and becomes practically zero before the upper limit $t = 2\pi/\omega_0$ is reached.

30. Inversion by Laguerre functions. Our first two solutions of the inversion problem are applicable to any Laplace transform. These two solutions made use of equidistant values of $\mathcal{L}(z)$ along the real axis. Our third solution was specifically shaped to the demands of network analysis. We took advantage of the fact that the range of integration, although theoretically extending to infinity, was in actual fact reducible to a finite time interval T_0, the "memory time" of the network. In the present section a method will be discussed which

is again of general applicability. It is based on Fourier's reciprocity
theorem, which solves the problem of inverting the Fourier transform. The inversion of the Laplace transform is reducible to Fourier's
problem.

Already in § 4 we have briefly touched on the relation between
Laplace and Fourier transforms. If we consider the Laplace trans-
form $\mathcal{L}(z)$ for purely imaginary values $i\omega$ of z, we obtain the Fourier
transform of the indicial function $f(t)$. Now we can use Fourier's
reciprocity theorem (17.16) and obtain $f(t)$ as the Fourier transform
of $\mathcal{L}(i\omega)$.

$$\mathcal{L}(i\omega) = \int_0^\infty f(t)e^{-i\omega t}\, dt \qquad (4\text{-}30.1)$$

$$f(t) = \frac{1}{2\pi}\int_{-\infty}^{+\infty} \mathcal{L}(i\omega)e^{i\omega t}\, d\omega = \frac{1}{2\pi i}\int_{-i\infty}^{+i\infty} \mathcal{L}(z)e^{zt}\, dz \quad (4\text{-}30.2)$$

Although $f(t)$ is thus obtained in the form of a definite integral,
the actual evaluation of this integral is by no means trivial, in view
of the infinite limits of integration. We will develop a method which
permits us to obtain numerically the Fourier transform of a function
which is given in either analytical or numerical form. For this
purpose we make use of a remarkable conformal mapping of the
complex plane on itself, encountered earlier in the chapter on
algebraic equations (cf. I, 18). This transformation mapped the
entire right complex half plane into the inside of the unit circle. The
unit circle itself became the image of the infinite imaginary axis.
Hence an integration along the imaginary axis is changed into an
integration along the unit circle.

We transform the complex variable z of the Laplace transform
into a new variable v, according to the following transformation:

$$z = \frac{1-v}{1+v} = -1 + \frac{2}{1+v}, \qquad dz = -\frac{2\,dv}{(1+v)^2}$$

$$= -\frac{(z+1)^2}{2}\, dv \quad (4\text{-}30.3)$$

In the new variable the integral (30.2) is extended over the unit circle

$$v = e^{i\theta} \qquad (4\text{-}30.4)$$

The limits of the angle variable θ are $[-\pi,+\pi]$.

$$f(t) = \frac{1}{2\pi i} \int_{-i\infty}^{+i\infty} \mathcal{L}(z)e^{tz}\, dz = -\frac{1}{4\pi} \int_{-\pi}^{+\pi} \mathcal{L}(z)(1+z)^2 e^{tz+i\theta}\, d\theta \quad (4\text{-}30.5)$$

Let us now introduce the function

$$\mathcal{L}_1(z) = \frac{(1+z)^2}{2}\left[\mathcal{L}(z) - \frac{p_{n-1}}{z+1}\right] \quad (4\text{-}30.6)$$

We have seen (cf. 25.2) that the order of $p(z)$ is generally one lower than the order of $q(z)$. Modification of $\mathcal{L}(z)$ by subtracting the last term of (6) has the purpose that the difference between the orders of numerator and denominator—if the two fractions are brought to one denominator—shall be increased to two. Then the multiplication by $(1+z)^2$ balances the orders of numerator and denominator and $\mathcal{L}_1(z)$ remains finite even at $z = \infty$. In compliance with the modification of $\mathcal{L}(z)$ we will put

$$f(t) = p_{n-1}e^{-t} + f_1(t) \quad (4\text{-}30.7)$$

The function $f_1(t)$ now starts from zero at $t = 0$.

We will now transform $\mathcal{L}_1(z)$ to the new variable v. This can be accomplished by multiplying the coefficients of the numerator and similarly the coefficients of the denominator by the pretabulated matrix B_n (cf. Table II). This transformation is quickly accomplished, since the coefficients of this matrix are all integers. If we write $\mathcal{L}_1(v)$, we mean that this operation has been performed already on the original $\mathcal{L}_1(z)$. (If it so happens that the numerator of $\mathcal{L}_1(z)$ is of lower order than the denominator, we consider the numerator nevertheless a polynomial of nth order, replacing the missing coefficients by zeros.)

Since $\mathcal{L}_1(e^{i\theta})$ is a periodic function of θ, it is natural that we shall expand it in a Fourier series.

$$\mathcal{L}_1(e^{i\theta}) = \sum_{k=0}^{\infty} (c_k e^{ik\theta} + c_{-k}e^{-ik\theta}) = \sum_{k=0}^{\infty} (c_k v^k + c_{-k}v^{-k}) \quad (4\text{-}30.8)$$

If the Fourier transform of an empirically given function is in question, the coefficients c_k, c_{-k} of this expansion can be obtained by trigonometric interpolation (cf. §§ 12, 13). Then the representation by a trigonometric series will not be exact, of course. It is enough, however, to know the maximum absolute error ε of the expansion at any point of the unit circle. We can now estimate the maximum error of the Fourier transform (5), if $f(t)$ is modified in the sense of (7). The integral (5) shows that $\mathcal{L}_1(z)$ is multiplied along the path of integration by a factor which has the absolute value 1, no matter what the value of t is (t being real). Hence we can estimate that the maximum absolute error of $f_1(t)$, calculated on the basis of (5), cannot surpass ε.

$$| \eta(t) | \leq \varepsilon \qquad (4\text{-}30.9)$$

In the case of inverting the Laplace transform, much more can be said. Here we know that $\mathcal{L}_1(v)$ is an analytical function of v which has no singularities inside or on the unit circle. We can expand $\mathcal{L}_1(v)$ in a Taylor series around the point $v = 0$.

$$\mathcal{L}_1(v) = c_0 + c_1 v + c_2 v^2 + \cdots \qquad (4\text{-}30.10)$$

This expansion converges everywhere inside and on the unit circle. Hence we are here in the fortunate situation that we possess the exact Fourier coefficients of $\mathcal{L}_1(v)$ along the unit circle, without any integration. The comparison with the general expansion (8) shows that the c_{-k} are not present at all. This corresponds to the fact that $f(t)$ is identically zero for all values of t which are less than zero.

In network problems the coefficients c_k can be obtained by a simple and elegant algorithm, called the "synthetic division of two polynomials," encountered earlier in studying the roots of an algebraic equation (cf. I, 14). The only difference is that in our earlier scheme that expansion proceeded in *reciprocal* powers of the variable, while now an expansion in direct powers of v is demanded. But this means only that the coefficients of the polynomials $p(v)$ and $q(v)$ are written in *reversed* order, starting with the absolute term and ending with the highest power. The absolute term q_0 of the denominator cannot vanish, since $q(v)$ can have no root inside the unit circle and certainly no root at the center $v = 0$. Hence we can always divide both numerator and denominator by q_0, thus making the absolute term of the denominator equal to 1. We put the successive coefficients of the numerator (in increasing order) on the "fixed strip," and the

successive coefficients of the denominator—going from the bottom
to the top and changing the sign of all coefficients beyond 1—on the
"movable strip." Then the regular movable strip technique (I, 8)
generates on the nascent strip the successive coefficients c_k. We may
continue with the algorithm until 15 to 20 coefficients are obtained.

Our problem is now reduced to inversion of the special Laplace
transform

$$\frac{2v^k}{(1+z)^2} = \frac{2(1-z)^k}{(1+z)^{k+2}} = \frac{2(-1)^k z_1^k}{(2+z_1)^{k+2}} \qquad (4\text{-}30.11)$$

We have put

$$z \cdot 1 + z_1 \qquad (4\text{-}30.12)$$

We will solve this problem in steps. In the first place, introducing
z_1 as a new variable, we obtain

$$\mathcal{L}(z_1) = \int_0^\infty f(t)e^{-t}e^{-z_1 t}\, dt = \int_0^\infty \varphi(t)e^{-z_1 t}\, dt \qquad (4\text{-}30.13)$$

where

$$\varphi(t) = f(t)e^{-t} \qquad (4\text{-}30.14)$$

Now we know that the indicial function of the transform $(2+z)^{-1}$ is

$$f(t) = e^{-2t}$$

We also know that *multiplication* by t means *negative differentiation*
in the transform plane. Hence the transform of the indicial function

$$t^{k+1}e^{-2t}$$

becomes

$$\mathcal{L}(z_1) = \frac{(k+1)!}{(2+z_1)^{k+2}}$$

Similarly, *multiplication* by z_1 in the transform plane means *differentiation* in the original plane. Hence the indicial function of

$$\mathcal{L}(z_1) = z_1^{k+1}\frac{(k+1)!}{(2+z_1)^{k+2}} \qquad (4\text{-}30.15)$$

becomes

$$(t) = \frac{d^{k+1}}{dt^{k+1}}\left(t^{k+1}e^{-2t}\right) \qquad (4\text{-}30.16)$$

An important class of orthogonal polynomials, called "Laguerre polynomials," are defined by the following operation.[1]

$$L_n(t) = e^t \frac{d^n}{dt^n} (t^n e^{-t}) \qquad (4\text{-}30.17)$$

Hence

$$L_n(2t) = e^{2t} \frac{d^n}{dt^n} (t^n e^{-2t}) \qquad (4\text{-}30.18)$$

On the basis of (15)–(18) the following statements can be made:

Indicial function: $\varphi(t) = \dfrac{e^{-2t}}{(k+1)!} L_{k+1}(2t)$

Laplace transform: $\mathcal{L}(z_1) = \dfrac{z^{k+1}}{(2+z_1)^{k+2}}$

Moreover,

Indicial function: $\varphi(t) = e^{-2t} \left[\dfrac{L_k(2t)}{k!} - \dfrac{L_{k+1}(t)}{(k+1)!} \right]$

Laplace transform: $\mathcal{L}(z_1) = \dfrac{z_1^k}{(2+z)^{k+1}} - \dfrac{z_1^{k+1}}{(2+z)^{k+2}} = \dfrac{2z_1^k}{(2+z_1)^{k+2}}$

Going back to the original variable z, we have to multiply $\varphi(t)$ by e^t. Hence we come to the conclusion that the inversion of the Laplace transform (11) becomes

$$(-1)^k e^{-t} \left[\frac{L_k(2t)}{k!} - \frac{L_{k+1}(2t)}{(k+1)!} \right] \qquad (4\text{-}30.19)$$

We will introduce the so-called "Laguerre functions" (cf. 5–20.16)

$$\varphi_k(t) = \frac{e^{-t/2} L_k(t)}{k!} \qquad (4\text{-}30.20)$$

which form an ortho-normal function system in the range $[0,\infty]$ of the variable t. The expression (19) can now be rewritten in the form

$$(-1)^k [\varphi_k(2t) - \varphi_{k+1}(2t)]$$

[1] Cf. {1}, p. 93.

and the inversion of the series (10) becomes

$$f_1(t) = c_0\varphi_0(t) - (c_0 + c_1)\varphi_1(t) + (c_1 + c_2)\varphi_2(t) - \cdots$$
$$= \sum_{k=0}^{\infty} (-1)^k (c_{k-1} + c_k)\varphi_k(2t) \qquad (4\text{-}30.21)$$

We have seen before that termination of this series to a finite expansion involves an error which for no value of t can surpass the error of the truncated series (10).

The Laguerre functions $\varphi_k(2t)$ can be pretabulated in reasonably small intervals of the variable t and up to a reasonable order k (for example, $k = 20$).[1] Then, by substituting for t a sufficiently dense set of values, we can plot the resulting expansion (21). We must not forget, however, to add the correction (7), in order to restore the complete pulse response of the network. This correction merely means that the coefficient of $\varphi_0(2t) = e^{-t}$ is changed as follows:

$$c_0' = c_0 + p_{n-1} \qquad (4\text{-}30.22)$$

In retrospect we can say that we have obtained four convenient numerical schemes for construction of the pulse response of a network, given by its structural elements, without evaluating the roots of the polynomial $q(z)$. While the first two methods utilized as key values equidistant values of $\mathcal{L}(z)$ along the real axis, the third method utilized equidistant values along the imaginary axis, and the present method made use of the infinitesimal neighborhood of the single point $z = 1$. The value of $\mathcal{L}(z)$ and all its derivatives at $z = 1$ uniquely determine the coefficients c_k of the expansion (11) and thus the indicial function $f(t)$.

In all these solutions one fundamental condition has to be fulfilled. The network must not contain a high-frequency vibration of small damping. Proper representation of such a vibration in our expansions would require an inordinate number of terms. However, a vibration of this kind corresponds to a root very near to the imaginary axis. Such roots can be spotted with no great difficulty

---

[1] Such a tabulation of the normalized Laguerre functions was prepared by Miss Fanny Gordon, staff member of the National Bureau of Standards, during the author's stay with the Institute for Numerical Analysis, National Bureau of Standards, Los Angeles, California. It is reproduced as Table X of the Appendix, by permission of the Bureau.

(cf. I, 19) and their effect determined separately. We have to subtract the contribution of these roots (cf. 25.5) before we start our numerical scheme of inverting the Laplace transform.

31. Interpolation of the Laplace transform. The Laplace transform, $\mathcal{L}(z)$ being an analytical function of the complex variable z, is uniquely determined if it is known on an arbitrarily small continuous arc of the complex plane. Of greater interest, however, is the method of prescribing an analytical function in an infinity of equidistant points and obtaining the values everywhere else by interpolation and extrapolation. Such an interpolation formula for the Laplace transform can actually be developed.

We have seen in § 27 that by giving the Laplace transform at the integer points $z = 1, 2, 3, \cdots$ along the real axis we can uniquely determine the "indicial function" $f(\xi)$, which is transformed into $\mathcal{L}(z)$ by the transformation (27.4). Now let us substitute the infinite expansion (27.11) for $f(\xi)$ and integrate term by term. For this purpose we need the definite integrals

$$w_k(z+1) = \int_0^1 \xi^z P_k^*(\xi)\, d\xi = \sum_{\alpha=0}^{k} \frac{p_k^\alpha}{z+\alpha+1} \qquad (4\text{-}31.1)$$

If these fractions are brought to one denominator, we obtain in the numerator a polynomial of the order k, while the denominator becomes the product

$$(z+1)(z+2)(z+3)\cdots(z+k+1)$$

The numerator can likewise be resolved into its root factors and we can see from the fact that any integer power $z = 0, 1, 2, \cdots, k-1$ is orthogonal to $P_k^*(\xi)$, that these root factors must be

$$z(z-1)(z-2)\cdots(z-k+1)$$

and what remains undetermined is a mere constant in front of these factors. But then, letting z go to infinity and remembering that $P_k^*(1) = 1$, we find that the undecided numerical constant becomes 1. Introducing the symbol

$$\begin{bmatrix} z \\ k \end{bmatrix} = \frac{z(z-1)(z-2)\cdots(z-k+1)}{(z+1)(z+2)\cdots(z+k+1)} = \frac{(z!)^2}{(z+k+1)!(z-k)!}$$

$$(4\text{-}31.2)$$

we obtain

$$w_k(z+1) = \left[\begin{array}{c} z \\ k \end{array} \right] \qquad (4\text{-}31.3)$$

and the interpolation formula for the Laplace transform becomes

$$\mathscr{L}(z+1) = \sum_{k=0}^{\infty} c_k \left[\begin{array}{c} z \\ k \end{array} \right] = \sum_{k=0}^{\infty} (2k+1) P_k^*(y) \left[\begin{array}{c} z \\ k \end{array} \right] \qquad (4\text{-}31.4)$$

The symbolic notation $P_k^*(y)$ (cf. § 27) has the following significance. We form the polynomial $P_k^*(y)$ and replace y^{α} by $\mathscr{L}(\alpha+1)$, i.e., the value of $\mathscr{L}(z)$ at the integer point $z = \alpha + 1$. The coefficients of the interpolation formula (4) are thus determined by the values of $\mathscr{L}(z)$ at the points 1, 2, 3, $\cdots$. If the condition

$$\sum_{k=0}^{\infty} |c_k| = \text{finite} \qquad (4\text{-}31.5)$$

is fulfilled, the formula (4) will converge for all z points to the right from the imaginary axis.

While this interpolation formula establishes the fact that the Laplace transform is uniquely determined if we know its value at all integer points, it would be a mistake to believe that these values can be prescribed at will. In the first place we know from the arbitrariness of the scale that the "unit distance" can be stretched out to any length. Hence we could thin out our originally given data to any degree we like.

Furthermore, the transformation

$$z = \alpha + z_1, \qquad f(x) = f_1(x)e^{\alpha x} \qquad (4\text{-}31.6)$$

shows that the points $z_1 = 1, 2, 3, \cdots$ correspond to the points $z = \alpha+1$, $\alpha + 2$, $\alpha + 3$, $\cdots$. Hence not only can we thin out our data to any degree, but we can start them at a point which is arbitrarily far from the origin. Our functional data are thus always redundant to an infinite degree and we can never reduce them to a basic system which is both necessary and sufficient. This seems in peculiar contradiction to the fact that the c_k can always be constructed to an arbitrarily given set of $\mathscr{L}(m)$ values; $(m = 1, 2, 3, \cdots)$. However, if these values are not properly given, the sum (5) will not converge, and our interpolation formula loses its significance.

Another interesting conclusion can be drawn from the interpolation formula (4). The Laplace transform associated with an arbitrary four-terminal network is always the ratio of two polynomials. This is only a very special class of Laplace transforms, characterized by those indicial functions which are defined as a finite linear combination of exponential functions with exponents whose real part is negative. Yet the convergence of the expansion (4) means that an arbitrary Laplace transform can be generated with any degree of accuracy with the help of electric networks, in fact with the help of purely "ballistic" networks, without any self-inductance L. Indeed, any absolutely integrable function of bounded variation can be approximated by a finite Legendre expansion of the form (27.11), with an error which can be made as small as we wish. The corresponding Laplace transform (4) can be written as the ratio of two polynomials, with the zeros of the denominator normalized to $z = -1, -2, -3, \cdots$. This shows that not only can we simulate an arbitrarily given pulse response $K(t)$ with the help of RC circuits, but we can normalize the constants of these circuits in such a way that the products RC shall be in the ratio $1 : \frac{1}{2} : \frac{1}{3} : \frac{1}{4} : \cdots$ to one another. Particularly instructive is the application of the interpolation formula (4)—truncating the number of terms to n—to the function

$$\frac{1}{(z + \beta)^2 + \gamma^2}$$

thus demonstrating explicitly that the output of a single RCL circuit is replaceable by the output of a sufficient number of properly coupled RC circuits, although the output of every one of these circuits is a monotonously decreasing function, without any vibrations.

We now come to the second inversion method, studied in § 28. Here the indicial function $f(\xi)$ was expanded in a Fourier sine series; the coefficients of this series were obtained with the help of the same $\mathscr{L}(m + 1)$ as before, but weighted by a different set of weight factors (the coefficients of the Chebyshev instead of Legendre polynomials). The integrals which take the place of (1) are now

$$w_k(z + 1) = \int_0^1 \xi^z \sin k\theta \, d\xi = \int_0^\pi \cos^{2z} \frac{\theta}{2} \sin k\theta \sin \theta \, d\theta \quad (4\text{-}31.7)$$

This integral is expressible in terms of gamma functions and we obtain

$$w_k(z + 1) = k\sqrt{\pi}\,\frac{(z + \frac{1}{2})!z!}{(z + k + 1)!(z - k + 1)!} \quad (4\text{-}31.8)$$

Once more we introduce a simplified notation.

$$\left\{\begin{matrix} z \\ k \end{matrix}\right\} = \frac{(z + \frac{1}{2})!z!}{(z + k + 1)!(z - k + 1)!} \quad (4\text{-}31.9)$$

and obtain the interpolation formula,

$$\mathcal{L}(z + 1) = \frac{4}{\sqrt{\pi}}\sum_{k=1}^{\infty} kb_k \left\{\begin{matrix} z \\ k \end{matrix}\right\} = \frac{4}{\sqrt{\pi}}\sum_{k=1}^{\infty} kU_{k-1}^{*}(y) \left\{\begin{matrix} z \\ k \end{matrix}\right\} \quad (4\text{-}31.10)$$

The poles of the functions (9) are at $z = -\frac{3}{2}, -\frac{5}{2}, -\frac{7}{2}, \cdots$ but these functions can no longer be written as the ratio of two polynomials, and are thus not realizable with the help of electric circuits.

Still another interpolation formula is derivable if the third method of inverting the Laplace transform is used (cf. § 29). This method is restricted to the case of a finite integration time T_0, as encountered in network problems. The points of interpolation are here equidistant points along the imaginary axis. The method of deriving the interpolation formula is the same as that before; we introduce the expansion obtained for $f(t)$ in the definition of the Laplace transform and integrate term by term. We thus obtain

$$\mathcal{L}(z) = 2e^{T_0 z/2}\sum_{k=-\infty}^{+\infty} \mathcal{L}(ik\omega_0)\,\frac{\sinh{(zT_0/2)}}{zT_0 - 2\pi ik} \quad (4\text{-}31.11)$$

The basic interpolating functions are here free of any poles and cannot be interpreted in network terms.

Finally in the fourth solution of the inversion problem we have expanded $\mathcal{L}(z)$ around the point $z = 1$, after transforming z into a new variable v. However, the resulting expansion can be formulated in the original variable z and written as follows:

$$\mathcal{L}(z) = 2\sum_{k=0}^{\infty}\frac{L_k(-2y)}{k!}\left(\frac{1-z}{1+z}\right)^k\frac{1}{1+z} \quad (4\text{-}31.12)$$

where the notation $L_k(-2y)$ is used operationally in the sense that in expanding the Laguerre polynomial $L_k(-2y)$, we should replace y^{α}

by $\mathcal{L}^{(\alpha)}(1)$. This expansion converges for all complex z points whose real part is positive.

If this interpolation formula is interpreted in network terms, we see that now an arbitrary Laplace transform is generated with the help of ballistic circuits whose products RC are crowded around the constant value 1. The interpolating functions have a pole at the point $= -1$, but this pole is of ever-increasing order k. Now a pole of the multiplicity k at $z = -1$ is equivalent to a superposition of k RC circuits of properly chosen strengths, and with products RC which differ from 1 by the arbitrarily small values $\varepsilon_1, \varepsilon_2, \cdots, \varepsilon_k$.

While in this interpolation (or rather extrapolation) formula the key values are taken from the infinitesimal neighborhood of $\mathcal{L}(z)$ around the point $z = 1$, we can obtain a still different interpolation formula by using as key values the entire set of $\mathcal{L}(i\omega)$ values along the imaginary axis. For this purpose we consider that the entire right half plane of the complex variable z is mapped into the inside of the unit circle of the variable v. In this new frame of reference we can make use of Cauchy's integral theorem:

$$\mathcal{L}(v) = \frac{1}{2\pi i} \oint \frac{\mathcal{L}(v_0)\, dv_0}{v - v_0} \tag{4-31.13}$$

where v_0 moves along the unit circle. Going back to the original variable z we obtain

$$\mathcal{L}(z) = \frac{1}{2\pi} (1 + z) \int_{-\infty}^{+\infty} \frac{\mathcal{L}(i\omega)\, d\omega}{(1 + i\omega)(z - i\omega)} \tag{4-31.14}$$

Since an integral is the limit of a sum, we observe that an arbitrary indicial function—and the Laplace transform generated by it—can also be simulated in terms of pure LC circuits, without any R, i.e., with the help of pure sinusoidal vibrations, without any damping. We have here the extreme counterpart of the ballistic circuits of the interpolation formula (4), as if the key values, distributed along the real axis, had swung around by $90°$. Yet there is a fundamental difference between the two cases. The strength of the simulating circuits is in the imaginary case continuous and without any extremes, because of the orthogonality of the basic functions. In the real case, however, the strength of the simulating ballistic circuits has to be chosen in an extreme fashion, because of the strongly nonorthogonal character of the basic interpolating functions.

Bibliographical References

[1] BUSH, V., *Operational Circuit Analysis* (Wiley, New York, 1929).

[2] CARSLAW, H. S., *Theory of Fourier Series and Integrals* (Macmillan, London, 1921).

[3] CHURCHILL, R. V., *Fourier Series and Boundary Value Problems* (McGraw-Hill, New York, 1941).

[4] FRANKLIN, PH., *Fourier Methods* (McGraw-Hill, New York, 1949).

[5] GUILLEMIN, E. A., *Mathematics of Circuit Analysis* (Wiley, New York, 1949).

[6] JACKSON, D., *Fourier Series and Orthogonal Polynomials* (Math. Assoc. of America, Oberlin, 1941).

[7] THOMSON, W. T., *Laplace Transformation* (Prentice-Hall, New York, 1950).

[8] WIDDER, D. V., *The Laplace Transform* (Princeton University Press, Princeton, 1941).

[9] WIENER, N., *The Fourier Integral and Its Applications* (Cambridge University Press, New York, 1933).

Article

[10] DANIELSON, G. C., and LANCZOS, C., "Improvements in Practical Fourier Analysis and Their Application to X-ray Scattering from Liquids," *J. Franklin Inst.*, **233**, 365, 435 (1942).

DATA ANALYSIS

1. Historical introduction. The limited accuracy of physical observations has been recognized since ancient days. Archimedes estimated the minimum size of the universe on the basis of the absence of any observed astronomical aberration within the error limits of the ancient instruments, assuming that Aristarchus' heliocentric theory is accepted as correct. Hand in hand with the ever-increasing accuracy of measuring instruments, a definite mathematical problem took shape. Let us assume that we have occasion to observe the same phenomenon repeatedly, under practically identical conditions. "Practically identical" means that the decisive factors which are responsible for a certain physical event remain the same, while uncontrollable side events change in a random fashion. For example, we may measure repeatedly the relation between displacement and time of the motion of a ball, falling freely toward the earth. The mass of the ball and the force of gravity, which determine the event, remain unchanged, while slight imperfections of the length- and time-measuring instruments will vary erratically. What these "erratic" or "random" variations mean in each case is often difficult to establish. But the demand arose for a general mathematical procedure by which surplus measurements could be handled, from the viewpoint of taking the best advantage of all measurements.

In the beginning of the nineteenth century C. F. Gauss (in 1809) and A. M. Legendre (in 1806) discovered independently a remarkable universal method by which surplus measurements could be adjusted. This method, known as the "method of least squares," gave an answer to the problem of surplus data, based purely on mathematical logic rather than on inductive physical reasoning. Later the method was further developed and became the cornerstone of all statistical considerations. The great success with which this method could be applied to an innumerable variety of physical and engineering problems leads to the conclusion that the method of least squares, originally discovered by sheer mathematical intuition, is nevertheless in fundamental harmony with the structure of the physical universe.

The problem of analyzing data has still another important aspect. Measurements always occur in a *discrete* set of points, while the functions we postulate as the basis of our measurements exist for a *continuous* range of the variable. Essentially we always *tabulate* functions, whether that tabulation is the result of mere calculations or of physical observations. This brings up the question of defining the unknown function *between* the points of tabulation. This is the problem of "interpolation" which fascinated mathematical research from the earliest times. The great analysts of the seventeenth century were well acquainted with the methods of equidistant interpolation and developed the theory of finite differences to a remarkable degree. They used already ordinary and also central differences. The pioneering work of James Gregory (1638–1675) was further advanced by Isaac Newton (1642–1727); J. Stirling (1692–1770) and later F. W. Bessel (1784–1846) added further formulas. The rigorous treatment of interpolation theory starts with D. Hahn and L. Fejér (1918). The dangers of equidistant polynomial interpolation were discovered independently by C. Runge (1901) and E. Borel (1903).

2. Interpolation by simple differences. One of the most fundamental formulas of analysis is the Taylor expansion which represents an analytical function $f(x)$ in terms of the derivatives of $f(x)$, all taken at the point $x = 0$.

$$f(x) = f(0) + f'(0)x + f''(0)\frac{x^2}{2!} + \cdots + f^{(n)}(0)\frac{x^n}{n!} + \cdots \quad (5\text{-}2.1)$$

This formula can be extended to finite differences instead of derivatives. Let $f(x)$ be given at the equidistant points

$$x = 0, h, 2h, 3h, \cdots, nh, \cdots \quad (5\text{-}2.2)$$

and let us form the successive "difference coefficients"

$$\frac{\Delta f(0)}{\Delta x} = \frac{f(h) - f(0)}{h}$$

$$\frac{\Delta^2 f(0)}{\Delta x^2} = \frac{f(2h) - 2f(h) + f(0)}{h^2} \tag{5-2.3}$$

$$\cdot \quad \cdot \quad \cdot \quad \cdot$$

These quantities take the place of the successive derivatives $f^{(k)}(0)$. The functions

$$\varphi_k(x) = \frac{x^k}{k!} \tag{5-2.4}$$

are characterized by the functional equation

$$\varphi_k'(x) = \varphi_{k-1}(x) \tag{5-2.5}$$

They have to be replaced by a class of functions which shall satisfy the equation

$$\frac{\Delta \varphi_k(x)}{\Delta x} = \frac{\varphi_k(x + h) - \varphi_k(x)}{h} = \varphi_{k-1}(x) \tag{5-2.6}$$

the solution of which is

$$\varphi_k(x) = \frac{x(x - h)(x - 2h) \cdots (x - kh + h)}{k!} \tag{5-2.7}$$

Our equations gain greatly in simplicity if we choose a scale for the independent variable x which makes $h = 1$. The kth difference coefficient can now be replaced by the kth difference itself and the auxiliary functions $\varphi_k(x)$ become

$$\varphi_k(x) = \frac{x(x - 1)(x - 2) \cdots (x - k + 1)}{k!} \tag{5-2.8}$$

We thus obtain the Gregory-Newton interpolation formula

$$f(x) = f(0) + \Delta f(0)x + \Delta^2 f(0) \frac{x(x - 1)}{2!} \tag{5-2.9}$$

$$+ \cdots + \Delta^k f(0) \frac{x(x - 1) \cdots (x - k + 1)}{k!} + \cdots$$

The Taylor series has the character of an "extrapolating" (predicting) series, since the value of $f(x)$ and all its derivatives *at the origin $x = 0$* serve to evaluate $f(x)$ *outside* the origin. The series (9), on the other hand, can serve for evaluation of $f(x)$ *between* the data points. Hence it is of an *interpolating* type.

The numerical application of formula (9) is greatly facilitated by setting up a "difference table," according to the following pattern.

x	y	Δy	$\Delta^2 y$	$\Delta^3 y$	$\Delta^4 y$
0	10	11	-5	3	-4
1	21	6	-2	-1	
2	27	4	-3		
3	31	1			(5-2.10)
4	32				

If, e.g., the value of $y = f(x)$ is requested at the point $x = 0.4$, application of (9) gives

$$y(0.4) = 10 + 11 \cdot 0.4 - 5 \frac{(0.4)(-0.6)}{2} + 3 \frac{(0.4)(-0.6)(-1.6)}{6}$$

$$- 4 \frac{(0.4)(-0.6)(-1.6)(-2.6)}{24} = 15.36$$

The "zero line" of formula (9) is flexible and can be shifted to any integer value of x, since the values of x are not more than an ordering scheme which may start at any point. If, for example, $y(2.1)$ is demanded, we would naturally interpolate from the point $x = 2$ on, instead of starting from $x = 0$.

$$f(2.1) = 27 + 4 \cdot 0.1 - 3 \frac{(0.1)(-0.9)}{2} = 27.54$$

We thus gain in convergence, although we may lose in the number of available terms if we are too far down in the scheme. This deficiency can be remedied, however, by filling in the holes downward. The last column remains -4, and thus the next unoccupied diagonal from the right to the left downward becomes $-4, -5, -8, -7, 25$. Hence at the point 2.1 we could have added the term

$$\frac{-5 \cdot (0.1)(-0.9)(-1.9)}{6} = -0.14$$

3. Interpolation by central differences. Simple differences are advisable only at the beginning or the end of a table of functional values. ("Beginning" and "end" are relative terms and become interchangeable by reverse ordering of the x values. Hence the "end" of a table is covered by our previous discussion.) In the middle of the table, however, we can take good advantage of the fact that every functional value $y(k)$ has a neighbor to the right *and to the left*. Only at the two ends of the table does it happen that we run out of neighbors, viz., neighbors to the left at the beginning and neighbors to the right at the end of the table. As soon as we have somewhat advanced in our table, we can set up a "central difference table" which tapers off toward the middle from both ends in the shape of two triangles, one being the reflection of the other.

The principle of a central difference table is once more that we take the difference of two consecutive values of the same column. Only the *arrangement* of these values is different. The result is not written parallel to the upper of the two values but *halfway between* the two values involved. For this reason we distinguish between "full lines" and "half lines". The original data are written on the "full lines". The arrangement is shown in the following scheme.

x	y	δy	$\delta^2 y$	$\delta^3 y$	$\delta^4 y$
0	10				
		11			
1	21		−5		
		6		3	
2	27		−2		−4
		4		−1	
3	31		−3		−4
		1		−5	
4	32		−8		−3
		−7		−8	
5	25		−16		
		−23			
6	2				

$$(5\text{-}3.1)$$

The table in its present form is full of gaps. On both the full lines and the half lines only every *second* term is present. We are now going to fill out these gaps by the following simple device which holds universally for *all* columns: *Take the arithmetic mean of the number above and below the gap.* Hence the completed central difference scheme will finally look as follows.

x	y	δy	$\delta^2 y$	$\delta^3 y$	$\delta^4 y$	
0	10					
0.5	15.5	11				
1	21	8.5	−5			
1.5	24	6	−3.5	3		
2	27	5	−2	1	−4	(5-3.2)
2.5	29	4	−2.5	−1	−4	
3	31	2.5	−3	−3	−4	
3.5	31.5	1	−5.5	−5	−3.5	
4	32	−3	−8	−6.5	−3	
4.5	28.5	−7	−12	−8		
5	25	−15	−16			
5.5	13.5	−23				
6	2					

The great advantage of this central difference scheme lies in its *increased convergence.* For example in the line 3 we find the second and third differences as −3, −3 while in the original arrangement these values would be −8, −8. Moreover, since we possess the half lines in addition to the full lines, we can interpolate from both kinds of lines, and thus the x of the interpolation formula need not go beyond ±0.25, compared with the ±0.5 of the previous table. This, too, is of great advantage by increasing the accuracy with the same number of differences. An interpolation with the help of three successive differences will almost always be sufficient in a reasonably spaced set of data, if central differences are used.

The fundamental $\varphi_k(x)$ functions associated with a central difference table are defined by the functional equation

$$\varphi_k(x + 1) + \varphi_k(x - 1) - 2\varphi_k(x) = \varphi_{k-2}(x) \qquad (5\text{-}3.3)$$

starting with $\qquad \varphi_0(x) = 1, \qquad \varphi_1(x) = x$

We have to differentiate between even and odd orders, according to the following formula.

$$\varphi_{2k}(x) = \frac{x^2(x^2 - 1)(x^2 - 4) \cdots (x^2 - (k-1)^2)}{(2k)!}$$

$$\varphi_{2k-1}(x) = \frac{x(x^2 - 1)(x^2 - 4) \cdots (x^2 - (k-1)^2)}{(2k-1)!} \qquad (5\text{-}3.4)$$

The functions for the half lines were found by Bessel. We thus obtain the interpolation formulas of Stirling and Bessel. Stirling's formula, using any *full line* as base line is

$$y(x) = y_0 + (\delta y)_0 x + (\delta^2 y)_0 \frac{x^2}{2} + (\delta^3 y)_0 \frac{x(x^2 - 1)}{6} \quad (5\text{-}3.5)$$

$$+ (\delta^4 y)_0 \frac{x^2(x^2 - 1)}{24} + (\delta^5 y)_0 \frac{x(x^2 - 1)(x^2 - 4)}{120} + \cdots$$

Bessel's formula, using any *half line* as base line is

$$y(x) = y_0 + (\delta y)_0 x + (\delta^2 y)_0 \frac{x^2 - \frac{1}{4}}{2} + (\delta^3 y)_0 \frac{x(x^2 - \frac{1}{4})}{6} \quad (5\text{-}3.6)$$

$$+ (\delta^4 y)_0 \frac{(x^2 - \frac{1}{4})(x^2 - \frac{9}{4})}{24} + (\delta^5 y)_0 \frac{x(x^2 - \frac{1}{4})(x^2 - \frac{9}{4})}{120} + \cdots$$

As an example for the use of these formulas let us apply them to the table (2), evaluating $y(3.18)$. We are near to the full line $x = 3$ and thus we will use Stirling's formula. We find on line 3 the value 31, 2.5, -3, -3 and thus obtain

$$y(3.18) = 31 + 2.5 \cdot 0.18 - 3 \frac{(0.18)^2}{2} - 3 \frac{(0.18)(0.18^2 - 1)}{6} = 31.49$$

On the other hand, let us obtain $y(3.41)$. Here we are near to the half line 3.5 and we can use Bessel's formula, with $x = -0.09$. The values found on line 3.5 are 31.5, 1, -5.5, -5, and we obtain

$$y(3.41) = 31.5 - 1 \cdot 0.09 - 5.5 \frac{(0.09)^2 - 0.25}{2}$$

$$+ 5 \cdot 0.09 \frac{(0.09)^2 - 0.25}{6} = 32.06$$

We will finally discuss the question of interpolation in the case of *nonequidistant arguments*. The operation with "divided differences" is very cumbersome. However, their use can frequently be avoided on the basis of the following consideration. We introduce an auxiliary variable t by some functional relation

$$x = u(t) \quad (5\text{-}3.7)$$

Then an equidistant tabulation in t corresponds to a nonequidistant tabulation in x. If the changeable scale of tabulation is sufficiently smooth, we can assume that we will interpolate in the variable t rather than in x. But then the previous interpolation formulas remain in force, in spite of the variable scale. The only change is that the x of the previous formulas has to be replaced by

$$t = \frac{x - x_0}{x_1 - x_0} \qquad (5\text{-}3.8)$$

where x_0 and x_1 are the two points between which we interpolate. This consideration shows that the assumption of a strictly *equidistant* tabulation of $f(x)$ is not very critical to the application of the Stirling or the Bessel formula of interpolation as long as the changing scale follows a smooth mathematical law.[1]

4. Differentiation of a tabulated function. The operation d/dx is not directly applicable to a function which is merely given in discrete equidistant data points. After the interpolation, however, we possess $f(x)$ everywhere and can now perform the differentiation. We are usually interested in the slope of the curve at the same points in which the observations were made. Hence we can differentiate our interpolation formulas at $x = 0$ and thus obtain a relation between the d/dx and the $\Delta/\Delta x$ operations.

Simple differences. In the case of simple differences we use the Gregory-Newton formula (2.8) and obtain

$$f'(0) = \Delta f(0) - \frac{\Delta^2 f(0)}{2} + \frac{\Delta^3 f(0)}{3} - \cdots \qquad (5\text{-}4.1)$$

which gives the following operational relation:

$$D = \Delta - \Delta^2/2 + \Delta^3/3 - \cdots = \log(1 + \Delta) \qquad (5\text{-}4.2)$$

The convergence of this formula is very slow.

---

[1] For the general polynomial interpolation at arbitrary points by means of the Lagrangian interpolation formula cf. VI, 10. For the ingenious method of Aitken (and its later modification by Neville), which reduces any polynomial interpolation to a succession of linear interpolations, cf. {4}, p. 84; [4], p. 76.

Central differences. Much better convergence is obtained in the case of central differences. The differentiation of Stirling's formula gives the relation

$$D = \delta - \delta^3/6 + \delta^5/30 - \cdots \tag{5-4.3}$$

The convergence compared with the simple difference formula (2) is greatly improved. Of great advantage also is the fact that only the differences of *odd* order enter the formula.

Even stronger convergence is obtained if we differentiate Bessel's formula (3.6). Here we get

$$D = \delta - \frac{1}{24}\,\delta^3 + \frac{3}{640}\,\delta^5 - \cdots \tag{5-4.4}$$

The factor of δ^3 is 4 times, the factor of δ^5 is 7 times smaller than on the full lines. We thus see that the slope of a curve can be ascertained with particularly great accuracy *halfway between* the points of observation.

5. The difficulties of a difference table. While the calculus of finite differences is an eminently important tool of applied analysis, we can employ it only with great caution if *observed* functions are involved. The accuracy of a calculated mathematical table is usually sufficiently good to set up a difference table. If, however, $y = f(x)$ is merely *observed* in equidistant intervals, the higher differences have the unpleasant quality of greatly magnifying the errors of the observations. Let us assume that in a sequence of observations our data are correct except for one observation where an error of 1 unit appears. Let us see how this isolated error will propagate through a central difference scheme.

				0
			0	
		0		1
	0		1	
0		1		−4
	1		−3	
1		−2		6
	−1		3	
0		1		−4
	0		−1	
		0		1
			0	
				0

We see from this table that a single peak spreads out more and more to a mountain of increasing base and increasing height. The increase occurs according to the law of the binomial coefficients. While this increase does not seem too rapid, it is actually sufficient to destroy the usefulness of a difference table in many instances. Let us keep in mind that the operation with higher differences presupposes a certain smoothness of the function. This means that the higher differences decrease rapidly and are quickly negligibly small. A function tabulated in small intervals—such as a customary logarithm table for example—has seldom differences higher than second order which are still significant. Here the interpolation is reduced to one or two terms. On the other hand, the use of higher differences can be very useful by making possible a tabulation which proceeds in much larger intervals. But then it is necessary that the functional values shall be given with *very great accuracy*. A closely spaced table of a few decimal places is thus replaceable by a widely spaced table, provided that the key values are given with excessive accuracy in order to make a differencing up to sufficiently high order possible. Such conditions, however, cannot be matched if the key values of the table are not mathematically calculated, but physically observed quantities. The interval in which the observations proceed is usually small enough to make an interpolation by two or three central differences sufficiently accurate. But the difficulty is that the magnification of the errors in the higher differences completely masks the true values of the higher differences because of two opposing trends: the true differences strongly decreasing, the errors strongly increasing. The result is that the difference table of an empirically observed function behaves quite differently from that of a mathematically evaluated function. The first and second differences are not too much modified, but the *relative errors* are already large. The higher differences show a completely erratic behavior and cannot be used for computational purposes.

Under these circumstances we have to discard the methods of higher differences if empirical and not mathematical functions are involved. We have to invent other methods for effective interpolation and differentiation of empirical functions, making use of the principles of "least squares." One particular feature of a difference table, however, deserves honorable mention: if there are *glaring single errors* among our observations which completely disrupt the smooth flow of a function, such errors stand out like illuminated spots against a generally dark background, if the test of a central difference table is applied. The fourth or fifth difference will now show some exceptionally high maxima, surrounded by terms of alternating sign. This locates the "bad" spots in our measurements and we should first adjust the data in a way that these anomalous peaks shall disappear. We can do that by dividing the peak value by the binomial coefficient

$$(-1)^k \binom{2k}{k}$$

if the peak appeared in the $2k$th difference, and subtracting this quantity from the y entry of the line in which the peak appeared.

6. The fundamental principle of the method of least squares. Let us assume that two variables are connected by a mathematical law whose form is known by hypothesis, although some of the constants of the law are unknown. If we have as many measurements as the number of unknown parameters demands, determination of these parameters is a purely algebraic problem. If, however, we have made more observations than necessary, we have a mathematical situation which we call "overdetermined" since the number of equations is greater than the number of unknowns. If the measurements were free of errors, we could simply discard the surplus equations, since they do not contribute anything new to the previous statements. Since, however, the measurements are not accurate, each new observation adds some new information of its own. The equations as they stand are not solvable and have to be "adjusted." We do that by taking the difference between the theoretical value and the observation, calling this quantity the "residual." While it is in general impossible to find values of the unknown parameters which would make each of the residuals equal to zero, we can always find values of these parameters which will make the *sum of the squares of the residuals a minimum*. The magnitude of this minimum gives us a measure of the closeness of our observations. Assuming that the postulated mathematical law is correct, the minimum "zero" would mean that each residual is zero and therefore all our observations are perfect. The larger the minimum, the more are our observations at error. The square root of the sum of the squares of the minimized residuals, divided by their number, can be considered the "average error" of our measurements, although it is sometimes advisable to examine the *distribution* of the residuals and see whether some residuals are not conspicuous by their magnitude. If we find residuals which are more than 3 times the average error, we will prefer to discard these measurements altogether and minimize the sum of the squares of the remaining residuals.

In many problems of physics and engineering the unknown parameters enter the given mathematical law in a *linear* fashion. The problem of minimizing the sum of the squares of the residuals is then equivalent to the minimizing of an algebraic expression of second order. The resultant equations are linear, with a symmetric and nonnegative (in most well-formulated problems even positive definite) matrix. These equations are called the "normal equations" of the given problem.

The same principle is applicable in all expansion problems which involve orthogonal function systems. The problem is here to expand a given function into a linear sequence of prescribed functions. If this sequence is *finite*, we cannot obtain a perfect answer. Yet we can always obtain a "best" answer by taking the difference between the given function and its expansion and considering this difference as the "residual" of our approximation. Since we are not interested in approximating the function in one particular point only, but everywhere within a given continuous domain, the method of least squares now requires that we *integrate* the square of the residual over the given domain. The resultant normal equations then give the "best" approximation of $f(x)$ in that domain, in terms of the prescribed functions. This procedure is the basis of the theory of orthogonal function systems and is of fundamental importance in almost all branches of applied analysis.

7. Smoothing of data by fourth differences. In a sequence of observations which scatter on account of accidental errors, the question of interpolation is quite different from the corresponding problem of a mathematically tabulated function. The data points are usually so close together that mere linear interpolation would already suffice. The difficulty is, however, that the basic values themselves are inaccurate and would give a very "bumpy" curve if we simply join them by straight lines. We have to "smooth" our data in order to draw further conclusions from them. This means that by some statistical considerations we try to reduce the influence of the random errors. One simple method, which is the analytical counterpart of the French curve type of smoothing, is based on the use of fourth differences.

We argue as follows. We assume that our data are sufficiently close together to justify the hypothesis that the second derivative does not change essentially during the course of a few measurements. In particular, we want to combine every measurement with its two neighbors to the left and to the right. This gives us altogether five consecutive observations, and we assume that these observations would lie very nearly on a parabola of second order, were it not for experimental errors. The theoretical course for these measurements is then given by the law

$$y = a + bx + cx^2 \tag{5-7.1}$$

where the coefficients a, b, c have to be adjusted to our data. However, these 3 parameters cannot be adjusted to 5 data, and thus we use the principle of least squares. We form the difference

$$\sum_{k=-2}^{+2} (y - y_k)^2 \tag{5-7.2}$$

and minimize this quantity with respect to a, b, c. In our case the data belong to the points $x = -2, -1, 0, 1, 2$, and thus the function to be minimized becomes

$$
\begin{aligned}
& (a - 2b + 4c - y_{-2})^2 \\
+ & (a - b + c - y_{-1})^2 \\
+ & (a \qquad\qquad - y_0)^2 \\
+ & (a + b + c - y_1)^2 \\
+ & (a + 2b + 4c - y_2)^2
\end{aligned}
\tag{5-7.3}
$$

The condition of minimum with respect to a and c gives the two "normal equations"

$$5a + 10c - \sum_{k=-2}^{+2} y_k = 0 \tag{5-7.4}$$

$$10a + 34c - \sum_{k=-2}^{+2} k^2 y_k = 0$$

The equation for b is of no interest to us at the present moment. Our goal is to correct the center value y_0, which belongs to $x = 0$. But then equation (1) shows that the theoretical value at $x = 0$ is

$y = a$. Hence we need only a. The solution of (7.4) for a yields

$$70a = -6y_{-2} + 24y_{-1} + 34y_0 + 24y_1 - 6y_2 \qquad (5\text{-}7.5)$$
$$= 70y_0 - 6(y_{-2} - 4y_{-1} + 6y_0 - 4y_1 + y_2)$$
$$= 70y_0 - 6\,\delta^4 y_0$$

where $\delta^4 y_0$ denotes the fourth central difference of the zero line. We thus find
$$a = y_0 - \tfrac{3}{35}\,\delta^4 y_0 \qquad (5\text{-}7.6)$$
and obtain the following simple method by which the "random scatter" of a sufficiently close set of data frequently can be effectively reduced. Construct a central difference table up to the order 4. Subtract from each ordinate $\tfrac{3}{35}$ of the fourth difference associated with that particular line. Since $\tfrac{3}{35}$ is very near to $\tfrac{1}{12}$, we can say that the correction is practically $= -\tfrac{1}{12}\delta^4 y_0$. Every single datum may be corrected by this amount. The new data will fit a much smoother curve. We can demonstrate that by constructing once more a difference table and showing that the fourth differences are now considerably reduced and less irregular than they were before. It is frequently advisable to give the corrected data to one more decimal place than the original data. This will assure a smoother fit. If the fourth differences still show a rather irregular pattern the method may be repeated a second time.[1]

The following numerical example shows the application of this method by applying it to the take-off performance measurements of an airplane. The observations were made from second to second,

[1] If the observations are so close together that 7 points, viz., 3 neighbors on both sides plus the central point, can be joined by a least-square parabola of second order, the correction formula becomes
$$\bar{y}_0 = y_0 - \frac{9\delta^4 y_0 + 2\delta^6 y_0}{21}$$

In both cases the operation with a central difference table can be avoided and replaced by a "movable strip technique" of the type (8.5). The code numbers of the movable strip are in the case of two neighbors,

$$\frac{-3, \quad 12, \quad \overset{\downarrow}{17}, \quad 12, \quad -3}{35}$$

and in the case of three neighbors,

$$\frac{-2, \quad 3, \quad 6, \quad \overset{\downarrow}{7}, \quad 6, \quad 3, \quad -2}{21}$$

The arrow indicates the "zero line" opposite to which the result of the weighting is written.

but in the table only every *second* measurement is recorded. This was done because the observations were divided into two independent groups, those belonging to the even and those to the odd seconds. Each group was independently adjusted by the same method, comparing the resulting graphs. This provides a valuable check on the results by obtaining an estimation of how much fluctuation is caused by the scattering of the data. The following table contains the calculation for one single group only. The other group was handled in identical fashion and gave consistent results.

t	y	δy	$\delta^2 y$	$\delta^3 y$	$\delta^4 y$	$-\frac{3}{35}\delta^4 y$	$\bar{y}$
0	0						0.8
		25.0					
2	25		17.4				20.7
		42.4		44.8			
4	67.4		62.2		−95.5	8.2	75.6
		104.6		−50.7			
6	172		11.5		22.1	−1.9	170.1
		116.1		−28.6			
8	288.1		−17.1		134.4	−11.5	276.6
		99.0		105.8			
10	387.1		88.7		−201.2	17.2	404.3
		187.7		−95.4			
12	574.8		−6.7		127.2	−10.9	563.9
		181.0		31.8			
14	755.8		25.1		−37.1	3.2	759.0
		206.1		−5.3			
16	961.9		19.8		17.4	−1.5	960.4
		225.9		12.1			
18	1187.8		31.9		−28.1	2.4	1190.2
		257.8		−16.0			
20	1445.6		15.9		46.5	−4.0	1441.6
		273.7		30.5			
22	1719.3		46.4		−59.8	5.1	1724.4
		320.1		−29.3			
24	2039.4		17.1		7.5	−0.6	2038.8
		337.2		−21.8			
26	2376.6		−4.7		31.2	−2.7	2373.9
		332.5		9.4			
28	2709.1		4.7		−14.2	1.2	2710.3
		337.2		−4.8			
30	3046.3		−0.1				3046.4
		337.1					
32	3383.4						3383.3

The first two and the last two data are not covered by the general procedure, since here we do not possess neighbors on both sides. We are thus forced to use neighbors on *one side only* and we will lay a least-square parabola of second order $a + bx + cx^2$ through five consecutive points. This still amounts to the same procedure as before, since we can agree that the value $x = 0$ shall belong to the third point, designating the first two points by $x = -2$ and $x = -1$. However, we now need the theoretical value (1) also for these two points, adopting

$$\bar{y}_{-2} = a - 2b + 4c \qquad \text{and} \qquad \bar{y}_{-1} = a - b + c$$

as the corrected values of the first two observations. The calculations give the following result:

$$\bar{y}_{-2} = y_{-2} + \tfrac{1}{5}\delta^3 + \tfrac{3}{35}\delta^4$$

$$\bar{y}_{-1} = y_{-1} - \tfrac{2}{5}\delta^3 - \tfrac{1}{7}\delta^4$$

with the understanding that for δ^3 and δ^4 we use the nearest available central differences, found along the upper slanted line. For example, the correction of $y(0)$, i.e., y_{-2}, becomes in our case

$$\bar{y}(0) = 0 + \tfrac{1}{5}(44.8) + \tfrac{3}{35}(-95.5) = 0.8$$

while the correction of $y(2)$, i.e., y_{-1}, becomes

$$\bar{y}(2) = 25 - \tfrac{2}{5}(44.8) - \tfrac{1}{7}(-95.5) = 20.7$$

The corresponding formulas for the other end of the table become

$$\bar{y}_2 = y_2 - \tfrac{1}{5}\delta^3 + \tfrac{3}{35}\delta^4$$

$$\bar{y}_1 = y_1 + \tfrac{2}{5}\delta^3 - \tfrac{1}{7}\delta^4$$

using the nearest differences of the lower slanted line. This means for our example

$$\bar{y}(32) = 3383.4 - \tfrac{1}{5}(-4.8) + \tfrac{3}{35}(-14.2) = 3383.2$$

$$\bar{y}(30) = 3046.3 + \tfrac{2}{5}(-4.8) - \tfrac{1}{7}(-14.2) = 3046.4$$

8. Differentiation of an empirical function. The definition of the derivative as the limit of a difference coefficient is of little value if our observations are not free of errors. The ratio $\Delta y / \Delta x$ gets excessively sensitive to even small errors if Δx becomes very small. The higher differences are even less reliable. Hence we cannot use the same methods for differentiation of equidistant data which were employed for mathematically accurate data. Once more we have to resort to least-square methods for solving the problem of differentiation.

The hypothesis we have made in the previous section, namely, that the acceleration changes little during five consecutive observations, will be valid in many situations. We can then lay a parabola of second order through these points, but this parabola has to be obtained by least-square minimization, since 5 points will generally not lie on a polynomial of second order. The procedure is exactly the same as the one followed in § 7. We combine every point with its two neighbors to the left and to the right. We again minimize the sum (7.3) and again obtain the normal equations (7.4). The only difference is that in our previous problem we wanted to obtain the corrected value of the *function* at the point $x = 0$, while now it is the corrected value of the *derivative* that we want to obtain at the same point. Hence it is now the constant b in which we are interested. The condition of minimum gives

$$10b = -2y_{-2} - y_{-1} + y_1 + 2y_2 \qquad (5\text{-}8.1)$$

or

$$b = \frac{-2y_{-2} - y_{-1} + y_1 + 2y_2}{10} \qquad (5\text{-}8.2)$$

If the interval between two observations is not 1 but h, then b has to be divided by h in order to get the derivative. Hence the final formula for the differentiation of an empirically given function with the help of five neighboring ordinates becomes

$$f'(x) = \frac{-2f(x - 2h) - f(x - h) + f(x + h) + 2f(x + 2h)}{10h} \qquad (5\text{-}8.3)$$

If not 2 but k neighbors are used on both sides, the formula becomes

$$f'(x) = \frac{\sum\limits_{\alpha=-k}^{+k} \alpha f(x + \alpha h)}{2 \sum\limits_{\alpha=1}^{k} \alpha^2 h} \qquad (5\text{-}8.4)$$

We apply this formula numerically by making use of the "movable strip" technique. We put on the movable strip the numbers -2, -1, 0, 1, 2, and obtain the smoothed derivative by performing the indicated multiplications and then moving the strip downward. The division by 10 merely changes the position of the decimal point.

The following data were obtained as the result of take-off measurements, giving the horizontal position of the airplane from second to second. We want to know the velocity of the airplane from second to second. The table indicates the method applied and the results of the calculation for a few data. The first two places are left blank since here we do not possess the neighbors on both sides.

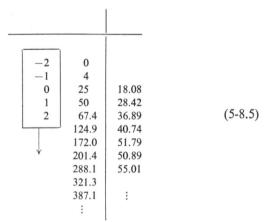

(5-8.5)

Procedure for the two first and last ordinates. At the beginning and at the end of our observations we lose out in neighbors, since only neighbors on *one side* are at our disposal. This destroys the symmetry of the process and reduces accuracy considerably. We are thus forced to lay a least-square parabola through the first few points and use its derivative at the points $t = 0$ and $t = 1$. It seems advisable to choose only *four* instead of five points at the beginning of the curve, since here the physical conditions are not settled yet and we cannot count on the smoothness of the curve to the extent that we can later on.

The solution of the normal equations for the present problem yields the following formulas for the first two missing velocities:

$$f'(0) = \frac{-21f(0) + 13f(h) + 17f(2h) - 9f(3h)}{20h}$$

$$f'(h) = \frac{-11f(0) + 3f(h) + 7f(2h) + f(3h)}{20h}$$

(5-8.6)

In our numerical example the first two missing velocities thus become

$$y'(0) = \tfrac{27}{20} = 1.35, \qquad y'(1) = \tfrac{237}{20} = 11.85$$

These formulas are applicable also at the *end* of a series of observations, in which case the ordinates $f(0), f(h), f(2h), f(3h)$ have to be changed to $f(x_n), f(x_n - h), f(x_n - 2h), f(x_n - 3h)$, and the formulas (6) give us $-f'(x_n)$ and $-f'(x_n - h)$.

It may happen that the sequence of our data is not rapid enough to justify the assumption that five consecutive data are practically on a parabola of second order. In such a situation we may feel safer if only *four* data are combined for one smoothing. We preserve symmetry if we agree that the velocities shall be obtained *halfway between the data*. The movable strip now contains four numbers only: $-3, -1, 1, 3$, and the zero line is halfway between ∓ 1. We now miss only *one* value at the beginning and at the end. That value is now obtained as the arithmetic mean of the two quantities of (6):

$$f'\left(\frac{h}{2}\right) = \frac{-8f(0) + 4f(h) + 6f(2h) - 2f(3h)}{10}$$

(5-8.7)

Applying this alternative method to the previous set of observations we now obtain the following set of velocity values:

−3	0	
		(6.6)
−1	4	
		17.10
1	25	
		21.52
3	50	
		31.71
	67.4	
		42.35
	124.9	
		44.91
	172.0	
		51.90
	201.4	
		53.46
	288.1	
		59.03
	321.3	
	387.1	

$$(5\text{-}8.8)$$

The first bracketed quantity was obtained on the basis of formula (7).

9. Differentiation by integration. It is of interest to observe that the method of differentiation here described is more an *integration* than a differentiation process. Let us assume that the sequence of our observations is very dense, i.e., that the h in the formula (8.4) is very small. Then in the limit the formula (8.4) goes over into the following integration process.

$$f'(x) = \frac{3}{2\varepsilon^3} \int_{-\varepsilon}^{+\varepsilon} t f(x+t)\, dt \qquad (5\text{-}9.1)$$

where ε is sufficiently small. We can see by expanding $f(x+t)$ into a Taylor series that the operation on the right side of (1) actually gives

$f'(x)$ with arbitrary accuracy at all points where $f(x)$ is analytic. We substitute on the right side

$$f(x+t) = f(x) + tf'(x) + \frac{t^2}{2} f''(x) + \frac{t^3}{6} f'''(x) + \cdots \quad (5\text{-}9.2)$$

and obtain as a result of the integration:

$$f'(x) \doteq f'(x) + \frac{1}{10} \varepsilon^2 f'''(x) + \cdots \quad (5\text{-}9.3)$$

Hence the error is of second order in ε. Moreover, formula (1) establishes a derivative even in points where a derivative in the ordinary sense does not exist. This is exactly what we want in the presence of "noise," since noise is a typically nonanalytical phenomenon which destroys the analytical nature of the true $f(x)$. This comes into strong profile in the behavior of the difference table of an empirically given table. Formulas (4.3) and (4.4) indicate that the central differences of odd order should be used for evaluation of the derivative. But these differences get invalidated practically from the beginning by the presence of noise, and the more so the smaller h is, since the differences decrease with h, while the noise remains on the same level, thus causing an intolerably large relative error. On the other hand, a difference table can be extended also to the left, in form of the differences of *negative* order, which actually mean *summation* instead of differencing. The damaging influence of the noise in these sums is greatly diminished, because of the randomness of the errors, which tend to balance them out in the process of summation.

The following formula of integrating by parts shows directly that the process indicated by the formula (8.4) is available if we possess the first two left columns of the difference table.

$$\int_a^b xu' \, dx = \left| xu \right|_a^b - \int_a^b u \, dx \quad (5\text{-}9.4)$$

If we put $u' = y$ and denote u by $D^{-1}y$ and the integral of u by $D^{-2}y$, we can write

$$\int_a^b xy \, dx = \left| xD^{-1} \right|_a^b - \left| D^{-2} \right|_a^b \quad (5\text{-}9.5)$$

We will thus show that the table (8.5) can be obtained in a numerically different manner which is sometimes even quicker than the previous algorithm. We now construct two columns of the differences of negative order, i.e., the sums and then the sums of the sums of the data. For the sake of simpler arrangements we write the columns to the *right* from the data column, although from the standpoint of the difference table they belong to the *left*:

x	y	Σy	$\Sigma^2 y$	
0	0	0	0	
1	4	4	4	
2	25	29	32	
3	50	79	112	
4	67.4	146.4	258.4	(5-9.6)
5	124.9	271.3	529.7	
6	172.0	443.3	973.0	
7	201.4	644.7	1617.7	
8	288.1	932.8	2550.5	
9	321.3	1254.1	3804.6	
10	387.1	1641.2	5445.8	

Let us consider the line $x = 8$, $y = 288.1$. Since two neighbors were used, we go to the line $x = 10$, while the lower line $x = 6$ is diminished by 1, thus giving us $x = 5$. In the column Σy we find the two numbers 1641.2 and 271.3. Applying the formula (5), the x in the first term of the right side has to be put equal to 2, and since at the lower limit $x = -2$, we form the *sum* $1641.2 + 271.3$ and multiply by 2.

$$(1641.2 + 271.3)2 = 3825.0$$

Now we come to the second term, which can be found in the column $\Sigma^2 y$. We go to the lines $10 - 1 = 9$ and $6 - 1 = 5$ and find there 3804.6 and 529.7, the difference of which is formed.

$$3804.6 - 529.7 = 3274.9$$

We now have two terms and obtain their difference.

$$3825.0 - 3274.9 = 550.1$$

This agrees—except for the position of the decimal point—with the value of $y'(x)$ found in table (8.5), opposite the entry $y = 288.1$. We have thus found an alternative method for construction of the velocity data.

In many problems of engineering we are interested in the integral of a function and also in the *moment* of the integral. The method discussed in the present section shows how both quantities can be obtained by constructing the Σy and the $\Sigma^2 y$ columns.

10. The second derivative of an empirical function. The difficulty of coping with the noise in the problem of differentiation is even more strongly emphasized if the *second* derivative of an empirically observed function has to be found. Yet such problems are frequent in analysis of tracking data, since we want to draw conclusions concerning the action of forces in aerodynamical problems; thus we are forced to obtain the acceleration of the displacement measurements. The difficulty can be illustrated by the fact that even a sudden change in the force will cause but a slight disturbance in the displacement. Smoothing our data causes a small error if the displacement itself is considered, but the second derivative may be altered by that process very considerably. Hence we cannot expect any great accuracy in the numerical evaluation of the second derivative.

Instead of trying to obtain the second derivative directly, it seems preferable to proceed in two steps, obtaining the first derivative by the process described before and then applying the same process once more to the first derivative. Since each single point of the derivative involves an interval of 5 observations, and again an interval of 5 derivative points contributes to one point of the derivative of the derivative, we find that *nine consecutive points* of the original curve determine *one* point of the acceleration curve. This gives a very good chance for averaging out accidental errors but works against us if the acceleration curve is in reality unsmooth because of more or less sudden changes of the force.

If we are not interested in the velocity curve but want to obtain the acceleration curve directly, we can combine the two operations into one movable strip operation, letting the movable strip $-2, -1, 0, 1, 2$ operate on $0, 0, \cdots, 1, 0, 0, \cdots$ and then repeating the operation on the result. We thus find that the code numbers of the second derivative process become

$$\frac{4, 4, 1, -4, -10, -4, 1, 4, 4}{100h^2} \qquad (5\text{-}10.1)$$

This process is numerically very convenient since all the multiplications by 4 can be combined into one operation. Once more we put the code numbers on a movable strip and let that strip glide down along the column of data. The result belongs to the time moment opposite to the code number -10.

As a numerical example, we carry through the process for the data of the table (8.5), obtaining now the second derivative $y''(x)$ from second to second. The h of our problem is 1, and the adjustment due to the denominator of (1) is merely a shift of the decimal point by 2 places to the left.

	y	y''	
4	0		
4	4		
1	25		
-4	50		
-10	67.4	7.97	(5-10.2)
-4	124.9	5.98	
1	172	4.64	
4	201.4	$\vdots$	
4	288.1		
	321.3		
	387.1		
	$\vdots$		

Procedure for the first four and last four ordinates. Since the general process requires 4 neighbors on both sides of each observation, we lose the first four acceleration values at the head of our table, and the same will happen again at the end. We can restore these values, however, although with diminished reliability, by using the separate technique (8.6) for evaluation of the first (and last) two ordinates of the velocity curve. The final result can be expressed in terms of weight factors which have to be applied to the first sequence of observations. We obtain a separate set of factors for the first, second, third, and fourth points of the acceleration curve. They are tabulated in the successive columns of the following table.

	$f''(0)$	$f''(h)$	$f''(2h)$	$f''(3h)$
y_0	115	85	53	26
y_1	-116	-76	-33	-4
y_2	-124	-84	-51	-18
y_3	118	58	13	-14
y_4	25	15	2	-8
y_5	-18	2	8	2
y_6			8	8
y_7				8

$$(5\text{-}10.3)$$

To be divided by $200h^2$.

In our numerical example (2), application of these weights leads to the following results:

$$y''(0) = 8.86$$
$$y''(1) = 8.78$$
$$y''(2) = 8.76$$
$$y''(3) = 7.66$$

$$(5\text{-}10.4)$$

We have pointed out already in the velocity procedure that the combination of 5 successive data may not be tolerable if the time marks of our observations do not follow each other closely enough. The table (8.8) was thus based on only *four* neighboring observations, and we have obtained the velocities halfway between the data points. Carrying through the same process twice, the original position of the points is once more restored. We now utilize only *three* instead of four neighbors on each side of every point. The code numbers for the acceleration curve now become

$$\frac{9, \quad 6, \quad -5, \quad -20, \quad -5, \quad 6, \quad 9}{100}$$

$$(5\text{-}10.5)$$

and we lose only 3 instead of 4 ordinates in the beginning and at the end of the curve.

Since the first velocity we possess is $f'(h/2)$, and the derivative of these data starts with $f''(h)$, we are unable to give $f''(0)$. Hence we have to leave a blank space opposite the first and the last observation. The weights for the next two accelerations are given in the following table which takes the place of the more elaborate table (3).

	$f''(h)$	$f''(2h)$
y_0	52	27
y_1	-54	-14
y_2	-44	-29
y_3	36	1
y_4	16	6
y_5	-6	9

(5-10.6)

To be divided by $100h^2$.

A comparison of the four-neighbor and three-neighbor technique of obtaining the acceleration of take-off measurements is given in the following table, which contains the first 20 acceleration data of a flight test, once evaluated by using four neighbors $\langle y'' \rangle$ and once by using three neighbors (y'') on both sides of each observation.

y	$\langle y'' \rangle$	(y'')		y	$\langle y'' \rangle$	(y'')
0	8.86			387.1	8.22	9.90
4	8.78	8.13		500.6	3.60	4.01
25	8.76	7.97		574.8	3.39	-1.59
50	7.86	8.59		650.1	3.81	3.84
67.4	7.97	8.08		755.8	6.26	7.46
124.9	8.98	6.31		815.8	10.17	11.70
172	4.64	4.03		961.9	8.25	8.31
201.4	6.12	4.39		1051.8	7.49	8.52
288.1	6.58	6.09		1187.8	5.21	3.91
321.2	8.76	11.33		1321.6	5.49	3.27

(5-10.7)

The general pattern of the acceleration course comes out similarly in both calculations. However, the amplitudes of the fluctuations are much larger in the second case where 3 instead of 4 neighbors were used. From the discrepancy of the two sets of results, the conclusion can be drawn that the second-to-second observations of this take-off were not sufficiently close together to make a satisfactory smoothing possible without disturbing at the same time the true course of the path. The first phase of an airplane flight is not smooth and far from the steady-state conditions which dominate the later development of the flight. The transient oscillations, caused by the coupling between horizontal and vertical motion, make themselves felt even during the time when the airplane is still on the runway. The condition that the acceleration does not change essentially during five or even four consecutive measurements was not fulfilled in the present problem. It would have been necessary to obtain four to five frames per second as the basis of our analysis if our aim is to get a reliable acceleration curve in which the random errors of the observations are eliminated and yet the true course of the acceleration is essentially preserved.

11. Smoothing in the large by Fourier analysis. In our previous discussions every observation was combined with its immediate neighbors to the left and to the right. We made use of the analytical nature of $f(x)$ in the neighborhood of a point and tried to eliminate the nonanalytical behavior of the noise by operations which did not leave the immediate neighborhood of a given point x. We can thus speak of "smoothing in the small," or the "neighbor technique," of eliminating noise. We will now consider an entirely different possibility. Instead of breaking our observations into a sequence of neighborhoods we may consider the *entire set of data* as one unified whole and try to find clues by which the true course of the function and the superimposed noise may be separated. This method of smoothing has the advantage that it is more independent of any special assumptions concerning the nature of the unknown $f(x)$. In our previous considerations we have assumed, for example, that in a certain finite neighborhood of a point, the second derivative of $f(x)$ is practically a constant. This means in physical terms that the force acting on a moving body changes but slowly within a specified time interval. Such assumptions may not always hold under actual aerodynamical conditions. A sudden gust, for example, represents a sudden and unforeseen change in the second derivative which does not satisfy our previous hypothesis of continuity. The neighbor technique of smoothing would tend to smooth out this discontinuity and thus change the true course of the second derivative by presupposing a smoothness which does not agree with the actual physical picture. Hence we may welcome a method of smoothing for which such assumptions need not be made.

Whenever approximation problems in the large are considered, the Fourier series appears automatically on the scene as one of the most powerful mathematical tools. We can thus attempt to analyze the problem of noise in terms of the Fourier series. We have observed already in our earlier discussions that the noise does not share with ordinary analytical functions the property of smoothness and differentiability. This property puts the noise into a special category with respect to the Fourier series.

The Fourier series is strictly speaking an infinite series, but the convergence of the series makes it possible to truncate the series to the first n terms. How large this n has to be chosen will depend decisively on the analytical nature of the function to which the Fourier series is applied. If the function is everywhere continuous but the derivative is discontinuous at some point, the terms of the Fourier series decrease with the speed n^{-2}. If the function itself becomes discontinuous at some point, the terms decrease with the speed n^{-1} only. A pulse of short duration endangers the convergence of the series, and the formal expansion of an infinitely sharp pulse ("delta function") becomes actually divergent. Now "noise" can be conceived as an irregular sequence of sharp pulses, and thus the harmonic analysis of noise will not have the tendency to converge. Here, then, is a chance to distinguish between the true course of a function and the noise superimposed on it. The harmonic analysis of the function will show *much faster convergence* than the harmonic analysis of the noise.

In order to take advantage of this characteristic difference in the convergence behaviour of function and noise, it is necessary to make the dividing line as sharp as possible. Given is a large number of observations at the points

$$x = 0, h, 2h, \cdots, nh = l \qquad (5\text{-}11.1)$$

If we do not use proper precautions, the Fourier series set up for the approximation of the unknown $f(x)$ will have very poor convergence even if $f(x)$ is everywhere well behaved, because the boundary conditions are not fulfilled. Since neither $f(x)$ nor $f'(x)$ returns at $x = l$ to the value it had at $x = 0$, the discontinuity at the boundary will determine the rate of convergence. We can improve the convergence by reflecting $f(x)$ as an even function for negative x, and thus expand $f(x)$ into a pure *cosine* series. This dispenses with the discontinuity in the function itself, but the *derivative* is still discontinuous at the boundary. However, we can go one step further. We subtract from $f(x)$ a properly chosen $\alpha + \beta x$, thus considering

$$g(x) = f(x) - (\alpha + \beta x) \tag{5-11.2}$$

We determine α and β by the boundary conditions

$$g(0) = 0, \qquad g(l) = 0 \tag{5-11.3}$$

Then we reflect $g(x)$ for negative x as an *odd* function.

$$g(-x) = -g(x) \tag{5-11.4}$$

The result is that we have obtained a function which, if made periodic with the period $2l$, has no discontinuity in either function or derivative. The first discontinuity appears in the *second* derivative. The asymptotic order of magnitude of the Fourier terms is now n^{-3}, and that is a practically satisfactory convergence.

Hence we are going to develop the function $g(x)$ into a *pure sine series* of the form

$$g(x) = b_1 \sin \frac{\pi}{l} x + b_2 \sin \frac{2\pi}{l} x + \cdots \tag{5-11.5}$$

Since we have at our disposal the values

$$y_k = f(kh), \qquad (k = 0, 1, 2, \cdots, n) \tag{5-11.6}$$

we first modify $f(x)$ to

$$g(x) = f(x) - f(0) - \frac{f(l) - f(0)}{l} x \tag{5-11.7}$$

and achieve the boundary conditions (3). Then we determine the coefficients b_k of the expansion (5) by the condition that at the data points $x = kh$ the series shall give the modified basic data $g(kh)$, that is, the original measurements corrected by $\alpha + \beta x$. This yields

$$b_k = \frac{2}{n} \sum_{\alpha=1}^{n-1} g(\alpha h) \sin k\alpha \frac{\pi}{n} \qquad (5\text{-}11.8)$$

Now smoothing is always based on the fact that we have many more measurements at our disposal than are needed by the smoothness of the function. The harmonic analysis of $f(x)$ does not contain overtones beyond a certain frequency, usually called the "cutoff frequency" ν_0. This means that beyond a certain predictable point the Fourier coefficients b_k are practically zero. The highest order $k = m$ which need be considered is determined by the condition

$$\frac{m\pi}{l} x = 2\pi\nu_0 x \qquad (5\text{-}11.9)$$

or

$$m = 2\nu_0 l = 2\nu_0 nh \qquad (5\text{-}11.10)$$

This gives the condition

$$m/n = 2\nu_0 h \qquad (5\text{-}11.11)$$

The product $2\nu_0 h$, i.e., the double cutoff frequency times the time interval between two consecutive measurements, is a pure number which is decisive for the effectiveness with which we can smooth our measurements. We will denote the reciprocal of this number by the symbol p and call it "smoothing parameter."

$$p = 1/(2\nu_0 h) \qquad (5\text{-}11.12)$$

If p is smaller than 1, this means that our measurements are so far apart that they are unable to determine the course of $f(x)$ even in the absence of noise. If $p = 1$, we have just the minimum number of observations for the determination of $f(x)$ and nothing is left over for smoothing. Hence p must be larger than 1 in order to smooth at all. Generally the significance of the smoothing parameter p is the *ratio of the actual number of observations to the minimum number required in the absence of noise.* Effective smoothing demands that p shall be at least 2, but we will hardly be satisfied if it is less than 4 to 5.

Let us first assume that for physical reasons we can establish the position of the cutoff frequency ν_0 in advance. Then we can argue as follows. In the absence of noise the Fourier coefficients $b_1, b_2, \cdots, b_m$ would have certain values but would be practically zero beyond b_m. Then the Fourier synthesis

$$g(x) = b_1 \sin \frac{\pi}{l} x + b_2 \sin 2 \frac{\pi}{l} x + \cdots + b_m \sin m \frac{\pi}{l} x \quad (5\text{-}11.13)$$

would properly interpolate our function, not only in the data points but at *all* points of the range.

The presence of noise changes the picture. The coefficients $b_{m+1}, \cdots, b_{n-1}$ are no longer zero nor have they the tendency to diminish. They represent the nonconvergent part of the Fourier series, caused by the nonanalytical nature of the noise. An ideally "random" noise would have a Fourier spectrum which has no preference for any frequency and would thus have an average amplitude with random fluctuations for all frequencies. If our analysis included the sine and cosine functions, we could speak of a "random distribution of phase," while the amplitudes would remain of the same constant order of magnitude. Since our analysis contains only sine functions, the randomness of the phase is replaced by a random sequence of plus and minus signs in the distribution of the b_k amplitudes; $(k > m)$.

Now the amplitudes b_k of the harmonic analysis are influenced by the noise in two ways. The amplitudes beyond b_m are *completely* caused by noise. In this part of the spectrum we can eliminate the noise altogether by simply *omitting* in the Fourier synthesis every term beyond $k = m$. Originally we assumed the Fourier series in the form

$$g(x) = \sum_{k=1}^{n-1} b_k \sin k \frac{\pi}{l} x \quad (5\text{-}11.14)$$

and determined the coefficients b_k by the condition that our data shall be exactly represented; now we *truncate* the series by forming the sum

$$\bar{g}(x) = \sum_{k=1}^{m} b_k \sin k \frac{\pi}{l} x \quad (5\text{-}11.15)$$

By this process we have eliminated all the high-frequency components of the noise. Now it is true that even the b_k components for $k < m$ are to some extent influenced by the noise. But in this part of the spectrum we are unable to separate noise and true component, since the randomness of the sign prevents us from subtracting the noise part of the component. This uncertainty is unavoidable. Yet we have succeeded in eliminating the *major portion* of the noise by disregarding that part of the Fourier series which is completely caused by noise.

In actual practice we will seldom possess the frequency ν_0 in advance. Even if we know that for physical reasons no frequency beyond a certain ν_0 can appear in the measured $g(x)$, there is no guarantee that *all* the frequencies up to ν_0 are genuine. The smoothness of $g(x)$ may be such that the genuine spectrum ends much sooner than ν_0, and all amplitudes beyond that point are spurious and should be omitted. The smaller the number of components of the genuine function $\bar{g}(x)$ is, the more we will succeed in eliminating the noise of our measurements.

Now the subtraction of a linear trend $\alpha + \beta x$ from our observations made it possible to reduce the unsmoothness at the boundary to a discontinuity in the *second* derivative. This is a mathematical condition which is frequently matched by the actual physical situation. In tracking problems, a discontinuity in the second derivative means a sudden change in the force. Such a sudden change in the force, necessitated mathematically by our desire to make our function periodic, can occur also physically during our observations on account of sudden gusts, or running out of fuel, or separation from the booster, and similar effects. Hence the physical unsmoothness of the problem is of the same character as the unsmoothness at the boundary, and the n^{-3} law of the coefficients could not be improved even if the function were genuinely periodic.

12. Empirical determination of the cutoff frequency ν_0. Under the given conditions we do much better if we do not try to determine ν_0 in advance but let our data themselves make the decision. The decrease of the Fourier components according to the n^{-3} law is sufficiently steep for a rather effective separation of the analytical and the nonanalytical parts of the spectrum. We make a complete Fourier analysis of our data and plot the components b_k obtained as ordinates at the abscissa values $k = 1, 2, 3, \cdots$. In the beginning, the b_k are large, and no particular law can be detected in their distribution. But then they diminish rather steeply to relatively small values which do not decrease any further. From here on we observe that the amplitudes remain within a certain band of plus-minus values. They have no tendency to become either much larger or much smaller than a certain average size. This average size can be ascertained by starting from the end of the spectrum and evaluating the sum

$$\beta^2 = \frac{1}{N} \left(b_{n-1}^2 + b_{n-2}^2 + \cdots + b_{n-N}^2 \right) \qquad (5\text{-}12.1)$$

As N increases, β will approach a fairly constant value. We then draw the horizontal lines $y = \pm\beta$ and determine the point $k = m$ where these lines intersect the low-frequency part of the spectrum. We keep all the b_k up to $k = m$ and omit all the b_k with $k > m$. *The mere truncation of the Fourier series to m terms performs the smoothing of our data.*

It is impossible to determine the *exact* point $k = m$ where we should terminate the series. A certain "twilight zone" is inevitable in which the b_k caused by the function and those caused by the noise are of the same order of magnitude. This uncertainty, however, is of no critical importance. We have the inevitable uncertainty caused by the low-frequency noise in the analytical components which could not be eliminated. In the face of this uncertainty it makes very little difference whether a few spurious Fourier coefficients have been added or not, or whether we have omitted a few Fourier coefficients which actually belong to the function but which have been interpreted as noise. It is important only that the twilight zone shall not be too extended. But this danger does not exist, since the n^{-3} law guarantees a sufficiently steep decrease of the genuine amplitudes to separate them from the nondecreasing noise components.

The following numerical example gives an actual demonstration of the method. A sequence of 68 take-off data was subjected to a Fourier sine analysis, after subtraction of the linear trend $\alpha + \beta x$. The Fourier components b_k were evaluated on the basis of formula

(11.8) (with $n = 67$), with the help of I.B.M. equipment. The analysis gave the following values for the 66 sine components of the given function $g(x)$.

k	$-b_k$	k	$-b_k$	k	$-b_k$	k	$-b_k$	
1	2024.40	17	2.33	33	4.50	49	1.27	
2	200.65	18	−5.05	34	−2.47	50	−3.54	
3	115.14	19	2.65	35	2.99	51	1.83	
4	20.35	20	−1.80	36	−2.85	52	−0.23	
5	16.08	21	2.53	37	−0.48	53	2.83	
6	5.62	22	−0.87	38	−0.07	54	1.71	
7	8.31	23	−1.58	39	0.02	55	−3.59	
8	6.89	24	0.60	40	2.29	56	2.50	(5-12.2)
9	0.95	25	−4.33	41	−1.14	57	−4.35	
10	7.62	26	4.26	42	0.49	58	2.93	
11	−2.30	27	−3.82	43	−2.42	59	0.34	
12	3.24	28	4.70	44	−0.46	60	−0.12	
13	1.97	29	−4.45	45	−0.04	61	0.62	
14	1.31	30	0.55	46	0.58	62	−2.57	
15	3.89	31	−1.13	47	2.18	63	0.12	
16	−2.86	32	−2.14	48	−1.91	64	−1.19	
						65	2.25	
						66	0.67	

Examination of this table shows that the "noise" in the present problem is not of a completely random character. The sequence of the $\pm$ signs shows a definite regularity, with many systematic $+$, $-$ alternations. Such regularities can occur if we have isolated "glaring errors" among our observations which fall out of the average order of magnitude of errors. We could have eliminated these "bad spots" right at the beginning by the device of differencing; this would have made the noise more incoherent. However, the method of the Fourier series has the advantage that these bad spots do no essential damage to the smooth part of the function, but merely increase the noise to some extent. Hence we get valid results even without removing the obviously bad observations from our data.

The "twilight zone" is not extensive in our problem. The ordinate $b_{10} = 7.62$ is definitely outside the noise band and has to be included in the analytical part of the function. We may terminate the series here, but we may also include b_{11} and b_{12} in our expansion and terminate our series at $k = 12$. The uncertainty is thus of no vital consequence. We shall decide on $k = 12$ and omit all overtones beyond the 12th. The number of observations is thus about 5.5 times as large as the needed number of components. Hence we can assume that the truncated series, apart from being smooth, has eliminated about 80% of the total noise. The remaining 20% is present in the form of a contamination of the retained first twelve b_k components.

The question of differentiating this series still remains. The fact that all overtones beyond the 12th are negligible for the displacement does not mean that the same will be true for the velocity and even less for the acceleration. The fact that we have a good approximation for $f(x)$ does not imply that we have a good approximation for $f'(x)$ or $f''(x)$ as well. The derivative of a good approximation is not necessarily a good approximation of the derivative of the function. The function $f''(x)$ is much less smooth than $f(x)$ itself and may require a considerably larger number of Fourier terms for its representation than $f(x)$. But these higher order terms are not available by differentiating the Fourier series of $f(x)$ because the b_k associated with these k are almost completely caused by noise and are not indicative of the behavior of $f(x)$ in the absence of noise.

Under these circumstances we have to return to the local operations by which a derivative can be defined. A derivative is by its very nature determined by the values of $f(x)$ in the neighborhood of a certain point. The law "mass times acceleration equals moving force" shows that in a sufficiently small time interval the acceleration cannot change too rapidly. We do not go wrong if we assume that five neighboring observations lie on a parabola of third order. Then the second derivative has still enough freedom to change linearly in this time interval. If this condition is not fulfilled, our observations are altogether too far apart from each other to allow any effective smoothing. A cubic parabola of the form $y = \alpha + \beta x + \gamma x^2 + \delta x^3$ has four degrees of freedom, and thus we determine four constants on the basis of five observations. The amount of overdetermination is thus slight and we did not violate the fidelity requirements of our problem. If we take the second derivative of this local least-square parabola at the point of symmetry, we obtain the following symmetric weighting of five successive ordinates, which take the place of the previous weight factors (10.1):

$$\frac{2, \quad -1, \quad -2, \quad -1, \quad 2}{7h^2} \qquad (5\text{-}12.3)$$

In the previous scheme every observation was combined with *four* of its neighbors on both sides. Hence *nine* consecutive data determined one point of the acceleration curve. This is justified if the observations follow each other in sufficiently close intervals. But if this condition is not fulfilled, the previous method (10.1) will lead to oversmoothing. We will blot out certain details of the acceleration curve which are real and not caused by noise. It is necessary to keep in mind that the problem of smoothing has two aspects. One is that we should eliminate noise as much as possible; the other is that we should not eliminate details which actually belong to the function to be observed. We have no a priori reasons to assume that the acceleration curve will be particularly smooth if the flight of an airplane is in question. Sudden gusts may interfere with the action of the regular aerodynamical forces, but even the ordinary aerodynamical forces alone can cause complicated deviations from steady-state flight. If we want to study the acceleration curve realistically, we will try to avoid oversmoothing the curve. The weighting (3) has better fidelity chances than the previous weighting (10.1), since it involves only five instead of nine consecutive data. On the other hand, the danger is now that we have not succeeded sufficiently with elimination of the noise. The acceleration curve thus obtained may be too unsmooth because of observational errors.

At this point the Fourier analysis technique may be invoked for additional smoothing. If n is the total number of observations and we evaluate $n - 1$ sine coefficients for the Fourier analysis of the acceleration data obtained by local least square parabolas, we can assume that not all these coefficients will actually be needed for representation of the true acceleration. The highest coefficients correspond to high-frequency oscillations which for aerodynamical reasons have to be considered highly implausible. An earlier termination of the series will thus properly smooth the acceleration curve, without violating essentially the fidelity requirements. A combination of slight local smoothing with additional smoothing by

Fourier truncation can thus be considered the most plausible solution of the problem of smoothing, if great caution is demanded by the fact that the data do not follow each other closely enough to allow effective smoothing by local parabolas alone.

The numerical example on page 342 shows the effect of smoothing by two different techniques. A sequence of 68 take-off observations was analyzed to obtain the acceleration at every point of the curve. The column s contains the actual displacement data, in feet, taken at intervals of $h = 0.96$ second. The column γ contains the acceleration, evaluated by the method of § 10. The first four and the last four γ values were obtained with the help of the table (10.3).

The column a has a different origin. The weighting was now done according to the scheme (12.3), omitting first the denominator $7h^2$. The first two and the last two acceleration data were obtained on the assumption that the local parabola constructed at the point $k = 2$ (and likewise $k = n - 2$) can be used for obtaining the acceleration at the missing points. This gives the following table, in correspondence to the previous table (10.3).

	$f''(0)$	$f''(h)$	
y_0	9	5.5	
y_1	-15	-8	
y_2	-2	-2	To be divided by $7h^2$ (5-12.4)
y_3	13	6	
y_4	-5	-1.5	

The values thus obtained (without the common denominator $7h^2$) are contained in column a. These values show considerable scatter and require additional smoothing. For this purpose the method of the Fourier analysis was employed. A Fourier sine series was applied in accordance with the principles of § 11. In our example the subtraction of a linear trend $\alpha + \beta x$ could be omitted because the acceleration could easily be extrapolated to zero at both ends of the series. The data were thus directly suited to a sine analysis. The series was then truncated to 30 terms. The synthesis of these 30 terms, evaluated for the points of observation, gave the column $\bar{a}$. Finally, dividing by $7h^2$, we obtained the last column $\bar{\gamma}$.

k	s	γ	a	$\bar{a}$	$\bar{\gamma}$	k	s	γ	a	$\bar{a}$	γ
0	117	−0.74	14	12.1	1.86	34	4042	2.85	19	16.3	2.51
1	144	2.38	22	17.7	2.72	35	4230	2.84	17	21.5	3.31
2	163	5.69	30	38.8	5.96	36	4425	2.57	23	21.9	3.37
3	192	8.34	74	67.1	10.31	37	4619	2.29	12	13.9	2.14
4	229	9.48	75	78.5	12.07	38	4819	1.82	12	6.0	0.92
5	281	8.70	67	62.9	9.67	39	5017	2.03	7	8.4	1.29
6	340	6.98	34	38.0	5.84	40	5218	2.52	11	18.5	2.84
7	407	5.55	26	26.8	4.12	41	5420	3.18	31	25.3	3.89
8	472	5.05	34	31.1	4.78	42	5623	3.48	25	23.7	3.64
9	545	5.11	39	35.8	5.50	43	5839	3.27	24	19.7	3.03
10	625	5.40	27	34.0	5.23	44	6047	3.53	5	20.5	3.15
11	706	6.02	35	34.8	5.35	45	6266	3.39	36	24.4	3.75
12	792	6.69	51	44.1	6.78	46	6479	3.51	28	25.5	3.92
13	887	7.01	49	52.7	8.10	47	6708	2.87	18	21.3	3.27
14	989	6.82	47	49.2	7.56	48	6933	1.95	5	13.7	2.11
15	1096	6.37	39	38.0	5.84	49	7157	1.37	16	5.5	0.84
16	1212	5.91	35	33.0	5.07	50	7389	0.90	3	3.6	0.55
17	1329	5.79	37	37.5	5.76	51	7618	1.68	−6	4.1	0.64
18	1453	5.66	41	41.3	6.35	52	7845	2.79	19	18.1	2.77
19	1584	5.33	35	36.9	5.67	53	8075	3.97	43	32.9	5.05
20	1718	5.24	32	30.5	4.69	54	8312	4.72	30	36.3	5.58
21	1858	5.10	32	30.8	4.73	55	8557	4.67	24	29.5	4.53
22	2002	5.04	36	35.5	5.46	56	8798	4.57	32	25.8	3.97
23	2150	4.91	32	35.5	5.46	57	9049	5.00	37	30.7	4.72
24	2306	4.51	30	29.1	4.47	58	9305	3.63	17	30.0	4.61
25	2462	4.17	25	23.6	3.63	59	9562	−0.07	12	8.5	1.31
26	2625	3.82	28	23.9	3.67	60	9821	−3.30	−16	−22.8	−3.51
27	2790	3.69	18	25.0	4.00	61	10082	−4.55	−35	−33.2	−5.10
28	2959	3.71	27	25.4	3.90	62	10330	−0.32	−18	−9.4	−1.44
29	3129	3.76	28	23.7	3.64	63	10578	2.04	32	25.3	3.89
30	3307	3.46	25	24.0	3.69	64	10830	2.00	42	38.4	5.90
31	3486	3.17	17	23.8	3.66	65	11092	0.99	18	25.2	3.87
32	3668	3.10	23	19.9	3.06	66	11356	0.64	11	7.7	1.18
33	3853	3.12	16	15.2	2.34	67	11616	0.30	4	0	0

For the sake of comparison, the two curves γ and $\bar{\gamma}$ are plotted in conjunction (cf. the accompanying figure: γ = dotted line, $\bar{\gamma}$ = solid line). We notice that the nine-point smoothing oversmooths the curve by blotting out certain details of the acceleration curve which belong to the data and are not the result of the smoothing procedure. The origin of these oscillations cannot be decided on the basis of the data alone. A careful examination of the physical situation would be needed to obtain more information about these peculiar details of the $\bar{\gamma}$ curve. The Fourier analysis reveals unmistakably the presence of a certain high-frequency component. But on the basis of the data alone we cannot decide whether these oscillations belong to the aerodynamical situation or to the measuring instruments.

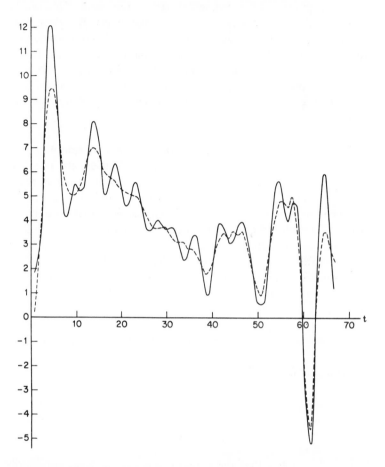

The present example illustrates the difficulties we may encounter in evaluation of data which are not obtained under optimum conditions. If the aerodynamical analysis is not interested in all the accidental details of the take-off phenomenon, then the stronger smoothing of the γ curve will be quite sufficient to give the essential features of the event. But the existence of additional oscillations of small amplitudes can be, if properly understood, of considerable interest. An adequate study of these finer features demands recording instruments of superior accuracy and a sufficiently close sequence of observations in order to eliminate the instrumental part of the noise. This can be done with the help of local least-square parabolas. But then, after obtaining a sequence of acceleration data which are considerably free of instrumental noise, we can subject these data to an additional Fourier analysis. The components which stand out above the noise level will reveal the existence of hidden periodicities, the study of which may contribute to a deeper understanding of aerodynamic phenomena.

13. Least-square polynomials. In the problem of smoothing we have encountered the process of laying a least-square parabola through four or five data points. Similar situations occur in many other problems of physics and engineering, and thus a general treatment of the subject is justified. Gauss, who first treated this problem, introduced an elegant notation which brings the resultant equations in particularly lucid form. He uses the bracket expression $[u]$ in the following sense. The quantity u shall be taken at all the data points and the sum of all these values shall be formed. Hence, if the data points are distinguished by the subscript $i = 1, 2, \cdots, n$, the notation $[u]$ shall mean

$$[u] = u_1 + u_2 + \cdots + u_n \tag{5.13-1}$$

consequently $[x^k]$ shall mean

$$[x^k] = x_1^k + x_2^k + \cdots + x_n^k$$

and similarly

$$[yx^k] = y_1 x_1^k + y_2 x_2^k + \cdots + y_n x_n^k$$

It is not assumed that the observations are made at necessarily *equidistant* values of the independent variable x.

The general problem of a least-square polynomial can be stated as follows: We have theoretical reasons to assume that a set of observations

$$y_1, y_2, \cdots, y_n \tag{5-13.2}$$

which belong to an unknown function $y = f(x)$ at some prescribed points of the independent variable

$$x = x_1, x_2, \cdots, x_n \tag{5-13.3}$$

can be fitted by a polynomial of the order m.

$$y = a_0 + a_1 x + \cdots + a_m x^m \tag{5-13.4}$$

We know the order m of the polynomial, while the coefficients $a_0, a_1, \cdots, a_m$ are at our disposal and to be determined by the measurements. We assume that the number of observations surpasses m, otherwise the problem has no unique solution.

The unknown coefficients a_i of the polynomial (13.4) are now determined by the following principle. At each point of observation we form the "residual"

$$a_0 + a_1 x_i + \cdots + a_m x_i^m - y_i \tag{5-13.5}$$

and take the sum of the squares of all these residuals.

$$Q = \sum_{i=1}^{m} (a_0 + a_1 x_i + \cdots + a_m x_i^m - y_i)^2 \tag{5-13.6}$$

The quantity Q is by nature positive or in the limit zero. The zero value is only possible if each one of the residuals vanishes, that is, if our measurements are all consistent and fit an mth order polynomial *exactly*. This cannot be expected. We can, however, find a_i values for which the sum Q becomes a *minimum*. We consider the polynomial associated with these a_i as the "best fit" of our measurements; (cf. also § 16).

This minimum principle has a unique solution in the form of a linear set of equations with nonvanishing determinant. According to the principles of calculus the condition of minimum requires that the partial derivative of Q with respect to any a_i shall vanish. This leads to the equations

$$a_0[x^0] + a_1[x] + \cdots + a_m[x^m] = [y]$$

$$a_0[x] + a_1[x^2] + \cdots + a_m[x^{m+1}] = [xy] \tag{5-13.7}$$

$$\vdots$$

$$a_0[x^m] + a_1[x^{m+1}] + \cdots + a_m[x^{2m}] = [x^m y]$$

These remarkable equations, called the "normal equations" of the least square problem, belong to a group of linear equations which have not only a symmetric but even a "recurrent" matrix, characterized by the property

$$a_{ik} = a_{i-1, \, k+1} \qquad (5\text{-}13.8)$$

The matrix of this system depends on only $2m + 1$ different quantities, instead of the usual full number $(m + 1) \cdot (m + 2)/2$. We have encountered this type of equations before (cf. 3-3.5) and seen that there exists a successive algorithm for their solution.

In the most important case of equidistant observations, the scale factor of x can be normalized in such a way that the constant interval between two observations shall become 1. Moreover, the origin $x = 0$ of the variable x can be put symmetrically into the mid-point of the total range. Then for every given m and every given n a numerical solution of the system (7) can be found, and the results can be tabulated in such a form that for every coefficient a_i the given data y_k are to be multiplied by a pretabulated set of numbers. This obviates the necessity of solving the system (7) separately in every case. Tables of this kind, for a reasonably small range of m and a reasonably large range of n, have actually been published.[1]

14. Polynomial interpolations in the large. If a function $f(x)$ is observed in equidistant intervals, the question arises what the values of $f(x)$ are *between* the data points. In § 2 was developed the Gregory-Newton formula, which interpolates the function between the data points by a polynomial in x. In § 3 were developed the Stirling and Bessel kind of interpolation formulas which operate with central differences and give much better convergence. In both kinds of interpolation we take it for granted that the function *can* be interpolated with the help of a power series. Although this assumption seems very reasonable, its validity is actually by no means guaranteed. Let us assume that a function $f(x)$ exists in an *infinite* range, from 0 to infinity, and is even analytical throughout this range. Let us give this function in the infinity of points $x = 0, 1, 2, \cdots$. Can we now interpolate this function between the data points with the help of a

---

[1] Cf. H. T. Davis, *Tables of the Higher Mathematical Functions*, Vol. I (The Principia Press, Bloomington, 1933).

Gregory-Newton interpolation? Will this interpolation formula converge and will it give the right answer even *if* it converges? The closer investigation of this problem shows that only a very definite *class* of functions, defined by a certain integral transform, allows the Gregory-Newton type of polynomial interpolation. Functions which do not belong to this class yield a divergent expansion, which means that the interpolation formula loses its significance.

This phenomenon shows that great caution is demanded when interpolating by powers. The difficulty is usually hidden by the fact that in tabulated functions the higher differences decrease so fast that in a few steps they are "off the board." This does not mean, however, that the interpolation formula, if pursued to arbitrarily high terms, would reduce the error to arbitrarily small amounts. In fact, in many cases something quite unexpected would happen. The error would get smaller and smaller by the correcting influence of the higher differences, but eventually a minimum would be reached, and beyond that the error would increase again and become arbitrarily large. (We have in mind mathematical functions which in principle could be evaluated to any degree of accuracy. In observed functions the "noise" rules out use of the differences of high order.)

How can we explain this puzzling phenomenon? Let us observe that the central differences found along a certain line of a difference table are determined purely by the *immediate neighborhood* of the data point at the head of the line. As we go to higher and higher differences, this "neighborhood" spreads out more and more. The use of a few terms or of many terms of an interpolation formula can thus be juxtaposed as follows. In one case we assume the validity of a polynomial approximation *in the small*, in the other case *in the large*. Now it so happens that a polynomial approximation in a sufficiently *small* neighborhood of a point is always safe and justified. But a polynomial approximation *in the large* is *not* always safe, and demands the proper safeguards.

Weierstrass proved in 1885 the fundamental theorem that any continuous function of a finite range can always be approximated to any degree of accuracy by powers. This theorem establishes thus the justification for a polynomial expansion, and it seems that we cannot go wrong if we interpolate our data by powers. Actually, however, the theorem of Weierstrass, while establishing the validity of a polynomial approximation, does *not* imply that the approximating polynomial is obtainable by fitting equidistant data. It was E. Borel in 1903 and O. Runge in 1901 who discovered the startling fact that we can take very simple analytical functions, such as for example,

$$y = \frac{1}{1 + 25x^2} \tag{5-14.1}$$

in the range $[-1,+1]$ and obtain quite wrong results by equidistant interpolation. As we put our data points closer and closer together, the interpolating polynomial which fits all our points actually converges to the given $f(x)$ unlimitedly in a large portion of the given range. But outside of a certain precalculable point—in the example (1) the point $x = \pm 0.726 \cdots$—up to the end of the range the interpolating polynomial does *not* converge to any limit, and in fact goes beyond all bounds at every point of the range. Since Runge investigated this phenomenon in great detail, it seems justifiable to call it the "Runge phenomenon."

Hence the strange fact holds that a polynomial which does *not* fit the data points may be in a much better position relative to over-all accuracy than a polynomial which fits the data points. The following fact for example is of interest. Let $f(x)$ satisfy the boundary condition $f(\pm 1) = 0$. Let us operate with a polynomial of the order $4n$. But instead of fitting $4n + 1$ equidistant points exactly, we fit only the $2n + 1$ key values $f(k/m)$, $(k = 0, \pm 1, \pm n)$. The data half-way between are calculated from these key values by the following interpolation formula.

$$f\left(\frac{k + \frac{1}{2}}{n}\right) = \frac{1}{\pi} \sum_{\alpha = -n}^{+n} (-1)^{\alpha - k} \frac{f(\alpha/n)}{\alpha - k + \frac{1}{2}} \tag{5-14.2}$$

Hence we have replaced the correct midpoint values by incorrect values. Nevertheless, the polynomial thus interpolated gives small errors all over the range, while the errors of the correctly interpolated polynomial go out of bound.

The difficulties examined by Runge are caused only by the *equidistant* character of the data. If the data are not equidistantly distributed but are placed into the zeros of the $(2n + 1)$st Chebyshev polynomial $T_{2n+1}(x)$, the difficulties disappear. The errors of the interpolation now oscillate with the same order of magnitude all over the range and converge at every point of the range to zero as the number of data points increases to infinity.

Since equidistant data are much more convenient from both the numerical and the observational standpoint, we may ask how we can obtain effective polynomial approximations *in spite* of the equidistant character of the data points. It so happens that in many problems of physics and engineering it is particularly desirable to replace a certain analytical function by a power series. The powers of x have great operational advantages, and we may want to use them even if the original (tabulated or observed) $f(x)$ is not a power series. We know from the theorem of Weierstrass that such a replacement is always possible. But we also know from Runge's investigation that we cannot obtain this polynomial by simple interpolation. The problem is not one in least squares, since we do not know in advance of what order the approximating polynomial will be, nor is it desirable to minimize the residuals, since small residuals in the data points can cause large errors between.

We return once more to the Fourier expansion of § 11 by which we tried to reduce the noise of our data. We subtracted a proper linear quantity $\alpha + \beta x$ from the given data and then employed a pure sine series. We then separated the analytical part of the function from the noise part by examining the trend of the Fourier components and truncating the series at the proper point. We thus obtained an analytical expression which not only smoothed our data points but interpolated the values of $f(x)$ between the data points. This method of trigonometric interpolation is free of the objections of equidistant polynomial interpolation. The trigonometric functions have orthogonality with respect to equidistant data and are thus in the same preferential position relative to such data as the powers of x are relative to data which are distributed according to the zeros of the Chebyshev polynomials. The trigonometric sine interpolation will automatically converge to $f(x)$ at every point of the interval if the data points get denser and denser.

Our aim is, however, to obtain a *polynomial* approximation of $f(x)$. For this reason we will now convert our sine expansion into a polynomial expansion. To use the Taylor expansion for each one of the sine functions and then collect terms would not serve our purpose, since the resulting series would have slow convergence and thus require a large number of terms. We know in advance that we fare best if we use the Chebyshev polynomials for expansion purposes, since this series will give fastest convergence (cf. VII, 6). Hence we will have to investigate the problem of converting a sine function into Chebyshev polynomials.

The sine functions of the expansion (11.13), for data normalized to the range [0,2], are the functions

$$\varphi_k(x_1) = \sin k\,\frac{\pi}{2}\,x_1 \qquad (5\text{-}14.3)$$

It will be more convenient, however, to put the origin of our reference system in the mid-point of the range, thus separating in advance the even and the odd parts of our function. This means that x_1 is to be replaced by $x = x_1 - 1$. The functions $\varphi_k(x)$ now become

$$\varphi_{2k}(x) = (-1)^k \sin k\pi x$$
$$\varphi_{2k+1}(x) = (-1)^{k+1} \cos (k + \tfrac{1}{2})\pi x \qquad (5\text{-}14.4)$$

The first group of functions gives the odd, the second group gives the even part of the function $g(x)$; the addition of the correction $\alpha + \beta x$ finally restores the original $f(x)$.

Now the expansion of the trigonometric functions (4) into Chebyshev polynomials is available in terms of the Bessel functions, $J_k(x)$. We obtain the following results.

$$\sin k\pi x = 2 \sum_{\alpha=1}^{\infty} (-1)^{\alpha} J_{2\alpha+1}(k\pi) T_{2\alpha+1}(x)$$
$$\cos (k + \tfrac{1}{2})\pi x = 2 \sum_{\alpha=0}^{\infty}{}' (-1)^{\alpha} J_{2\alpha}((k + \tfrac{1}{2})\pi) T_{2\alpha}(x) \qquad (5\text{-}14.5)$$

The Σ' in the second formula refers to the fact that the first term of the sum must be halved.

The method of interpolating a set of equidistant data by powers can thus be described as follows. After applying a linear correction which makes the two extreme data zero, we expand the data into a Fourier sine series, according to the method discussed in IV, 12. If the data are free of noise (mathematical data), we leave this expansion as it is. If the data have noise superimposed on them, we truncate the series at a properly chosen point. We now have a Fourier sine series of m

terms, with the coefficients $b_1, b_2, \cdots, b_m$. We convert this series into an infinite expansion of Chebyshev polynomials.

$$g(x) = \sum_{k=0}^{\infty} c_k T_k(x) \qquad (5\text{-}14.6)$$

the coefficients of which are evaluated as follows.

$$c_{2k} = 2(-1)^k \left[-J_{2k}\left(\frac{\pi}{2}\right) b_1 + J_{2k}\left(3\frac{\pi}{2}\right) b_3 - J_{2k}\left(5\frac{\pi}{2}\right) b_5 + \cdots \right] \qquad (5\text{-}14.7)$$

$$c_{2k+1} = 2(-1)^k \left[-J_{2k+1}(\pi)b_2 + J_{2k+1}(2\pi)b_4 - J_{2k+1}(3\pi)b_6 + \cdots \right]$$

The values of the Bessel functions of even order at the odd multiples of $\pi/2$, and the values of the Bessel functions of odd order at the multiples of π can be pretabulated (cf. Table XI). Evaluation of the expansion coefficients c_k is then reduced to multiplication of the b_k coefficients by a numerical matrix.

In the absence of noise we can go one step further. Evaluation of the Fourier coefficients b_k and subsequent evaluation of the c_k can be combined into one single step. We can take our equidistant data, after separating the even and the odd parts of the function $g(x)$, and directly multiply them by a preassigned numerical matrix, thus obtaining in one step the even and the odd c_k, without any preliminary Fourier analysis (cf. Table XII).

Theoretically the expansion (6) into the Chebyshev polynomials $T_k(x)$ is an infinite expansion. However, the good convergence of the series makes an early termination possible. We evaluate the finite sum of $\nu + 1$ terms,

$$\bar{g}(x) = \sum_{k=0}^{\nu} c_k T_k(x) \qquad (5\text{-}14.8)$$

at the data points and see how much error is caused by the truncation of the series. We can stop at an order $k = \nu$ at which the residuals become sufficiently small. In the presence of noise, a natural termination is effected by the criterion that the accuracy of our interpolation need not go much beyond the accuracy of our data. Hence we will truncate the series (8) at an order at which the maximum residual in any of the data points remains well within the average error of the data.

This method of obtaining a well-convergent polynomial expansion for a set of equidistant data has the advantage that we need not know in advance what order polynomial will be the most suitable for our purposes. The data themselves decide the most appropriate polynomial for a given accuracy. The difficulties of the Runge phenomenon are avoided and we obtain a close approximation comparable with that obtainable by the theoretically more desirable but practically much less accessible unequal distribution of data points corresponding to the zeros of the first neglected Chebyshev polynomial $T_{n+1}(x)$. In our case the data points are equidistant and we still have the benefit of an expansion into Chebyshev polynomials, which will give a practically uniform approximation throughout the range.[1]

15. The convergence of equidistant polynomial interpolation. The previous section gave an answer to the problem of polynomial interpolation in equidistant points which solved the two basic difficulties: the problem of the noise and the problem of the non-uniformity of the error which can cause oscillations of harmful amplitudes around the two ends of the interval. Both phases of the problem were beneficially influenced by the temporary interjection of the Fourier sine functions, which make an effective separation of function and noise possible and which distribute the errors with a uniform order of magnitude throughout the given interval. The resultant truncated sine series was finally converted into an infinite Chebyshev expansion, which again could be terminated at a properly designated point. By this procedure we know in advance that our interpolation must converge to the given $f(x)$ for any finite, single-valued, sectionally continuous function which does not oscillate infinitely many times in the given interval.

However, in spite of these results, the investigation of Runge concerning the convergent or divergent behavior of the simple equidistant interpolation remains of fundamental interest. Since certain functions give convergent, certain others divergent expansions, the question arises whether we can decide *in advance* what type

---

[1] This last section is a brief summary of an elaborate investigation of equidistant polynomial interpolation, with many numerical examples, of which a multilithed laboratory report came out under the title: "Analytical and Practical Curve Fitting of Equidistant Data," *Nat. Bur. Standards, Report 1591,* 1952.

of functions will lead to one and what type to the other kind of expansion. This decision can actually be made if we know the analytical nature of the function $f(x)$. We will show that if we can consider $u = f(z)$ a function of the complex variable $z = x + iy$, the behavior of $f(z)$ in a certain region around the x-axis uniquely determines the convergence behavior of this function with respect to equidistant interpolation, without any detailed investigation of the remainder. What we have to know is merely whether or not a certain explicitly given oval-shaped region around the x axis is free of analytical singularities. If the function $f(z)$ behaves throughout this region including the boundary analytically, we know in advance that the equidistant polynomial interpolation of this function will converge uniformly in the interval $[-1, +1]$. If at any point a singularity occurs, the convergence will hold only within a certain subinterval of the total range $[-1, +1]$, while outside of this range the interpolation diverges as n grows to infinity.

This convergence behavior is very similar to that of the Taylor series, which is likewise determined by the analytical nature of $f(z)$ outside of the x axis, even if our interest is completely restricted to the real range. If we draw a circle from the center of expansion which reaches up to the nearest point of singularity, the Taylor series will definitely converge inside of this circle and definitely diverge outside of it, while on the circle itself the behavior is dubious.

The difference in our case is only that the critical region is much more complicated than a circle, although mathematically available. We start with the following fundamental function.

$$\frac{F_m(z) - F_m(x)}{z - x} \qquad (5\text{-}15.1)$$

where $F_m(x)$ is the fundamental polynomial, i.e., the polynomial composed of the root factors

$$F_m(x) = (x - x_1)(x - x_2) \cdots (x - x_m) \qquad (5\text{-}15.2)$$

where $x_1, x_2, \cdots, x_m$ are the points of interpolation.

Now the numerator of (1) is divisible by the denominator, and thus the function (1), considered as a function of x, must be a polynomial of the order $m - 1$. The same remains true even if we divide by $F_m(z)$ and consider the function

$$\phi(x,z) = \frac{F_m(z) - F_m(x)}{F_m(z)(z - x)} = \frac{1}{z - x} - \frac{F_m(x)}{F_m(z)} \frac{1}{z - x} \qquad (5\text{-}15.3)$$

Consequently the function $\phi(x,z)$, considered as a function of x, allows an *exact* polynomial interpolation, no matter how the x_i are located. The resulting expansion terminates after m terms. Moreover the second term on the right side of (3) has no influence on the interpolation, since it vanishes at all points x_i. This term can be considered the *remainder* of the interpolation.

We thus see that the special function

$$v(x) = \frac{1}{z - x} \qquad (5\text{-}15.4)$$

has the property that, if interpolated by powers in arbitrarily chosen points, the remainder of the interpolation is explicitly available.

We restrict ourselves to the case of *equidistant* interpolation and put

$$x_k = \frac{k}{n} \qquad (k = 0, \pm 1, \cdots, \pm n)$$

$$\varphi_{2k+1}(x) = kx \frac{(k^2x^2 - 1)(k^2x^2 - 4) \cdots (k^2x^2 - k^2)}{(2k + 1)!} \qquad (5\text{-}15.5)$$

Then $m = 2n + 1$ and $F_m(x)$ becomes proportional to the $(2n + 1)$st Stirling function:

$$F_{2n+1}(x) = x \left(x^2 - \frac{1}{n^2} \right) \left(x^2 - \frac{4}{n^2} \right) \cdots \left(x^2 - \frac{n^2}{n^2} \right)$$

$$= \frac{(2n + 1)!}{n^{2n+1}} \varphi_{2n+1}(x) \qquad (5\text{-}15.6)$$

The resulting expansion becomes

$$\frac{1}{z - x} = \frac{n}{\varphi_1(z)} + \frac{n\varphi_1(x)}{2\varphi_2(z)} + \cdots + \frac{n}{2n + 1} \frac{\varphi_{2n}(x)}{\varphi_{2n+1}(z)} + \eta_{2n+1}$$

$$= n \sum_{k=0}^{2n} \frac{\varphi_k(x)}{(k + 1)\varphi_{k+1}(z)} + \eta_{2n+1}(x,z) \qquad (5\text{-}15.7)$$

where

$$\eta_{2n+1}(x,z) = \frac{\varphi_{2n+1}(x)}{\varphi_{2n+1}(z)} \frac{1}{z-x} \qquad (5\text{-}15.8)$$

We will now make use of Cauchy's fundamental integral theorem.

$$f(x) = \frac{1}{2\pi i} \oint \frac{f(z)\,dz}{z-x} \qquad (5\text{-}15.9)$$

where the integration extends over a closed loop which encloses the point $z = x$ and likewise the data points $z = x_i$, but is free of any analytical singularities of the function $f(z)$. If in this theorem the function (4) is replaced by its expansion (7), we obtain on the right side the Stirling expansion of $f(x)$, with a remainder. This remainder appears in the following form.

$$\eta_{2n+1}(x) \doteq \frac{\varphi_{2n+1}(x)}{2\pi i} \oint \frac{f(z)\,dz}{(z-x)\varphi_{2n+1}(z)} \qquad (5\text{-}15.10)$$

Now the function $\varphi_{2n+1}(x)$ remains bounded by ± 1 in the entire range $[-1,+1]$. Hence we will focus our attention on the function

$$\varphi_{2n+1}(z) = \frac{(nz+n)(nz+n-1)\cdots(nz-n)}{(2n+1)!}$$

$$= \frac{(nz+n)!}{(nz-n-1)!\,(2n+1)!} \qquad (5\text{-}15.11)$$

and investigate its properties as n grows to infinity.

First we make use of the reflection theorem of the factorial function.[1]

$$\frac{1}{(-\nu-1)!} = -\nu!\,\frac{\sin \pi\nu}{\pi} \qquad (5\text{-}15.12)$$

because of which the critical function becomes

$$\varphi_{2n+1}(z) = \frac{(-1)^n \sin n\pi z}{\pi} \frac{[n(1+z)]!\,[n(1-z)]!}{(2n+1)!} \qquad (5\text{-}15.13)$$

[1] Cf. {15}, p. 239.

We agree that the complex variable z shall stay within the positive half plane. Then we can effectively approximate the factorial function by Stirling's formula.

$$n! = \sqrt{2\pi}\, n^{n+1/2} e^{-n} \tag{5-15.14}$$

This formula becomes arbitrarily accurate as n grows to infinity, but it is remarkably accurate even in the realm of small n.

In terms of this approximation our function becomes

$$\varphi_{2n+1}(z) = \frac{(-1)^n}{\sqrt{\pi}}\, \frac{\sqrt{n}}{2n+1}\, \sqrt{1-z^2}\, \left[\frac{(1+z)^{1+z}(1-z)^{1-z}}{4}\right]^n \sin n\pi z$$

The last factor requires closer attention.

$$\sin n\pi(x+iy) = \frac{1}{2i}\left(e^{-n\pi y}e^{in\pi x} - e^{n\pi y}e^{-in\pi x}\right) \tag{5-15.16}$$

Let us assume that y is positive. Then the first term becomes negligible as n grows to infinity, while the second term can be combined with the previous factor to one single expression, raised to the power n.

For our purposes it suffices to investigate the *absolute value* of this expression.

$$A(z) = \tfrac{1}{4}\left|(1+z)^{1+z}(1-z)^{1-z}\right| e^{\pi y} \tag{5-15.17}$$

The nth power of this number, as n grows to infinity, behaves in an extreme fashion. If $A(z)$ is less than 1, the nth power goes to zero, if greater than 1, to infinity. Correspondingly the integrand of the remainder (10) converges in the first case to infinity, in the second case to zero. The boundary $A(z) = 1$ is characterized by a closed ellipse-like curve (see the accompanying drawing), determined by the transcendental equation

$$\tfrac{1}{2}(1+x)\log[(1+x)^2+y^2] + \tfrac{1}{2}(1-x)\log[(1-x)^2+y^2]$$
$$+ \pi|y| - y\left(\arctan\frac{y}{1+x} + \arctan\frac{y}{1-x}\right) = 2\log 2$$

The following table tabulates $y(x)$ in intervals of 0.1 of the independent variable. The neighborhood of the singular point $x = 1$, $y = 0$, is tabulated in intervals of 0.01.

x	y	x	y	x	y
0.0	0.5255	0.6	0.3855	0.95	0.0963
0.1	0.5219	0.7	0.3283	0.96	0.0813
0.2	0.5110	0.8	0.2556	0.97	0.0652
0.3	0.4925	0.9	0.1598	0.98	0.0475
0.4	0.4660	1.0	0.0	0.99	0.0273
0.5	0.4307			1.00	0.0

For any closed path which lies completely outside of this transcendental curve C, the integral on the right side of (10) converges to zero as n goes to infinity. This means that the interpolation converges to the true $f(x)$ throughout the interval of interpolation.

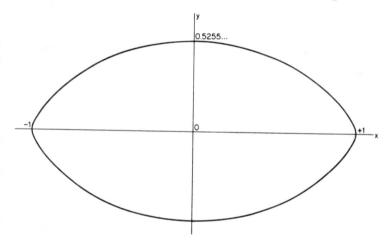

However, such a closed path (9) can be chosen only if the interior of C is free of any singularity of the analytical function $f(z)$. If $f(z)$ has one or more singular points within the enclosed region, the remainder goes to infinity instead of zero and the interpolation cannot remain everywhere convergent.

We have thus found the necessary and sufficient conditions for the convergent or divergent behavior of equidistant interpolation. Runge demonstrated the divergent character of equidistant polynomial interpolation with the help of the example

$$y = \frac{1}{1 + 25x^2}$$

The singularity occurs here at the points

$$z = \pm\tfrac{1}{5}i$$

which are clearly inside the critical region.

16. Orthogonal function systems. A much deeper insight into the nature of interpolation problems came from a different field, developed during the nineteenth century, but fully understood in all its fundamental implications in more recent times. It was an ingenious *geometrical* interpretation of the method of least squares which opened a new and tremendously fertile field of research. In this geometrical picture the concept of a "function" is translated into the concept of a "vector," placed in a space of infinitely many dimensions.

We will assume that a certain continuous domain of the variable x is given, limited by $x = a$ and $x = b$ (domains of more than one dimension, however, can be treated in an entirely similar way). In this domain we consider a function $f(x)$ which shall be conceived as everywhere finite and single-valued, and at least sectionally continuous. Such a function, although it comprises an infinity of values, can nevertheless be tabulated in the given interval with any degree of accuracy. Instead of giving all the values of $f(x)$, we record its values in the everywhere dense set of discrete points $x_1, x_2, \cdots, x_n$. The sectionally continuous character of $f(x)$ makes it possible that the discrete set of values

$$y_1 = f(x_1), \quad y_2 = f(x_2), \quad \cdots, \quad y_n = f(x_n) \qquad (5\text{-}16.1)$$

comes arbitrarily near to any value of $f(x)$ and that this discrete set can replace the original $f(x)$ for all analytical operations. Although a certain error is committed by this replacement, we can make the error as small as we wish, by making n sufficiently large.

Let us now assume that we plot the functional values as rectangular coordinates of an imaginary Euclidian space of n dimensions. More specifically we want to plot along the successive coordinate axes the values

$$f_1 = y_1 \sqrt{\varepsilon_1}, \quad f_2 = y_2 \sqrt{\varepsilon_2}, \quad \cdots, \quad f_n = y_n \sqrt{\varepsilon_n} \quad (5\text{-}16.2)$$

where

$$\varepsilon_k = x_k - x_{k-1} \qquad (x_0 = a)$$

All the ε_k converge to zero as n increases to infinity.

We now have a definite point

$$P = (f_1, f_2, \cdots, f_n)$$

of an n-dimensional space, representing the given function $f(x)$. We can also say that we have constructed the vector $\overrightarrow{OP}$ whose projections on the coordinate axes are proportional to the given functional values. By going with n higher and higher we will obtain an increasingly adequate representation of any finite, single-valued, and sectionally continuous function of the given interval. The analytical concept of a function can thus be dropped in favor of the more vivid picture of a *vector* in a many-dimensional space. The length-square of this vector is given by the sum

$$f^2 = \sum_{i=1}^{n} f_i^2 = \sum_{i=1}^{n} y_i^2 \varepsilon_i \qquad (5\text{-}16.3)$$

and we notice that in the limit, as n goes to infinity, this sum is replaceable by the integral

$$f^2 = \int_a^b f^2(x)\, dx \qquad (5\text{-}16.4)$$

If *two* functions $f(x)$ and $g(x)$ are considered, they represent two vectors, whose mutual orientation can be characterized by their "scalar product"

$$f \cdot g = \sum_{i=1}^{n} f_i g_i = \sum_{i=1}^{n} f(x_i) g(x_i) \varepsilon_i \qquad (5\text{-}16.5)$$

which again becomes an integral as n grows to infinity:

$$f \cdot g = \int_a^b f(x)g(x)\,dx \qquad (5\text{-}16.6)$$

In this picture the problem of approximating a function by a given set of functions—such as the powers of x, for example, in interpolation problems—appears likewise in new light. A given set of functions represents a set of vectors which can be conceived as a given *frame of reference* within our imaginary space. It is in fact a *partial* frame only if the number of approximating functions is finite while the number of dimensions goes to infinity, The problem of approximating a function as a linear combination of given functions $u_i(x)$ can be conceived as the geometrical problem of *analyzing a vector in a given frame of reference*. But then it is evident that we will prefer as particularly convenient the *orthogonal* frames of reference, characterized by the fact that any two of the base vectors u_i are orthogonal to each other:

$$u_i \cdot u_k = \int_a^b u_i(x)u_k(x)\,dx = 0 \qquad (5\text{-}16.7)$$

while the length of any one of the vectors is normalized to 1:

$$u_i^2 = \int_a^b u_i^2(x)\,dx = 1 \qquad (5\text{-}16.8)$$

Functions which satisfy these conditions are called "orthonormal." The matter of normalization is of smaller importance. The condition (7) alone defines the *orthogonality* of a given set of functions. In such an orthogonal frame of reference the problem of analyzing a given vector is immediately solvable. The mere *projection* of the vector $\vec{v}$ on the orthogonal axes gives the components of $\vec{v}$ in that particular frame of reference:

$$\vec{v} = c_1\vec{u_1} + c_2\vec{u_2} + \cdots + c_m\vec{u_m}$$

where

$$c_i = \frac{\vec{u_i} \cdot \vec{v}}{(\vec{u_i})^2}$$

This means that a function $f(x)$, analyzed in the reference system of the $u_i(x)$, appears in the form

$$f(x) = \sum_{i=1}^{m} c_i u_i(x) \tag{5-16.9}$$

where

$$c_i = \frac{\int_a^b f(x) u_i(x)\, dx}{\int_a^b u_i^2(x)\, dx} \tag{5-16.10}$$

This requires, however, that $f(x)$ shall lie inside the space included by the m functions $u_1(x), \cdots, u_m(x)$. If $f(x)$ lies partly outside that space, we obtain by the above construction a function which can be conceived as an effective *approximation* of $f(x)$, in terms of the functions $u_i(x)$. This approximation is the *projection* of $f(x)$ into the subspace of the base vectors $u_1(x), \cdots, u_m(x)$. This projection can be considered as that particular linear combination of the given vectors $u_i(x)$, which comes nearest to the given $f(x)$, inasmuch as the error of the approximation:

$$\eta_m(x) = f(x) - f_m(x) \tag{5-16.11}$$

has the smallest possible length.

It is conceivable that we may possess function systems which constantly satisfy the orthogonality condition (7) and yet newer and newer functions can be added to the previous set, without end. Such an infinite set of functions may include the entire function space. This means that if $f(x)$ is an arbitrary finite, single-valued, and sectionally continuous function of the given interval—and we will add the condition that $f(x)$ shall not have an infinite number of maxima or minima in that interval—we can form approximations of increasingly high order and, although it will never happen that we obtain *exactly*

$$f(x) = \sum_{i=1}^{n} c_i u_i(x)$$

no matter how far we go with n, yet we may obtain *in the limit*:

$$f(x) = \lim_{n \to \infty} \sum_{i=1}^{n} c_i u_i(x) \qquad (5\text{-}16.12)$$

Such function systems are called "complete orthogonal function systems." Expansions of the form (12), with coefficients determined according to (10), are called "orthogonal expansions." They play a superior role in the physical and mathematical problems of our days. They are not restricted to one single variable but exist equally in any number of variables. Nor is the domain of expansion necessarily finite. With the proper precautions even infinite domains find their place within the function space.

The Fourier functions, $\sin kx$, $\cos kx$, connected with the range $[-\pi, +\pi]$, were the first example of a complete orthogonal function system. At the time of their discovery the concept of the function space and the wider implications of orthogonality were not yet recognized. Today we know that the Fourier functions represent just *one* particularly interesting orthogonal frame of reference within the function space. But there are infinitely many other such frames, obtainable by a rigid rotation of the original axes, which all share the spectacular properties of the Fourier functions in approximating highly capricious functions. All these function systems are analytically equivalent, in the sense that all orthogonal reference systems of an n-dimensional Euclidian space share the same metrical properties, because of the homogeneity of space in every direction.

17. Self-adjoint differential operators. We have seen in Chapter II that every symmetric matrix with noncollapsing eigenvalues establishes an orthogonal frame of reference by its principal axes which are always present in sufficient number. In function space a correspondingly abundant source of complete orthogonal function systems exists through the medium of a certain class of linear differential operators called "self-adjoint."

The important matrix identity

$$y \cdot Ax - x \cdot \tilde{A}y \equiv 0 \qquad (5\text{-}17.1)$$

has a counterpart in the theory of linear differential operators. Let D be an arbitrary linear differential operator, ordinary or partial.

Then there exists a uniquely determined operator $\tilde{D}$ obtainable by purely algebraic and differential operations, which has the following property:

$$v\, Du - u\, \tilde{D}v \equiv \frac{\partial p_1}{\partial x_1} + \cdots + \frac{\partial p_n}{\partial x_n} \qquad (5\text{-}17.2)$$

Then, integrating over an arbitrary closed volume of the variables $x_1, x_2, \cdots, x_n$, and transforming on the right side the volume integral into a surface integral by the Gaussian theorem, we obtain a relation called "Green's identity" which is the counterpart of (1):

$$\int (v\, Du - u\, \tilde{D}v)\, d\tau = \text{surface integral} \qquad (5\text{-}17.3)$$

So far the functions u and v are completely arbitrary—except for their differentiability to the extent demanded by the operators D and $\tilde{D}$. We now assume that the function $u(x)$ is subjected to some more or less stringent "boundary conditions" on the boundary surface. Then we can always prescribe some properly chosen boundary conditions for $v(x)$, called the "adjoint boundary conditions," which will make the right side of (3) vanish:

$$\int (v\, Du - u\, \tilde{D}v)\, d\tau = 0 \qquad (5\text{-}17.4)$$

In the interior of the region of integration the functions u and v are still arbitrary.

Now it may happen that $\tilde{D} = D$, in which case we speak of a "self-adjoint differential operator." Moreover, we may have subjected u to such boundary conditions that the boundary conditions for v become *identical* with those for u. We then have a "self-adjoint differential operator with self-adjoint boundary conditions." Such an operator, conceived as an operator in function space, is a counterpart of a symmetric matrix. Its principal axes define a complete orthogonal function system. These principal axes are defined by the differential equation

$$Du = \lambda u \qquad (5\text{-}17.5)$$

together with the prescribed boundary conditions.

Equation (5) is solvable only for a definite set of λ-values, called the eigenvalues of D. However, since the function space has infinitely many dimensions, the eigenvalues are present in infinite number. They are always real, however, and always discrete, if the differential operator D is free of any singularities and the domain of integration is finite. They can be arranged according to their magnitude, starting with the absolutely smallest $\lambda = \lambda_1$ and continuing to larger and larger $\lambda_2, \lambda_3, \cdots$.

The orthogonality of two solutions, belonging to two different eigenvalues λ_i and λ_k, follows from (4) if we substitute for u and v the two solutions u_i and u_k:

$$(\lambda_i - \lambda_k)\int u_i u_k \, d\tau = 0 \qquad (5\text{-}17.6)$$

Since the first factor cannot vanish if λ_i and λ_k are different, the second factor must vanish, expressing the orthogonality of the obtained functions with respect to the realm of integration. In the case of multiple eigenvalues the associated eigensolutions are not automatically orthogonal to each other (although they are orthogonal to all the other u_i of the set) but we can orthogonalize them by choosing the proper linear combinations; (cf. II, 9).

18. The Sturm-Liouville differential equation. Although an infinite variety of self-adjoint differential operators can be constructed, together with an appropriate set of boundary conditions, the most important orthogonal function systems of applied analysis arise from differential operators of *second* order. This is the lowest order for self-adjoint differential operators, since differential operators of first order, (at least with real coefficients) cannot be self-adjoint. We will restrict ourselves to one single variable and thus deal with ordinary differential operators only. The most general ordinary linear and self-adjoint differential operator of second order—first investigated by Sturm and by Liouville and thus frequently named after them—has the following form:

$$Dy = \frac{d}{dx}(py') + qy \qquad (5\text{-}18.1)$$

The functions $p(x)$ and $q(x)$ are still arbitrary, although $p(x)$ has to be differentiable and we usually assume that it does not change its sign throughout the given interval.

Green's identity (17.3) associated with this differential operator becomes

$$\int_a^b \left\{ v \left[\frac{d}{dx}(pu') + qu \right] - u \left[\frac{d}{dx}(pv') + qv \right] \right\} dx$$
$$= \left| p(vu' - uv') \right|_a^b \qquad (5\text{-}18.2)$$

Any boundary condition is permitted which, if equally applied to u and v, makes the boundary term vanish, e.g.,

$$u(a) = u(b) = 0$$

or

$$u'(a) = u'(b) = 0$$

As a simple example let us choose $a = -\pi, b = \pi, p = -1, q = 0$. The boundary conditions shall be chosen as

$$u(-\pi) = u(\pi), \qquad u'(-\pi) = u'(\pi)$$

The eigenvalues and eigenfunctions are defined by the differential equation

$$-u'' = \lambda u$$

which is solved by

$$u = A \cos \sqrt{\lambda} x + B \sin \sqrt{\lambda} x$$

The boundary conditions determine λ to

$$\lambda_k = k^2 \qquad (k = 0, 1, 2, \cdots)$$

and the arbitrariness of A and B shows that every eigenvalue belongs to *two* functions. We can separate them by the choice

$$u_k = \cos kx, \qquad \bar{u}_k = \sin kx$$

These are the orthogonal functions of the Fourier series, here obtained as the solution of a simple Sturm-Liouville problem.

Let us extend, however, our considerations to the eigenvalue problem associated with the most general linear differential operator of second order:

$$A(x)u''(x) + B(x)u'(x) + [C(x) + \lambda]u(x) = 0 \qquad (5\text{-}18.3)$$

We can multiply this equation by a certain $\rho(x)$, chosen in such a way that the operator (1) shall be obtained again. This demands

$$\rho(x)B(x) = [\rho(x)A(x)]'$$

which gives the condition

$$\frac{\rho'}{\rho} = \frac{B - A'}{A} \tag{5-18.4}$$

We can now put

$$p(x) = \rho(x)A(x) \tag{5-18.5}$$

and write our equation in the self-adjoint form

$$\frac{d}{dx}[p(x)u'(x)] + \rho(x)[C(x) + \lambda]u(x) = 0 \tag{5-18.6}$$

If we apply Green's identity, the boundary term becomes again

$$\left| p(x)(vu' - uv') \right|_a^b \tag{5-18.7}$$

and again we assume that this term vanishes on account of the boundary conditions. The only difference compared with the previous case is that the orthogonality of two solutions belonging to λ_i and λ_k appears in the form

$$\int_a^b \rho(x)u_i(x)u_k(x)\, dx = 0 \tag{5-18.8}$$

An orthogonality condition of this type is called "weighted orthogonality," since the function $\rho(x)$—which must remain everywhere positive inside the interval of integration—can be interpreted as a weight factor. Function systems associated with weighted orthogonality are not essentially different from ordinary orthogonal functions, as we can see if a new independent variable ξ is introduced by the condition

$$d\xi = \rho(x)\, dx \tag{5-18.9}$$

The weighted orthogonality in x is now changed to ordinary orthogonality in ξ.

Weighted orthogonality is particularly useful if the range of integration becomes infinite in one or both directions. A properly chosen $\rho(x)$ may prevent the integrals (8) from becoming divergent on account of the infinite limits. By this artifice the validity of function space operations can be extended to an infinite interval.

19. The hypergeometric series. One of the many far-sighted discoveries of Gauss was the introduction of an infinite series, called the "hypergeometric series." It defines a function of x which at the same time depends on three constants α, β, γ, which can assume arbitrary real or complex values, except that γ must not be a negative integer. The hypergeometric function, usually denoted by $F(\alpha, \beta, \gamma; x)$, is defined by the following infinite series which converges for all $|x| < 1$, and diverges for all $|x| > 1$, except if it so happens that the infinite series terminates after a finite number of terms:

$$F(\alpha, \beta, \gamma; x) = 1 + \frac{\alpha\beta}{\gamma \cdot 1} x + \frac{\alpha(\alpha + 1)\beta(\beta + 1)}{\gamma(\gamma + 1) \cdot 1 \cdot 2} x^2$$
$$+ \frac{\alpha(\alpha + 1)(\alpha + 2)\beta(\beta + 1)(\beta + 2)}{\gamma(\gamma + 1)(\gamma + 2) \cdot 1 \cdot 2 \cdot 3} x^3 + \cdots$$
(5-19.1)

Almost all the special functions of mathematical physics—with the exception of the gamma function—are in some relation to the hypergeometric series, which includes a large class of functions.

The hypergeometric function satisfies the following differential equation of second order, called the differential equation of Gauss:

$$x(1 - x)u'' + [\gamma - (\alpha + \beta + 1)x]u' - \alpha\beta u = 0 \qquad (5\text{-}19.2)$$

This differential equation has the general form (18.3) of a linear differential equation of second order, with

$$A(x) = x(1 - x), \quad B(x) = \gamma - (\alpha + \beta + 1)x, \quad C(x) = 0, \quad \lambda = -\alpha\beta$$

In order to transform it into the self-adjoint form (18.6) we have to obtain $\rho(x)$ by the condition (18.4), which gives

$$\rho(x) = x^{\gamma - 1}(1 - x)^{\alpha + \beta - \gamma} \qquad (5\text{-}19.3)$$

and thus

$$p(x) = x^{\gamma}(1 - x)^{\alpha + \beta + 1 - \gamma} \qquad (5\text{-}19.4)$$

20. The Jacobi polynomials. An eigenvalue problem which shall lead to an orthogonal set of functions requires a self-adjoint differential operator with self-adjoint boundary conditions. In the previous section we have introduced the weight factor which made the Gaussian differential equation self-adjoint. Now we will investigate the question of boundary conditions. The realm of integration shall be limited to the interval [0,1]. Then the boundary term becomes, according to (18.7):

$$\left| x^{\gamma}(1-x)^{\delta}(vu'-uv') \right|_0^1 \qquad (5\text{-}20.1)$$

where we have put

$$\alpha + \beta + 1 - \gamma = \delta \qquad (5\text{-}20.2)$$

We will assume that γ and δ are given *positive* constants:

$$\gamma > 0, \qquad \delta > 0 \qquad (5\text{-}20.3)$$

In this case the boundary term (1) seems to vanish all by itself and it seems that we do not get any boundary conditions for $u(x)$. In actual fact the points $x = 0$ and $x = 1$ are *singular points* of our differential equation at which the solution goes generally to infinity. The demand that $u(x)$ shall remain *finite* at these two points, is in itself a boundary condition at the two endpoints of the interval. The finiteness at $x = 0$ ruled out already one of the solutions of the Gaussian differential equation and reduced our problem to the hypergeometric series. Now the condition "finiteness at $x = 1$" demands additional restrictions.

Let us observe that for any given γ and δ we still have either α or β freely at our disposal. Now the hypergeometric series (19.1) has the remarkable property that it automatically *terminates* with the power x^n if α is chosen as the negative integer $-n$. Then

$$\alpha = -n, \qquad \beta = n + \gamma + \delta - 1 \qquad (5\text{-}20.4)$$

and we obtain as the solution of our eigenvalue problem the polynomials

$$P_n^{(\gamma,\delta)}(x) = F(-n, n + \gamma + \delta - 1, \gamma; x) \qquad (5\text{-}20.5)$$

called "Jacobi polynomials." They have the remarkable property that they are orthogonal with respect to the weight factor (19.3):

$$\rho(x) = x^{\gamma-1}(1-x)^{\delta-1} \qquad (5\text{-}20.6)$$

establishing a complete orthogonal function system for any choice of γ and δ which is in harmony with the condition (3). For applications certain special choices of γ and δ are of particular interest.

The ultraspherical polynomials. The choice

$$\gamma = \delta \qquad (5\text{-}20.7)$$

leads to *symmetric* weighting with respect to the mid-point $x = \frac{1}{2}$ of the interval. In this case it is frequently preferable to put the origin of the reference system into the mid-point of the range by the transformation

$$x = \frac{1 - \xi}{2} \qquad (5\text{-}20.8)$$

The new variable ξ runs between -1 and $+1$. The polynomials thus obtained, called "ultraspherical," are now alternatively even and odd polynomials, according to the even or odd character of n. If the new variable is again called x, we obtain the definition

$$P_n^{(\gamma)}(x) = F\left(-n,\ n + 2\gamma - 1,\ \gamma;\ \frac{1 - x}{2}\right) \qquad (5\text{-}20.9)$$

The weight factor of orthogonality now becomes

$$\rho(x) = (1 - x^2)^{\gamma - 1} \qquad (5\text{-}20.10)$$

Of particular interest are the following special cases:

(a) *The Legendre polynomials:* $\gamma = 1$. Here the weight factor becomes 1 and weighted orthogonality changes to ordinary orthogonality:

$$P_n(x) = F\left(-n,\ n + 1,\ 1;\ \frac{1 - x}{2}\right) \qquad (5\text{-}20.11)$$

We will encounter these important polynomials later on, in quadrature problems (cf. VI, 10 and 19).

(b) *The Chebyshev polynomials:* $\gamma = \frac{1}{2}$. Here the weight factor becomes $(1 - x^2)^{-1/2}$ and the transformation $x = \cos \theta$ removes weighting. Then the polynomial

$$T_n(x) = F\left(-n,\ n,\ \tfrac{1}{2};\ \frac{1 - x}{2}\right) \qquad (5\text{-}20.12)$$

is transformed into $\cos n\theta$ and we obtain the Fourier cosine functions. These polynomials have the widest field of applications (see Chapter VII) and have the greatest efficiency in approximating arbitrary functions.

(c) *The Chebyshev polynomials of the second kind:* $\gamma = \frac{3}{2}$:

$$U_n(x) = (n + 1)F\left(-n, n + 2, \frac{3}{2}; \frac{1-x}{2}\right) \quad (5\text{-}20.13)$$

$$= \frac{\sin (n + 1)\theta}{\sin \theta} \quad (\text{if } x = \cos \theta)$$

They are well suited for the polynomial representation of a function which assumes large values in the neighbourhood of $x = \pm 1$ and remains small everywhere else (cf. III, 7, see also IV, 28).

(d) *The case* $\gamma = \infty$. This case is of interest on account of its relation to the Taylor series which can thus be conceived as the limit of an orthogonal expansion. Moreover, if x is simultaneously changed to

$$\xi = \sqrt{\gamma}x$$

the interval of ξ is extended from $-\infty$ to $+\infty$ and the weight factor (10) becomes $e^{-\xi^2}$. The resulting polynomials are called "Hermitian":

$$H_n(x) = \lim_{\gamma \to \infty} \sqrt{4\gamma}^n F\left[-n, n + 2\gamma - 1, \gamma; \frac{1}{2}\left(1 - \frac{x}{\sqrt{\gamma}}\right)\right] (5\text{-}20.14)$$

The Laguerre polynomials. Among the Jacobi polynomials of unsymmetric weighting ($\gamma \neq \delta$) the case $\delta \to \infty$ is of special interest This corresponds to letting β go to infinity. But then the expansion (19.1) of the hypergeometric series shows that we can introduce $\xi = \beta x$ as a new variable and obtain in the limit, as β goes to infinity, the function

$$\phi(\alpha, \gamma; \xi) = 1 + \frac{\alpha}{\gamma \cdot 1} \xi + \frac{\alpha(\alpha + 1)}{\gamma(\gamma + 1) \cdot 1 \cdot 2} \xi^2 + \cdots \quad (5\text{-}20.15)$$

which converges for all (real or complex) values of ξ. From the standpoint of orthogonality the range of ξ is now $[0, \infty]$ and the weight factor (6) becomes

$$\rho(\xi) = \xi^{\gamma - 1}e^{-\xi}$$

The choice $\gamma = 1$ yields the Laguerre polynomials:

$$L_n(x) = n!\phi(-n, 1; x) \qquad (5\text{-}20.16)$$

They are orthogonal with respect to the weight factor e^{-x}, in the infinite interval $[0, \infty]$. We have encountered this class of polynomials in the inversion problem of the Laplace transform, in connection with the transient response of an electric network; (cf. IV, 30).

21. Interpolation by orthogonal polynomials. In §§ 14 and 15 we have discussed the dangers of equidistant polynomial interpolation. We have seen that the error of an arbitrary interpolation by powers depends on the ratio of two polynomials:

$$Q_n(x, z) = \frac{F_n(x)}{F_n(z)} \qquad (5\text{-}21.1)$$

Here $F_n(x)$ is the fundamental polynomial, formed out of the root factors $x - x_i$, if x_i are the points of interpolation. The point x is some point of the interval $[-1, +1]$, while z is some point of the complex plane. If the ratio (1) approaches zero with n growing to infinity, the convergence of the interpolation is assured for all x values. But in the case of equidistant interpolation only such z values could be admitted which stayed outside of a certain oval-shaped region surrounding the interval of interpolation. This meant that not only had $f(x)$ to be analytical in the given interval, but this analytical behavior had to be demanded everywhere within the oval-shaped domain.

Very different is the behavior of orthogonal polynomials. We can expand an arbitrary $f(x)$, which satisfies much less than analytical conditions within the interval of interpolation and need not even be defined outside the interval, into a complete orthogonal set of functions, according to the equation (16.12). However, the coefficients c_i of this expansion demand the evaluation of the definite integrals (16.10) which are in actual fact but seldom at our disposal. Hence it is of great practical advantage that we can obtain an equivalent expansion with modified coefficients c_i' which also converges to $f(x)$ with increasing n, and whose accuracy even for finite n is not essentially worse than the expansion formed with the help of the c_i coefficients. This expansion is explicitly at our disposal on the basis of an *interpolation procedure*, without any integrations.

We can link up this procedure with the "classical" series

$$f(x) = \sum_{i=1}^{\infty} c_i u_i(x) \qquad (5\text{-}21.2)$$

by truncating the series to n terms and considering the remainder of this expansion:

$$\eta_n(x) = f(x) - f_n(x) = \sum_{i=n+1}^{\infty} c_i u_i(x) \qquad (5\text{-}21.3)$$

If we assume that this series has quick convergence, we may estimate the remainder by keeping only the *first term*. In this case

$$\eta_n(x) = c_{n+1} u_{n+1}(x) \qquad (5\text{-}21.4)$$

and this means that the error is zero at the roots of the first neglected orthogonal function. But this again means that we shall obtain the coefficients of the finite expansion

$$f_n(x) = \sum_{i=1}^{n} c_i' u_i(x) \qquad (5\text{-}21.5)$$

by fitting the functional values $f(x)$ at the zeros of the first neglected orthogonal function $u_{n+1}(x)$ (provided that n such points can be found inside the realm of orthogonality):

$$\sum_{i=1}^{n} c_i' u_i(\lambda_k) = f(\lambda_k), \quad u_{n+1}(\lambda_k) = 0, \qquad (k = 1, 2, \cdots, n) \quad (5\text{-}21.6)$$

This gives a simultaneous system of n linear equations for the determination of the c_i'. Although these coefficients will generally not coincide with the classical coefficients (16.10), obtained by integration, yet the error of the approximation will not be essentially worse, while the numerical procedure is now simple and straightforward.

The price we have to pay for convergence is that the key values of $f(x)$ have to be given at certain prescribed *nonequidistant* points

$x = \lambda_k$, defined as the zeros of the first neglected orthogonal poly-nomial.[1] The fundamental polynomial in (1) has to be replaced by the nth orthogonal polynomial $p_n(x)$ [the enumeration of these polynomials starts with $n = 0$ and thus $u_{n+1}(x)$ is actually $p_n(x)$]. The greatly increased convergence is demonstrated if we form the ratio (1). No longer is z confined to an oval-shaped region around the x-axis. We can choose for z *any* complex value $x + iy$, arbitrarily near to the x-axis, and yet $Q_n(x, z)$ converges to zero.

We can prove this statement explicitly in an elementary way for the special case of the Chebyshev polynomials. Here

$$p_n(x) = \cos n\theta$$

if x is transformed into θ according to

$$x = \cos \theta$$

Now for any complex value of x, in fact for any x outside the range ± 1, the angle θ must become complex. But then we can put

$$\theta = \varphi + i\psi$$
$$\cos n\theta = \tfrac{1}{2}(e^{in\varphi - n\psi} + e^{-in\varphi + n\psi})$$

and this quantity tends to infinity with increasing n for any $\psi \neq 0$.

Hence we see that the analytical nature of $f(x)$ outside the given interval is no longer demanded. In actual fact the interpolation converges to $f(x)$ for any x of the given interval, without demanding analyticity for $f(x)$ even *in* the given interval, as long as $f(x)$ belongs to the class of functions of "bounded variation."

The interpolation of functions with the help of orthogonal poly-nomials (not necessarily of the Jacobi type) has some further properties which greatly facilitate the numerical procedure. One of

---

[1] That $p_n(x)$ must have n zeros in the interval of orthogonality can easily be demonstrated. For, assuming that this is not the case, we would have

$$p_n(x) = (x - \lambda_1) \cdots (x - \lambda_m)q(x) \qquad (m < n)$$

where $q(x)$ does not change its sign between a and b. But then

$$\int_a^b (x - \lambda_1) \cdots (x - \lambda_m)p_n(x)\rho(x)\,dx$$

cannot vanish since the integrand does not change its sign anywhere. And yet, $p_n(x)$ being orthogonal to any polynomial of lower order (cf. text, later), the integral should be zero. This contradiction establishes the theorem.

the remarkable properties of orthogonal polynomials is that they satisfy a *recurrence relation* which connects three consecutive polynomials. This recurrence relation is of the following general form:

$$p_{n+1}(x) = (c_{n+1}x - a_n)p_n(x) - b_n p_{n-1}(x) \qquad (5\text{-}21.7)$$

The existence of such a relation is a direct consequence of the fact that an arbitrary $p_n(x)$ is orthogonal to all the previous polynomials and therefore also to any power x^α, $\alpha < n$, since such power is merely a linear combination of the $p_k(x)$, of an order less than n. But then we can conclude that more generally $p_n(x)$ must be orthogonal to *any polynomial* of an order less than n. Let us denote the highest coefficient of $p_n(x)$, i.e., the coefficient of x^n, by μ_n. Then the difference

$$p_{n+1}(x) - \frac{\mu_{n+1}}{\mu_n}xp_n(x)$$

eliminates the power x^{n+1} and what remains is a polynomial of the order n. This polynomial can certainly be obtained as a linear combination of $p_0(x)$, $p_1(x)$, $\cdots$, $p_n(x)$:

$$p_{n+1}(x) - \frac{\mu_{n+1}}{\mu_n}xp_n(x) = \gamma_0 p_0(x) + \gamma_1 p_1(x) + \cdots + \gamma_n p_n(x) \quad (5\text{-}21.8)$$

Now, multiplying on both sides by $\rho(x)p_m(x)$, $m < n - 1$, we obtain in consequence of orthogonality:

$$\gamma_m \int_a^b \rho(x)p_m^2(x)\, dx = -\frac{\mu_{n+1}}{\mu_n}\int_a^b \rho(x)p_n(x)xp_m(x)\, dx$$

But $xp_m(x)$ is a polynomial of a degree less than n, to which $p_n(x)$ is orthogonal. Hence all $\gamma_m(m < n - 1)$ must drop out on the right side of (8) and the only non-zero coefficients are γ_n and γ_{n-1}. The existence of a "three-term recurrence relation" of the form (7) is thus demonstrated and c_{n+1} is explicitly obtained:

$$c_{n+1} = \frac{\mu_{n+1}}{\mu_n} \qquad (5\text{-}21.9)$$

We will now assume that the orthogonal polynomials are normalized in length:

$$\int_a^b \rho(x)p_k^2(x)\, dx = 1 \qquad (5\text{-}21.10)$$

If $p_k^2(x)$ is written in the form

$$p_k(x)(\mu_k x^k + \cdots)$$

and the fact is taken into account that the dots represent a polynomial of not higher than $(n-1)$st order, we obtain

$$\mu_k \int_a^b \rho(x)p_k(x)x^k \, dx = 1 \qquad (5\text{-}21.11)$$

Let us now multiply (7) by $p_{n-1}(x)\rho(x)$ and integrate on both sides, taking into account that $xp_{n-1}(x)$ gives $\mu_{n-1}x^n$ plus a polynomial of not higher than $(n-1)$st order:

$$b_n = c_{n+1}\mu_{n-1}\int_a^b x^n p_n(x)\rho(x) \, dx = c_{n+1}\frac{\mu_{n-1}}{\mu_n}$$

We have thus obtained an explicit expression even for b_n:

$$b_n = \frac{\mu_{n-1}\,\mu_{n+1}}{\mu_n^2} \qquad (5\text{-}21.12)$$

and now we can write (7) in slightly different form, multiplying the equation by μ_n/μ_{n+1}:

$$\beta_{n+1}p_{n+1}(x) = (x - \alpha_n)p_n(x) - \beta_n p_{n-1}(x) \qquad (5\text{-}21.13)$$

where we have put

$$\alpha_k = a_k\frac{\mu_k}{\mu_{k+1}}, \qquad \beta_k = \frac{\mu_{k-1}}{\mu_k}$$

These recurrence relations can be conceived as a sequence of linear equations:

$$
\begin{aligned}
(\alpha_0 - x)y_0 + \beta_1 y_1 && &= 0 \\
\beta_1 y_0 + (\alpha_2 - x)y_1 + \beta_2 y_2 && &= 0 \quad (5\text{-}21.14) \\
\beta_2 y_2 + (\alpha_3 - x)y_2 + \beta_3 y_3 &= 0
\end{aligned}
$$

This never-ending sequence terminates, however, if we consider those particular values $x = \lambda_k$ for which $p_n(x)$ vanishes:

$$y_n = p_n(\lambda_k) = 0, \qquad (k = 1, 2, \ldots, n)$$

We then have a homogeneous set of n linear equations whose determinant must vanish:

$$\begin{vmatrix} \alpha_0 - \lambda & \beta_1 & & & & \\ \beta_1 & \alpha_1 - \lambda & \beta_2 & & & \\ & \beta_2 & \alpha_3 - \lambda & \beta_3 & & \\ & & & \cdot & & \\ & & & & \cdot & \\ & & & & \cdot & \\ & & & & \beta_{n-1} & \alpha_{n-1} - \lambda \end{vmatrix} = 0 \quad (5\text{-}21.15)$$

We see that we obtain a regular *eigenvalue problem* of a symmetric matrix of the order n. The eigenvalues $\lambda = \lambda_i$ are the roots of the equation $p_n(\lambda) = 0$. The eigensolutions are the principal axes of a symmetric matrix which are automatically *orthogonal* to each other. These eigensolutions are

$$u_i = p_0(\lambda_i), \quad p_1(\lambda_i), \quad \ldots, \quad p_{n-1}(\lambda_i)$$

Thus we obtain the orthogonality relations

$$\sum_{\alpha=0}^{n-1} p_\alpha(\lambda_i)\, p_\alpha(\lambda_k) = 0 \qquad (i \neq k) \qquad (5\text{-}21.16)$$

We can make the $p_k(\lambda_i)$ matrix to a truly orthogonal matrix by normalizing the length of each column to 1. This means that we put

$$q_{ki} = \rho_i p_k(\lambda_i)$$

where

$$\rho_i = \frac{1}{\sqrt{\displaystyle\sum_{\alpha=0}^{n-1} p_\alpha^2(\lambda_i)}} \qquad (5\text{-}21.17)$$

Since, however, an orthogonal matrix has the property that the orthogonality relations hold between the rows not less than between the columns, we obtain

$$\sum_{\alpha=1}^{n} \rho_\alpha^2 p_i(\lambda_\alpha) p_k(\lambda_\alpha) = 0 \qquad (i \neq k) \qquad (5\text{-}21.18)$$

This relation is a remarkable counterpart of the orthogonality condition

$$\int_a^b \rho(x)p_i(x)p_k(x)\,dx = 0 \qquad (i \neq k)$$

It shows that the orthogonal polynomials possess a *second* orthogonality property. They are orthogonal with respect to *integration*. But they are also orthogonal with respect to *summation*, if the key points are chosen as the zeros of the first neglected polynomial (in both cases the orthogonality is of the weighted type, but the two weight factors are quite different).

This second orthogonality property greatly reduces the labor of interpolating with the help of orthogonal polynomials. Our aim was to solve the linear equations (6):

$$c_0'p_0(\lambda_1) + c_1'p_1(\lambda_1) + \dots c_{n-1}'p_{n-1}(\lambda_1) = f(\lambda_1)$$
$$\vdots$$
$$c_0'p_0(\lambda_n) + c_1'p_1(\lambda_n) + \dots c_{n-1}'p_{n-1}(\lambda_n) = f(\lambda_n)$$

The orthogonality condition (18), together with the normalization condition

$$\sum_{\alpha=1}^n \rho_\alpha^2 p_i^2(\lambda_\alpha) = 1 \qquad (5\text{-}21.19)$$

allows us to solve the linear set explicitly and obtain the coefficients c_i' in the form:

$$c_i' = \sum_{\alpha=1}^n \rho_\alpha^2 f(\lambda_\alpha)p_i(\lambda_\alpha) \qquad (5\text{-}21.20)$$

These equations can be conceived as natural counterpart of the determining equations (16.10) of the coefficients c_i. The great advantage of the new coefficients is that they are numerically available by a simple *summation* process (the coefficients of which can be pretabulated), without any integration.

Here again the Chebyshev polynomials are endowed with superior properties. For them the weight factors ρ_α^2 become all *equal* and thus weighted orthogonality becomes ordinary orthogonality. Moreover, the coefficients $p_k(\lambda_\alpha)$ are here easily available since they are the simple and well-tabulated trigonometric functions $\cos k\theta$ at angles θ_i which are easily accessible (cf. IV, 16).

Other special cases of potential interest are the Legendre polynomials and the Laguerre polynomials. Although the zeros of these polynomials, together with the weights ρ_i, have been calculated,[1] an elaborate tabulation of the matrices $p_k(\lambda_i)$ is not available at the present time.

[1] The weights ρ_i^2 are in a remarkable relation to the weights of the Gaussian quadrature. Let the interpolation of $f(x)$ by orthogonal polynomials serve the purpose of obtaining a parexic value of the definite integral

$$A = \int_a^b \rho(x) f(x) \, dx$$

("Gaussian quadrature," cf. VI, 10). Then the orthogonality of all $p_i(x)$ to $p_0(x) = p_0 = $ const. shows that every term of the expansion (5), except the first vanishes in the process of integration and we obtain

$$\bar{A} = c_0' p_0 \int_a^b \rho(x) \, dx$$

But then, in view of (20),

$$\bar{A} = p_0^2 \int_a^b \rho(x) \, dx \sum_{\alpha=1}^n \rho_\alpha^2 \ (\ _\alpha) \tag{5-21.21}$$

The factor in front of the summation sign is 1, considering the normalization of $p_0(x)$. On the other hand, the weights w_i of the Gaussian quadrature

$$\bar{A} = \sum_{\alpha=1}^n w_\alpha f(\lambda_\alpha)$$

are tabulated. Comparison with (21) gives

$$\rho_i^2 = w_i$$

Bibliographical References

[1]   Cf. Ref. {4}, Chapters IV and V;   [2] Cf. Ref. {8}, Chapters VI and IX.

[3]   FORT, T., *Finite Differences and Difference Equations* (Oxford University Press, New York, 1948).

[4]   MILNE-THOMPSON, *Calculus of Finite Differences* (Macmillan, London, 1933).

[5]   STEFFENSON, J. F., *Interpolation* (Williams & Wilkins, Baltimore, 1927).

[6]   WHITTAKER, E. T., and ROBINSON, G., *A Short Course in Interpolation* (Blackie & Son, London, 1923).

   Article

[7]   SCHOENBERG, I. J., "Some Analytical Aspects of the Problem of Smoothing" (*Courant Anniversary Volume* (Interscience Publishers, New York, 1948).

VI

QUADRATURE METHODS

1. Historical notes. The problem of areas challenged the imagination of scientific thinking from earliest dates. The "problem of Dido," usually formulated as the problem of enclosing a maximum area by a chain of given length, originated in prehistoric times. The determination of areas of more or less complicated shape played an important part in agricultural civilizations, and both the old Babylonians and the old Egyptians were acquainted with a variety of formulas, expressed, of course, in verbal rather than algebraic form, by which areas included by straight lines could be evaluated. The Greeks with their advanced knowledge of geometry went into much greater details. The area of a circle offered a particularly challenging problem and led to establishment of the rigorous methods of limit theory, called in ancient times the "method of exhaustion." True integrations were performed by Archimedes, who used the method of inscribed and circumscribed polygons, thus obtaining upper and lower bounds which approached each other indefinitely. With the advent of infinitesimal calculus many areas could be evaluated by the discovery that integration and differentiation are inverse processes. Moreover, the simple trapezoidal procedure of the ancients became refined by using polynomials of second and higher orders for interpolation of equidistant data. An eminently useful formula, based on parabolas of second order, was introduced by the English mathematician Th. Simpson (in 1743) and is usually quoted as "Simpson's rule." Later Gauss (1814) invented a particularly ingenious and important method for obtaining areas, based on the properties of the Legendre polynomials.

The Swedish mathematician H. Fredholm introduced in 1900 his "integral equations" which in later decades became of fundamental importance in the solution of boundary-value problems, eigenvalue problems, and many problems of advanced statistics. If the integrals here encountered are evaluated by the simple trapezoidal rule, an integral equation becomes replaceable by a large system of ordinary linear algebraic equations. But frequently more convenient solutions are available if more advanced quadrature methods are applied to the definite integral which characterizes the left side of an integral equation.

2. Quadrature by planimeters. The problem of determining the area under a given curve is frequently referred to as "mechanical quadrature," although the implication does not mean that a mechanical instrument shall be used for the solution of the problem. In actual fact, however, mechanical instruments do exist which perform the quadrature operation by mechanical or electrical means. They are called "planimeters" if their basic principle is that we trace by a pointer the circumference of the area to be evaluated. The reading of the instrument gives the area directly in square inches.

Carefully constructed planimeters are rather expensive instruments and their accuracy is limited. They require careful drawing and tracing of the contour of the unknown area. Frequently a simple calculation can give more accurate results than the planimeter, particularly if an empirical curve is not observed as a continuous curve but as a sequence of discrete ordinates. The calculation utilizes the observed ordinates only and does not require that the points of observation shall be interpolated by a more or less arbitrary graphical procedure.

3. The trapezoidal rule. The oldest method of approximating the area under a continuous curve is the method of inscribed polygons, known today as the "trapezoidal rule." We connect the observed ordinates by straight lines and replace the area under the curve by the area under the polygon. If the observed ordinates

$$y = y_0, y_1, y_2, \cdots, y_n \qquad (6\text{-}3.1)$$

belong to the abscissa values

$$x = x_0, x_1, x_2, \cdots, x_n \qquad (6\text{-}3.2)$$

the elementary formula for the area of a trapezoid gives the following result for the area under the polygon.[1]

$$\bar{A} = \tfrac{1}{2}(y_0 + y_1)(x_1 - x_0) + \cdots + \tfrac{1}{2}(y_{n-1} + y_n)(x_n - x_{n-1}) \qquad (6\text{-}3.3)$$

$$= \tfrac{1}{2}[y_0(x_1 - x_0) + y_1(x_2 - x_0) + y_2(x_3 - x_1) + \cdots +$$

$$y_{n-1}(x_n - x_{n-2}) + y_n(x_n - x_{n-1})]$$

For equidistant spacing,

$$x_1 - x_0 = x_2 - x_1 = x_3 - x_2 = \cdots = x_n - x_{n-1} = h \qquad (6\text{-}3.4)$$

the formula (3) assumes the simpler form

$$\bar{A} = h(\tfrac{1}{2}y_0 + y_1 + y_2 + \cdots + y_{n-1} + \tfrac{1}{2}y_n) \qquad (6\text{-}3.5)$$

which means that, apart from the factor h, we simply add up all the observed ordinates, but apply the weight factor $\tfrac{1}{2}$ to the two limiting ordinates.

We know from the fundamental theorem of integral calculus that the area under the curve characterized by $y = f(x)$ can be defined as the limit to which $\bar{A}$ tends as h approaches zero.

$$A = \int_a^b f(x)\,dx = \lim_{h \to 0} h(\tfrac{1}{2}y_0 + y_1 + \cdots + y_{n-1} + \tfrac{1}{2}y_n) \quad (6\text{-}3.6)$$

For theoretical purposes this limit process is quite satisfactory since for mathematically given functions we can frequently obtain the limit by analytical tools. This was the procedure of Archimedes in evaluating the center of masses and center of buoyancies of many complicated figures.

From the viewpoint of practical computation the trapezoidal rule gives quite satisfactory results if we possess a sufficient number of ordinates. The calculation is extremely simple since straightforward addition of a given set of ordinates on the comptometer is a simple and quick process. Very often, however, it is cumbersome to ascertain the large collection of ordinates which are demanded by the trapezoidal method for a sufficiently close approximation.

4. Simpson's rule. Straight lines are too rigid for a really satisfactory approximation of curves. If we want to imitate a curve by

---

[1] We use the notation "bar" consistently to indicate an "approximation." Hence $\bar{A}$ is an approximation of the true area A; similarly $\bar{\eta}$ is an approximate value of η.

drawing a succession of straight lines, we need a great many small lines for this purpose. We can obviously fare much better if we use *parabolas of second order* for the approximation. With the help of such parabolas we can approximate a curve to a remarkable degree without changing the approximating parabolas too often. A large number of very short straight sections is thus replaced by a small number of much longer parabolic sections.

Hence we will increase the accuracy of our quadrature formula very considerably if, instead of connecting two consecutive ordinates by a straight line, we connect *three* consecutive ordinates by a parabola. We can always find a parabola of the form

$$y = a + bx + cx^2 \qquad (6\text{-}4.1)$$

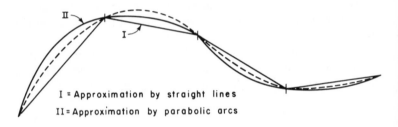

I = Approximation by straight lines

II = Approximation by parabolic arcs

which will fit three given points. By this method we come much closer to the actual curve than by mere straight lines. The error committed is thus much smaller.

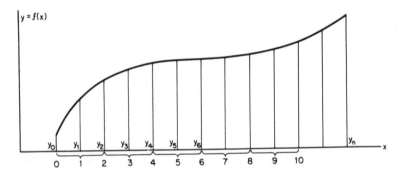

The resulting procedure is known as Simpson's method, and the resulting formula is called Simpson's rule. We divide the total area into an even number of equal panels, thus reading an odd number of ordinates because the two end ordinates are included in our reading. For the sake of convenience the width of each panel is chosen as 1. In each double panel we approximate the curve by a parabola of second order.

Let us consider for example the first double panel, composed of the three ordinates y_0, y_1, y_2. We can expand $y = f(x)$ around the point $x = 1$ into a local power series, making use of the method of central differences (cf. V, 3). According to Stirling's formula we have

$$f(1 + t) = f(1) + \delta f(1)t + \frac{\delta^2 f(1)}{2} t^2 \qquad (6\text{-}4.2)$$

Here

$$\begin{aligned} f(1) &= y_1 \\ \delta f(1) &= \tfrac{1}{2}(y_1 - y_0) \\ \delta^2 f(1) &= y_2 - 2y_1 + y_0 \end{aligned} \qquad (6\text{-}4.3)$$

Here we approximated the curve between $x = 0$ and $x = 2$ by a parabola of second order which coincides with the actual curve at the three points of interpolation $x = 0, 1, 2$.

The area under the approximating parabola can be obtained by integrating between the points $x = 0$ and $x = 2$.

$$\bar{A}_{02} = \int_{-1}^{+1} f(1 + t)\, dt \qquad (6\text{-}4.4)$$

$$= \left| f(1)t + \delta f(1) \frac{t_2}{2} + \frac{\delta^2 f(1)}{2} \frac{t^3}{3} \right|_{-1}^{+1}$$

$$= 2f(1) + \tfrac{1}{3} \delta^2 f(1)$$

$$= 2y_1 + \tfrac{1}{3}(y_2 - 2y_1 + y_0)$$

$$= \tfrac{1}{3} y_0 + \tfrac{4}{3} y_1 + \tfrac{1}{3} y_2$$

We repeat exactly the same process for the areas $\bar{A}_{24}, \bar{A}_{46}, \cdots$ until the total area is exhausted. We then get

$$\begin{aligned} \bar{A} &= \tfrac{1}{3} y_0 + \tfrac{4}{3} y_1 + \tfrac{1}{3} y_2 \qquad (6\text{-}4.5) \\ &\quad + \tfrac{1}{3} y_2 + \tfrac{4}{3} y_3 + \tfrac{1}{3} y_4 \\ &\quad + \cdot \quad \cdot \quad \cdot \quad \cdot \quad \cdot \quad \cdot \quad \cdot \\ &= \tfrac{2}{3}[\tfrac{1}{2} y_0 + y_2 + y_4 + \cdots + \tfrac{1}{2} y_{2n}] \\ &\quad + \tfrac{4}{3}[y_1 + y_3 + y_5 + \cdots + y_{2n-1}] \end{aligned}$$

This is *Simpson's formula*. We separate even and odd ordinates and apply the even and odd ordinates with different weights, instead of the same weights as we have done in the earlier trapezoidal method. This discrimination of weights greatly increases the accuracy of the result.

If the width of each panel is not 1 but h, we merely multiply by h, and the final formula becomes

$$\bar{A} = \tfrac{2}{3}h[\tfrac{1}{2}y_0 + y_2 + \cdots + \tfrac{1}{2}y_{2n} + 2(y_1 + y_3 + \cdots + y_{2n-1})] \quad (6\text{-}4.6)$$

The necessity of an even number of panels is sometimes an inconvenient limitation for the use of Simpson's formula. We can avoid this difficulty by using a different construction for the *last three* (or first three) panels, if the number of panels happens to be odd. Consider the four ordinates y_0, y_1, y_2, y_3. We put the origin of our reference system into the point $x = 1.5$ and approximate the function

$$\tfrac{1}{2}[f(1.5 + t) + f(1.5 - t)]$$

by a parabola $y = a + ct^2$

since we know in advance that the area under the curve between $t = \pm1.5$ is influenced only by the *even* part of the function. The conditions at $t = 0.5$ and 1.5 determine the coefficients a and c:

$$c = \tfrac{1}{4}[y_0 + y_3 - (y_1 + y_2)]$$

$$a = \tfrac{1}{16}[9(y_1 + y_2) - (y_0 + y_3)]$$

and we obtain for the area of the first three panels.

$$\bar{A}_{03} = \left| at + \frac{c}{3}t^3 \right|_{-1.5}^{+1.5} = \frac{3}{2}2\left(a + \frac{c}{3}\frac{9}{4}\right)$$

$$= \frac{3}{8}[3(y_1 + y_2) + (y_0 + y_4)] \quad (6\text{-}4.7)$$

The corresponding formula for the width h of the panels becomes

$$\bar{A}_{03} = \frac{3h}{8}[3(y_1 + y_2) + (y_0 + y_4)] \quad (6\text{-}4.8)$$

Hence if the number of panels is odd, we apply formula (8) to the first three, or last three, panels. Then the remaining number of panels is even, and here Simpson's formula (6) comes into operation.

5. The accuracy of Simpson's formula. We can estimate the accuracy of the parabolic approximation by the following consideration. If we integrate Stirling's formula between the limits ± 1, all the terms with odd differences drop out, since the area under an odd function, taken between symmetric limits, is zero. Now, assuming that the sequence of ordinates is sufficiently dense, the Stirling expansion will be sufficiently convergent to estimate the truncation error by the first neglected term. The first term we have neglected in the integration (4.4) is

$$\frac{\delta^4 f(1)}{24} \int_{-1}^{+1} t^2(t^2 - 1)\, dt = -\frac{\delta^4 f(1)}{90} \tag{6-5.1}$$

The same consideration holds for every double panel, from $k = 1$ to $k = 2n - 1$. We thus obtain for the complete area

$$\int_0^{2n} f(x)\, dx \doteq \bar{A} - \frac{1}{90} \sum_{k=0}^{n-1} \delta^4 f(2k + 1) \tag{6-5.2}$$

This expression for the error is not very convenient, since it requires setting up an elaborate difference table. However, if $f(x)$ is sufficiently smooth, the difference coefficient may be replaced by the derivative and the sum by an integral. Then the estimated error $\bar{\eta}$ of the quadrature appears in the form:

$$\bar{\eta} = -\frac{h^4}{180}[f'''(b) - f'''(a)] \tag{6-5.3}$$

where a is the lower and b the upper limit of the quadrature, and the width of the panels is no longer 1 but h. (For a more general estimation of the error, not assuming smoothness, cf. 17.6.)

More convenient and more reliable is another method of error estimation which will be discussed later (cf. § 12). If this method is applied to Simpson's quadrature procedure, we obtain the following error estimate:

$$\bar{\eta} = \frac{4h}{15}\left[\left(\tfrac{1}{2}y_0 + y_2 + \cdots + \tfrac{1}{2}y_{2n}\right) - (y_1 + y_3 + \cdots + y_{2n-1})\right]$$
$$- \frac{h^2}{15}[f'(b) - f'(a)] \tag{6-5.4}$$

6. The accuracy of the trapezoidal rule. In order to estimate the accuracy of the trapezoidal rule, we make use of Bessel's formula which interpolates on the half lines.

$$f(\tfrac{1}{2} + t) = \delta^0 f(\tfrac{1}{2}) + \delta f(\tfrac{1}{2})t + \frac{\delta^2 f(\tfrac{1}{2})}{2}(t^2 - \tfrac{1}{4}) + \cdots \quad (6\text{-}6.1)$$

Integrating between the limits $t = \pm\tfrac{1}{2}$, we obtain the area of one single panel.

$$\bar{A}_{01} = \int_{-1/2}^{+1/2} f(\tfrac{1}{2} + t)\, dt$$

$$= \left| \delta^0 f(\tfrac{1}{2})t + \delta f(\tfrac{1}{2})\frac{t^2}{2} + \frac{\delta^2 f(\tfrac{1}{2})}{2}\left(\frac{t^3}{3} - \frac{t}{4}\right) \right|_{-1/2}^{+1/2} \quad (6\text{-}6.2)$$

Adding up the area of every panel and neglecting higher order terms, we obtain

$$A \doteq \sum_{k=0}^{n-1} \delta^0 f(k + \tfrac{1}{2}) - \frac{1}{12}\sum_{k=0}^{n-1} \delta^2 f(k + \tfrac{1}{2})$$

$$\doteq \bar{A} - \frac{1}{12}[\delta f(n) - \delta f(0)] \quad (6\text{-}6.3)$$

The formula for the panel width h becomes

$$A \doteq \bar{A} - \frac{h}{12}[\delta f(nh) - \delta f(0)] \quad (6\text{-}6.4)$$

and replacing again differences by derivatives we finally obtain the formula which corresponds to (5.3).

$$\bar{\eta} = -\frac{h^2}{12}[y'(b) - y'(a)] \quad (6\text{-}6.5)$$

The comparison of (5.3) and (5) shows how much more accurate Simpson's rule is than the trapezoidal rule. The second power of h is changed to the fourth power, while the numerical factor 12 in the denominator is changed to the much larger factor 180. On the other hand, the first derivative is changed to the third derivative, which is often much less smooth than the first derivative.

7. The trapezoidal rule with end correction. We can perceive the significance of the trapezoidal formula from still a different viewpoint by introducing the Fourier series in our analysis. Let us assume that $f(x)$ is given in the range $[0,1]$ and let us expand it into a Fourier cosine series.

$$f(x) = \tfrac{1}{2}a_0 + a_1 \cos \pi x + a_2 \cos 2\pi x + \cdots \qquad (6\text{-}7.1)$$

If we now substitute for x the values

$$0, \frac{1}{n}, \frac{2}{n}, \cdots, \frac{n}{n}$$

and perform the summation

$$\bar{A} = \frac{1}{n}\left[\tfrac{1}{2}f(0) + f\left(\frac{1}{n}\right) + \cdots + f\left(\frac{n-1}{n}\right) + \tfrac{1}{2}f\left(\frac{n}{n}\right)\right] \qquad (6\text{-}7.2)$$

we obtain on the right side

$$\bar{A} = \tfrac{1}{2}a_0 + a_{2n} + a_{4n} + \cdots \qquad (6\text{-}7.3)$$

Now by the general definition of the Fourier coefficients,

$$a_k = 2 \int_0^1 f(x) \cos k\pi x \; dx \qquad (6\text{-}7.4)$$

Hence $\tfrac{1}{2}a_0$ is the true area A under the curve and we obtain

$$A = \bar{A} - (a_{2n} + a_{4n} + \cdots) \qquad (6\text{-}7.5)$$

We thus see that the trapezoidal rule will hold the better the more convergent the Fourier series (1) is. Now we know that the convergence of the Fourier series is decided by the analytical behavior of the function. The more continuous the function is in itself, and in its derivatives, the more convergent will the Fourier series be. In our present problem of a pure cosine series we have in the full range $[-1,1]$ an even function which remains continuous at the boundary points, while the derivative will generally have a discontinuity at the points $x = 0$ and $x = 1$. The order of magnitude of the coefficients a_k will be determined by this discontinuity. If in (4) we integrate by parts, we obtain

$$a_k = 2 \left|\frac{f(x) \sin k\pi x}{k\pi}\right|_0^1 = \frac{2}{k\pi}\int_0^1 f'(x) \sin k\pi x \; dx$$

$$\qquad (6\text{-}7.6)$$

$$= \frac{2}{(k\pi)^2}\left|f'(x) \cos k\pi x\right|_0^1 - \frac{2}{(k\pi)^2}\int_0^1 f''(x) \cos k\pi x \; dx$$

Hence the coefficients of even order become

$$a_{2k} = \frac{1}{2k^2\pi^2} \left[f'(1) - f'(0) \right] + \varepsilon \qquad (6\text{-}7.7)$$

where ε becomes small in comparison with the first term if k is large. If it so happens, however, that the boundary condition

$$f'(0) = f'(1) \qquad (6\text{-}7.8)$$

is satisfied, we can repeat the method of integrating by parts once more and obtain

$$a_{2k} = -\frac{1}{8k^4\pi^4} \left[f'''(1) - f'''(0) \right] \qquad (6\text{-}7.9)$$

In the first case, equation (5) gives

$$A = \bar{A} - \left[f'(1) - f'(0) \right] \frac{1}{2n^2\pi^2} \left[1 + \frac{1}{2^2} + \frac{1}{3^2} + \cdots \right] + \varepsilon \quad (6\text{-}7.10)$$

while in the second case,

$$A = \bar{A} + \frac{f'''(1) - f'''(0)}{8n^4\pi^4} \left[1 + \frac{1}{2^4} + \frac{1}{3^4} + \cdots \right] + \varepsilon \quad (6\text{-}7.11)$$

The infinite sums appearing in these equations can be evaluated in terms of the "Bernoulli numbers" B_{2k} on account of the relation

$$B_{2k} = \frac{2(2k)!}{(2\pi)^{2k}} \left(1 + \frac{1}{2^{2k}} + \frac{1}{3^{2k}} + \cdots \right) \qquad (6\text{-}7.12)$$

The first Bernoulli numbers are

$$B_2 = \frac{1}{6}, \quad B_4 = \frac{1}{30}, \quad B_6 = \frac{1}{42}, \quad \cdots$$

Hence

$$\frac{1}{\pi^2}\left(1 + \frac{1}{2^2} + \frac{1}{3^2} + \cdots \right) = \frac{1}{6} \qquad (6\text{-}7.13)$$

$$\frac{1}{\pi^4}\left(1 + \frac{1}{2^4} + \frac{1}{3^4} + \cdots \right) = \frac{1}{90}$$

We thus obtain from (10) as an estimate of the error of the trapezoidal rule:

$$A \doteq \bar{A} - \frac{1}{12n^2} \left[f'(1) - f'(0) \right] \doteq \bar{A} - \frac{h^2}{12} \left[f'(1) - f'(0) \right] (6\text{-}7.14)$$

in agreement with our previous result (6.5).

However, by a proper modification of $f(x)$ we can eliminate the jump in the first derivative and put formula (11) in operation. Let us consider the function

$$g(x) = f(x) - \frac{f'(1) - f'(0)}{2} x^2 \qquad (6\text{-}7.15)$$

For this function the boundary condition (8) is satisfied (replacing f by g) and we obtain the estimate [cf. (11) and (13)],

$$\int_0^1 g(x)\, dx \doteq \bar{A}_g + \frac{g'''(1) - g'''(0)}{720n^4}$$

$$\doteq \bar{A}_g + \frac{h^4}{720} [f'''(1) - f'''(0)] \qquad (6\text{-}7.16)$$

Now by the definition of $g(x)$,

$$\int_0^1 g(x)\, dx = \int_0^1 f(x)\, dx - \frac{1}{6} [f'(1) - f'(0)] \qquad (6\text{-}7.17)$$

while the operation $\bar{A}_g$ is composed of the following two parts:

$$\bar{A}_g = \frac{1}{n} \left[[\tfrac{1}{2} f(0) + f\left(\frac{1}{n}\right) + \cdots + f\left(\frac{n-1}{n}\right) + \tfrac{1}{2} f\left(\frac{n}{n}\right) \right]$$

$$- \frac{f'(1) - f'(0)}{2n^3} \sum_{k=0}^{n}{}' k^2 \qquad (6\text{-}7.18)$$

$$\sum_{k=0}^{n}{}' k^2 = \frac{n(n+1)(2n+1)}{6} - \frac{n^2}{2} = \frac{n^3}{3} + \frac{n}{6} \qquad (6\text{-}7.19)$$

and thus we obtain

$$\int_0^1 f(x)\, dx \doteq \frac{1}{n} \left[\frac{1}{2} f(0) + f\left(\frac{1}{n}\right) + \cdots \frac{1}{2} f(1) \right]$$

$$- \frac{1}{12n^2} [f'(1) - f'(0)] + \frac{1}{720n^4} [f'''(1) - f'''(0)] \qquad (6\text{-}7.20)$$

We return to our original notations,

$$\bar{A} = h[\tfrac{1}{2} y_0 + y_1 + \cdots + y_{n-1} + \tfrac{1}{2} y_n] - \frac{h^2}{12} [f'(b) - f'(a)] \qquad (6\text{-}7.21)$$

with the error estimate

$$\bar{\eta} = \frac{h^4}{720} [f'''(b) - f'''(a)] \qquad (6\text{-}7.22)$$

This result shows that the correction

$$-\frac{h^2}{12}\,[f'(b) - f'(a)] \qquad (6\text{-}7.23)$$

(which requires the knowledge of the derivative at the two endpoints of the range) greatly increases the accuracy of the simple trapezoidal rule. The new error, if compared with the estimated error (5.3) of Simpson's rule, is only $\frac{1}{4}$ of that value and of opposite sign.

Use of the trapezoidal rule with end correction is particularly advocated if the given ordinates are the result of observations and thus afflicted by accidental errors. The simple arithmetic mean after halving the two extreme ordinates will tend to minimize the influence of these errors. Moreover, we will be able to lay a least-square parabola of second order through a suitable number of points at the one and the other end of the range. The derivative of this parabola at the end point will then provide us with the values $f'(b)$ and $f'(a)$ which can be used for the end correction (23). By this device we can considerably increase the accuracy of the simple trapezoidal method.

8. Numerical examples. *Problem I.* As a numerical demonstration of the operation of the various formulas we choose a simple example which permits us to follow the general analytical procedures with little technical complications. We choose the simple function

$$y = e^x$$

and assume $h = 0.5$. The range of integration shall extend from $x = 0$ to $x = 4$. Taking the ordinates from a table, we have

x	y
0	1
0.5	1.64872
1	2.71828
1.5	4.48169
2	7.38906
2.5	12.18249
3	20.08554
3.5	33.11545
4	54.59815

The theoretical value of the area under the curve is in this problem

$$A = \int_a^b f(x)\, dx = \int_0^4 e^x\, dx = \left| e^x \right|_0^4 = e^4 - 1$$

which gives $A = 53.59815$

The trapezoidal formula gives

$\bar{A} = 0.5(\tfrac{1}{2} \cdot 1 + 1.64872 + 2\cdot71828 + \cdots + \tfrac{1}{2} \cdot 54.59815) = 54.71015$

The error of this result is

$$53.59815 - 54.71015 = -1.11200$$

The theoretical error estimate (6.5) gives

$$-\frac{h^2}{12}[y'(b) - y'(a)] = -\frac{1}{48}(54.59815 - 1) = -1.11663$$

We now apply Simpson's formula to the same problem.

$\bar{A} = \tfrac{2}{3}0.5[\tfrac{1}{2} \cdot 1 + 2.71828 + 7.38906 + 20.08554 + \tfrac{1}{2} \cdot 54.59815]$
$\quad + 2(1.64872 + 4.48169 + 12.18249 + 33.11545) = 53.61622$

The new error is $53.59815 - 53.61622 = -0.01807$

The estimated error (5.3) becomes

$$-\frac{h^4}{180}[y'''(b) - y'''(a)] = -\frac{1}{2880}(54.598 - 1) = -0.01861$$

while (5.4) yields $\bar{\eta} = -0.01816$.

The trapezoidal rule with end correction becomes

$$54.71015 - 1.11663 = 53.59352$$

This value is slightly *less* than the correct value, while Simpson's rule gave a value slightly *more* than the correct value. The new error is

$$53.59815 - 53.59352 = 0.00463$$

The estimated error is $-\tfrac{1}{4}$ of the error of Simpson's formula, that is,

$$0.25 \cdot 0.01861 = 0.00465$$

In all these cases the predictions and the numerical results agree very closely. We will now consider a problem which operates under less favorable circumstances.

Problem II. The function of Problem I was a "smooth" function, i.e., a function whose successive differences decreased satisfactorily. We will now choose a function whose Stirling expansion converges much less satisfactorily because of the nearness of a singular point. The given function shall be

$$y = \frac{1}{\sqrt{x}}$$

between the limits $x = 0.1$ and $x = 1.7$; the equidistant ordinates shall follow each other at the distance $h = 0.2$.

x	y
0.1	3.16228
0.3	1.82574
0.5	1.41421
0.7	1.19523
0.9	1.05409
1.1	0.95346
1.3	0.87706
1.5	0.81650
1.7	0.76696

Theoretical value:

$$\int_{0.1}^{1.7} x^{-1/2}\, dx = 2 \left| x^{1/2} \right|_{0.1}^{1.7} = 2(\sqrt{1.7} - \sqrt{0.1}) = 2 \cdot 0.98756$$

$$= 1.97512$$

Application of the trapezoidal rule:

$$\bar{A} = 0.2(\tfrac{1}{2} \cdot 3.16228 + 1.82574 + \cdots + \tfrac{1}{2}\, 0.76696)$$
$$= 0.2 \cdot 10.10092 = 2.02018 \qquad [\text{error: } -0.045]$$

Application of Simpson's formula:

$$\bar{A} = \tfrac{2}{3}0.2[\tfrac{1}{2} \cdot 3.16228 + 1.41421 + 1.05409 + 0.87706 + \tfrac{1}{2} \cdot 0.76696$$
$$+ 2(1.82574 + 1.19523 + 0.95346 + 0.81605)]$$
$$= 1.98558 \qquad [\text{error: } -0.010]$$

Trapezoidal rule with end correction:

$$\bar{A} = 2.02018 - \frac{0.04}{12}\left(-\frac{0.76696}{3 \cdot 4} + \frac{3.16228}{0.2}\right) = 1.96823$$

$$[\text{error: } 0.0069]$$

We observe in this example that the refinements of the simple trapezoidal method have now much less effect on the result than in the previous example. The order of magnitude of the error remained the same in all three methods. This is because the effect of the higher powers of h is counteracted by the strong increase of the higher derivatives. The nearness of the singular point $x = 0$ greatly reduces the effectiveness of the difference calculus by putting a much larger weight on the higher terms of the Stirling series, as we can see if we evaluate the estimated errors of the three formulas.

Estimated error of the trapezoidal formula:

$$A - \bar{A} = -\frac{0.04}{12}\left(-\frac{0.767}{3.4} + \frac{3.162}{0.2}\right) = -0.051$$

Estimated error of Simpson's formula (cf. 5.3):

$$A - \bar{A} = -\frac{0.0016}{180}\left(-\frac{15}{8}\frac{0.767}{(1.7)^3} + \frac{15}{8}\frac{3.162}{(0.1)^3}\right) = -0.053$$

Estimated error of the corrected trapezoidal formula:

$$A - \bar{A} = \frac{0.053}{4} = 0.013$$

The unreliable estimation of the error in the case of Simpson's formula is caused by the unsmoothness of the function, which has the consequence that the third derivative and the third central difference do not agree even approximately in the neighborhood of the lower limit. The formula (5.4) gives more reliable results. Its application yields $\bar{\eta} = -0.0139$, which overestimates the true error $\eta = -0.010$, but to no undue degree.

9. Approximation by polynomials of higher order. In Simpson's formula we terminated the Stirling formula with the *quadratic* term. We can obviously go further and terminate the series with a term of higher order. Correspondingly the number of panels involved will increase. Since we do not want to lose the great advantage of symmetric limits, the next step after the limits ±1 will be the limits ±2, involving *four* neighboring panels. This means five consecutive data:

$$y_0, y_1, y_2, y_3, y_4 \tag{6-9.1}$$

from which central differences up to the fourth order can be formed. Integrating between the limits ± 2 we obtain

$$\bar{A} = \int_{-2}^{+2} f(2 + t)\, dt$$

$$= \left| f(2)t + \frac{\delta^2 f(2)}{2}\frac{t^3}{3} + \frac{\delta^4 f(2)}{24}\left(\frac{t^5}{5} - \frac{t^3}{3}\right)\right|_{-2}^{+}$$

$$= 4f(2) + \frac{8}{3}\delta^2 f(2) + \frac{14}{45}\delta^4 f(2) \qquad (6\text{-}9.2)$$

$$= \frac{2}{45}\left[90 f(2) + 60\,\delta^2 f(2) + 7\,\delta^4 f(2)\right]$$

$$= \frac{2}{45}\left[7y_0 + 32y_1 + 12y_2 + 32y_3 + 7y_4\right]$$

If the width of the panels is h, the formula has to be multiplied by h, and thus the final five-point formula becomes

$$\bar{A} = \frac{2h}{45}(7y_0 + 32y_1 + 12y_2 + 32y_3 + 7y_4) \qquad (6\text{-}9.3)$$

The error of this approximation can again be estimated by the first neglected term of the Stirling expansion.

$$\bar{\eta} = \frac{h\,\delta^2 f(2h)}{6!}\int_{-2}^{+2} t^2(t^2 - 1)(t^2 - 4)\, dt$$

$$= \frac{h}{720}\delta^6 f(2)\left|\frac{t^7}{7} - 5\frac{t^5}{5} + 4\frac{t^3}{3}\right|_{-2}^{+2} = -\frac{8h}{945}\delta^6 f(2) \qquad (6\text{-}9.4)$$

If the 6th difference coefficient is replaced by the 6th derivative, the estimated error of the four-panel formula becomes

$$\bar{\eta} = -\frac{8h^7}{945}f^{(6)}(2h) \qquad (6\text{-}9.5)$$

In our previous numerical example the entire region was divided into 8 panels of equal width. Applying Simpson's formula, we grouped these panels in the form of 4 double panels. We will now group them as a double group of 4 panels. Hence the weight factors of the successive ordinates become, considering the fact that y_5 is the extreme right ordinate of the first group but simultaneously the extreme left ordinate of the second group,

$$\frac{2h}{45} (7, 32, 12, 32, 14, 32, 12, 32, 7) \qquad (6\text{-}9.6)$$

Applying these weights to the tabulated ordinates of Example I we obtain

$$\bar{A} = \frac{2 \cdot 0.5}{45} \, 2411.98706 = 53.59971$$

The error of this result is

$$A - \bar{A} = 53.59815 - 53.59971 = -0.00156$$

If we estimate the error on the basis of the formula (5), we have to remember that *two* sets of panels were employed. Accordingly the second factor has to be taken at the points $2h = 1$ and $6h = 3$ and their sum formed.

$$f^{(6)}(1) + f^{(6)}(3) = 2.72 + 20.09 = 22.81$$

This gives

$$A - \bar{A} = -\frac{8 \cdot (0.5)^7}{945} \, 22.81 = -0.00151$$

The accuracy of this estimation is very satisfactory.

We notice that the fourth-order approximation decreased the error by the factor 12 if compared with the second-order approximation of Simpson's rule. The gain is caused by the higher power of h, which is here not counteracted by an unduly large increase of the higher derivatives. Quite different is the situation in Problem II, where the higher derivatives go up very rapidly for the small values of x. If the weights (6) are applied to this problem, we obtain

$$\bar{A} = \frac{0.4}{45} \, [14(1.581139 + 1.054093 + 0.383482)$$

$$+ \, 32(1.825742 + 1.195229 + 0.953463 + 0.816497)$$

$$+ \, 12(1.414214 + 0.877058)] = 1.982818$$

The error is now -0.0077, which is only slightly less than the previous error, -0.010.

Here we do not succeed with a reliable estimation of the error, in contrast to the previous example. The reason is that the error estimation (4) was based on the convergence of the Stirling expansion, and this presumes the smooth behavior of the higher differences. In our present example the nearness of the singular point at $x = 0$ precludes the forming of the sixth difference. Consequently we cannot expect valid results from the application of the formula (5). On the other hand, the method of § 12, which does not require higher than first derivatives, operates again satisfactorily. Its application to the present problem gives the error estimate $\bar{\eta} = -0.0127$, which slightly, but not unduly, overestimates the true error $\eta = -0.0077$.

Generally we can say that a higher-order approximation is helpful only if at the same time h is chosen sufficiently small. We can greatly gain in accuracy if we operate with a sufficiently small h and a polynomial approximation of not too low order. It would be a mistake, however, to believe that we will always obtain great accuracy by approximating the *entire* region by one polynomial of the order N if the total number of ordinates is $N + 1$. This is prevented by the generally divergent behavior of equidistant polynomial approximation, as discussed before (cf. V, 15). In practice the excellent accuracy of Simpson's formula is usually satisfactory since it combines a relatively high power of h, viz., the fourth, with a derivative of relatively low order, viz., the third [cf. (5.3)]. In the case of smooth functions the four-panel formula (3) deserves attention and we want to add a useful six-panel formula, known as "Weddle's rule." This formula operates with a sixth-order parabola, but makes a slight error in the sixth difference, for the purpose of simpler weighting.

$$\bar{A} = \frac{3h}{10}(y_0 + 5y_1 + y_2 + 6y_3 + y_4 + 5y_5 + y_6) \qquad (6\text{-}9.7)$$

Of considerable value is also the simple trapezoidal rule, augmented by end correction [cf. (7.21)]. The error of this formula is only $-\frac{1}{4}$ of that of Simpson's formula. It can be employed for checking purposes and is particularly useful if the given ordinates are not the result of calculation but of observation.

10. The Gaussian quadrature method. The eminent mathematician Gauss injected an entirely new and exceptionally ingenious idea into the customary theory of quadratures. This idea in its wider implications was quite fundamental for many fields of practical analysis. We assume that a certain integrable function $y = f(x)$ is not given at every point of the continuous variable x but only at certain selected points $x_1, x_2, \cdots, x_n$ which shall lie inside of a given interval. Since we are going to deal with a *finite* range only, we can immediately *normalize* the range of interest. We will put the origin of the variable x in the middle of the range considered and choose a scale factor which makes the two end points of the range to the points $x = \pm 1$. Hence we will now deal with the range

$$-1 \leq x \leq +1 \qquad (6\text{-}10.1)$$

and assume that the points $x_1, x_2, \cdots, x_n$ in which the function $y = f(x)$ is given belong to this range. The ordinates

$$\begin{aligned} y_1 &= f(x_1) \\ y_2 &= f(x_2) \\ &\ \ \vdots \\ y_n &= f(x_n) \end{aligned} \qquad (6\text{-}10.2)$$

are generally not enough for a determination of the function $f(x)$, no matter how large n may be. But we can try to *interpolate* $f(x)$ for intermediate points. For this purpose we may use the powers of x. We can find a definite polynomial $p_{n-1}(x)$ of the order $n - 1$ which has the property that it assumes the given values y_k at the given points x_k.

In the usual calculus of finite differences we assume that the chosen points $x = x_k$ are *equidistantly* spaced. Gauss conceived the idea that we could possibly get much greater accuracy with the same number of ordinates if we did not fix their position in advance but utilized the distribution of the data points in some suitable fashion to our greatest advantage. By this procedure Gauss succeeded in obtaining not only a quadrature formula of extraordinary accuracy but also a procedure which is free of the dangers of equidistant polynomial interpolation, although these dangers were entirely unknown in his time.

Let us assume that we leave the points of interpolation $x = x_k$ entirely free and want to determine the polynomial $u = p_{n-1}(x)$ which will fit the given ordinates $y_1, y_2, \cdots, y_n$. The resulting formula is known as "Lagrange's interpolation formula."[1] It is based on constructing the fundamental polynomial

$$F_n(x) = (x - x_1)(x - x_2) \cdots (x - x_n) \qquad (6\text{-}10.3)$$

[1] Cf. {8}, p. 84, {11}, p. 86.

and dividing it by synthetic division by the n root factors $(x - x_1)$, $\cdots$, $(x - x_n)$. We thus obtain a set of polynomials,

$$Q_i(x) = \frac{1}{F_n'(x_i)} \frac{F_n(x)}{x - x_i} \qquad (i = 1, 2, \cdots, n) \qquad (6\text{-}10.4)$$

which have the following properties: $Q_i(x)$ vanishes at all points $x = x_k$ except $x = x_i$ where $Q_i(x)$ becomes 1. If we introduce "Kronecker's delta" δ_{ik} which is defined as 1 if $i = k$ and 0 if $i \neq k$, we can write

$$Q_i(x_k) = \delta_{ik} \qquad (6\text{-}10.5)$$

But then we see that a polynomial $p_{n-1}(x)$ constructed by the sum

$$p_{n-1}(x) = y_1 Q_1(x) + y_2 Q_2(x) + \cdots + y_n Q_n(x) \qquad (6\text{-}10.6)$$

satisfies the condition that it assumes at any point $x = x_k$ the prescribed ordinates $y = y_k$. The uniqueness of $p_{n-1}(x)$ follows from the fact that the difference between $p_{n-1}(x)$ and a hypothetical second polynomial $\bar{p}_{n-1}(x)$ would assume the values 0 at all n points $x = x_k$. But the difference $p_{n-1}(x) - \bar{p}_{n-1}(x)$ is again a polynomial of the order $n - 1$, and such a polynomial cannot have more than $n - 1$ roots except by vanishing identically, which means $\bar{p}_{n-1}(x) \equiv p_{n-1}(x)$.

Now, if we consider $p_{n-1}(x)$ a sufficiently close approximation of the given function $y = f(x)$, we can obtain the area of the unknown function $f(x)$ parexically by evaluating

$$\bar{A} = \int_{-1}^{+1} p_{n-1}(x)\, dx = \sum_{k=1}^{n} y_k \int_{-1}^{+1} Q_k(x)\, dx \qquad (6\text{-}10.7)$$

For any given distribution of the points $x = x_i$ the $Q_k(x)$ are uniquely determined, and thus the definite integrals

$$\int_{-1}^{+1} Q_k(x)\, dx = w_k \qquad (6\text{-}10.8)$$

will have some definite numerical values which can be tabulated. These values are entirely independent of the nature of the function $y = f(x)$ in whose area we are interested.

Now the ingenious Gaussian quadrature method can be introduced as follows. Let us add an additional point

$$x = x_{n+1}$$

to the previous points, without changing in any way the previous points x_i. This will now introduce an additional root factor $x - x_{n+1}$ and generate an additional $Q_{n+1}(x)$. We can see from the definition (4) of $Q_i(x)$ that this $Q_{n+1}(x)$ will be proportional to the previous $F_n(x)$, since the new root factor $(x - x_{n+1})$ drops out. Hence the weight factor w_{n+1} by which the new ordinate y_{n+1} has to be multiplied will be proportional to the definite integral

$$\int_{-1}^{+1} F_n(x)\, dx \tag{6-10.9}$$

Similarly, if m new points

$$x = x_{n+1}, \quad x_{n+2}, \quad \cdots, \quad x_{n+m} \tag{6-10.10}$$

are introduced together with their ordinates, the corresponding weights $w_{n+1}, w_{n+2}, \cdots, w_{n+m}$ are determined by a definite integral of the type

$$w_{n+i} = \int_{-1}^{+1} F_n(x) G_{m-1}^i(x) \tag{6-10.11}$$

where these $G_{m-1}^i(x)$ are some polynomials of the order $m - 1$. Now all these weights will become *automatically zero* if we let $F_n(x)$ satisfy the following integral conditions:

$$\int_{-1}^{+1} F_n(x)\, dx = 0, \cdots, \int_{-1}^{+1} F_n(x) x^{m-1}\, dx = 0 \tag{6-10.12}$$

in view of the fact that an arbitrary polynomial $G_{m-1}(x)$ is a linear superposition of the powers $1, x, x^2, \cdots, x^{m-1}$.

In actual fact we can go up to $m = n$ by requiring the integral conditions

$$\int_{-1}^{+1} F_n(x) x^\alpha\, dx = 0 \qquad (\alpha = 0, 1, 2, \cdots, n - 1) \tag{6-10.13}$$

The result is that we can add freely any n points to our originally given n points, and yet none of the new ordinates will change anything on the result obtained before. Hence *in effect* we operated with $2n$

ordinates and yet *in fact* we used only n ordinates, since all the additional ordinates contributed nothing to the area to be evaluated. By this procedure we save n terms in the sum

$$\bar{A} = \sum_{k=1}^{2n} y_k w_k$$

but even more important is the fact that we *need not even know* the additional ordinates $y_{n+1}, y_{n+2}, \cdots, y_{2n}$. The sum

$$\bar{A} = \sum_{k=1}^{n} y_k w_k \qquad (6\text{-}10.14)$$

gives the area with the help of n ordinates and yet with an accuracy as if $2n$ ordinates had been used.

Integral conditions of the type (13) are called "orthogonality conditions." We say that the polynomial $F_n(x)$ is "orthogonal" to the powers $1, x, x^2, \cdots, x^{n-1}$. We have encountered such conditions earlier when dealing with the "orthogonal functions systems" (cf. V, 16). We have studied the "Jacobi polynomials" (cf. V, 20), which have the property that they are orthogonal to all powers of lower order, exactly in the sense of the conditions (13). However, generally the orthogonality involves a weight factor $\rho(x)$ in the integrand. Only in the special case of the "Legendre polynomials" (cf. 5-20.11) does it happen that the weight factor becomes 1 and thus weighted orthogonality changes into simple orthogonality. The choice of $F_n(x)$ is thus decided; the Gaussian program requires that $F_n(x)$ shall be identified with the nth Legendre polynomial $P_n(x)$. The zeros of these polynomials give us the points at which the function $f(x)$ has to be prescribed. They have been tabulated with great accuracy, together with the numerical values of the coefficients w_i which can be calculated by evaluating the definite integrals (8) (cf. also § 13).[1]

11. Numerical example. We want to apply the Gaussian method to the same numerical examples which we considered earlier in § 8.

---

[1] Cf. [2]. The labor of calculating the coefficients w_i is reduced to one-half by the symmetry properties of the Legendre polynomials. The roots appear in pairs $\pm \xi_k$; the weights belonging to two such points are equal.

The first example represents a very smooth, the second a very unsmooth function. The rapidity of convergence with increasing n is thus very different in the two cases. In our previous equidistant procedures, 9 equidistant ordinates were used. We will now replace them by only 5 nonequidistant ordinates. Since the tabulation of the Gaussian zeros assumes the interval $[-1,1]$, we have to adjust an arbitrary interval to this normalization. We do that by the transformation

$$x = \frac{b+a}{2} + \frac{b-a}{2}\,\xi \qquad (6\text{-}11.1)$$

$$\int_a^b f(x)\,dx = \frac{b-a}{2}\int_{-1}^{+1} f(\xi)\,d\xi \qquad (6\text{-}11.2)$$

In our first problem $a = 0$, $b = 4$. Hence the ordinates have to be read at the points

$$x_k = 2 + 2\xi_k = 2(1 + \xi_k) \qquad (6\text{-}11.3)$$

For the choice $n = 5$ we obtain the five zeros,

$$2\,(1 \pm 0.906179846)$$
$$2\,(1 \pm 0.538469310)$$
$$2$$

The points of interpolation, together with the associated weights w_k, thus become (using 8 decimal place accuracy)

$$
\begin{aligned}
x_1 &= 0.18764031 & w_1 &= 0.47385377 \\
x_2 &= 0.92306138 & w_2 &= 0.95725734 \\
x_3 &= 2 & w_3 &= 1.13777778 \qquad (6\text{-}11.4) \\
x_4 &= 3.07693862 & w_4 &= 0.95725734 \\
x_5 &= 3.81235969 & w_5 &= 0.47385377
\end{aligned}
$$

These weights were obtained by multiplying the Gaussian weights by $(b-a)/2 = 2$. The ordinates of the function $y = e^x$ at these points can be taken from the "Tables of the Exponential Function" of the

Mathematical Tables Project, New York, after making the proper interpolations. These ordinates become

$$y_1 = 1.20639950$$
$$y_2 = 2.51698405$$
$$y_3 = 7.38905610$$
$$y_4 = 21.69189349$$
$$y_5 = 45.25710562$$

Multiplying these ordinates by the weight factors of the table (4) and summing, we obtain

$$\bar{A} = 53.59813663$$

against the true value,

$$A = e^4 - 1 = 53.59815003$$

The error of the approximation is

$$\eta = 0.0000134$$

The previous approximation (9.6) by 9 ordinates gave the much larger error

$$\eta = -0.0016$$

We see that the Gaussian method gives an admirable accuracy. In spite of operating with 5 instead of 9 ordinates, the error decreased by a factor of more than 100. We thus get the impression that the operation with uneven intervals performs even *more* than it promises. It seems to have a benefical effect on the error, even beyond the saving of ordinates. This is indeed the case. Equidistant interpolation is generally not a well-convergent process, and for functions which have singularities inside the unit circle, (although they may be entirely smooth between -1 and $+1$), the convergence is generally not even guaranteed (cf. V, 15). The Gaussian quadrature process uses the zeros of the Legendre polynomials, i.e., the zeros of an *orthogonal* set of functions, as points of interpolation. The convergence of this process is guaranteed by the general nature of orthogonal expansions, (cf. V, 21).

The Gaussian quadrature method is thus superior to the ordinary equidistant methods for two reasons. One is that n ordinates are comparable in effectiveness to $2n$ equidistant ordinates. The other is that interpolation by Legendre polynomials is much more convergent than interpolation by Lagrangian polynomials.

In our second example of § 8 even the Gaussian quadrature has slow convergence. But once more we can demonstrate the power of the Gaussian method in the saving of ordinates. This time we want to use but *four* ordinates, instead of the original nine. The limits now are

$$a = 0.1, \qquad b = 1.7$$

Hence
$$x_k = 0.9 + 0.8\xi_k$$

The four points of interpolation become

$$0.9 \pm 0.8 \cdot 0.33998104$$
$$0.9 \pm 0.8 \cdot 0.86113631$$

and we can set up the table

$x_1 = 0.21109095$	$w_1 = 0.34785484$
$x_2 = 0.62801516$	$w_2 = 0.65214515$
$x_3 = 1.17198483$	$w_3 = 0.65214515$
$x_4 = 1.58890905$	$w_4 = 0.34785484$

This time we have copied the weight factors w_k unchanged, since it is numerically simpler to obtain the result and then multiply by $(b - a)/2 = 0.8$ than to multiply every weight by that factor.

The ordinates y_k of the function $y = x^{-1/2}$; at the points of interpolation are now

$$y_1 = 2.17653268$$
$$y_2 = 1.26187092$$
$$y_3 = 0.92371714$$
$$y_4 = 0.79332380$$

Multiplying by the corresponding weight factors w_k and summing gives

$$2.45839962$$

Hence
$$\bar{A} = 0.8 \cdot 2.45839962 = 1.96671970$$

The correct value of the area is here

$$A = 1.97512$$

which gives the error $\qquad \eta = 0.0084$

This is only slightly more than the error obtained in § 9 by using 9 ordinates:

$$\eta = -0.0077$$

In the previous case we divided the range of integration into two panels and used in each panel an approximating polynomial of fourth order. In the present case the entire range seems to be approximated by a polynomial of only third order. In actual fact a polynomial of *seventh order* is used because each of our points counts actually as a *double point*. We lay a parabola of seventh order through eight specially chosen points. The points of interpolation are not merely the four zeros of $P_4(\xi)$. They are in fact four *pairs* of points, but each pair lies close together and in the limit collapses into one point. The *four* zeros of $P_4(\xi)$ actually stand for the *eight* zeros of $P_4^2(\xi)$.

12. The error of the Gaussian quadrature. The more effective a certain method of parexic analysis is, the more difficult it usually is to obtain a satisfactory estimate of the accuracy obtained. In the case of the Gaussian quadrature, estimation of the error is not easy. The traditional formula which estimates the error of the Gaussian quadrature requires knowledge of the $2n$th derivative of $f(x)$ throughout the interval of integration.[1] This formula is

$$\eta = \left[\frac{(n!)^2}{(2n)!} \right]^2 \frac{2^{2n+1}}{2n+1} \frac{f^{(2n)}(\theta)}{(2n)!} \ (\theta = some\ unknown\ point\ between\ \pm 1)$$

$$(6\text{-}12.1)$$

This estimate has several drawbacks. From the theoretical standpoint we can object to the assumption that $f(x)$ possesses $2n$ derivatives in the interval of integration since the Gaussian quadrature converges to the proper value even in the case of nonanalytical functions, such as for example $\sqrt{|x|}$, whose first derivative becomes already

---

[1] Cf. [2], p. 740.

infinite at $x = 0$ and yet is perfectly amenable to Gaussian quadrature. From the practical standpoint it is usually difficult (except in simple cases) to evaluate the $2n$th derivative of a function, even if its analytical form is given. But frequently $f(x)$ is given only in tabulated form, and the analytical expression of $f(x)$ is unknown.

The following procedure is free of these objections. If the Gaussian quadrature is applied to $f'(x)$, we know that the result should be

$$\int_{-1}^{+1} f'(x)\, dx = f(1) - f(-1)$$

Hence in this case we can check up explicitly on the error of the Gaussian quadrature. For the proper exploitation of this idea we have to consider, however, that the quadrature between the limits -1 and $+1$ involves only the *even* part of the function, viz., $f(x) + f(-x)$, while the quadrature of the derivative would involve an entirely different function, viz., the *odd* part of $f(x)$, which is $f(x) - f(-x)$. We avoid this difficulty by taking the derivative of $xf(x)$ which has the same symmetry character as $f(x)$ itself:

$$\int_{-1}^{+1} [xf(x)]'\, dx = f(1) + f(-1) \qquad (6\text{-}12.2)$$

Since $(xf)' = xf' + f$, we see that the numerical procedure is simple. We multiply the previous weights w_i by $\xi_i f'(\xi_i)$ instead of $f(\xi_i)$. The error of the new quadrature is now given as follows:

$$\eta' = f(1) + f(-1) - \bar{A} - \sum_{\alpha=1}^{n} w_\alpha \xi_\alpha f'(\xi_\alpha) \qquad (6\text{-}12.3)$$

In the case of general limits a and b (cf. 11.1),

$$\eta' = \frac{b-a}{2}[f(b) + f(a)] - \bar{A} - \left(\frac{b-a}{2}\right)^2 \sum_{\alpha-1}^{n} w_\alpha \xi_\alpha f'(x_\alpha) \qquad (6\text{-}12.4)$$

Now a closer investigation of the error of the Gaussian quadrature reveals that for functions which are not too unsmooth between -1 and $+1$ the point θ of the formula (1) is near to the origin $\xi = 0$. But then the last factor of the formula is practically equal to the coefficient of ξ^{2n} in the Taylor expansion around the origin. Now the expansion of $(\xi f')$ is identical with the original expansion except that a_{2n} is multiplied by $2n + 1$. Under these conditions we get the error estimate

$$\bar{\eta} = \frac{1}{2n + 1} \eta' \qquad (6\text{-}12.5)$$

The great advantage of this estimate is that it requires the *first derivative* of $f(x)$ only, instead of the 2nth derivative of the traditional formula. We would think that the 2nth derivative of $f(x)$ is necessary, considering the fact that the quadrature is accurate for any polynomial whose order is less than $2n$. In order to eliminate such a polynomial, we must differentiate $2n$ times. But the formula (5), although we have differentiated only once, does not lose out on this account either. If $f(x)$ is any polynomial of an order less than $2n$, the function $(xf)'$ is again a polynomial of the same type. Hence the second quadrature gives the error zero, and the estimate (5) becomes likewise zero.

If $f(x)$ does not have the smoothness required by the above argument, we can assume that the differentiation will increase rather than decrease the unsmoothness of the function. Hence we can assume that the estimation according to (5) will tend to *over*estimate the error and thus we will be on the safe side, even if our estimate is not too realistic. We cannot be sure, however, in the case that $f'(x)$ changes its sign in the given interval.[1]

We will now apply this method of estimating the error of the Gaussian quadrature to the two numerical examples of § 11. In the first example we obtain

$$\eta' = 2 \cdot 55.59815003 - 53.59813663 - 57.59801484 = 0.0001487$$

Dividing by $2n + 1 = 11$ we get

$$\eta = 0.0000135$$

which agrees perfectly with the actual error. Here the function was very smooth. We now come to the second example where the function is much less smooth inasmuch as the higher derivatives

---

[1] The exact safety limits of this procedure are not yet established.

increase strongly in a certain range of the critical interval. We now obtain

$$0.8 \cdot 3.929242650 - 1.96671970 - 1.02713980 = 0.149535$$

Here the number of points was $n = 4$ and thus we have to divide by 9:

$$\eta = 0.01661$$

The actual error is only *one-half* of this number. But this is still a satisfactory estimate and it is fortunate that we have *over*estimated the error.

13. The coefficients of a quadrature formula with arbitrary zeros. The coefficients of an arbitrary quadrature formula (with or without a weight factor $\rho(x)$) can be evaluated by a simple numerical scheme. We know that if we interpolate $f(x)$ at the n points $x = \xi_1, \xi_2, \cdots , \xi_n$, the interpolation will be exact for any polynomial whose order is lower than n. In particular the successive powers $1, x, x^2, \cdots , x^{n-1}$ will be interpolated without any error. Hence the quadrature associated with all these powers will also be exact. We assume that we have evaluated the n definite integrals:

$$u_k = \int_a^b \rho(x) x^k \, dx \tag{6-13.1}$$

Now the general form of a quadrature formula is

$$\bar{A} = \sum_{\alpha=1}^n w_\alpha f(\xi_\alpha) \tag{6-13.2}$$

and since in our special case $\bar{A}$ coincides with the exact value of the area A, we obtain the following n equations:

$$
\begin{aligned}
w_1 \quad &+ w_2 \quad + \cdots + w_n \quad\quad = u_0 \\
w_1\xi_1 \quad &+ w_2\xi_2 \quad + \cdots + w_n\xi_n \quad = u_1 \\
&\vdots \\
w_1\xi_1^{n-1} &+ w_2\xi_2^{n-1} + \cdots + w_n\xi_n^{n-1} = u_{n-1}
\end{aligned}
\tag{6-13.3}
$$

These n linear equations are sufficient for a unique determination of the w_i. We have dealt with this problem of "weighted moments" earlier (cf. IV, 23) and obtained a simple numerical algorithm for its solution (cf. 4-23.11, 12, 13]. This algorithm is applicable to our problem and yields a simpler method for the evaluation of the w_i than the explicit construction of the definite integrals (10.8].

14. Gaussian quadrature with rounded-off zeros. From the practical angle the Gaussian quadrature suffers from one serious drawback. The function has to be evaluated at irrational points. This requires heavy interpolation, which is a cumbersome procedure. In the case of tabulated functions it could easily happen that it would take much more effort to obtain n interpolated ordinates than to read off directly $2n$ equidistant ordinates. For this reason the Gaussian method is usually employed only if the evaluation of every y_k requires an independent calculation, because of absence of any tabulation.

This drawback of the Gaussian method can be remedied by the following modification of the original procedure. We *round off* the Gaussian zeros to a small number of decimal places, perhaps two or three, and evaluate the weight factors associated with these shifted zeros. The process of interpolation is then greatly simplified or even obviated if we possess tables of $f(x)$ which proceed in units of 0.01 or perhaps 0.001 of the argument. It is true that the full accuracy of the Gaussian procedure is not available in this manner, but the accuracy is still high. In fact, by an additional correction scheme, considered in the next section, the *full accuracy* of the Gaussian method may be maintained.

The coefficients w_i associated with the shifted zeros can be evaluated according to the general method of § 13. We first construct the fundamental polynomial $F_n(\xi)$ with the help of the root factors. Let us choose for example $\eta = 5$. Here the five Gaussian zeros are

$$\xi = \pm 0.53846 \cdots, \quad 0, \quad \pm 0.90617 \cdots$$

We round them off to two decimal places:

$$\xi = \pm 0.54, \quad 0, \quad \pm 0.91$$

and construct the fundamental polynomial out of the root factors:

$$F_5(\xi) = \xi(\xi^2 - 0.54^2)(\xi^2 - 0.91^2)$$
$$= \xi^5 - 1.1197\xi^3 + 0.24147396\xi$$

Comparison with the fifth Legendre polynomial

$$P_5(\xi) = \tfrac{1}{8}(63\xi^5 - 70\xi^3 + 15\xi)$$

requires multiplication by 63 and the factor $\tfrac{1}{8}$ in front:

$$\tfrac{1}{8}(63\xi^5 - 70.5411\xi^3 + 15.21285948\xi)$$

We see that the two polynomials have nearly equal coefficients. The y_k quantities of the equations (4-23.8) have now the meaning of the definite integrals (13.1) with $\rho(x) = 1$:

$$\int_{-1}^{+1} x^k \, dx = \frac{2}{k+1} \qquad (k = 0, 2, 4, \cdots)$$
$$= 0 \qquad (k = 1, 3, 5, \cdots)$$

giving rise to the reciprocal polynomial

$$2\xi^{-1} + \tfrac{2}{3}\xi^{-3} + \tfrac{2}{5}\xi^{-5}$$

by which $F_5(\xi)$ has to be multiplied, according to the scheme (4-23.11):

$$\begin{array}{rrrrr}
0.4829479200, & 0, & -2.2394000000, & 0, & 2 \\
-0.7464666667, & 0, & 0.6666666667, & & \\
0.4 & & & & \\
\hline
0.1364812533, & 0, & -1.5727333333, & 0, & 2
\end{array}$$

Hence

$$G_4(\xi) = 0.1364812533 - 1.5727333333\xi^2 + 2\xi^4$$

and

$$\frac{G_4(\xi)}{F_5'(\xi)} = \frac{0.1364812533 - 1.5727333333\xi^2 + 2\xi^4}{0.24147396 \quad - 3.3591\xi^2 \quad + 5\xi^4}$$

Substitution of $\xi = \pm 0.91, \pm 0.54, 0$, yields the five weights w_i of the quadrature formula:[1]

$$\begin{array}{cc}
\xi = \pm 0.91 & w = 0.231387878 \\
\pm 0.54 & 0.486011767 \\
0 & 0.565200708
\end{array}$$

We apply these weights to our standard example of $y = e^x$ treated before (cf. § 11) with the exact Gaussian zeros. The new ordinates become

$x = 0.18$	$y = 1.19721736$
0.92	2.50929039
2	7.38905610
3.08	21.75840240
3.82	45.60420832

The weights have to be multiplied by 2, because of the double range of x.[1] The sum of weighted ordinates becomes

$$\bar{A} = 53.59910051$$

against the true value

$$A = 53.59815003$$

The error of the approximation is thus

$$\eta = -0.00095048$$

Compared with the Gaussian error, the error has increased by the factor 71, which shows the great sensitivity of the Gaussian method to even small shifts of the zeros. Nevertheless, the accuracy is still considerable.

The estimation of the error on the basis of the method discussed in § 12 operates again satisfactorily. We now obtain

$$\eta' = 111.19630006 - 53.59910051 - 57.6093084 = -0.012109$$

Division by 11 gives

$$\bar{\eta} = -0.00110$$

which is only slightly more than the true error given above.

15. The use of double roots. In § 11 the remark was made that the great efficiency of the Gaussian method is explainable on the basis that the fundamental polynomial is not $P_n(\xi)$ but actually $P_n^2(\xi)$. Every point of the interpolation may be counted as a double point, for it is in the nature of Gaussian quadrature that n points can be added freely to the points of interpolation without changing

---

[1] A systematic table for the operation with rounded-off zeros was prepared by the Mathematical Tables Project, New York City, and is reprinted as Table XIV of the Appendix, by permission of the Project.

anything, since the new points enter the quadrature formula with the weight zero. Let us now assume that we choose an *arbitrary* set of n points within the range but choose every point as a double root of $F_n(\xi)$. Since double roots are equivalent to two close points which in the limit collapse into one, interpolation by n double points has the significance that at every point of interpolation the functional values $f(x_k)$ *and its derivative* $f'(x_k)$ are given. The Gaussian points are now chosen in such particular fashion that the weights of the derivatives shall become zero. The quadrature formula is thus reduced to n instead of $2n$ terms.

If now the Gaussian zeros are slightly out of focus, the weight factors of the derivatives will not vanish any more but they will remain *small*. Hence it will not be necessary to know the derivatives with great accuracy. If the function is tabulated in sufficiently close intervals, the mere difference coefficient between two neighboring tabular values can take the place of the derivative. In this fashion we can round off the Gaussian zeros to convenient numbers, thus avoiding the inconvenience of interpolation, and still maintain the full accuracy of the Gaussian procedure.

As a numerical example we return once more to our previous example of the exponential function, using five Gaussian points (cf. § 11). We round off these points to $\xi = \pm 0.54$ and $\xi = \pm 0.90$ (sacrificing the slightly better 0.91 in favor of a more convenient value). The coefficients w_i and w_i' can again be evaluated according to the numerical scheme of the previous section, but raising the order of $F_n(\xi)$ to 10 by squaring. Acccordingly the polynomial $G_{n-1}(\xi)$ will be of the order 9, but in actual fact we get a polynomial of the order 4 in ξ^2 since all the odd powers of ξ drop out. As a result we obtain the five weights w_i of the ordinates $f(\xi_i)$, augmented by five weights w_i' of the derivatives $f'(\xi_i)$:[1]

$\xi =$	$w =$	$w' =$
$-0.90,$	$0.23640530,$	-0.00155377
$-0.54,$	$0.47899553,$	0.00058042
0	$0.56919830,$	0
0.54	$0.47899553,$	-0.00058042
0.90	$0.23640530,$	0.00155377

[1] A systematic table of the weights w_i and w_i' for the operation with the ordinates and their derivatives at the rounded off Gaussian zeros is not available at the present time.

In our example the change of the limits has the consequence that the points of interpolation become

$$x = 0.2, \quad 0.92, \quad 1, \quad 3.08, \quad 3.8$$

The successive ordinates and their derivatives become

$$y_1 = y_1' = 1.22140276$$
$$y_2 = y_2' = 2.50929039$$
$$y_3 = y_3' = 7.38905610$$
$$y_4 = y_4' = 21.75840240$$
$$y_5 = y_5' = 44.70118449$$

The weights w_i of the ordinates have to be multiplied by 2, the weights w_i' of the derivatives by $2^2 = 4$. The result of the weighting and summing is

$$\bar{A} = 53.37259512 + 0.22554004 = 53.59813516$$

The error is now

$$\eta = 0.00001488$$

and we see that the *full accuracy* of the Gaussian procedure is preserved.

Once more we can estimate the error on the basis of the method described in § 12. Once more we obtain the formula (12.4) but with the following modifications. The correction in $\bar{A}$, caused by the weights w_i', is multiplied by 2. Moreover, we have to add one more term of the following form:

$$-\left(\frac{b-a}{2}\right)^3 \sum_{\alpha=1}^{n} w_\alpha' \xi_\alpha f''(x_\alpha)$$

In our numerical example we obtain

$$\eta' = 111.19630006 - 53.59813516 - 0.22554004$$
$$-4 \cdot 14.229894611 - 8 \cdot 0.0566116792 = 0.000153$$

Division by $2n + 1 = 11$ gives $\bar{\eta} = 0.0000139$. The agreement with the actual error is again satisfactory.

16. Engineering applications of the Gaussian quadrature method. The Gaussian quadrature method is characterized by very high accuracy. Even a small number of ordinates gives usually a very accurate evaluation of a definite integral. In problems of engineering, excessive accuracy is seldom required. The Gaussian quadrature method has its place, however, as an excellent device for economizing in the number of ordinates. It happens rather frequently that the average value of a function of unknown structure has to be established on the basis of very few observations. In this case it is strongly advocated that the points where the ordinates are measured shall follow the Gaussian pattern.

For example, in our standard numerical problem of evaluating the definite integral

$$A = \int_0^4 e^x \, dx$$

the use of Simpson's rule, employing nine equidistant ordinates, gave an error of 0.02 in 54 units, i.e., an accuracy of 0.04%. Such accuracy will seldom be required in an engineering problem. Let us now cut down the number of ordinates to *three*. The Gaussian procedure requires that these ordinates shall be placed at the following x values and taken into account with the following weight factors:

$x = 0.45$	$w = 1.11$
2	1.78
3.55	1.11

The calculation gives $\bar{A} = 53.535$

compared with the true value

$$A = 53.598$$

The error is $\eta = 0.063$

This error is three times as large as the error of nine ordinates but still very acceptable. Yet the number of ordinates was only three. Hence use of the Gaussian quadrature method is strongly indicated if for some reasons we have to economize on the number of ordinates employed for establishment of the average value of an unknown function. Had we used three *equidistant* ordinates in the above example, we would have obtained the value

$$\bar{A} = 56.77$$

The error is now $\qquad \eta = 3.2$

Hence the error is 50 times as big as when using the Gaussian method.

The pressure tubes in an airduct will give much more favorable results if they are not uniformly distributed over the cross section of the airduct but in conformity with the Gaussian zeros. The same holds for temperature measurements along a wall or for temperature measurements spread over a certain time interval if the purpose of these measurements is to establish average values.

The Gaussian zeros and the associated weight factors have been calculated with great accuracy and are available in tabular form. The Table XIII of the Appendix containing these data is taken from the calculations of the Mathematical Tables Project in New York. The same project evaluated the weight factors which belong to the rounded off values of the Gaussian zeros. Part of this table is included in the Appendix (cf. Table XIV), with the permission of the Project.

17. Simpson's formula with end correction. We have seen that the error caused by a small shift of the Gaussian zeros could be counteracted by adding the knowledge of the derivatives to the knowledge of the functional values. The weights of the derivatives at symmetrically placed points entered with $\pm$ signs. We can take advantage of this property of the weights w'_k for a modification of Simpson's rule which greatly increases its accuracy at the cost of a small additional calculation. In § 4 we discussed Simpson's method. It consisted in dividing the entire range into an even number of panels and approximating every double panel by a parabola of second order. Let us concentrate on such a double panel and assume that both the mid-point and the two end points shall be taken as *double points*. Then in effect we approximate by a parabola of fifth order, and the error will be proportional to the sixth derivative, instead of the previous fourth derivative. For sufficiently smooth functions, the accuracy of the formula is thus greatly increased. On the other hand, we now have to know function *and derivative* at every panel point. In actual fact, however, since every panel point (with the exception of the two end points) is the terminal point of one panel and at the same time the starting point of the next panel, the derivatives enter with the weights $w' - w' = 0$; only the two end points behave differently, and thus we have to know the derivatives only at these two points.

The evaluation of the weights is here so simple that we need not take recourse to the general procedure of § 13. We can solve the linear equations for the weights directly, making use of the fact that for the powers $1, x, \cdots, x^5$ we must get exact results. The odd powers can be neglected, since for them the equations balance automatically, because of symmetry. Hence only $y = 1$, x^2, x^4 have to be tried. We have 3 unknowns, viz., the two weights

$$\begin{array}{c|ccc} x = & -1 & 0 & 1 \\ \hline & w_1 & w_0 & w_1 \end{array}$$

and the third weight $-w_1' \quad 0 \quad w_1'$

The weight w_0' can be equated to zero in advance, because of symmetry. Now we have for the three trial functions

$$y = 1 \Big\} \quad \begin{aligned} y(-1) &= 1 \\ y'(-1) &= 0 \end{aligned} \quad y(0) = 1 \quad \begin{aligned} y(1) &= 1 \\ y'(1) &= 0 \end{aligned}$$

$$y = x^2 \Big\} \quad \begin{aligned} y(-1) &= 1 \\ y'(-1) &= -2 \end{aligned} \quad y(0) = 0 \quad \begin{aligned} y(1) &= 1 \\ y'(1) &= 2 \end{aligned}$$

$$y = x^4 \Big\} \quad \begin{aligned} y(-1) &= 1 \\ y'(-1) &= -4 \end{aligned} \quad y(0) = 0 \quad \begin{aligned} y(1) &= 1 \\ y'(1) &= 4 \end{aligned}$$

This gives the three conditions:

$$\begin{aligned} 2w_1 + w_0 &= 2 \\ 2w_1 + 4w_1' &= \tfrac{2}{3} \\ 2w_1 + 8w_1' &= \tfrac{2}{5} \end{aligned} \qquad (6\text{-}17.1)$$

from which

$$w_1' = \tfrac{1}{15}, \qquad w_1 = \tfrac{7}{15}, \qquad w_2 = \tfrac{16}{15} \qquad (6\text{-}17.2)$$

and the resulting formula becomes

$$\bar{A} = \tfrac{1}{15}[7(f(-1) + f(1)) + 16f(0) + f'(-1) - f'(1)] \qquad (6\text{-}17.3)$$

If the distance between neighboring ordinates is not 1 but h, the formula has to be modified as follows.

$$\bar{A} = \frac{h}{15}[7(y_0 + y_2) + 16y_1 + h(y_0' - y_2')] \qquad (6\text{-}17.4)$$

where y_0, y_1, y_2 indicate the three successive ordinates.

Example. The function

$$y = \sin x$$

between $x = 0$ and π is roughly of a parabolic shape. Hence we get a satisfactory approximation of this function by giving it in the three equidistant points,

$x =$	0	$\pi/2$	π
$y =$	0	1	0

and approximating it by a parabola of second order. In this problem

$$h = \frac{\pi}{2}, \quad y_0 = 0, \quad y_1 = 1, \quad y_2 = 0$$

Moreover: $\qquad\qquad y_0' = 1, \qquad y_2' = -1$

Application of Simpson's rule gives

$$\bar{A} = \frac{\pi}{6}(y_0 + 4y_1 + y_2) = \frac{2\pi}{3} = 2.094$$

while the true area is

$$A = \int_0^\pi \sin x \, dx = -\left| \cos x \right|_0^\pi = 2$$

The error is thus only 4.7%.

$$\eta = -0.094$$

The new formula gives for the same area,

$$\bar{A} = \frac{\pi}{30}(16 + \pi) = 2.0045$$

The error $\qquad\qquad \eta = -0.0045$

is 20 times smaller than before.

We can motivate this increase of accuracy by an estimate of the error. In Simpson's case the estimated error becomes, according to the mean value theorem of integral calculus:

$$\eta = -\frac{f'''(\theta)}{4!} \int_{-1}^{+1} x^2(1-x^2)\, dx = -\frac{f'''(\theta)}{90} \qquad (6\text{-}17.5)$$

and generally (panel width h, limits a, b),

$$\eta = -\frac{f'''(\theta)}{180} h^4(b-a) \qquad (6\text{-}17.6)$$

In the case of the formula (3), however, we obtain

$$\eta = \frac{f^{(6)}(\theta)}{6!} \int_{-1}^{+1} x^2(1-x^2)^2\, dx = \frac{f^{(6)}(\theta)}{4725} \qquad (6\text{-}17.7)$$

and in the general case

$$\eta = \frac{f^{(6)}(\theta)}{9450} h^6(b-a) \qquad (6\text{-}17.8)$$

This gives, applied to our numerical example, the estimates

$$\bar{\eta} = -0.106 \qquad \text{(Simpson)}$$

$$\bar{\eta} = -0.00499 \quad \text{(modified Simpson)}$$

The great accuracy of these error estimates is explainable by the fact that in our example, the point of maximum of $f^{(n)}(\theta)$ and the midpoint of the range *coincide*.

Let us now divide a given range into an even number of panels, as we have done before in applying Simpson's parabolic approximation. We apply our formula to each double panel and put these areas together. The correction term in y' cancels out at all inside points, since it comes in with alternate sign from both sides. The only correction which remains is that at the two *end points* of the range. The resulting formula becomes

$$\bar{A} = \frac{h}{15} \, [14(\tfrac{1}{2}y_0 + y_2 + y_4 + \cdots + \tfrac{1}{2}y_{2n})$$

$$+ \, 16(y_1 + y_3 + y_5 + \cdots y_{2n-1})$$

$$+ \, h(y_0' - y_{2n}')] \qquad (6\text{-}17.9)$$

This formula shows that at a relatively little sacrifice, namely, adding the derivatives at the two end points of the range, we gain very considerably in accuracy. Our approximation is now of fifth order. That means our formula is exact for any $f(x)$ which can be represented by an arbitrary power expansion of fifth order. Simpson's rule gives exact results for a power expansion of third order only. The two additional powers mean a large increase in accuracy in the case of smooth functions.

Numerical example. We go back to the numerical example of § 8 and apply the formula (9) to the nine ordinates of that problem. At present $h = \frac{1}{2}$ and

$$y' = e^x$$

If we substitute the numerical values in (9) we obtain

$$\bar{A} = 53.5980641$$

compared with the true value

$$A = 53.5981500$$

The error is thus $\qquad \eta = 0.000086$

Simpson's formula gave the error -0.0181. The new error is 200 times smaller. This shows the great effectiveness of the end correction. The estimated increase of accuracy is given by the ratio

$$-\frac{f^{(6)}(\theta)}{f^{(4)}(\theta_1)} \frac{90h^2}{4725}$$

This gives in our example the fraction $1/210$; i.e., the error of the corrected formula is an estimated 210 times smaller than that of the uncorrected formula, in good agreement with the facts.

18. Quadrature involving exponentials. In many problems of applied analysis a certain "integral transform" of the following form is encountered:

$$F(p) = \int_a^b f(x)e^{px} \, dx \qquad (6\text{-}18.1)$$

We assume that $f(x)$ is given in tabulated form, x proceeding in equidistant intervals $\Delta x = h$. We assume that this interval is so small that linear interpolation is sufficient for functional values which lie between the tabulated points. In this case we could use the simple trapezoidal rule (3.5) for the numerical evaluation of the integral (1), were it not for the exponential factor e^{px}. If p is small enough, the trapezoidal rule would still hold. But we may need the value of the integral (1) for larger values of p. Hence it is of advantage to know how to evaluate an integral of the form (1) for sufficiently closely tabulated $f(x)$, but without making any restrictions concerning p, which may assume any real or imaginary or complex values.

We interpolate $f(x)$ linearly from panel to panel, starting at the mid-point $x_k + \frac{1}{2}h$ of each panel and proceeding to the two points $(x_k + \frac{1}{2}h) \pm \frac{1}{2}h$. We now perform the integration in each panel and form the sum. The result is the trapezoidal formula but with certain corrections. Let the result of the trapezoidal summation procedure be $S(p)$. Then

$$F(p) = \left[\sigma\left(\frac{p}{2}\right) \right]^2 S(p) + \frac{\sigma(p) - 1}{p} \left[f(a)e^{pa} - f(b)e^{pb} \right] \qquad (6\text{-}18.2)$$

where

$$\sigma(p) = \frac{\sinh ph}{ph} \qquad (6\text{-}18.3)$$

This formula has many applications and is particularly useful if the Fourier coefficients of an empirically given function are to be determined (in which case p is purely imaginary). The uncorrected sum $S(p)$ gives the coefficients of the finite trigonometric series which passes through the given points, (cf. IV, 11–15). Hence the formula (2) can be conceived as an expression of the relation between the true Fourier coefficients (demanded for example in the acoustical analysis of an empirically given function), and the coefficients obtained by trigonometric interpolation.

19. Quadrature by differentiation. Many functions of applied analysis are defined by a certain differential equation. If a function of this kind has to be integrated, we might consider it desirable to base the integration on the knowledge of the function and its derivatives at the two end points of the range, since the successive derivatives are easily calculable from the defining differential equation if we know the boundary values at the two end points. Our problem is then to obtain an effective quadrature formula which uses no inside ordinates but only the two end ordinates and its derivatives.

A formula of this kind makes use of the end information not for increased accuracy, but for complete evaluation of a definite integral. We may also say that our ordinates are now distributed in an extreme fashion inasmuch as they crowd infinitely near to the two end points of the given range. The ordinary Taylor expansion corresponds to the case when all the given ordinates are infinitely near to *one* point of the range. But we want to assume that *both* end points are equally represented.

We start with an elementary formula of integral calculus, based on the method of integrating by parts.

$$\int_a^b u(x)v^{(n)}(x)\,dx - \int_a^b (-1)^n v(x)u^{(n)}(x)\,dx$$

$$= \left| uv^{(n-1)} - u'v^{(n-2)} + \cdots \right|_a^b \qquad (6\text{-}19.1)$$

$$= \left| \sum_{k=0}^{n-1} u^{(k)}(x)v^{(n-k-1)}(x)(-1)^k \right|_a^b$$

We will make the following use of this formula. The range of integration shall be normalized to $[0, 1]$. Moreover, we choose

$$u = f(x), \qquad v = \frac{p_n(x)}{\gamma_n^n n!} \qquad (6\text{-}19.2)$$

where the polynomial

$$p_n(x) = \gamma_n^n x^n + \gamma_{n-1}^n x^{n-1} + \cdots + \gamma_0^n \qquad (6\text{-}19.3)$$

is freely at our disposal.

With this choice of the functions $u(x)$ and $v(x)$ the formula (1) may be written as follows.

$$\int_0^1 f(x)\,dx = \frac{1}{\gamma_n^n n!} \left| \sum_{k=0}^{n-1} f^{(k)}(x)p_n^{(n-k-1)}(x)(-1)^k \right|_0^1 + \eta_n \quad (6\text{-}19.4)$$

where η_n stands for the definite integral

$$\eta_n = (-1)^n \int_0^1 \frac{p_n(x)}{\gamma_n^n n!} f^{(n)}(x)\,dx \qquad (6\text{-}19.5)$$

Formula (4) can be conceived as a *quadrature formula* which obtains the area under the curve solely in terms of boundary values, given at the two end points $x = 0$ and $x = 1$ of the range,

$$\bar{A} = \frac{1}{\gamma_n^n n!} \sum_{k=0}^{n-1} f^{(k)}(x) p_n^{(n-k-1)}(x)(-1)^k \Bigg|_0^1 \qquad (6\text{-}19.6)$$

whereas η_n, given in the form (5), represents the *remainder* of our quadrature formula.

We first concentrate on this remainder. Our aim will be to dispose of $p_n(x)$ in such a way that the remainder (5) shall become particularly small. One choice is of particular interest here because it translates the outstanding features of the Gaussian quadrature method to our present problem. In discussing the Gaussian method (cf. § 10) we have seen that while an arbitrary distribution of the points of interpolation led to a quadrature formula which gives exact results for an arbitrary polynomial of not higher than $(n-1)$st order, the Gaussian points of interpolation gave a quadrature formula which yields exact results for any polynomial of $(2n-1)$st order.

In our case we have no choice concerning the points of interpolation, since we have decided already that our quadrature will be based on the boundary values of the given function and its derivatives at both end points of the range. But the polynomial $p_n(x)$ is still freely at our disposal. The form (5) of the remainder shows that our quadrature formula will be exact for any polynomial of not higher than $(n-1)$st order, but the remainder will not vanish generally for a polynomial of still higher order. For one particular choice of $p_n(x)$, however, we can make the quadrature formula (4) exact for any polynomial up to the order $2n-1$.

Let us consider the Legendre polynomials $P_n(x)$ (cf. V, 20), but renormalized to the range $[0, 1]$, instead of the traditional range $[-1, 1]$. We will denote these polynomials by $P_n^*(x)$. They are directly expressible in terms of the Gaussian hypergeometric function $F(\alpha, \beta, \gamma, x)$ [cf. 5-20.11].

$$P_n^*(x) = F(-n, n+1, 1; x) \qquad (6\text{-}19.7)$$

$$= 1 - \frac{n(n+1)}{1 \cdot 1} x + \frac{n(n-1)(n+1)(n+2)}{1 \cdot 2 \cdot 1 \cdot 2} x^2 - \cdots$$

$$= \sum_{k=0}^{n} \frac{(n+k)!}{(n-k)!(k!)^2} (-x)^k$$

The polynomial $P_n^*(x)$ has the following remarkable property. It can be written as the nth derivative of a function which vanishes together with all its derivatives up to the order $n-1$ at the two end points $x=0$ and $x=1$ of the range.

$$P_n^*(x) = \frac{1}{n!}\frac{d^n[x(1-x)]^n}{dx^n} \qquad (6\text{-}19.8)$$

But then we can make use of the integral transformation (1), identifying $u(x)$ with $f^{(n)}(x)$ and $v(x)$ with $[x(1-x)]^n$. The boundary terms on the right side drop out entirely, on account of the properties of $v(x)$. What remains can be written

$$\eta_n = \frac{(-1)^n}{\gamma_n^n(n!)^2}\int_0^1 f^{(n)}(x)\frac{d^n[x(1-x)]^n}{dx^n}\,dx \qquad (6\text{-}19.9)$$

$$= \frac{1}{\gamma_n^n(n!)^2}\int_0^1 f^{(2n)}(x)[x(1-x)]^n\,dx$$

By definition, γ_n^n denotes the coefficient of the highest power of $P_n^*(x)$. According to (7) we obtain

$$\gamma_n^n = (-1)^n\frac{(2n)!}{(n!)^2} \qquad (6\text{-}19.10)$$

Moreover, the Legendre polynomials $P_n^*(x)$ possess the symmetry property

$$P_n^*(x) = (-1)^n P_n^*(1-x) \qquad (6\text{-}19.11)$$

Because of this property, the sum on the right side of (6) becomes reducible from $2n$ to n terms.

$$\bar A = \frac{1}{\gamma_n^n n!}\sum_{k=0}^{n-1}(-1)^{k+1}P^{*(n-k-1)}(0)y_k \qquad (6\text{-}19.12)$$

where

$$y_k = f^{(k)}(0) + (-1)^k f^{(k)}(1) \qquad (6\text{-}19.13)$$

Now the expansion (7) gives

$$P_n^{*(k)}(0) = (-1)^k\frac{(n+k)!}{(n-k)!k!} \qquad (6\text{-}19.14)$$

and we will denote

$$(-1)^{n-k}P_n^{*(n-k)}(0) = C_k^n = \frac{(2n-k)!}{(n-k)!k!} \qquad (6\text{-}19.15)$$

With this notation the resulting quadrature formula becomes

$$\bar{A}_n = \frac{1}{C_0^n} \sum_{k=0}^{n-1} C_{k+1}^n y_k \qquad (6\text{-}19.16)$$

For arbitrary limits $[a, b]$ the formula (16) becomes modified as follows. We define

$$y_k = f^{(k)}(a) + (-1)^k f^{(k)}(b) \qquad (6\text{-}19.17)$$

and obtain

$$\bar{A}_n = \frac{1}{C_0^n} \sum_{k=0}^{n-1} C_{k+1}^n (b-a)^{k+1} y_k \qquad (6\text{-}19.18)$$

The coefficients C_k^n are tabulated in Table XV of the Appendix.

As an example, let us assume that the function and its first and second derivatives are given at both end points of the range. Then $n = 3$, and we obtain

$$\bar{A} = \tfrac{1}{2} f(a)h + \tfrac{1}{10} f'(a)h^2 + \tfrac{1}{120} f''(a)h^3$$

$$+ \tfrac{1}{2} f(b)h - \tfrac{1}{10} f'(b)h^2 + \tfrac{1}{120} f''(b)h^3$$

with $h = b - a$.

The error η_n of the quadrature formula (16) can be estimated as follows. Since the weight function $[x(1-x)]^n$ does not change its sign throughout the interval $[0, 1]$, we obtain by the theorem of weighted means,

$$\frac{\int_0^1 f^{(2n)}(x)[x(1-x)]^n dx}{\int_0^1 [x(1-x)]^n dx} = f^{(2n)}(\theta) \qquad (6\text{-}19.19)$$

where θ is some point within the interval $[0,1]$. Now

$$\int_0^1 [x(1-x)]^n \, dx = \frac{\sqrt{\pi}}{2^{2n+1}} \frac{n!}{(n+\frac{1}{2})!} \qquad (6\text{-}19.20)$$

and thus

$$\eta_n = \frac{(-1)^n \sqrt{\pi}}{2 \cdot 4^n} \frac{n!}{(n+\frac{1}{2})!(2n)!} f^{(2n)}(\theta) \qquad (6\text{-}19.21)$$

and more generally for the case of arbitrary limits,

$$\eta_n = \sqrt{\pi} \, \frac{(-1)^n n!}{(n + \frac{1}{2})!(2n)!} \, f^{(2n)}(\theta) \left(\frac{b-a}{2}\right)^{2n+1} \quad (6\text{-}19.22)$$

The uncertainty in the position of the point θ can be alleviated if we can assume that $f^{(2n)}(x)$ does not change too violently within the interval $[a, b]$. The function $[x(1 - x)]^n$ cuts out a narrow "window" around the point $x = \frac{1}{2}$ because, if n is large enough, the function falls off rapidly on both sides of that point. The weighted mean (19) is thus heavily loaded in favor of the immediate neighborhood of $x = \frac{1}{2}$. If $f^{(2n)}(x)$ is sufficiently smooth, we can identify θ with the point $x = \frac{1}{2}$ and in the general case with the point

$$\theta = \frac{b-a}{2}$$

Furthermore, Stirling's formula for the factorial function shows that for estimation purposes we may put

$$\frac{n!}{(n + \frac{1}{2})!} = n^{-1/2} \quad (6\text{-}19.23)$$

Under these conditions we obtain the following realistic estimation of the error of the quadrature formula (18).

$$\bar{\eta}_n = (-1)^n \sqrt{\frac{\pi}{n}} \, \frac{f^{(2n)}\left(\dfrac{b-a}{2}\right)}{(2n)!} \left(\frac{b-a}{2}\right)^{2n+1} \quad (6\text{-}19.24)$$

The general character of this formula is quite similar to the one valid in the Gaussian quadrature;[1] [cf. (12.1)]. The numerical factor is quite different, however, in the one and in the other case. The ratio of the two factors is

$$\mu_n = \frac{(-4)^n}{\sqrt{\pi n}} \quad (6\text{-}19.25)$$

---

[1] For this reason the error estimation method of § 12 is applicable again. As test function we use the function $[(2x - 1)f(x)]'$ (assuming the range $[0, 1]$), for which the result of the quadrature is $f(1) + f(0)$. Hence η' is again explicitly at our disposal and once more the estimated error of the original quadrature becomes $\bar{\eta} = \eta'/(2n + 1)$.

in favor of the Gaussian quadrature; i.e., the estimated error of the quadrature formula based on $2n$ boundary values is μ_n times larger than the estimated error of the Gaussian quadrature, using n interior points.

This is understandable, however, if we realize how badly handicapped we are by taking our information completely from the two boundaries of the interval, instead of using judiciously chosen ordinates of the inside. The importance of the method lies in the fact, however, that it is frequently so much easier to get the function and its derivatives at the two end points of the interval than to evaluate the functional values at some inside points. This is particularly true if an untabulated and unknown function is defined by a *differential equation*. Then application of the quadrature formula (16) can lead to an effective method of *solving* the given differential equation (cf. § 21).

20. The exponential function. A particular example which is well adapted to demonstrate the power of the method is the exponential function

$$f(x) = e^{\alpha x} \tag{6-20.1}$$

integrated between 0 and 1. Here we know that the result of the integration is

$$\int_0^1 e^{\alpha x}\,dx = \frac{1}{\alpha}(e^\alpha - 1) \tag{6-20.2}$$

Collecting all the terms with e^α and replacing α by x we obtain a rational approximation of the exponential function e^x which appears in the following form.[1]

$$e^x = \frac{\displaystyle\sum_{k=0}^n C_k^n x^k}{\displaystyle\sum_{k=0}^n C_k^n (-x)^k} \tag{6-20.3}$$

[1] The author is indebted to his friend Charles Davis, numerical analyst, North American Aviation for pointing out to him that this approximation was found earlier by P. M. Hummel and C. L. Seebeck, "A Generalization of Taylor's Theorem," *Association Monthly*, **56**, 243–247 (1949).

The first four approximations ($n = 1, 2, 3, 4$) are given as follows.

$$\frac{2 + x}{2 - x}, \quad \frac{12 + 6x + x^2}{12 - 6x + x^2}, \quad \frac{120 + 60x + 12x^2 + x^3}{120 - 60x + 12x^2 - x^3}, \quad (6\text{-}20.4)$$

$$\frac{1680 + 840x + 180x^2 + 20x^3 + x^4}{1680 - 840x + 180x^2 - 20x^3 + x^4}$$

The osculation with e^x is of the order $2n$; that means that by expanding these ratios into powers of x, the agreement with the coefficients of the Taylor expansion extends up to the term of the order $2n$, although the number of coefficients at our disposal is only n.

If we put $x = 1$, we obtain successive rational convergents of the transcendental number e which are of astonishing precision.

$$e = \quad \frac{3}{1} \quad = 3 \qquad (6\text{-}20.5)$$

$$\frac{19}{7} \quad = 2.714 \qquad (\eta = 4 \cdot 10^{-3})$$

$$\frac{193}{71} \quad = 2.71831 \qquad (\eta = -3 \cdot 10^{-5})$$

$$\frac{2721}{1001} \quad = 2.7182817 \qquad (\eta = 1 \cdot 10^{-7})$$

$$\frac{49171}{18089} \quad = 2.7182818287 \qquad (\eta = -3 \cdot 10^{-10})$$

$$\frac{1084483}{398959} = 2.7182818284586 \qquad (\eta = 4 \cdot 10^{-13})$$

It is also of interest to apply our quadrature formula to the same numerical example that we employed in demonstrating the power of the Gaussian quadrature; (cf. § 11). Here we had $n = 5$, $a = 0$, $b = 4$, and the given function was $y = e^x$. The application of the formula (19.18), taking the coefficients C_k^5 from Table XV of the Appendix, gives

$$\bar{A} = \frac{1}{30240} \, [15120 \cdot 4(1 + e^4) + 3360 \cdot 16(1 - e^4)$$

$$+ \, 420 \cdot 64(1 + e^4) + 30 \cdot 256(1 - e^4) + 1 \cdot 1024(1 + e^4)]$$

$$= 53.60174$$

Comparison with the true value $A = 53.59815$ shows that

$$\eta = -0.0036$$

while the Gaussian error was only $\eta = 0.000013$. The change of sign is explained by the factor $(-1)^n$ of formula (19.24) which gives the minus sign in the case of $n = 5$. Moreover, the factor μ_n [cf. (19.25)], which is the estimated magnification of the error compared with the Gaussian error, is now

$$|\mu_5| = \frac{4^5}{\sqrt{5\pi}} = 258$$

in close agreement with the facts.

21. Eigenvalue problems. In V, 17 and 18 we encountered a class of problems which play a fundamental and increasingly important role in all types of vibration problems associated with elasticity, flutter analysis, wave guides, atomic physics. They are called "eigenvalue problems." The general situation encountered in such problems can be described as follows. Given a certain linear differential operator which contains an unknown constant parameter usually called the "eigenvalue" λ and given certain homogeneous boundary conditions which are such that without the proper choice of λ no solution outside the trivial solution $y = 0$ is possible, the problem is to find the smallest λ_1, or a few of the smallest λ_i, which make a solution possible.

In such eigenvalue problems the quadrature method of § 19 may be of considerable help, since it is based on the knowledge of the function and its derivatives at the two end points of the interval, and these quantities are available on the basis of the given differential equation and the given boundary conditions.

In order to show the operation of the method, we choose a simple example, but the method is applicable under much more complicated conditions. It is our aim, however, to study the essential features of the method, unhampered by technical difficulties. For this reason we choose a simple differential operator of second order, with constant coefficients:

$$y'' + y' + \lambda y = 0 \qquad (6\text{-}21.1)$$

with the boundary conditions

$$y(0) = 0, \qquad y'(1) = 0 \qquad (6\text{-}21.2)$$

The given interval is thus normalized to [0, 1].

Since a linear homogeneous differential equation leaves an amplitude factor undetermined, we can assign arbitrarily the value

$$y'(0) = 1$$

to the derivative at $x = 0$. Then the differential equation (1) determines uniquely the values of all the derivatives at $x = 0$. We obtain these by successive differentiation, or by substituting into the differential equation the power expansion

$$y = a_0 + a_1 x + a_2 x^2 + \cdots$$

We collect terms and put the resulting coefficient of x^k equal to zero. In our simple problem we get

$$a_0 = 0, \qquad a_1 = 1, \qquad a_2 = -\tfrac{1}{2}, \qquad a_3 = \frac{1 - \lambda}{6}$$

$$\tag{6-21.3}$$

$$y(0) = 0, \qquad y'(0) = 1, \qquad y''(0) = -1, \qquad y'''(0) = 1 - \lambda$$

The higher we go with the evaluation of the successive coefficients, the greater accuracy can we expect. For our present purposes we will stop with a_3.

At the other end point we will put similarly

$$x = 1 + \xi, \qquad y = b_0 + b_1 \xi + b_2 \xi^2 + b_3 \xi^3$$

Using the same method of substituting in the differential equation and collecting terms, we obtain no value for b_0, but the later coefficients all appear as linear functions of b_0.

$$b_0, \quad b_1 = 0, \qquad b_2 = -\frac{\lambda b_0}{2}, \qquad b_3 = \frac{\lambda b_0}{6} \qquad (6\text{-}21.4)$$

$$y(1) = b_0, \quad y'(1) = 0, \quad y''(1) = -\lambda b_0, \quad y'''(1) = \lambda b_0$$

We will now apply our quadrature formula, considering $y''(x)$ as our function $f(x)$; hence $f(x) = y''(x)$ and $f'(x) = y'''(x)$ are given at the two end points

$$f(0) = -1, \qquad f(1) = -\lambda b_0$$
$$f'(0) = 1 - \lambda, \qquad f'(1) = \lambda b_0$$

The quadrature formula for $n = 2$ gives

$$\int_0^1 y''(x)\,dx = y'(1) - y'(0) = \frac{6(-1-\lambda b_0) + 1(1 - \lambda - \lambda b_0)}{12}$$

$$-1 = \frac{-5 - 7\lambda b_0 - \lambda}{12}$$

This gives the relation

$$b_0 = \frac{7 - \lambda}{7\lambda} \qquad (6\text{-}21.5)$$

We now use the quadrature formula once more, considering $y'(x)$ as $f(x)$

$$f(0) = 1, \qquad\qquad f(1) = 0$$
$$f'(0) = -1, \qquad\qquad f'(1) = -\lambda b_0$$
$$f''(0) = 1 - \lambda \qquad\qquad f''(1) = \lambda b_0$$

At present we have 3 data on both ends and thus use the formula for $n = 3$

$$\int_0^1 y'(x)\,dx = y(1) - y(0)$$

$$= \frac{60(1 + 0) + 12(-1 + \lambda b_0) + (1 - \lambda + \lambda b_0)}{120}$$

$$b_0 = \frac{49 - \lambda + 13\lambda b_0}{120}$$

This gives the new relation

$$b_0 = \frac{49 - \lambda}{120 - 13\lambda} \qquad (6\text{-}21.6)$$

Equating the right sides of (5) and (6) we obtain a quadratic equation for determination of the eigenvalue λ

$$20\lambda^2 - 554\lambda + 840 = 0$$

The two roots of this equation are

$$\lambda_1 = 1.6095, \quad \lambda_2 = 26.0905 \tag{6-21.7}$$

Only the smaller root has significance, since the larger root is enormously sensitive to higher-order corrections. The smaller root, however, does not change much if we continue our differentiation process. We have stopped with $y'''(x)$. If we go one step further and include $y'''(0)$ and $y'''(1)$ in our calculations, the first application of the quadrature formula involves $n = 3$ and the second involves $n = 4$. We then get the two relations

$$b_0 = \frac{71 - 10\lambda}{\lambda(73 - \lambda)} \quad \text{and} \quad b_0 = \frac{679 - 18\lambda}{1680 - 201\lambda + \lambda^2}$$

which now yield the following cubic equation for λ

$$28\lambda^3 - 4074\lambda^2 + 80638\lambda - 119280 = 0$$

This is still an essentially quadratic equation, since the cubic term is only a small correction. We can first neglect the cubic term and obtain a preliminary λ_0, then add the cubic term evaluated with this λ_0 as a correction to the absolute term. This yields for the smallest root

$$\lambda_1 = 1.608467 \tag{6-21.8}$$

The small change of λ_1 compared with the corresponding value found in (7) shows that the first rather crude approximation was very close, the error being 1 unit in the third decimal, which is an accuracy of 0.07%. In the present simple problem we can check our results, because the exact value of λ is theoretically available. We find

$$\lambda = \tfrac{1}{4} + \theta^2 \tag{6-21.9}$$

where θ is a solution of the transcendental equation

$$\tan \theta = 2\theta \tag{6-21.10}$$

The smallest root of this equation is

$$\theta_1 = 1.1655618$$

which gives

$$\lambda_1 = 1.608534 \tag{6-21.11}$$

Hence the error of the $n = 2, 3$ approximation is $\eta = -0.0010$, while the error of the $n = 3, 4$ approximation is only $\eta = 0.000067$. We see that the convergence of the method if applied to a smooth function is rapid. Of interest is also the approach from above and below in the two successive cases.

The so-called "Rayleigh-Ritz method" of obtaining eigenvalues is based on minimization of a certain integral. Hence it is applicable only to *self-adjoint* differential problems. The present method does not require that either the differential operator or the boundary conditions shall be self-adjoined. It operates under more general conditions, and in principle even linearity of the differential equation need not be demanded for application of the method.

If we pursue these ideas one step further, we come to the conclusion that this quadrature procedure can be applied not only to evaluation of eigenvalues, but also to actual *solution* of differential equations. Let us make the substitution

$$x = \alpha x_1 \qquad (6\text{-}21.12)$$

considering x_1 the new independent variable and α a given constant parameter. The point $x_1 = 1$ in the new variable corresponds to the point $x = \alpha$ in the old variable. Hence $y(1)$ in the new variable gives actually the original $y(\alpha)$, and replacing α by x, we have obtained the unknown function at the variable point x.

We show the operation of this method by using the previous example. The differential equation (1) now becomes

$$y'' + \alpha y' + \alpha^2 \lambda y = 0$$

The table (3) of the initial values has to be replaced as follows:

$$a_0 = 0, \qquad a_1 = \alpha, \qquad a_2 = -\tfrac{1}{2}\alpha^2 \qquad a_3 = \frac{1 - \lambda}{6}\,\alpha^3$$

$$y(0) = 0, \quad y'(0) = \alpha, \quad y''(0) = -\alpha^2, \quad y'''(0) = (1 - \lambda)\alpha^3$$

while the table of end values (4) becomes modified as follows (the boundary condition $y'(1) = 0$ is no longer valid).

$$y(1) = b_0, \quad y'(1) = b_1, \quad y''(1) = -\alpha b_1 - \alpha^2 \lambda b_0$$

$$y'''(1) = (1 - \lambda)\alpha^2 b_1 + \alpha^3 \lambda b_0$$

In the present problem, the unknowns are b_0 and b_1, while λ is already known. The procedure, however, is once more the same. The first relation which before led to (5) becomes now

$$b_1 - \alpha = \frac{1}{12} [6(-\alpha^2 - \alpha b_1 - \alpha^2 \lambda b_0)$$
$$+ (1 - \lambda)\alpha^3 - (1 - \lambda)\alpha^2 b_1 - \alpha^3 \lambda b_0]$$

This gives the linear relation

$$(6\alpha^2 \lambda + \alpha^3 \lambda)b_0 + [12 + 6\alpha + (1 - \lambda)\alpha^2]b_1$$
$$= 12\alpha - 6\alpha^2 + (1 - \lambda)\alpha^3$$

For the second relation we will simplify matters at the sacrifice of accuracy. We will once more use the quadrature formula for $n = 2$ only, applied to the first three boundary values, and not make use of the values $y'''(0)$ and $y'''(1)$. We now get

$$b_0 = \frac{1}{12} [6(\alpha + b_1) + (-\alpha^2 + \alpha b_1 + \alpha^2 \lambda b_0)]$$

which gives the new linear relation

$$(12 - \alpha^2 \lambda)b_0 - (6 + \alpha)b_1 = 6\alpha - \alpha^2$$

Solving the two linear equations for b_0 and b_1, we obtain

$$b_0 = \frac{\alpha(144 - 12\lambda\alpha^2)}{144 + 72\alpha + 12(1 + \lambda)\alpha^2 + 6\lambda\alpha^3 + \lambda^2\alpha^4}$$

$$b_1 = \alpha \frac{144 - 72\alpha + 12(1 - 5\lambda)\alpha^2 + 6\lambda\alpha^3 + \lambda^2\alpha^4}{144 + 72\alpha + 12(1 + \lambda)\alpha^2 + 6\lambda\alpha^3 + \lambda^2\alpha^4}$$

We now go back to the original variable x; in this variable the significance of b_0 and b_1 becomes $b_0 = y(\alpha)$ and $b_1 = \alpha y'(\alpha)$. We thus see that we have obtained two independent approximations for $y(x)$ and $y'(x)$ in the form

$$\bar{y}(x) = \frac{x(144 - 12\lambda x^2)}{144 + 72x + 12(1 + \lambda)x^2 + 6\lambda x^3 + \lambda^2 x^4}$$

$$\bar{y}'(x) = \frac{144 - 72x + 12(1 - 5\lambda)x^2 + 6\lambda x^3 + \lambda^2 x^4}{144 + 72x + 12(1 + \lambda)x^2 + 6\lambda x^3 + \lambda^2 x^4}$$

The second derivative $y''(x)$ and all the higher derivatives are then determined by the defining differential equation (1).

To test the accuracy of our solution, we apply it to the special value $\lambda = -2$, in which case the explicit solution of our problem becomes

$$y(x) = \tfrac{1}{3}(e^x - e^{-2x}), \qquad y'(x) = \tfrac{1}{3}(e^x + 2e^{-2x})$$

while

$$\bar{y}(x) = \frac{x(144 + 24x^2)}{144 + 72x - 12x^2 - 12x^3 + 4x^4}$$

$$\bar{y}'(x) = \frac{144 - 72x + 132x^2 - 12x^3 + 4x^4}{144 + 72x - 12x^2 - 12x^3 + 4x^4}$$

At the point $x = 1$ the correct solutions are

$$y(1) = 0.8610, \qquad y'(1) = 0.9963$$

while the approximation gives

$$\bar{y}(1) = 0.8571, \qquad \bar{y}'(1) = 1$$

We see that the accuracy is very satisfactory.

Generally, if this quadrature method is applied to an analytical solution of a given boundary value problem in the realm of ordinary differential equations, our procedure can be described as follows. First the given realm is normalized to [0,1] by a proper scale transformation. Then we make a list of the boundary values

$$y(0), y'(0), y''(0), \ldots$$

and

$$y(1), y'(1), y''(1), \ldots$$

up to the point where the higher derivatives are already determined by the given differential equation. Some of these boundary values are prescribed on account of the given boundary conditions. The others are replaced by letter symbols. If the differential equation is homogenous, one of the boundary values can be normalized to 1. Now the quadrature formula (16) comes into operation, establishing a linear relation between the unknown boundary data. Repeating the quadrature formula for derivatives of lower and lower order, eventually all the boundary data become determined, on the basis of a system of linear equations. If an eigenvalue λ is involved, elimination of the unknown boundary data results in an algebraic equation for λ, whose absolutely smallest root we retain.

At this point we have an *initial value problem*, since $y(0)$ and all its higher derivatives at $x = 0$ are already known. Now the transformation $x = \alpha\xi$ follows, considering α a given constant parameter. Applying the quadrature formula in the previous manner to the new differential equation, we eventually obtain $y(\alpha)$, and independently $y'(\alpha)$, $y''(\alpha)$, ... , from the given initial values.

As a valuable check we can proceed similarly from the *other* end point $x = 1$ and compare the new $y(\alpha)$, obtained in terms of the *end values* $y(1)$, $y'(1)$, $\cdots$, with the previous $y(\alpha)$ obtained in terms of the *initial* values $y(0)$, $y'(0)$, $\cdots$. The degree of agreement will give a good measure of the accuracy of the solution.

22. Convergence of the quadrature based on boundary values. We will now investigate the general convergence properties of this quadrature procedure. For this purpose we will use a method which is very similar to that utilized before, when dealing with the convergence of equidistant polynomial interpolation, (cf. V, 15). We assume the analytical character of $f(x)$ in the interval $[0,1]$ and a certain domain of the complex plane surrounding this interval. Then we can represent $f(x)$ with the help of Cauchy's loop integral (5-15.9). This makes it possible that the entire investigation shall be restricted to the special function $(z_0 - x)^{-1}$ where z_0 is some fixed point of the complex plane. If we can show that the quadrature method converges for this particular function, provided that z_0 stays outside of a certain well circumscribed domain of the complex plane, then the convergence is assured for any analytical function $f(z)$ which stays analytic inside and on the boundaries of that domain.

In (19.9) the following form of the remainder was developed:

$$\eta_n = \frac{(-1)^n}{(2n)!} \int_0^1 f^{(2n)}(x)[x(1-x)]^n \, dx \qquad (6\text{-}22.1)$$

For the special function:

$$f(x) = \frac{1}{z_0 - x} \qquad (6\text{-}22.2)$$

we get

$$\frac{f^{(2n)}(x)}{(2n)!} = \frac{1}{(z_0 - x)^{n+1}} \tag{6-22.3}$$

and thus

$$\eta_n = (-1)^n \int_0^1 \frac{1}{z_0 - x} \left[\frac{x(1 - x)}{(z_0 - x)^2} \right]^n dx \tag{6-22.4}$$

Our aim is to show that with n increasing to infinity, η_n converges to zero. We notice that the decisive quantity is the absolute value of the ratio whose nth power appears in the integrand:

$$A(x) = \left| \frac{x(1 - x)}{(z_0 - x)^2} \right| \tag{6-22.5}$$

If $A(x)$ stays everywhere smaller than 1, the gradual decrease of η_n to zero is assured.

Now it is possible that z_0 is so far away from the interval [0,1] of the x-axis that $A(x)$ remains smaller than 1 throughout the given interval, in which case the convergence is already guaranteed and no further discussion is demanded. But it is also possible that z_0 approaches the x-axis to a degree that $A(x)$ grows beyond 1 in a certain portion of the critical interval. Let us assume that z_0 has the form

$$z_0 = \alpha - i\beta \tag{6-22.6}$$

where β is positive. Hence z_0 is some complex point below the x-axis. We start from the point $x = 0$ where $A(x)$ is zero and proceed up to the point P, where $A(x)$ becomes 1. Similarly we start symmetrically from the point $x = 1$ (where $A(x)$ is also zero) and go backward until a point P_2 is reached where again $A(x)$ becomes 1. We have difficulty only between the points P_1 and P_2, since the contribution of the interval OP_1 and $P_2 1$ converges to zero. But now we will make use of the well-known property of analytic functions that the path of integration can be deformed in any way we like, as long as singular points of the integrand are avoided. Hence we will replace x by the complex variable $z = x + iy$ and choose our path somewhere in the upper half of the complex plane in which the integrand is free of any singularities. We first establish the geometrical locus of all the points in which $A(z)$ becomes 1. This gives the condition

$$[x(1 - x) + y^2]^2 + (1 - 2x)^2 y^2 = [(\alpha - x)^2 + (\beta + y)^2]^2 \tag{6-22.7}$$

which may be conceived as a cubic equation in y with the cubic term $4\beta y^3$ and the absolute term

$$[(\alpha - x)^2 + \beta^2]^2 - [x(1 - x)]^2$$

In the critical interval between P_1 and P_2 this quantity becomes *negative*. But then the cubic equation must have a real positive root for every x of the critical interval. We thus obtain in the upper half-plane a definite convex curve between P_1 and P_2, inside of which $A(z)$ is larger than 1, outside of which $A(z)$ is smaller than 1. By choosing our path of integration outside of the critical curve, we now have a finite path along which $A(z) < 1$ and thus we have demonstrated that η_n converges to zero.

The only case we have not covered yet is the choice $\beta = 0$. Then z_0 lies on the x-axis and the integrand is real. We consider the ratio

$$A(x) = \frac{x(1 - x)}{(\alpha - x)^2}$$

The maximum occurs at the point

$$x = \frac{\alpha}{2\alpha - 1}$$

At this point

$$A(x) = \frac{1}{4\alpha(\alpha - 1)}$$

The condition that this ratio shall remain smaller than 1 gives two bounds for α, viz., on the positive side:

$$\alpha > \frac{1 + \sqrt{2}}{2} = 1.20705 \tag{6-22.8}$$

and on the negative side:

$$\alpha < \frac{1 - \sqrt{2}}{2} = -0.20705 \tag{6-22.9}$$

The result of our investigation can be summarized as follows. We extend the x-axis symmetrically on both sides by the amount of 0.207 beyond the terminal points 0, 1. This line can be surrounded by a fence of *arbitrarily small width*. If $f(z)$ is analytical in this infinitesimal domain, the convergence of the quadrature formula is secured.

$$
\begin{array}{ccccc}
\vdash & & & & \dashv \\
-0.207 & 0 & & 1 & 1.207
\end{array}
$$

Bibliographical References

[1]   Cf. Ref. {11}, Chapter IX.

[2]   Lowan, A. N., Davids, N., and Levenson, A., Table of the zeros of the Legendre Polynomials of order 1–16 and the weight coefficients for Gauss' mechanical quadrature formula, *Bull. Amer. Math. Soc.*, **48**, 739 (1942).

[3]   Milne, W. E., *Numerical Solution of Differential Equations* (Wiley, New York, 1953).

VII

POWER EXPANSIONS

1. Historical introduction. The powers of a variable x appeared originally purely in algebraic problems. With the development of calculus the great importance of power expansions became evident. The expansion discovered by Taylor (1715) and by Maclaurin (1742) enables us to predict the course of a function if we know the value of the function and all its derivatives in one particular point. The "Taylor series" thus became one of the cornerstones of analytical research and was particularly useful in establishing the existence of solutions of differential equations. From the middle of the nineteenth century on, a more cautious attitude toward power expansions became noticeable. The mere existence of a Taylor expansion does not prove that this series has any inner affinity to the function it represents. If, on the other hand, the series has no other purpose than numerical evaluation of the function, the degree of *convergence* has to be investigated. The Taylor expansion may converge in the entire complex plane or within a given circle only, and it may diverge even at every point. It was recognized, however, that a more liberal formulation of the question of convergence greatly increases the usefulness of an expansion. One can make good use, for example, of "semiconvergent expansions" which actually diverge if we increase the number of terms to infinity, but converge *in the beginning*, thus allowing evaluation of the function with a certain *limited accuracy* which cannot be surpassed, since the error of the truncated series decreases to a certain minimum and then increases again. Much attention was paid also to the problem of inventing methods of summing a series in such a way that it shall become convergent, although the original series, if added term by term, increased to infinity.

438

With the development of the theory of orthogonal expansions the realization came that occasionally power expansions whose coefficients are not determined according to the scheme of Taylor can operate much more effectively than the Taylor series itself. Such expansions are not based on the process of successive differentiation but on integration. A large class of functions which are not sufficiently analytic to allow a Taylor expansion can be represented by such orthogonal expansions. The realm of power expansions is thus extended far beyond the family of analytical functions. But even for analytical functions we may gain in convergence if we do not employ the powers directly but in the form of polynomials which are members of an orthogonal set of functions. These expansions belong to a given definite real realm of the variable x, and our aim is to approximate a function in such a way that the error shall not become too small or too large at any particular point of the range, but rather of the same order of magnitude all over the range. The gain in comparison with the Taylor series arises from the fact that we sacrifice in accuracy at the point where the Taylor series gave very accurate results but reduce the error in the peripheral regions where the error of the Taylor expansion became intolerably large.

The theory of orthogonal expansions is based on the theory of least squares developed by Gauss and Legendre. Many orthogonal function systems were known through the study of the Laplacian operator (potential equation) which played a fundamental role not only in astronomy but also in electricity and magnetism. The Fourier series was the first classical example of an orthogonal expansion. However, the nature of orthogonal expansions was not recognized until the great acoustical investigations of Lord Rayleigh (*Theory of Sound*, 1894), who has to be considered as the true founder of the theory of orthogonal expansions. Fredholm's integral equation (1900) and the subsequent investigations of Hilbert (1910) gave a tremendous impetus to the broader understanding of the problems of orthogonality and their relation to metrical geometry.

In the present chapter we will be particularly interested in the question of *rapidly convergent power expansions*. Mere convergence of an expansion, valuable as it is from the purely analytical standpoint, is of little practical use if the number of terms demanded for a reasonable accuracy is very large. Methods can be developed, however, by which a long power expansion may be "telescoped" into a much shorter series of only a few terms. A finite power series cannot be replaced by any other series if *absolute* accuracy is required. If, however, we are satisfied with a given limited accuracy of, let us say, 5%, it can easily happen that a long power expansion of, let us say, 50 terms may be replaceable by another expansion of only 5 terms. This "telescoping" of a power series is not obtained by merely *omitting* the terms from a certain point on, but by *rearranging* the series according to a certain pattern. Another possibility is that we make use of the differential equation which defines a certain function and take care in advance that in solving this differential equation by a finite power expansion the errors shall be evenly distributed over the entire range. In these investigations a certain outstanding class of polynomials, introduced by the Russian mathematician Chebyshev (in 1864) and called "Chebyshev polynomials," play a fundamental role.

2. Analytical extension by reciprocal radii. The Taylor expansion of an analytical function $w = f(z)$ has a certain radius of convergence which is determined by the analytical nature of the function $f(z)$. If $f(z)$ becomes infinite at a certain point $z = z_0$ or has some other "singularity" at that point which is not in harmony with the general conditions of an analytical behaviour, the infinite series

$$w = a_0 + a_1 z + a_2 z^2 + \cdots \tag{7-2.1}$$

cannot converge beyond the circle $|z| = |z_0|$. But the series (1) will converge *within* that circle if we know that $z = z_0$ is the *nearest* point where the analytical behavior of $f(z)$ is violated. *On* the circle itself the series may or may not converge, and requires special investigation.

As an example we consider the two functions

(a) $$w = e^{z^2} \tag{7-2.2}$$

(b) $$w = \frac{1}{1 + e^z}$$

The first function remains finite and analytical at *all* points of the complex plane $z = x + iy$. Hence the Taylor expansion around the origin must remain convergent for any finite value of z. A function of this kind is called an "entire function." The second function becomes infinite at the points

$$z = \pm \pi i, \quad \pm 3\pi i, \quad \pm 5\pi i, \cdots$$

The nearest singularity is at the point $z = \pm i\pi$. This decides that the Taylor expansion (2.1) of this function converges only inside the circle $|x + iy| = \pi$.

It is frequently possible to obtain the coefficients of the Taylor expansion on the basis of the defining equation, without going through the process of successive differentiations. For example, the equation (2b) may be written as follows.

$$(1 + e^z)w = 1$$

and expanding both factors in a Taylor series, we have

$$\left(2 + z + \frac{z^2}{2!} + \frac{z^3}{3!} + \cdots \right)\left(a_0 + a_1 z + a_2 z^2 + \cdots \right) = 1$$

If we carry through the multiplications term by term, we obtain successive recursions for the determination of the a_i:

$$2a_0 = 1 \qquad\qquad (a_0 = \tfrac{1}{2})$$

$$2a_1 + a_0 = 0 \qquad\qquad (a_1 = -\tfrac{1}{4})$$

$$2a_2 + a_1 + a_0 = 0 \qquad\qquad (a_2 = 0)$$

$$2a_3 + a_2 + \frac{a_1}{2} + \frac{a_0}{6} = 0 \qquad (a_3 = \tfrac{1}{48})$$

. . . .

Since, however, our series is not valid outside of a definite circle, the question can be raised whether by some method we could not go *beyond* the circle of convergence. One method of great analytical but little practical significance is the method of "analytical continuation" by local expansions. In this method we choose a point still inside the radius of convergence, e.g., the point $z = \tfrac{2}{3}\pi$. On the basis of the Taylor series we obtain the function and all its derivatives at that point and now choose this point as the center of a new Taylor expansion. The radius of the new circle of convergence is again determined by the nearest singularity, i.e., the points $z = \pm i\pi$, and we see that we now cover ground which was not included in the previous circle. By repeated applications of this method we can extend the definition of a function, originally given in terms of an infinite expansion (1), and restricted to a definite circle, to larger and larger portions of the complex plane.

A numerically much more convenient method can be derived from the "transformation by reciprocal radii." Many of the important transcendental functions of applied analysis have the property that they are analytical in the entire right complex half plane, although they have a singularity at the point $z = 0$. Let us assume that we possess the Taylor expansion of this function about a certain point $z = p$, where p is a positive real number.

$$w = f(z) = a_0 + a_1(z - p) + a_2(z - p)^2 + \cdots \qquad (7\text{-}2.3)$$

This expansion converges for all z which are inside the circle $|z| = p$.

We now apply the transformation

$$z = p\,\frac{1 + \xi}{1 - \xi} \qquad (7\text{-}2.4)$$

This transformation has the property that the entire right half plane of the variable z is mapped inside the unit circle $|\xi| = 1$ of the new variable ξ. Hence the function $w = f(z)$, considered as a function of the variable ξ, has no singularity inside the unit circle, and allows there an expansion of the form

$$w = b_0 + b_1\xi + b_2\xi^2 + \cdots \qquad (7\text{-}2.5)$$

The radius of convergence of this expansion is $|\xi| = 1$. Hence the series (5) allows evaluation of $w(\xi)$ with arbitrary accuracy at any point ξ whose absolute value is less than 1, and that means any point z which is in the right complex half plane.

Now the series (3) and the series (5) are in close relation to each other. We have

$$z - p = 2p\,\frac{\xi}{1 - \xi} \qquad (7\text{-}2.6)$$

We write this transformation in two steps:

$$z - p = 2pt \quad \text{and} \quad t = \frac{\xi}{1 - \xi} \qquad (7\text{-}2.7)$$

The first transformation gives

$$w = a_0 + 2pa_1t + (2p)^2a_2t^2 + \cdots = a_0' + a_1't + a_2't^2 + \cdots$$

$$(7.2\text{-}8)$$

where

$$a_k' = (2p)^k a_k \qquad (7\text{-}2.9)$$

The second transformation means that the term $a_k't^k$ is to be replaced by an infinite series:

$$a_k't^k = a_k'\frac{\xi^k}{(1-\xi)^k} = a_k'\xi^k \sum_{\alpha=0}^{\infty} \binom{k+\alpha-1}{\alpha}\xi^\alpha$$

$$= a_k' \sum_{m=k}^{\infty} \binom{m-1}{k-1}\xi^m \qquad (7\text{-}2.10)$$

This shows that the rearrangement of the series (3) into a series in ξ gives the following values for the coefficients b_k.

$$b_{k+1} = \sum_{\alpha=0}^{k} \binom{k}{\alpha}a_{\alpha+1}' \qquad (7\text{-}2.11)$$

or written out in detail

$$\begin{aligned} b_0 &= a_0' \\ b_1 &= a_1' \\ b_2 &= a_1' + a_2' \\ b_3 &= a_1' + 2a_2' + a_3' \\ b_4 &= a_1' + 3a_2' + 3a_3' + a_4' \end{aligned} \qquad (7\text{-}2.12)$$

$$\cdot \ \cdot \ \cdot \ \cdot \ \cdot \ \cdot \ \cdot \ \cdot \ \cdot \ \cdot$$

The coefficients of the expansion (5) are thus available by a *simple binomial weighting* of the coefficients a_k'.

We have thus extended the validity of a local expansion to a much larger range. The original expansion (3) converged only within the circle $|z| = p$ and was unsuitable for evaluation of $f(z)$ at any point z which lies outside of that circle. We now transformed these coefficients by two simple operations. First we multiplied the successive coefficients by the successive powers of $2p$. Then we weighted these coefficients, starting with a_1', binomially. The new coefficients give us a new series by which $f(z)$ can be evaluated at any point z which lies somewhere in the right half plane. The variable of this new series is not the original z but

$$\xi = \frac{z - p}{z + p} \qquad (7\text{-}2.13)$$

which for all permissible z remains in absolute value smaller than (or in the limit equal to) unity.

This method can be useful in many problems. It so happens that almost all the fundamental transcendentals encountered in the mathematical description of physical events—such as the Bessel functions, the Gaussian error function, the gamma function, the exponential integral—have the property that the bulk of the function can be approximated by a simple exponential function, and what remains is an amplitude factor which does not change radically for any point z of the complex right plane. This amplitude factor is still analytical throughout the complex half plane, and becomes singular only at $z = 0$ and $z = \infty$. Moreover, the defining differential equation or functional equation permits us to obtain the successive derivatives of this amplitude function at a suitably chosen point $z = p$. We now have a method by which that function can be analytically represented and numerically evaluated for all the z points of interest, in terms of the coefficients of that single expansion.

3. Numerical example. We demonstrate the operation of the method by applying it to a simple but characteristic example, viz., to the "exponential integral":

$$w(z) = \int_z^\infty \frac{e^{-t}}{t}\, dt \qquad (7\text{-}3.1)$$

The defining differential equation here is

$$w'(z) = -\frac{e^{-z}}{z} \qquad (7\text{-}3.2)$$

The method of integrating by parts reveals that for large values of z we will have practically

$$w(z) = \frac{e^{-z}}{z}$$

Hence we will put

$$w(z) = \frac{e^{-z}}{z} u(z) \qquad (7\text{-}3.3)$$

and consider $u(z)$ our unknown function:

$$u(z) = ze^z w(z) \qquad (7\text{-}3.4)$$

By substitution into (2) we obtain the defining equation for $u(z)$:

$$u' - \frac{u}{z} - u = -1 \qquad \text{or} \qquad -zu' + (1 + z)u = z$$

We want to expand about the point $p = 1$. Hence we put

$$z = 1 + t, \qquad -(1 + t)u' + (2 + t)u = 1 + t \qquad (7\text{-}3.5)$$

If we substitute in this equation for u the expansion

$$u = a_0 + a_1 t + a_2 t^2 + \cdots \qquad (7\text{-}3.6)$$

and tabulate the coefficients, we obtain in succession

$$
\begin{aligned}
&a_1 = -1 + 2a_0, && a_5 = \tfrac{1}{5}(a_3 - 2a_4) \\
&a_2 = \tfrac{1}{2}(-1 + a_0 + a_1), && a_6 = \tfrac{1}{6}(a_4 - 3a_5) \\
&a_3 = \tfrac{1}{3}a_1, && a_7 = \tfrac{1}{7}(a_5 - 4a_6) \\
&a_4 = \tfrac{1}{4}(a_2 - a_3) && \qquad (7\text{-}3.7)
\end{aligned}
$$

All coefficients are thus uniquely determined if a_0 is given. We take the value of a_0 from a table.[1]

$$a_0 = e \int_1^\infty \frac{e^{-t}}{t}\, dt = 0.596347361$$

Successive substitutions give

$$
\begin{aligned}
&a_1 = 0.192694722, && a_5 = 0.029817368 \\
&a_2 = -0.105478958, && a_6 = -0.021979956 \\
&a_3 = 0.064231574, && a_7 = 0.016819599 \\
&a_4 = -0.042427633 && \cdot \ \cdot \ \cdot \ \cdot \ \cdot \ \cdot \ \cdot \qquad (7\text{-}3.8)
\end{aligned}
$$

The next step is to multiply by the successive powers of $2p$, i.e., in our

---

[1] See footnote to IV, 21.

case by powers of 2. These a_i' coefficients are then weighted by the binomial coefficients:

$$
\begin{array}{rrrrrrrrr}
a_1' = & 0.385389444 & 1 & 1 & 1 & 1 & 1 & 1 & 1 \\
a_2' = & -0.421915834 & & 1 & 2 & 3 & 4 & 5 & 6 \\
a_3' = & 0.513852592 & & & 1 & 3 & 6 & 10 & 15 \\
a_4' = & -0.678842130 & & & & 1 & 4 & 10 & 20 \\
a_5' = & 0.954155779 & & & & & 1 & 5 & 15 \\
a_6' = & -1.406717197 & & & & & & 1 & 6 \\
a_7' = & 2.152908672 & & & & & & & 1
\end{array} \qquad (7\text{-}3.9)
$$

The resultant series (truncated to 8 terms) becomes

$$
\begin{aligned}
u(\xi) = & \; 0.5963 + 0.3854\xi - 0.0365\xi^2 + 0.0554\xi^3 \\
& - 0.0176\xi^4 + 0.0196\xi^5 - 0.0100\xi^6 + 0.00978\xi^7
\end{aligned} \qquad (7\text{-}3.10)
$$

This series has slow convergence if we approach the unit circle $|\xi| = 1$. An approximation of this kind is not characterized by excessive accuracy, and we possess other methods for approximation of the exponential integral, which give much higher precision. But these other methods make use of the differential equation which defines the function $u(z)$, while the expansion by reciprocal radii is available under much more general conditions. The value of the method lies in the fact that by a simple weighting of a local Taylor expansion a new expansion is obtained which represents the desired function $u(z)$ in a large domain of z by a simple analytical expression of moderate accuracy.

The maximum error of the expansion (10) can be estimated as follows. The singularity of $u(z)$ lies at the point $z = 0$, i.e., $\xi = -1$. Here $u(z)$ does not go to infinity but to zero with the strength $z \log z$. We can thus expect that the series does not fail even on the limiting circle $|\xi| = 1$, which corresponds to the imaginary axis of the variable z. The convergence will be slowest for the singular point $\xi = -1$. Here the series (10) gives

$$
u(-1) = 0.0619
$$

Hence we can estimate that the absolute error of $u(\xi)$ will nowhere surpass 0.062.

As a numerical experiment let us obtain from our approximation the value of $w(z)$ at the point $z = i$. The corresponding value of ξ is

$$\xi = \frac{i-1}{i+1} = i$$

We are thus on the limiting circle. Substituting in (10), we obtain

$$u(i) = 0.6253 + 0.3398i$$

and going back to the original function $w(z)$ according to (3):

$$
\begin{aligned}
w(i) &= -Ci(1) - i\left(\frac{\pi}{2} - Si(1)\right) \\
&= \frac{1}{i}\,(\cos 1 - i\sin 1)u(z) \\
&= -(0.6253 + 0.3398i)(0.8415 + 0.5403i) \\
&= -0.3425 - 0.6238i
\end{aligned}
$$

from which $Ci(1) = 0.3425, \qquad Si(1) = 0.9470$

while the actual values are

$$Ci(1) = 0.3374, \qquad Si(1) = 0.9461$$

We thus get an accuracy of 1%.

4. The convergence of the Taylor series. The coefficients of the Taylor series are obtained by successive differentiations at a definite point $z = z_0$, the center of expansion. Since the derivatives of a differentiable function are all limits of difference coefficients, taken for an h which converges to zero, we see that the coefficients of the Taylor series are uniquely determined if the function is known in an *arbitrarily small* neighborhood of the point $z = z_0$. We thus obtain the remarkable result that the course of an analytical function can be *predicted* by knowing the function only in the small. The Taylor series is thus an *extrapolating* and not an interpolating series. Since interpolation is always safer than extrapolation, we would think that a series based on interpolation would give much better results, due to stronger convergence, than the Taylor series. While this viewpoint is correct, as the following discussions will demonstrate, it is true only if we are interested in a *one-dimensional* range of the variable z; the situation is different, however, if we pay our attention to the complete *two-dimensional* area included by the circle of convergence. In this range the validity of the Taylor series is derived from Cauchy's integral theorem:

$$f(z) = \frac{1}{2\pi i} \oint \frac{f(t)}{t - z} \, dt \qquad (7\text{-}4.1)$$

where the contour integral is extended over the boundary values of $f(z)$ along the limiting circle. In this formulation the values of $f(z)$ are known *inside* a circle if the boundary values *on* the circle are prescribed. It is only because of the analytical nature of $f(z)$ that this integral is replaceable by an infinite convergent power expansion whose coefficients are obtainable by the successive derivatives of $f(z)$ at the center of the circle. If $z - z_0$ is written in polar form:

$$z - z_0 = re^{i\theta} \qquad (7\text{-}4.2)$$

the Taylor series becomes

$$f(z) = \sum_{k=0}^{\infty} a_k r^k e^{ik\theta} \qquad (7\text{-}4.3)$$

Let us fix our attention on the values of $f(z)$ along a circle of fixed radius $r = r_0$. Then $f(z)$ appears in the form of a *Fourier series* in the variable θ. But this series is already an orthogonal expansion which approximates in the "best" way in the sense of least squares. Hence we cannot hope to improve on the convergence of the Taylor series, if our aim is to obtain a valid expansion everywhere inside of a circle. Given the condition that $f(z)$ shall be approximated by a power series of nth order, where n is a prescribed positive integer, and given a circular analytical domain in which our expansion shall hold, we can find no series which would be preferable to the truncated Taylor series by giving smaller errors throughout the given domain.

5. Rigid and flexible expansions. The great analytical value of the Taylor series and the admirable properties of the Fourier series led to overemphasis of a definite kind of limit procedure. In this procedure we consider an infinite series of the following kind.

$$f(x) = a_1 \varphi_1(x) + a_2 \varphi_2(x) + \cdots \qquad (7\text{-}5.1)$$

The meaning of this "dot-dot-dot writing" is not immediately clear, since addition of an infinite number of terms cannot be taken literally. By tacit assumption we agree that neither the "equal" sign nor the summation on the right side shall be meant in the algebraic sense. The right side points out a successive *procedure*, while the equal sign refers to a certain relation of this procedure to the given $f(x)$. The procedure is as follows. Take n terms of the series for a given x.

$$s_n = a_1 \varphi_1(x) + a_2 \varphi_2(x) + \cdots + a_n \varphi_n(x) \qquad (7\text{-}5.2)$$

Then add one more term:

$$s_{n+1} = s_n + a_{n+1} \varphi_{n+1}(x)$$

Then again add one more term:

$$s_{n+2} = s_{n+1} + a_{n+2} \varphi_{n+2}(x)$$

Proceed in this fashion, always adding one more term. This defines the procedure for any x we may choose.

The equal sign between the left and the right side expresses two independent facts:

1. The sums $s_n, s_{n+1}, s_{n+2}, \cdots$ approach a definite limit. This means that there exists a certain number $l(x)$ which is approached by the $s_n(x)$ in such a way that eventually, for sufficiently large n, the difference between $l(x)$ and $s_n(x)$ can be made in absolute value as small as we wish, although generally not zero. This $l(x)$ is called the "limit of $s_n(x)$, as n increases to infinity."

2. In a certain given interval of x this $l(x)$ coincides with $f(x)$.

While an infinite limit process of this kind is very valuable and received historically overwhelming eminence, it is nevertheless characterized by one serious drawback which ties our hands considerably. In going from n to $n + 1, n + 2, \cdots$ we constantly add one more term, *without changing anything on the sum we have obtained before.* This puts a heavy responsibility on us when we write the term $a_k \varphi_k(x)$, because we are not going to use that term only for the $s_k(x)$ when we first encounter it, but for *all the later* $s_{k+\alpha}(x)$, to all eternity. It is conceivable that we could gain much more in effective approximations if we were entitled to do *two* things in going from n to $n + 1$: add one more new term *and change all the coefficients we had before.* In this case the one-dimensional "rigid" sequence of coefficients

$$a_0, a_1, \cdots, a_n, \cdots$$

would be changed to a *two*-dimensional sequence, since now the coefficients of the expansion (1) will depend on n, the number of terms. Hence now we obtain a matrix scheme of the following structure.

$$
\begin{array}{llll}
a_{11} & & & \\
a_{21}, & a_{22} & & \\
a_{31}, & a_{32}, & a_{33} & \\
\cdot & \cdot & \cdot & \cdot
\end{array}
\tag{7-5.3}
$$

The sum $s_n(x)$ is now to be formed as follows.

$$
s_n(x) = a_{n1}\varphi_1(x) + a_{n2}\varphi_2(x) + \cdots + a_{nn}\varphi_n(x) \tag{7-5.4}
$$

Again we proceed from n to $n + 1$, to $n + 2$, $\cdots$, and again our aim is that $s_n(x)$ shall approach $f(x)$ with ever-increasing accuracy. Again we can say that

$$
f(x) = \lim_{n \to \infty} s_n(x) \tag{7-5.5}
$$

This statement is identical with the statement of equation (1) but the simple shorthand by which the previous statement could be put in symbols is no longer at our disposal.

The use of such "flexible" coefficients extends the power of infinite expansions very noticeably. By adjusting the coefficients to the number of terms we may reduce the error of the finite series for the same number of terms. Moreover, we may expand functions into infinite series which are not expandable with the help of rigid coefficients.

The value of such expansions was already demonstrated when we were dealing with the differentiation of a Fourier series (cf. IV, 6). We have seen how advantageous it is to smooth out the Gibbs oscillations by replacing $f(x)$ by a modified $\overline{f(x)}$, obtained by taking the arithmetic mean of $f(x)$ between the points $x - \pi/n$ and $x + \pi/n$. The width of this integration process changes with n. Hence the function we are expanding changes constantly with the number of terms, and thus the coefficients of the expansion also change. Nevertheless, as n increases to infinity, $\overline{f_n(x)}$ approaches the given $f(x)$ more and more, and so does the Fourier expansion, which comes nearer and nearer to $\overline{f_n(x)}$. A function like $y = \tan x$, which is not integrable and thus not expandable by the method of rigid coefficients, becomes immediately expandable if flexible coefficients are used.

The greatest importance of flexible coefficients for parexic purposes lies, however, in the fact that the coefficients of an orthogonal expansion require evaluation of certain definite integrals which are frequently not at our disposal. Hence it is of great value to know that such an expansion of n terms may be replaced by another expansion of n terms, using the same n functions but with different coefficients which can be evaluated more easily. And yet the error of this modified expansion may not be essentially worse than the error of the classical expansion, with practically unavailable coefficients.

6. Expansions in orthogonal polynomials. In V, 20 we have discussed an important family of polynomials, called Jacobi polynomials. These doubly infinite family of polynomials differ from each other according to the weight function with respect to which the orthogonality holds. But whatever this weight function is (its positiveness being assured), an arbitrary function, restricted by very few conditions (absolute integrability suffices if Fejér's summation method is employed, cf. IV, 2), can be expanded in these polynomials and thus they all share the same analytical properties. Is there any reason why we should make a choice in favor of a special weight factor when all these polynomials have the same capacity in representing arbitrary functions?

The differences reveal themselves at once if we inquire deeper into the meaning of the word "represent." We can be easily deceived if we endow the concept of a "limit" with magic significance. We may be inclined to believe that the expression "the limit of an infinite expansion becomes $f(x)$" means that a proper combination of given functions eventually yields $f(x)$. But in actual fact this "eventually" never occurs because it is not possible to add up an infinity of terms. What is meant is only that adding up more and more terms we can approach $f(x)$ *as near as we wish*, provided that we do not say: "I want to get $f(x)$ exactly." Hence an error is admittedly tolerated and what we claim is only that this error can be diminished to an arbitrarily small amount.

Now, if somebody prescribes an arbitrarily small error limit $\pm\varepsilon$, the pure analyst is satisfied in showing that the sum of a sufficiently large number of terms, constructed according to a certain scheme, will match the given limit. He can be lavish with the number of terms and it makes no difference to him whether 20 or 10,000 terms are needed for his purpose. The viewpoint of the applied analyst is markedly different. Given a certain error limit $\pm\varepsilon$, his aim will be to *economize* as much as possible in the number of terms which can match that limit.

Here is where the various classes of orthogonal polynomials diverge sharply in their behavior. It is entirely possible that a certain kind of polynomials will attain with 20 terms the accuracy for which another kind of polynomials may need 10,000 terms. This is understandable if we investigate the role of the weight factor $\rho(x)$. Let us assume that $\rho(x)$ is such that it is large in the immediate neighborhood of zero but then drops to almost negligible amounts. This means that we take the immediate neighborhood of zero seriously, at the expense of the rest of the interval. Now a finite sum of such an expansion will show the following behavior. It will give $f(x)$ with great accuracy around the origin but in the rest of the interval the accuracy will be greatly diminished. Since, however, the error limit $\pm\varepsilon$ shall mean that the deviation of the series from $f(x)$ shall not surpass $\pm\varepsilon$ *at any point of the interval* ±1, we see at once that the unfortunate nature of the weight factor compels us to pile up a large number of terms, since it will take a long time before the accuracy ε is reached at the two ends ±1 of the interval.

The Taylor expansion around the point $x = 0$ actually operates in this uneconomical fashion since we take all our information from the functional values infinitely near to $x = 0$. This explains why the Taylor series, in spite of its tremendous analytical significance, is frequently of little use in applied problems and can be replaced by much more useful expansions.

Let us now focus our attention on the Jacobi polynomials. We have a doubly infinite family of polynomials (cf. 5-20.5), characterized by the two parameters γ and δ. However, we will seldom be interested in a "lopsided" weighting in which the left and the right sides of the interval $[-1, +1]$ are weighted differently. If we restrict ourselves to symmetric weighting, the Jacobi polynomials are at once reduced to the "ultraspherical polynomials" (5-20.7) which depend on *one* parameter γ only. This parameter can still vary between 0 and ∞. The case $\gamma = \infty$ corresponds to the aforementioned overemphasis of the point $x = 0$ at the cost of all other points and thus leads to the Taylor series. The case $\gamma = 1$

leads to the Legendre polynomials for which the most natural weighting $\rho(x) = 1$ is realized. Here every point of the interval ± 1 is taken equally seriously. And yet this is not necessarily the best choice.

If we want to investigate the error of a truncated series, we do not go far wrong if we estimate the error by the first neglected term of the series. Let us expand $f(x)$ in Legendre polynomials $P_k(x)$. Then the error of $\eta_n(x)$ of a finite expansion of n terms will be approximately

$$\bar{\eta}_n(x) = c_n P_n(x)$$

Now what is the nature of $P_n(x)$? It is a function which oscillates back and forth around zero, with variable amplitudes. The amplitudes steadily (although slowly) *increase* as we leave the mid-point of the range and reach the maximum 1 at the two endpoints $x = \pm 1$. Hence the error is not quite evenly distributed over the range. In order to assure that η_n does not surpass $\pm \varepsilon$ even at the endpoints ± 1, we have to pay a price by a too large number of terms, because we have obtained *more* than we wanted in the inside of the range. It is similarly of no advantage if the amplitudes of the error oscillations constantly *decrease*, as we go from $x = 0$ towards $x = 1$. In this case we get more than what we want at the *periphery* of the interval and again we have to pay the price.[1]

Now the ultraspherical polynomials $P_n^\gamma(x)$ have the following behavior. If γ has any value between ∞ and $\frac{1}{2}$, the amplitudes of the successive oscillations steadily *increase* as we proceed from $x = 0$ to $x = 1$; if γ has any value between $\frac{1}{2}$ and 0, they steadily *decrease*. The limit $\gamma = \frac{1}{2}$ is thus of special interest. Here the amplitudes remain *constant* throughout the interval. Hence we have distributed the errors in the most advantageous fashion, namely *uniformly* over the entire range. The number of terms now needed for a certain accuracy $\pm \varepsilon$ cannot be diminished. We have obtained *maximum convergence* inasmuch as the smallest number of terms is reached by which an approximation of $f(x)$ can be achieved which does not deviate from the true value by more than $\pm \varepsilon$ at any point of the given range $[-1, +1]$. These polynomials, called after the Russian mathematician Chebyshev, are actually the most elementary (and historically oldest) set of polynomials, because of their simple relation to the elementary trigonometric functions $\cos n\theta$.

[1] See [4], p. VIII.

7. The Chebyshev polynomials. Apart from the feature of maximum convergence, the polynomials $T_k(x)$ (cf. 5-20.12) have many other valuable properties which make them a particularly interesting class of functions. Their most valuable property is the fact that they are expressible in terms of elementary trigonometric functions. They are nothing but the simple trigonometric functions $\cos k\theta$, but expressed in the variable

$$x = \cos \theta \tag{7-7.1}$$

Consider an arbitrary integrable function of bounded variation $y = f(x)$. Substituting $\cos \theta$ for x, we get the function

$$y = f(\cos \theta) \tag{7-7.2}$$

which is now a periodic function of the angle variable θ in the interval $[-\pi, +\pi]$. Since a change of θ to $-\theta$ does not modify the value of the function, $f(\cos \theta)$ is an *even function* of θ which may be expanded in a Fourier series

$$f(\cos \theta) = \tfrac{1}{2}a_0 + a_1 \cos \theta + a_2 \cos 2\theta + \cdots \tag{7-7.3}$$

with

$$a_k = \frac{2}{\pi} \int_0^\pi f(\cos \theta) \cos k\theta \, d\theta \tag{7-7.4}$$

This same series, viewed from the standpoint of the variable x, appears in the form:

$$f(x) = \tfrac{1}{2}a_0 + a_1 T_1(x) + a_2 T_2(x) + \cdots \tag{7-7.5}$$

with

$$a_k = \frac{2}{\pi} \int_{-1}^{+1} f(x) T_k(x) \frac{dx}{\sqrt{1 - x^2}} \tag{7-7.6}$$

This is a special case of the expansion into ultraspherical polynomials (cf. § 6), for the choice $\gamma = \tfrac{1}{2}$, which characterizes the Chebyshev polynomials.

Expansion of a function into Chebyshev polynomials is thus a mere reinterpretation of the expansion of an even function into a cosine series. This fundamental relation, which translates the outstanding properties of the Fourier series into the realm of power expansions, is the most important property of the Chebyshev polynomials.

For numerical applications, the *integer* nature of the Chebyshev polynomials and their simple relation to the binomial coefficients are convenient. The trigonometric formula

$$\cos (k + 1)\theta + \cos (k - 1)\theta = 2 \cos \theta \cos k\theta$$

gives in translation the *recursion formula* for the Chebyshev polynomials[1]

$$T_{k+1}(x) = 2xT_k(x) - T_{k-1}(x) \qquad (7\text{-}7.7)$$

starting with

$$T_0(x) = 1, \qquad T_1(x) = x \qquad (7\text{-}7.8)$$

It leads throughout to integer coefficients. In fact, the relation (7) shows that the coefficient of x^m must be divisible by 2^{m-1} and still remain an integer. If we write

$$T_k(x) = c_k^0 + c_k^1 x + \cdots + c_k^k x^k \qquad (7\text{-}7.9)$$

we obtain for the coefficients the expression,

$$c_k^m = 2^{m-1} \left[2 \binom{\frac{1}{2}(k + m)}{\frac{1}{2}(k - m)} - \binom{\frac{1}{2}(k + m) - 1}{\frac{1}{2}(k - m)} \right] (-1)^{\frac{k-m}{2}} \quad (7\text{-}7.10)$$

with the understanding that any coefficient for which $k + m$ is odd is put equal to zero. The first ten $T_k(x)$ are tabulated in Table V of the Appendix.

8. The shifted Chebyshev polynomials. In many problems of applied analysis, normalization of the mid-point of the range to $x = 0$ is highly inconvenient. We frequently want to expand a function around the point $x = 0$, but are interested only in the *positive* range. Or we may want to expand about the point $x = \infty$, using reciprocal powers, which is again reduced to the previous problem if $\xi = 1/x$ is introduced as a new variable. Here again the analytical natures of $f(\xi)$ to the left and to the right of $\xi = 0$ are usually completely incongruous and should not be treated together. Hence the normalization of the range to [0,1] is frequently much more convenient than the previous normalization [−1,+1]. This

---

[1] For further algebraic and functional properties of the $T_k(x)$, cf. [4], Introduction.

is particularly true if we want to utilize the "parametric method," encountered earlier in VI, 21 [cf. (6-21.12)].
We put

$$x = \alpha x_1 \tag{7-8.1}$$

and make the point $x = \alpha$ correspond to the point $x_1 = 1$ in the new variable. Although α is treated as a constant during the problem, it can be identified with any complex value z. Then, by taking the function at the end point $x_1 = 1$ of the range, we have actually obtained $f(z)$ in the original variable. Here again it is necessary that the variable x_1 shall be normalized to the interval [0,1].

The renormalization of the $T_k(x)$ to the new range is quite simple. We now put

$$\cos \theta = 2x - 1 \tag{7-8.2}$$

or

$$x = \frac{1 + \cos \theta}{2} = \cos^2 \frac{\theta}{2} \tag{7-8.3}$$

While θ varies from 0 to π, x varies from 0 to 1. The Chebyshev polynomials are again defined by

$$T_k^*(x) = \cos k\theta \tag{7-8.4}$$

but now expressed in the new variable (3). The new polynomials have entirely different coefficients, although they would be obtainable from the previous $T_k(x)$ by a simple substitution.

$$T_k^*(x) = T_k(2x - 1) \tag{7-8.5}$$

These $T_k^*(x)$ are now, apart from a sign, directly equal to a hypergeometric function [cf. 5-20.12].

$$T_k^*(x) = (-1)^k F(-k, k, \tfrac{1}{2}; x) \tag{7-8.6}$$

If $T_k^*(x)$ is again written in the form (7.9), the coefficients c_k^m now become

$$c_k^m = 2^{2m-1} \left[2 \binom{k + m}{k - m} - \binom{k + m - 1}{k - m} \right] (-1)^{m+k} \tag{7-8.7}$$

These polynomials are no longer alternately even and odd functions. The coefficients of *all* orders are present between 0 and k, with alternating $\pm$ signs. The first ten $T_k^*(x)$ are tabulated in Table VII of the Appendix.

The polynomial $T_0^*(x)$ requires special attention. According to the general definition (4) we must put $T_0^*(x) = 1$. It so happens, however, that in many summation formulas involving Chebyshev polynomials the polynomial $T_0^*(x)$ enters with *half weight* only and has to be treated separately. In order to avoid this inconvenience, we will use the following convention. We write $T_0^*(x) = 1$, but we put

$$T_0^* = \tfrac{1}{2} \qquad (7\text{-}8.8)$$

9. Telescoping of a power series by successive reductions. The discussions of § 6 have shown that the same function $f(x)$ may be expanded into a whole spectrum of different power series. They all represent the same function, but with very different degrees of convergence. If our aim is absolute accuracy, all these expansions are essentially equivalent. We have to take in more and more terms in order to reduce the error below an arbitrarily small ε. If, however, our aim is *limited* accuracy, these expansions are vastly different. We may expand directly in the powers of x, as we do in the case of the Taylor series. The convergence is then the slowest, i.e., the error, if we terminate the series after the nth term, is the largest. On the other end of the spectrum are the Chebyshev polynomials. If we expand into these polynomials, the convergence is the fastest, i.e., the error, if we terminate the series after the nth term, is the smallest. In fact we have obtained a numerical comparison of the two expansions and found that the error of the latter series is reduced in order of magnitude by the factor 2^{n-1} (in the range [0,1] the same factor becomes 2^{2n-1}). This makes it possible to obtain great accuracy with even a small number of terms.

To obtain directly an expansion in Chebyshev polynomials by evaluating the coefficients c_k on the basis of (7.6) will be only rarely possible. In all probability evaluation of the definite integrals (7.6) would become too cumbersome. But there are frequently other procedures available which require much less labor. We may start for example, with the Taylor series, which is frequently at our disposal, and "telescope" that series into a much shorter series, without losing essentially in accuracy. In order to see the operation of the method we will assume that a certain finite polynomial of the order n is given. We ask the question whether this polynomial is replaceable by a polynomial of lower order without losing too much in accuracy.

Suppose the following function is given in the interval [0,1]:

$$y = 1 - x + x^2 - x^3 + x^4 - x^5 + x^6 \qquad (7\text{-}9.1)$$

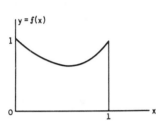

The graph of this function is quite smooth. Yet it is associated with a polynomial of 6th order. The last term is certainly not negligible without committing a large error. Hence it seems that we need all these powers for the representation of y. Yet this is not the case. By a proper technique our long expansion can be simplified.

Let us rewrite the defining equation for $T_6^*(x)$ in the following form.

$$x^6 = \frac{6144x^5 - 6912x^4 + 3584x^3 - 840x^2 + 72x - 1}{2048} + \frac{T_6^*(x)}{2048}$$

$$(7\text{-}9.2)$$

This equation shows a remarkable fact. The power x^6 is algebraically independent of the lower powers and is certainly not expressible as a linear combination of lower powers. But this independence, if the range [0,1] is considered, is astonishingly weak. In this range x^6 is *almost* equal to a definite linear combination of the lower powers. If we write

$$x^6 = \frac{6144x^5 - 6912x^4 + 3584x^3 - 840x^2 + 72x - 1}{2048} \qquad (7\text{-}9.3)$$

which means

$$x^6 = 3x^5 - 3.375x^4 + 1.75x^3 - 0.410156x^2$$
$$+ 0.035156x - 0.000488 \qquad (7\text{-}9.4)$$

we commit an error which is nowhere larger than $1/2048 = 0.000488$, in view of the fact that $T_6^*(x)$ oscillates between ± 1 throughout the range. Moreover, this accuracy cannot be surpassed, since the coefficient 2048 in $T_6^*(x)$ is the largest possible coefficient for any polynomial which in the given interval oscillates between ± 1.

We see that if we do not neglect x^6 but reduce it to the lower powers according to the equation (4), we commit an error which is very small and may well be below the accuracy we demand. Substituting (4) into the given $y = f(x)$ we obtain the approximation

$$\bar{y} = 0.999512 - 0.964844x + 0.589844x^2 + 0.75x^3$$
$$- 2.375x^4 + 2x^5 \qquad (\pm 0.0005) \qquad (7\text{-}9.5)$$

The new y is a polynomial of only *fifth* order.

We can obviously continue this process. The table of the Chebyshev polynomials $T_k^*(x)$ shows that x^5 can be expressed by the lower powers as follows.

$$x^5 = \frac{1280x^4 - 1120x^3 + 400x^2 - 50x + 1}{512} \qquad \left(\pm \frac{1}{512}\right) \quad (7\text{-}9.6)$$

$$= 2.5x^4 - 2.1875x^3 + 0.78125x^2 - 0.09765x$$

$$+ 0.001953 \ (\pm 0.00195)$$

If we substitute this expression into our previous $\bar{y}$, we obtain a new approximation which uses not more than *four* powers:

$$\bar{y} = 1.003418 - 1.160156x + 2.152344x^2 - 3.625x^3 \qquad (7\text{-}9.7)$$

$$+ 2.625x^4 \quad (\pm 0.0044)$$

The error has increased somewhat but it may still lie substantially below the accuracy we expect. Hence we can try our luck once more. According to the table x^4 is expressible by the lower powers with an accuracy slightly less than 0.01.

$$x^4 = \frac{256x^3 - 160x^2 + 32x - 1}{128} \qquad \left(\pm \frac{1}{128}\right)$$

$$(7\text{-}9.8)$$

$$= 2x^3 - 1.25x^2 + 0.25x - 0.007812 \quad (\pm 0.00781)$$

Substituting back in (9.7) we now obtain the approximation

$$\bar{y} = 0.982910 - 0.503906x - 1.128906x^2 + 1.625x^3 \ (\pm 0.0249)$$

$$(7\text{-}9.9)$$

The error is now much larger than before but still sufficiently small for most practical purposes.

If we go still further by one step, we can substitute

$$x^3 = \frac{48x^2 - 18x + 1}{32}$$

$$= 1.5x^2 - 0.5625x + 0.03125 \quad (\pm 0.03125) \quad (7\text{-}9.10)$$

and obtain

$$\bar{y} = 1.0337 - 1.4180x + 1.3086x^2 \qquad (\pm 0.0757) \qquad (7\text{-}9.11)$$

The original long polynomial of *sixth* order has been reduced to a new polynomial of but *second* order with an accuracy which may still suffice in many cases.

10. Telescoping of a power series by rearrangement. If we analyze closely what we have done in the previous section, we come to a new formulation of our procedure. Suppose we neglect nothing but write the first step of our reduction process in the form

$$y = y_5 + 0.000488T_6^*$$

Then the next step can be written as follows.

$$y_5 = y_4 + 0.003906T_5^*$$

The next step gave $\quad y_4 = y_3 + 0.020508T_4^*$

and the last step, $\quad y_3 = y_2 + 0.050781T_3^*$

All these steps combined give

$$y = y_2 + 0.050781T_3^* + 0.020508T_4^* + 0.003906T_5^* + 0.000488T_6^*$$

Suppose that we continued this reduction process to the very end, arriving finally at a constant. We would have obtained [with $T_0^* = \frac{1}{2}$, cf. (8.8)]

$$y = 1.630859T_0^* - 0.054687T_1^* + 0.163574T_2^* + 0.050781T_3^*$$
$$+ 0.020508T_4^* + 0.003906T_5^* + 0.000488T_6^*$$

What we have here is no longer an ordinary power series, but an *expansion in Chebyshev polynomials*. To be sure, we have exactly the same function with which we started. It is only the *form* in which that function appears that is different. This change of form, however, is exceedingly beneficial for parexic purposes.

Instead of going from step to step, we can proceed in a more systematic manner as follows. Just as the $T_n^*(x)$ are expressible as linear combinations of the powers of x, we can reverse the process and express the powers of x as linear combinations of the $T_n^*(x)$. This can easily be done if we remember the original definition of the polynomials $T_n^*(x)$:

$$T_n^*(x) = \cos n\theta, \qquad x = \cos^2 \frac{\theta}{2} \qquad (7\text{-}10.1)$$

Hence

$$x^n = \cos^{2n}\frac{\theta}{2} = \left(\frac{e^{i\theta/2} + e^{-i\theta/2}}{2}\right)^{2n} \tag{7-10.2}$$

$$= \frac{2}{4^n}\left[\cos n\theta + \binom{2n}{1}\cos(n-1)\theta\right.$$

$$\left. + \binom{2n}{2}\cos(n-2)\theta + \cdots + \binom{2n}{n}\frac{1}{2}\right]$$

$$= \frac{2}{4^n}\left[T_n^*(x) + \binom{2n}{1}T_{n-1}^*(x) + \binom{2n}{2}T_{n-2}^*(x) + \cdots + \binom{2n}{n}T_0^*\right]$$

We see that, apart from a constant factor, a very simple *binomial weighting* of the $T_k^*(x)$ gives the powers x^n.

We can thus add to the table of the Chebyshev polynomials $T_k^*(x)$ a new table which solves the *inverse* problem; it expresses the powers of x in terms of the $T_i^*(x)$ (cf. Table VIII of the Appendix), for example:

$$1 = 2T_0^*$$
$$x = \tfrac{1}{2}(2T_0^* + T_1^*)$$
$$x^2 = \tfrac{1}{8}(6T_0^* + 4T_1^* + T_2^*)$$

. . . .

This conversion process is numerically not too cumbersome, particularly if we multiply the given series by a proper factor which cancels out all the denominators. The rest is straight multiplication of the given coefficients by integers and additions, which can be done on the machine cumulatively. In our problem for example we will weight the given coefficients, from the lowest to the highest, by the factors

$$\frac{4096, \quad 1024, \quad 256, \quad 64, \quad 16, \quad 4, \quad 1}{2048}$$

This gives, leaving aside the denominator,

$$4096, \quad -1024, \quad 256, \quad -64, \quad 16, \quad -4, \quad 1$$

This row is now multiplied by the successive columns of the following matrix, obtained directly from Table VIII:

$$
\begin{array}{ccccccc}
1 \\
2 & 1 \\
6 & 4 & 1 \\
20 & 15 & 6 & 1 \\
70 & 56 & 28 & 8 & 1 \\
252 & 210 & 120 & 45 & 10 & 1 \\
924 & 792 & 495 & 220 & 66 & 12 & 1
\end{array}
$$

This gives

$$2048y = 3340T_0^* - 112T_1^* + 335T_2^* + 104T_3^* + 42T_4^* + 8T_5^* + T_6^*$$

Hence, dividing by 2048,

$$y = 1.630859T_0^* - 0.054687T_1^* + 0.163574T_2^* + 0.050781T_3^*$$
$$+ 0.020508T_4^* + 0.003906T_5^* + 0.000488T_6^*$$

What we gain by this rearrangement of the original power expansion is the *increased convergence* of the series. The coefficients of the original series had the values

$$1, \quad -1, \quad 1, \quad -1, \quad 1, \quad -1, \quad 1$$

The coefficients of the new series have the values

$$1.6308, \quad -0.0547, \quad 0.1636, \quad 0.0508, \quad 0.0205, \quad 0.0039, \quad 0.0005$$

Without omissions, the new series is merely a modified form of the original series and nothing is gained. But in fact the much more rapid convergence of the new series will allow us to *drop terms*. The error caused by neglecting terms is simply given by the coefficients of the neglected terms. We get an upper bound of the error at any point of the range by adding up the absolute values of all the neglected coefficients. In our example, all the last coefficients happen to be accidentally positive numbers. This, however, is quite immaterial, since the error committed by neglecting a certain term is always oscillatory in nature, and thus it is imperative that the *sum of the absolute values of all the omitted coefficients* are to be taken for a realistic estimation of the maximum error.

In our problem, for example, we obtain the successive error estimates,

$$0.0005, \quad 0.0044, \quad 0.0249, \quad 0.0757$$

Depending on the accuracy desired, we can tell at once how many terms we can drop without committing an error which is above the permissible limit.

Whether we shall prefer the successive reduction process of the previous section or the systematic rearrangement process of the present section will depend on the numerical nature of the problem. Both processes lead to the same result, namely the *telescoping of a given power expansion due to increased* (and maximized) *convergence*.

11. Power expansions beyond the Taylor range. The methods of the two previous sections are applicable only if we start with a given power expansion. We have found a procedure by which this expansion is replaceable by a shorter expansion if a certain error can be tolerated. Hence we could start, for example, with a given Taylor expansion and reduce it to a more economical expansion of fewer terms. But this procedure may become very cumbersome if we come near to the convergence radius of the Taylor series, since then we would need a very large number of terms at the start, and the numerical work might become prohibitive.

Even worse is the situation if we want a power expansion in an interval which goes beyond the convergence of the Taylor series. For example, the function

$$y = \frac{1}{1+x} \tag{7-11.1}$$

possesses a convergent Taylor series only up to $x = 1$, while we may want to approximate this function in the range [0,4]. Here the rearrangement procedure loses its significance since we have no guarantee that operating with a divergent series will give convergent results.

It is thus necessary to look for methods which will give practically well-convergent expansions even if we cannot rely on any primary series which may be rearranged into a more effective form. The existence of such series is guaranteed from the theory of orthogonal expansions, which gives the proof that any quadratically integrable function of bounded variation may be expanded into a complete orthogonal function system, such as the Legendre polynomials or the Chebyshev polynomials. Hence the possibility of expanding our given $y = f(x)$ beyond the convergence interval of the Taylor series cannot be doubted; the difficulty is only that the coefficients (7.6) of such an expansion are not at our disposal, since we do not know how to evaluate conveniently the definite integrals by which these coefficients are defined. We have to search for other more convenient methods which may not give us the coefficients c_k but some other coefficients of practically equivalent effectiveness.

12. The τ method. We are in the fortunate position that at least for a limited, but practically very frequent, class of functions a satisfactory solution of the problem can be found. Let us consider a function which is defined by a linear differential equation (or even purely algebraic equation), whose coefficients are *rational functions of* x. Bessel's differential equation

$$y'' + \frac{y'}{x} + \left(1 - \frac{p^2}{x^2}\right) y = 0 \qquad (7\text{-}12.1)$$

and many other differential equations of mathematical physics are examples for this class of functions. In fact, the majority of the important functions encountered in the advanced chapters of physics and engineering belong to this category, if we add those functions which are not *directly* defined by such a law but which can be *conceived* as the solution of such a differential equation. For example, the function

$$y = \sqrt[3]{1 - x} \qquad (7\text{-}12.2)$$

does not belong directly to this category but in actual fact appears immediately in the proper form if by logarithmic differentiation we find

$$\frac{y'}{y} = -\frac{1}{3(1 - x)} \qquad (7\text{-}12.3)$$

and write this equation in the form

$$3(1 - x)y' + y = 0 \qquad (7\text{-}12.4)$$

with the boundary condition

$$y(0) = 1 \qquad (7\text{-}12.5)$$

which makes the solution unique. Also Bessel's differential equation can be written without any denominator if we multiply through by x^2.

$$x^2 y'' + x y' + (x^2 - p^2)y = 0 \qquad (7\text{-}12.6)$$

Generally we will assume that our equation is already written in this form. For example, the algebraic definition of the function (11.1) may be replaced by the equation

$$(1 + x)y - 1 = 0 \qquad (7\text{-}12.7)$$

All these equations have something in common. We can *formally* substitute a power expansion,

$$y = b_0 + b_1 x + b_2 x^2 + \cdots \qquad (7\text{-}12.8)$$

in these equations and obtain *recurrence relations* for the coefficients. If we know the proper initial conditions, these coefficients become uniquely determined and are easily calculable. The expansion (8) is almost always an *infinite* expansion. This expansion may or may not converge. If it converges, it may converge in an infinite range or in a limited range. But it is also possible—as it happens in the case of the so-called "asymptotic expansions"—that the formal series (8) exists but diverges for every point outside $x = 0$.

We will now introduce a new viewpoint in the study of these expansions by *truncating* the series (8) to a finite number of terms. This means that we try to satisfy our equation by an expansion of the form

$$y = b_0 + b_1 x + b_2 x^2 + \cdots + b_n x^n \qquad (7\text{-}12.9)$$

where n may be given. We will undoubtedly not succeed since the recurrence relation which demands a definite nonvanishing b_{n+1}, cannot be satisfied. We may say that we get an "overdetermined" system of equations since we have $n + 1$ coefficients at our disposal, but the number of equations we have to satisfy is larger than $n + 1$. In the case of a *homogeneous* equation we know in advance that a common factor of the coefficients must remain undetermined. Hence one of the coefficients, e.g., a_0, can be normalized to 1. The number of disposable coefficients is then not higher than n, but the number of equations to be satisfied will be larger than n.

The degree of overdetermination can easily be established in advance by simple enumeration. For example, equation (4) does not increase the order of the nth power, and thus the number of recurrence equations will be $n + 1$. This surpasses the number of permissible equations by only *one*. In the case of Bessel's equation (6) the last term raises the nth power to $n + 2$, and thus the number of coefficient equations will become $n + 3$. This surpasses the number of permissible equations by *three*. In the case of the equation (7) the nth power is raised to $n + 1$, hence the number of coefficient equations becomes $n + 2$. Here the equation is *inhomogeneous* and $n + 1$ coefficients are at our disposal. Hence the degree of overdetermination is only *one*.

In all these cases we see that the degree of overdetermination is not serious since it is a small *constant* number, while the number of coefficients to be determined can increase arbitrarily.

We will now remove this overdetermination by putting something on the right side of the differential equation. This means that we are reconciled to the fact that we cannot solve the given equation exactly. The right side can be considered an *error term* which we introduce intentionally in order to make our equation solvable by a finite power series.

Our first thought will be to introduce the Taylor expansion (8) in our equation, only truncated to $n + 1$ terms. Since the coefficients were determined by successive recurrences, the only equations which are not fulfilled are the *last* ones. Hence the error term we have to put on the right side will be of the form τx^n in the case of (4), or τx^{n+1} in the case of (7), where τ is an a priori undetermined constant.

In the case of Bessel's differential it would be necessary to put *three* terms of the form

$$\tau_1 x^n + \tau_2 x^{n+1} + \tau_3 x^{n+2}$$

on the right side of the differential equation (6). In actual fact, the proper study of this differential equation shows that for large x we obtain as good approximation

$$y = \frac{e^{ix}}{\sqrt{x}}$$

If now we *take out* this asymptotic solution as a factor and write

$$y = \frac{e^{ix}}{\sqrt{x}} u(x) \tag{7-12.10}$$

we have defined a new function $u(x)$ which is much smoother than the original function $y(x)$ has been. For this $u(x)$ the determining differential equation becomes

$$x^2u'' + 2ix^2u' - (p^2 - \tfrac{1}{4})u = 0 \qquad (7\text{-}12.11)$$

This equation requires only *two τ terms* on the right side, since the power x^{n+2} disappeared. But a further simplification takes place if the new variable

$$\xi = \frac{1}{x} \qquad (7\text{-}12.12)$$

is introduced and we expand in powers of ξ, i.e., in reciprocal powers of x. The new differential equation in the variable ξ becomes

$$\xi^2u'' + 2(\xi - i)u' - (p^2 - \tfrac{1}{4})u = 0 \qquad (7\text{-}12.13)$$

This equation does not go beyond the nth power, and only the single term τx^n is needed to remove the overdetermination.

If we are interested in *small* values of x, a similar simplification can be implanted on Bessel's differential equation but by somewhat different tools. A study of the Taylor expansion reveals that x^p can be taken out as a universal factor in front of the expansion. Moreover, the remaining series proceeds in *even* powers of x only. It will thus be reasonable to introduce a new variable

$$t = \left(\frac{x}{2}\right)^2 \qquad (7\text{-}12.14)$$

and to take out an amplitude factor. We thus put

$$y(2\sqrt{t}) = t^{p/2}u(t) \qquad (7\text{-}12.15)$$

This substitution gives for $u(t)$ the following differential equation.

$$tu'' + (1 + p)u' + u = 0 \qquad (7\text{-}12.16)$$

Again we notice that the differential equation in the new form does not overstep the power x^n. Hence *one* term of the form τx^n will remove the overdetermination.

This example shows that by the proper transformations we may greatly simplify our problem and reduce the degree of overdetermination. In actual practice a *single τ term* covers much ground, and more than *two* τ terms are almost never required.

A further examination of our problem reveals that we need not choose our τ term in the form τx^n. We can remove the overdetermination equally well by a right side of the form

$$\rho(x) = \tau p_n(x) \qquad (7\text{-}12.17)$$

where $p_n(x)$ may be an *arbitrary polynomial*. This provides us with a powerful opportunity to improve on the accuracy of the Taylor expansion. The error term

$$\rho(x) = \tau x^n \qquad (7\text{-}12.18)$$

which would lead to the truncated Taylor expansion, has the property that it remains exceedingly small in the neighborhood of $x = 0$ but then increases very strongly in the neighborhood of $x = 1$. Hence we have satisfied the given differential equation with excessive accuracy in the neighborhood of the origin $x = 0$ but the price we have to pay is that the error increases practically exponentially as we come near to the point $x = 1$.

Let us agree that we have introduced already a scale factor which normalizes our range of interest to [0,1]. We will undoubtedly fare much better if we replace the error term (18) by the term

$$\rho(x) = \tau T_n^*(x) \qquad (7\text{-}12.19)$$

because the Chebyshev polynomial $T_n^*(x)$ *oscillates* in the range [0,1] with equal amplitudes. Hence we have adjusted our error throughout our interval to a balanced quantity which is neither too small nor too large at any point of the range. At no neighborhood will we get excessive accuracy. But at no neighborhood will we get a very large error which would make our series diverge. The accuracy we obtain is equivalent to the accuracy obtained by the "ideal" Fourier coefficients (7.6), but the new coefficients are evaluated by *simple algebraic recursions*, without any integrations.

The coefficients of the expansion are of course no longer "rigid" coefficients like the coefficients of the Taylor series which form a single row of numbers. If we change n, the order of the approximating polynomial, we also change completely the coefficients of $T_n^*(x)$ since $T_n^*(x)$ is now replaced by $T_{n+1}^*(x)$. But this means that in the successive recursions we get an entirely new set of coefficients for every n (cf. § 5). We adjust our coefficients to the range in which we want our function *and* to the number of powers with which we want to operate. The result of this flexibility is threefold:

1. We can increase the convergence of the Taylor expansion, that is, we can make the error of our approximation much smaller than the error of the Taylor expansion with the same number of terms.

2. We can obtain a convergent expansion in cases when the Taylor expansion *diverges* in the given range, either from the beginning or from a certain point on. All the so-called "asymptotic" expansions of the customary transcendentals of mathematical physics can thus be transformed from divergent to convergent expansions.

3. We can obtain a convergent expansion in cases when the Taylor series does not exist at all.

13. The canonical polynomials. If we operate with "flexible" coefficients, the following objection can be raised. Let us decide that we are not satisfied with a certain order of approximation. Then, if we go from n to $n + 1$, we have to throw away the previous results completely and start the entire calculation again. We will now develop a method which obviates this objection.

Instead of putting an error term of the form $\tau T_n^*(x)$ on the right side, we will put the single power x^m on the right side and solve the resultant coefficient equations, assuming a finite expansion. But then, enumerating the number of equations to be solved, we find that the number of conditions surpasses the number of available parameters by μ. We relieve this overdetermination by *dropping the first μ equations* for the time being. We consider them as a kind of "initial conditions" which will be taken care of later.

To illustrate the situation, let us examine the differential equation

$$x^3 y' - 2y = 0 \qquad (7\text{-}13.1)$$

We would encounter this equation if our aim were to obtain a power expansion for the function

$$y = e^{-1/x^2} \qquad (7\text{-}13.2)$$

The recurrence relations for the coefficients demand $n + 3$ equations for $n + 1$ coefficients. The number of surplus equations is thus *two*. Hence we will drop two equations in the beginning in order to make our system consistent. If we substitute the formal expansion

(12.8) in our equation and tabulate the resulting coefficients, we obtain in succession

$$-2b_0 = 0$$

$$\boxed{\begin{aligned}-2b_1 &= 0\\ -2b_2 &= 0\end{aligned}}$$

$$b_1 - 2b_3 = 0$$
$$2b_2 - 2b_4 = 0 \qquad (7\text{-}13.3)$$
$$3b_3 - 2b_5 = 0$$
$$4b_4 - 2b_6 = 0$$

. . . . . .

We relieve the overdetermination by omitting the two boxed equations.

Now the error term x^m on the right side of the original equation means the appearance of a single "one" on the right side of the equation (3). This 1 appears in the mth equation. By assigning to m the successive values 0, 1, 2, $\cdots$, we let this solitary 1 glide down from the top to the bottom. Since, however, in our problem the second and third equations have been obliterated, we avoid the indices $m = 1$ and $m = 2$.

What we obtain in this manner is a *definite set of polynomials*, inherently associated with the given differential operator. We will call them "canonical polynomials" and denote them by $Q_m(x)$. This $Q_m(x)$ is not necessarily of the order m. The index m refers merely to the fact that if the operation indicated by the left side of the given differential equation is performed on the polynomial $Q_m(x)$ (omitting the surplus equations), the result is x^m. In our problem for example we will refrain from defining $Q_1(x)$ and $Q_2(x)$, but the remaining $Q_k(x)$ are uniquely determined

$$Q_0(x) = -\tfrac{1}{2}$$
$$Q_3(x) = x$$
$$Q_4(x) = \tfrac{1}{2}x^2$$
$$Q_5(x) = \tfrac{1}{3}(2x + x^3) \qquad (7\text{-}13.4)$$
$$Q_6(x) = \tfrac{1}{4}(x^2 + x^4)$$
$$Q_7(x) = \tfrac{1}{15}(4x + 6x^3 + 3x^5)$$

. . . .

Another interesting example is the differential equation

$$xy' - y - x = 0 \tag{7-13.5}$$

which defines the function

$$y = x \log x \tag{7-13.6}$$

apart from the additive term cx. Here the mere enumeration of the equations shows no overdetermination, since we get $n + 1$ equations for $n + 1$ coefficients, and the system is inhomogeneous. But the second coefficient equation expresses the contradictory statement $-1 = 0$. We consider this equation as overdetermination and drop it from our system.

In this example the canonical polynomials become

$$Q_0(x) = -1$$
$$Q_2(x) = x^2$$
$$Q_3(x) = \frac{x^3}{2} \tag{7-13.7}$$
$$\vdots$$
$$Q_m(x) = \frac{x^m}{m - 1}$$

In all cases, step-by-step construction of the canonical polynomials can occur without difficulty, even if we cannot always write down their general expression in such simple manner as in the last example.

These canonical polynomials put us in the position to obtain the solution of the differential equation $Dy = 0$, if an error term of the form (12.19) is put on the right side. We make use of the superposition principle of linear operators. Since $T_n^*(x)$ can be written out in its actual coefficients,

$$T_n^*(x) = \sum_{m=0}^{n} c_n^m x^m \tag{7-13.8}$$

the solution of

$$Dy = \tau T_n^*(x) \tag{7-13.9}$$

becomes

$$y = \tau \sum_{m=0}^{n} c_n^m Q_m(x) \qquad (7\text{-}13.10)$$

The solution appears in explicit form as a linear superposition of a *rigid set of polynomials*, viz., the canonical polynomials which are uniquely associated with the given differential operator. The coefficients of the Chebyshev polynomials appear as mere *weight factors* of these polynomials. Step-by-step solution of the coefficient equations separately for each n is now avoided. The solution (10) may be put in the following operational form

$$y = \tau T_n^*(Q(x)) \qquad (7\text{-}13.11)$$

with the understanding that in the formal expansion of $T_n^*(x)$ the power Q^m is replaced by $Q_m(x)$.

The freedom of τ can now be used for satisfying the surplus equation which was originally omitted. If we had to omit more than one equation, we need more than one τ term on the right side. In the case of *two* surplus equations, for example, we will need the error terms

$$\rho(x) = \tau_1 T_n^*(x) + \tau_2 T_{n+1}^*(x)$$

which gives rise to *two* τ factors, to be determined from the two remaining conditions. In actual practice it will be more convenient, and practically equally effective, to operate with only *one* set of Chebyshev coefficients. We can do that by using an error term of the following form

$$\rho(x) = T_n^*(x)(\tau_1 + \tau_2 x) \qquad (7\text{-}13.12)$$

The explicit solution (13.10) now becomes

$$y_n = \sum_{m=0}^{n} c_n^m [\tau_1 Q_m(x) + \tau_2 Q_{m+1}(x)] \qquad (7\text{-}13.13)$$

The remaining two equations give two simultaneous linear equations for τ_1 and τ_2 which can be solved without difficulty.

We carry through the program for the case of the differential equation (5). Here the solution becomes

$$y_n = -\tau + \tau \sum_{m=2}^{n} c_n^m \frac{x^m}{m-1} + cx \qquad (7\text{-}13.14)$$

where c is arbitrary. The equation we have omitted becomes

$$-1 = \tau c_n^1 \qquad (7\text{-}13.15)$$

from which

$$\tau = -\frac{1}{c_n^1} = \frac{(-1)^n}{2n^2} \qquad (7\text{-}13.16)$$

The constant c can be determined from the boundary condition

$$y(1) = 0 \qquad (7\text{-}13.17)$$

This solution gives ever-improving approximations of the function $y = x \log x$ in the form of a power expansion with flexible coefficients. No Taylor series exists in this case, in view of the analytical singularity at the point $x = 0$. But the expansion (14) converges to $y = x \log x$ uniformly throughout the domain [0,1]. Moreover, the convergence is satisfactory even for small n, since the maximum error is bounded by $|\tau|$, i.e., by $1/2n^2$. For example, the choice $n = 4$ gives the expansion

$$y_4 = -\frac{1}{32} + \frac{1}{32}\left(160x^2 - \frac{256}{2}x^3 + \frac{128}{3}x^4\right) - \frac{221}{96}x \qquad (7\text{-}13.18)$$

$$= \frac{1}{96}(-3 - 221x + 480x^2 - 384x^3 + 128x^4)$$

with a maximum error of ± 0.03.

Use of the canonical polynomials $Q_m(x)$ is advocated from still another viewpoint. It makes it possible to adjust the solution to *arbitrary ranges*. Let us assume that we want a solution in the range [0,α], instead of [0,1]. This α may be any given real or *complex* number, provided that the ray [0,α] does not include any singular point of the function $y(x)$. Now this readjustment of the range is immediately accomplished by merely replacing $T_n^*(x)$ by $T_n^*\left(\frac{x}{\alpha}\right)$.

This means that the coefficients c_n^m are to be replaced by $c_n^m \alpha^{-m}$. Hence the solution (10) will now appear in the form

$$y(x) = \tau \sum_{m=0}^{n} c_n^m \alpha^{-m} Q_m(x) \qquad (7\text{-}13.19)$$

where x is any point along the complex ray $[0,\alpha]$. If we go to the end point of the range $x = \alpha$, we obtain a solution of the given differential equation *at an arbitrary point z of the complex plane*, compatible with the general regularity conditions. The solution appears in the form

$$y(z) = \tau \sum_{m=0}^{n} c_n^m z^{-m} Q_m(z) = \tau T_n^* \left(\frac{Q(z)}{z} \right) \qquad (7\text{-}13.20)$$

and we have overcome the restriction that x has to lie between 0 and 1. However, this solution will generally not be a polynomial any more but the *ratio of two polynomials* (cf. 14, Example 5).

14. Examples for the τ method. The τ method has such wide fields of applications that it will be of value to discuss a few characteristic examples which demonstrate the power of the method. The problem of obtaining suitable error bounds will be left to the next section.

The first example deals with a function whose Taylor expansion is well convergent. We show how the application of the τ method greatly increases the degree of convergence. In the second example we show how the realm of convergence of the Taylor series can be extended to a wider range. The third example demonstrates how a completely divergent "asymptotic" expansion can be changed into a convergent expansion. The fourth example deals with a function whose Taylor expansion does not exist. The fifth example illustrates the extension of the method to the complex domain.

Example 1. We consider the function

$$y = e^x \qquad (7\text{-}14.1)$$

defined by the differential equation

$$y - y' = 0 \qquad (7\text{-}14.2)$$

with the boundary condition

$$y(0) = 1 \qquad (7\text{-}14.3)$$

We have $n + 2$ equations for $n + 1$ coefficients. The equations become compatible by merely dropping the boundary condition.

The canonical polynomials $Q_m(x)$ of our problem become

$$Q_0(x) = 1$$

$$Q_1(x) = 1 + x$$

$$Q_2(x) = 2\left(1 + x + \frac{x^2}{2}\right) \qquad (7\text{-}14.4)$$

$$Q_3(x) = 3!\left(1 + x + \frac{x^2}{2!} + \frac{x^3}{3!}\right)$$

$$\vdots$$

$$Q_m(x) = m!\left(1 + x + \frac{x^2}{2!} + \cdots + \frac{x^m}{m!}\right)$$

$$= m!\, S_m(x)$$

where
$$S_m(x) = 1 + x + \frac{x^2}{2!} + \cdots + \frac{x^m}{m!}$$

represents the successive "partial sums" of the Taylor series. Hence the solution of our problem becomes, according to (13.10),

$$y_n(x) = \tau \sum_{m=0}^{n} c_n^m m!\, S_m(x) \qquad (7\text{-}14.5)$$

We now satisfy the condition (3) and obtain

$$y_n = \frac{\displaystyle\sum_{m=0}^{n} c_n^m m!\, S_m(x)}{\displaystyle\sum_{m=0}^{n} c_n^m m!} \qquad (7\text{-}14.6)$$

The solution appears as a *weighted arithmetic mean of the partial sums of the Taylor series.* The successive convergents of e^x in the range [0,1] thus become

$$y_0 = 1$$

$$y_1 = \frac{1 + 2x}{1}$$

$$y_2 = \frac{9 + 8x + 8x^2}{9}$$

(7-14.7)

$$y_3 = \frac{113 + 114x + 48x^2 + 32x^3}{113}$$

$$y_4 = \frac{1825 + 1824x + 928x^2 + 256x^3 + 128x^4}{1825}$$

The corresponding e values, putting $x = 1$, become

$$1, \quad 3, \quad 2.77\underset{.}{7}, \quad 2.71\underset{.}{6}8, \quad 2.718\underset{.}{3}6, \quad 2.7182806\underset{.}{8}, \quad \cdots \quad (7\text{-}14.8)$$

(A dot under a digit indicates the first decimal in error.) The convergence is strong in comparison with the much slower convergence of the Taylor series:

$$1, \quad 2, \quad 2.5, \quad 2.667, \quad 2.70867, \quad 2.71667, \quad \cdots$$

yet the results are no match to the spectacular e approximations found in VI, 20. The previous approximations are analogous to the present ones and can be obtained by the present procedure, provided that the coefficients c_n^m of the Chebyshev polynomials are replaced by the corresponding coefficients of the Legendre polynomials $P_n^*(x)$. Since we have decided that the Chebyshev polynomials give a smaller error than the Legendre polynomials, it seems strange that the previous e values are so much superior to the new ones. However, we have to consider the *entire* range of x between 0 and 1. The smallness of the error at the end point $x = 1$ does not guarantee that the error will be small *everywhere*. Moreover, the choice of $T_n^*(x)$ as error term of the differential equation, while securing good convergence, will not give necessarily the *best* possible convergence. What we want to achieve is an even oscillation of the error in the *function* $y = f(x)$. The operation Dy may considerably alter this uniform character of the error. By proper study of the given differential operator we may construct an error term $\rho(x)$ which is better than the choice $\tau T_n^*(x)$.

For example, in our problem we can see without difficulty that the dominant term in the given differential equation is $-y'$, since the differentiation of a high-frequency oscillation of the type $\cos n\theta$ magnifies the amplitude by the factor n. Hence in this problem it would be preferable to put

$$\rho(x) = \tau T_{n+1}^{*}{}'(x) \qquad (7\text{-}14.9)$$

because, assuming that the polynomial approximation $y_n(x)$ is of the form

$$y_n(x) \doteq e^x - \tau T_{n+1}^*(x) \qquad (7\text{-}14.10)$$

the operation $y - y'$ will generate (in good approximation) the error term (9). The solution (5) will now hold again, but the c_n^m have to be replaced by the coefficients of $T_{n+1}^{*}{}'(x)$, called "Chebyshev polynomials of the second kind." Furthermore, the estimated solution (10) indicates that we should not satisfy the boundary condition $y(0) = 1$ exactly. We should rather determine τ from the condition

$$y_n(0) = 1 - \tau T_{n+1}^*(0) = 1 - \tau(-1)^{n+1} \qquad (7\text{-}14.11)$$

This gives

$$\tau \left[(-1)^{n+1} + \sum_{m=0}^{n} \bar{c}_n^m m! \right] = 1 \qquad (7\text{-}14.12)$$

and thus the final solution becomes

$$y_n(x) = \frac{\displaystyle\sum_{m=0}^{n} \bar{c}_n^m m! S_m(x)}{(-1)^{n+1} + \displaystyle\sum_{m=0}^{n} \bar{c}_n^m m!} \qquad (7\text{-}14.13)$$

This is again a weighted arithmetic mean of the partial sums of the

Taylor series, but the weight factors have changed. The first five convergents of e^x in the range $[0,1]$ now become

$$y_0 = \frac{2}{1}$$

$$y_1 = \frac{8 + 16x}{9}$$

$$y_2 = \frac{114 + 96x + 96x^2}{113} \tag{7-14.14}$$

$$y_3 = \frac{1824 + 1856x + 768x^2 + 512x^3}{1825}$$

$$y_4 = \frac{36690 + 36640x + 18720x^2 + 5120x^3 + 2560x^4}{36689}$$

The estimated maximum error at any point of the range does not exceed the reciprocal of the denominator. Hence the quadratic approximation is bounded by ± 0.01, and the cubic approximation is better than ± 0.0006. The first 6 e values become

$$2, \quad 2.\dot{6}67, \quad 2.70\dot{7}96, \quad 2.71\dot{7}808, \quad 2.718\dot{2}53, \quad 2.7182807\dot{5}$$

The "best" nature of this approximation is not invalidated by the superiority of the remarkable approximations (6-20.4) and (6-20.5) obtained on the basis of the quadrature method described in V, 22. These are *end point approximations*, and one can show that in our problem the Legendre polynomials are superior to the Chebyshev polynomials from the standpoint of minimizing the error of $y(1)$ at the end point of the range.

Example 2. We consider the function

$$y = \frac{1}{a + x} \tag{7-14.15}$$

defined by the algebraic equation

$$(a + x)y - 1 = 0 \tag{7-14.16}$$

Instead of using the canonical polynomials, we will deal with this problem in a different manner, in order to illustrate a method which in some problems is more adequate. Instead of expanding $y(x)$ into powers, we will immediately arrange our expansion in *Chebyshev polynomials*

$$y = \tfrac{1}{2}c_0 T_0^*(x) + c_1 T_1^*(x) + \cdots + c_{n-1} T_{n-1}^*(x) \quad (7\text{-}14.17)$$

As usual, we put a τ term on the right side, to insure compatibility

$$(a + x)y = 1 + \tau T_n^*(x) \qquad (7\text{-}14.18)$$

In this problem we can obtain τ in advance. If we substitute $x = -a$, we get zero on the left side. Hence

$$\tau = -\frac{1}{T_n^*(a)} \qquad (7\text{-}14.19)$$

We now substitute the expansion (17) on the left side and perform the multiplication by $(a + x)$. For this purpose we make use of the recurrence relation (7.7), which for the shifted polynomials $T_n^*(x)$ becomes

$$2(2x - 1)T_k^*(x) = T_{k+1}^*(x) + T_{k-1}^*(x) \qquad (7\text{-}14.20)$$

In order to use this relation we multiply our equation (18) by 4 and put

$$4a + 2 = 2b \qquad (7\text{-}14.21)$$

$$[2b + (4x - 2)]y = 4 + 4\tau T_n^*(x) \qquad (7\text{-}14.22)$$

We tabulate the coefficients of the expansion, obtaining in succession:

Left side:	bc_0	$2bc_1$	$2bc_2$	$2bc_3$	
		c_0	c_1	c_2	
	c_1	c_2	c_3	c_4	$\cdots$
Right side:	4	0	0	0	

This gives the equations

$$bc_0 + c_1 = 4$$
$$c_0 + 2bc_1 + c_2 = 0$$
$$c_1 + 2bc_2 + c_3 = 0 \qquad (7\text{-}14.23)$$
$$\cdot \quad \cdot \quad \cdot \quad \cdot \quad \cdot \quad \cdot$$

Let us disregard for the time being the first equation. The remaining homogeneous equations are solvable by an assumption of the following form

$$c_k = Cp^k \tag{7-14.24}$$

All the equations are then reduced to one quadratic equation for p

$$1 + 2bp + p^2 = 0 \tag{7-14.25}$$

which gives the two roots

$$p = -b \pm \sqrt{b^2 - 1} \tag{7-14.26}$$
$$= -(2a + 1) \pm 2\sqrt{a(a + 1)}$$

In view of the superposition principle of the solution of linear system, we thus get

$$c_k = C_1 p_1^k + C_2 p_2^k \tag{7-14.27}$$

where p_1 and p_2 are the two roots, corresponding to the $\pm$ sign in (26). The two constants C_1 and C_2 are so far arbitrary. But the condition $c_n = 0$ gives

$$C_1 p_1^n + C_2 p_2^n = 0 \tag{7-14.28}$$

while the first equation of the system (23) demands

$$C_1 p_1 + C_2 p_2 + b(C_1 + C_2) = 4 \tag{7-14.29}$$

The quadratic equation (25) shows that the product of the two roots p_1 and p_2 give 1. We will call p_1 the absolutely smaller and p_2 the absolutely larger root. Now the relation (28) gives

$$C_2 = -C_1 \left(\frac{p_1}{p_2}\right)^n \tag{7-14.30}$$

If n grows very large, C_2 converges to zero. We simplify matters without making the error essentially larger if we drop C_2 altogether. This means that we establish immediately the *infinite* expansion into the $T_n^*(x)$ i.e., the Fourier series if we think in terms of the variable θ [cf. (8.3)] and then truncate this series to n terms. Our solution is then simply

$$\gamma_k = \frac{4}{b + p} p^k = \frac{2}{\sqrt{a(a + 1)}} p^k \tag{7-14.31}$$

with

$$p = 2\sqrt{a(a + 1)} - (2a + 1) \tag{7-14.32}$$

assuming that a is positive.

Generally a may be any real or complex constant. The solution of our problem is quite generally the infinite expansion

$$y = \frac{1}{a+x} = \frac{2}{\sqrt{a(a+1)}} [\tfrac{1}{2}T_0^*(x) + pT_1^*(x) + p^2T_2^*(x) + \cdots \quad (7\text{-}14.33)$$

where p is the absolutely smaller of the two roots (26). If this series is terminated after n terms, the remainder of the series can be estimated on the basis of the fact that all $T_k^*(x)$ oscillate between ± 1.

$$|\eta_n| \le \frac{2|p|^n}{\sqrt{a(a+1)}} (1 + |p| + |p|^2 + \cdots)$$

$$|\eta_n| \le \frac{2|p|^n}{\sqrt{a(a+1)}} \frac{1}{1-|p|} \quad (7\text{-}14.34)$$

Since $p_1 p_2 = 1$, one of the roots must always remain smaller than 1 in absolute value except in the limiting case when both p_1 and p_2 lie on the unit circle and p_2 is the complex conjugate of p_1. This is possible only if a is a *real negative number* between 0 and -1. In that case $y(x)$ has a singularity in the critical interval [0,1]. In all other cases the series (33) converges, and the convergence is even absolute.

Compare this result with the convergence of the Taylor series which extends only up to $x = |a|$ and becomes divergent beyond that point. In both cases the expansion has the character of a geometric series, and the estimated maximum error η_n at any point of the range has the form (34). But the p of this formula has in the two cases widely different values. The following numerical table makes a comparison between the two p values for a wide range of the parameter a, assuming that a is a real positive number.

a	Taylor $-p$	$T_n^*(x)$ $-p$	
3	0.333	0.072	
2	0.5	0.101	
1	1	0.172	(7-14.35)
0.5	2	0.263	
0.333	3	0.333	
0.2	5	0.420	
0.1	10	0.537	
0.01	100	0.819	

Of particular interest is the case

$$y = \frac{1}{1+x} = 1 - x + x^2 - x^3 + \cdots$$

$$= \sqrt{2}\left[\tfrac{1}{2} - 0.1716 T_1^*(x) + (0.1716)^2 T_2^*(x) \right. \quad (7\text{-}14.36)$$

$$\left. - (0.1716)^3 T_3^*(x) + \cdots \right]$$

The quotient of the second expansion is deduced from

$$3 - 2\sqrt{2} = 0.1716 \cdots$$

Compared with the Taylor series we have gained the large factor 5.83. The rough overall estimate[1] suggests the factor 4, and that value is actually correct for *large* values of a where the Taylor series is still well convergent, as we can demonstrate by the case $a = 3$. For decreasing a the gain increases slowly and reaches at the limit of convergence of the Taylor series the maximum value

$$3 + 2\sqrt{2} = 5.8284 \cdots$$

Beyond $a = 1$ the Taylor series does not hold in the entire interval [0,1], while the $T_n^*(x)$ expansion retains its convergence.

Example 3. This example is chosen because it illustrates the manner in which the τ method transforms a completely divergent expansion into a strictly convergent one. We consider the "exponential integral"

$$w(z) = \int_z^\infty \frac{e^{-t}}{t}\, dt \quad\quad (7\text{-}14.37)$$

encountered in (3.1). As before, we write

$$w(z) = \frac{e^{-z}}{z}\, u(z) \quad\quad (7\text{-}14.38)$$

and obtain for the new function $u(z)$ the differential equation

$$-zu' + (1 + z)u = z \quad\quad (7\text{-}14.39)$$

We now introduce the reciprocal of z as a new variable

$$x = \frac{1}{z} \quad\quad (7\text{-}14.40)$$

---

[1] See footnote to p. 453.

and obtain the new differential equation,

$$x^2 y' + (1 + x)y = 1 \qquad (7\text{-}14.41)$$

If we put for $y(x)$ a formal power expansion, we obtain the following infinite series.

$$y(x) = 1 - x + 2!x^2 - 3!x^3 + 4!\,x^4 - \cdots \qquad (7\text{-}14.42)$$

The variable x of this expansion is in actual fact the *reciprocal* of the original variable z. Hence we can conceive the expansion (42) as the formal antipode of the expansion

$$e^{-z} = 1 - z + \frac{z^2}{2!} - \frac{z^3}{3!} + \cdots \qquad (7\text{-}14.43)$$

inasmuch as every term is the exact reciprocal of the latter expansion. While the series (43) is exceedingly well convergent—the convergence being preserved for *any* value of z—the series (42) is exceedingly well *divergent*; the convergence is lost for even arbitrarily small values of x, except for the trivial value $x = 0$. Such a series is nevertheless of great value, since one can show that the error of the truncated series is smaller than the first neglected term. Hence we use only the *decreasing* part of the series, stopping at the proper term. We choose this term by the condition that the first neglected term shall be smaller than the term which follows it. By this method we obtain good approximation for sufficiently small values of x but with increasing x the series gradually loses its value.

We will now treat this problem by the τ method. We drop the first equation for the purpose of compatibility and establish the canonical polynomials $Q_m(x)$, starting with $m = 1$. We thus find

$$Q_m(x) = \frac{(-1)^{m-1}}{m!}\, S_{m-1}(x) \quad (m = 1, 2, \cdots) \qquad (7\text{-}14.44)$$

and the τ solution becomes

$$y_{n-1}(x) = \tau \sum_{m=1}^{n} c_n^m (-1)_n^{m-1} \frac{S_{m-1}(x)}{m!} \qquad (7\text{-}14.45)$$

We now satisfy the temporarily omitted first equation.

$$b_0 = 1 + \tau c_n^0 \qquad (7\text{-}14.46)$$

This gives

$$\tau = \frac{1}{\sum_{m=0}^{n} c_n^m (-1)^{m-1}/m!} \tag{7-14.47}$$

and the complete solution becomes

$$y_{n-1}(x) = \frac{\sum_{m=1}^{n} (-1)^m c_n^m (S_{m-1}(x)/m!)}{\sum_{m=0}^{n} (-1)^m c_n^m (1/m!)} \tag{7-14.48}$$

It is of interest to make a comparison between this solution and the solution (6) found in the case of the exponential function. In both cases the solution appears as a weighted arithmetic mean of the partial sums of the Taylor series. The essential difference is that the factor $m!$ appeared previously in the *numerator* of each term, while now it appears in the *denominator*. Previously the partial sums of high order were strongly emphasized; now they are strongly de-emphasized. Use of the truncated Taylor expansion of the order n corresponds to the extreme weighting $0, 0, \cdots, 0, 1$. In the case of the well-convergent e^x function this weighting is not damaging and not too far from the weighting of the more effective τ series, which is much less extreme than the weighting of the Taylor series but still not too dissimilar to it, by putting the center of importance out to the *large m*. In the second case the situation is quite different. Here the influence of the high-order terms is cut down by small weight factors and the center of importance is on the partial sums of *low* order. This corresponds to the customary method of going with the partial sums only up to a point and be satisfied with an error which we cannot further diminish. Here, however, we need not stop with any definite n. The higher-order partial sums are not thrown away, but their damaging influence is cut down by a proper factor. The higher-order sums serve as correction terms and in actual fact we come closer and closer to the correct functional value at any point of the range, as n increases more and more. Whether the original $S_n(x)$ converge in themselves or not, is irrelevant. The *proper weighting establishes their convergence in every case.*

The first five convergents of our problem become

$$y_0 = \frac{2}{1+2} = \frac{2}{3} \qquad\qquad (7\text{-}14.49)$$

$$y_1 = \frac{8 + \frac{8}{2}(1-x)}{1 + 8 + \frac{8}{2}} = \frac{12 - 4x}{13}$$

$$y_2 = \frac{18 + \frac{48}{2}(1-x) + \frac{32}{6}(1-x+2x^2)}{1 + 18 + \frac{48}{2} + \frac{32}{6}} = \frac{142 - 88x + 32x^2}{145}$$

$$y_3 = \frac{32 + \frac{160}{2}(1-x) + \frac{256}{6}(1-x+2x^2) + \frac{128}{24}(1-x+2x^2-6x^3)}{1 + 32 + \frac{160}{2} + \frac{256}{6} + \frac{128}{24}}$$

$$= \frac{160 - 128x + 96x^2 - 32x^3}{161}$$

$$y_4 = \frac{50 + \frac{400}{2}(1-x) + \frac{1120}{6}(\quad) + \frac{1280}{24}(\quad) + \frac{512}{120}(\quad)}{1 + 50 + \frac{400}{2} + \frac{1120}{6} + \frac{1280}{24} + \frac{512}{120}}$$

$$= \frac{7414 - 6664x + 7328x^2 - 5184x^3 + 1536x^4}{7429}$$

The estimated accuracy (cf. § 15) of the solution is such that the maximum error at any point of the range cannot surpass τ. Hence

$$|\eta| \le 0.33,\ 0.077,\ 0.021,\ 0.0062,\ 0.0021,\ 0.00070,\ \cdots$$

At the end point $x = 1$ of the range we obtain the following convergents for $y(1) = 0.596358414$.

$$0.667,\quad 0.6154,\quad 0.5931,\quad 0.59627,\quad 0.596312,\quad 0.5963666,\quad \cdots \qquad (7\text{-}14.50)$$

The asymptotic series tells us at this point only that the functional value must lie between 0 and 1. The τ solution, on the other hand, gives the functional value, as τ increases, with arbitrary accuracy.

Example 4. We will now obtain an expansion for the function

$$y = \sqrt{x} \qquad\qquad (7\text{-}14.51)$$

Since all derivatives of the function become infinite at the point $x = 0$, the Taylor series does not exist in this case.

By logarithmic differentiation we find

$$\frac{y'}{y} = \frac{1}{2x} \tag{7-14.52}$$

Hence we see that we can characterize our function by the differential equation

$$2xy' - y = 0 \tag{7-14.53}$$

but we change this equation to

$$2xy' - y = \tau T_n^*(x) \tag{7-14.54}$$

The canonical polynomials $Q_m(x)$ of our problem become

$$
\begin{aligned}
Q_0(x) &= -1 \\
Q_1(x) &= x \\
Q_2(x) &= \frac{x^2}{3} \\
&\vdots \\
Q_m(x) &= \frac{x^m}{2m - 1}
\end{aligned}
\tag{7-14.55}
$$

and thus

$$y_n(x) = \tau \sum_{m=0}^{n} \frac{c_n^m}{2m - 1} x^m \tag{7-14.56}$$

The factor τ can be determined by the boundary condition

$$y(1) = 1 \tag{7-14.57}$$

Hence

$$y_n(x) = \frac{\displaystyle\sum_{m=0}^{n} \frac{c_n^m}{2m - 1} x^m}{\displaystyle\sum_{m=0}^{n} \frac{c_n^m}{2m - 1}} \tag{7-14.58}$$

The error analysis of this problem is particularly interesting. The critical point is the point $x = 0$ where the function has an analytical singularity. Hence the largest errors will occur in the neighborhood of the origin. We will operate with the angle variable θ of § 8, which is the adequate variable whenever the Chebyshev polynomials are involved. For the sake of convenience, we replace θ by $\pi - \theta$ and have accordingly

$$x = \sin^2 \frac{\theta}{2}$$

$$T_n^*(x) = (-1)^n \cos n\theta$$

For small x we can put

$$x = \frac{\theta^2}{4} \tag{7-14.59}$$

and rewrite the differential equation (54) in terms of θ.

$$\theta y' - y = \bar{\tau} \cos n\theta \tag{7-14.60}$$

where $\bar{\tau} = (-1)^n \tau$. We solve this differential equation by the method of the "variation of the constants." We put

$$y = C\theta \tag{7-14.61}$$

$$C' = \bar{\tau} \frac{\cos n\theta}{\theta^2} \tag{7-14.62}$$

Integrating by parts,

$$C = -\frac{\bar{\tau}}{\theta} \cos n\theta - n\bar{\tau} \int \frac{\sin n\theta}{\theta} \, d\theta \tag{7-14.63}$$

Hence

$$y_n = -\bar{\tau} \cos n\theta - n\bar{\tau}\theta \left[\int_0^\theta \frac{\sin \theta}{\theta} \, d\theta + A \right] \tag{7-14.64}$$

where A is a constant of integration. Since, however, $y(\theta)$ must be a polynomial in θ^2, a term of the form $A\theta$ cannot occur. This shows that A must be put equal to zero, and we obtain

$$y_n(\theta) = -\bar{\tau} \cos n\theta - \bar{\tau}n\theta Si(n\theta) \tag{7-14.65}$$

The function $Si(n\theta)$ oscillates around the asymptotic value $\pi/2$ with decreasing amplitudes. This shows that $y(\theta)$ oscillates around the function

$$-\bar{\tau}n \frac{\pi}{2} \theta = -\bar{\tau}n\pi\sqrt{x}$$

Since the function we want to obtain is $\sqrt{x}$, we have to choose $\bar{\tau}$ in such a way that the factor of $\sqrt{x}$ shall become 1. This yields

$$\tau = \frac{(-1)^{n+1}}{n\pi} \tag{7-14.66}$$

The error of the approximation $y_n(\theta)$ thus becomes

$$\eta_n(\theta) = \sqrt{x} - y_n(\theta) = \frac{-1}{n\pi}\left(\cos n\theta + n\theta\left[Si(n\theta) - \frac{\pi}{2}\right]\right)$$
$$\tag{7-14.67}$$

In order to obtain the points of maxima of the oscillatory error $\eta_n(\theta)$, we put the derivative equal to zero. This gives the condition

$$Si(n\theta) = \frac{\pi}{2}$$

But at these points

$$\eta_n(\theta) = -\frac{1}{n\pi}\cos n\theta$$

and thus we see that

$$\eta_n = \frac{1}{n\pi}$$

is an absolute error-bound for the entire interval. With increasing n this η_n decreases slowly to zero. But even $n = 6$ gives an accuracy of ± 0.053; that is, a certain polynomial of 6th order approximates $\sqrt{x}$ with a maximum error of ± 0.053 in the entire range between 0 and 1. By substituting in (56) this polynomial becomes

$$y_6(x) = \frac{1}{6\pi}\left[1 + 72x - \frac{840}{3}x^2 + \frac{3584}{5}x^3 - \frac{6912}{7}x^4 + \frac{6144}{9}x^5 - \frac{2048}{11}x^6\right]$$

$$= 0.053 + 3.820x - 14.854x^2 + 38.028x^3 - 52.385x^4$$

$$+ 36.217x^5 - 9.877x^6$$

Before we came to the discussion of the error, we had already a method for obtaining τ; we made use of the condition (57). Since the error at the point $x = 1$ is small, we can expect that the two τ determinations will differ but slightly. This gives the following approximate relation which holds even for small n with remarkable accuracy, and becomes exact for $n = \infty$.

$$\sum_{m=0}^{n} \frac{c_n^m}{2m-1} \doteq (-1)^{n+1} \pi n \qquad (7\text{-}14.68)$$

For example in the case $n = 6$ the number π thus determined becomes 3.1427, which means an accuracy of 1 : 1000. We will return to this problem once more in the next section.

Example 5. The following example is chosen to illustrate how the τ method can serve for the precision approximation of functions even in a complex range. The function

$$y = \text{arc tan } z \qquad (7\text{-}14.69)$$

can be defined by the differential equation

$$y' = 1 + z^2$$

We put

$$y = zu(z) \qquad (7\text{-}14.70)$$

This $u(z)$ is now an *even* function of z and can thus be conceived as a function of z^2. Hence we introduce the new variable

$$z^2 = \xi \qquad (7\text{-}14.71)$$

and obtain for $u(\xi)$ the differential equation

$$2\xi u' + u = \frac{1}{1+\xi} \qquad (7\text{-}14.72)$$

The equation we want to solve is thus

$$(1 + \xi)(2\xi u' + u) = 1 + \tau T_n^*(\xi) \qquad (7\text{-}14.73)$$

The infinite Taylor expansion, obtained by substituting for $u(\xi)$ a polynomial of infinite order (without any τ term) becomes

$$S(\xi) = 1 - \frac{\xi}{3} + \frac{\xi^2}{5} - \cdots + (-1)^k \frac{\xi^k}{2k+1} + \cdots \qquad (7\text{-}14.74)$$

We now derive the $Q_m(\xi)$ polynomials, omitting the first equation, in order to relieve overdetermination. Hence the $Q_m(\xi)$ start with $m = 1$, and we obtain

$$Q_m(\xi) = (-1)^{m-1} S_{m-1}(\xi) \qquad (7\text{-}14.75)$$

Consequently the solution of (73), apart from the equation of zeroth order, becomes

$$u(\xi) = \tau \sum_{m=1}^{n} (-1)^{m-1} c_n^m S_{m-1}(\xi) \qquad (7\text{-}14.76)$$

The omitted equation was

$$u(0) = 1 + \tau c_n^0 \qquad (7\text{-}14.77)$$

which yields for τ the condition

$$\tau = \frac{1}{\displaystyle\sum_{m=0}^{n} (-1)^{m-1} c_n^m} = \frac{1}{T_n^*(-1)} \qquad (7\text{-}14.78)$$

and thus

$$u_{n-1}(\xi) = \frac{\displaystyle\sum_{m=1}^{n} (-1)^m c_n^m S_{m-1}(\xi)}{T_n^*(-1)} \qquad (7\text{-}14.79)$$

The estimated maximum error is

$$\eta_{n-1} = |\tau| = \frac{1}{|T_n^*(-1)|} \qquad (7\text{-}14.80)$$

If we substitute for ξ the value 1, the Taylor series gives the celebrated Leibniz-series for $\pi/4$.

$$1 - \tfrac{1}{3} + \tfrac{1}{5} - \tfrac{1}{7} + \tfrac{1}{9} - \cdots$$

which is very slowly convergent. Weighting the partial sums by the Chebyshev coefficients according to (79) makes the series rapidly convergent. The first five convergents of $\pi/4$ become

Taylor:

$$1, \ \tfrac{2}{3} = 0.667, \ \tfrac{13}{15} = 0.867, \ \tfrac{76}{105} = 0.724, \ \tfrac{789}{945} = 0.835$$

τ method:

$$\frac{2 \cdot 1}{1+2} = \frac{2}{3} = 0.667$$

$$\frac{8 + 8 \cdot \frac{2}{3}}{1 + 8 + 8} = \frac{40}{51} = 0.7843$$

$$\frac{18 + 48 \cdot \frac{2}{3} + 32 \cdot \frac{13}{15}}{1 + 18 + 48 + 32} = \frac{1166}{1485} = 0.78518$$

$$\frac{32 + 160 \cdot \frac{2}{3} + 256 \cdot \frac{13}{15} + 128 \cdot \frac{76}{105}}{1 + 32 + 160 + 256 + 128} = \frac{47584}{60585} = 0.785409$$

$$\frac{50 + 400 \cdot \frac{2}{3} + 1120 \cdot \frac{13}{15} + 1280 \cdot \frac{76}{105} + 512 \cdot \frac{789}{945}}{1 + 50 + 400 + 1120 + 1280 + 512}$$

$$= \frac{2496018}{3178035} = 0.7853966 \quad (7\text{-}14.81)$$

$$\left(\frac{\pi}{4} = 0.785398163\right)$$

However, our aim is to extend our solution to arbitrary *complex* values of the variable ξ. For this purpose we make use of the "parametric method," discussed at the end of § 13. According to the formula (13.20) the only difference is that the coefficient c_n^m has to be replaced by $c_n^m \xi^{-m}$. Hence the solution (76) is now transformed into

$$u(\xi) = \tau \sum_{m=1}^{n} (-1)^{m-1} c_n^m \xi^{-m} S_{m-1}(\xi) \quad (7\text{-}14.82)$$

The omitted equation remains unchanged: $u(0) = 1 + \tau c_n^0$. This determines τ in the form

$$\tau = \frac{1}{\sum_{m=0}^{n} (-1)^{m-1} c_n^m \xi^{-m}} = \frac{-1}{T_n^*(-1/\xi)} \quad (7\text{-}14.83)$$

and the final solution becomes

$$u_{n-1}(\xi) = \frac{1}{T_n^*(-1/\xi)} \sum_{m=1}^{n} (-1)^m \frac{c_n^m}{\xi^m} S_{m-1}(\xi) \quad (7\text{-}14.84)$$

Going back to our original variable z, we finally obtain the following sequence of approximations of the function $y = \text{arc tan } x$.

$$y_{n-1} = \frac{z}{T_n^*(-1/z^2)} \sum_{m=1}^{n} (-1)^m \frac{c_n^m}{z^{2m}} S_{m-1}(z^2) \qquad (7\text{-}14.85)$$

This is no longer a simple polynomial approximation, but the *ratio of two polynomials*. For example, putting $n = 4$ we obtain the following approximation.

$$\text{arc tan } z = \frac{32z}{105} \frac{420 + 700z^2 + 329z^4 + 38z^6}{128 + 256z^2 + 160z^4 + 32z^6 + z^8} \qquad (7\text{-}14.86)$$

We demonstrate this approximation by applying it to the values $z = 2$ and $z = i/2$. In the first case we obtain

$$\text{arc tan } 2 = \frac{64}{105} \frac{10916}{6016} = 1.10598 \qquad (7\text{-}14.87)$$

while the correct value is 1.10715.

In the second case we obtain

$$\text{arc tan } \frac{i}{2} = \frac{16i}{105} \frac{67832}{18817} = 0.54930673i \qquad (7\text{-}14.88)$$

Now the significance of the arc tan z function for imaginary arguments is

$$\text{arc tan } ip = \frac{i}{2} \log \frac{1+p}{1-p} \qquad (7\text{-}14.89)$$

Hence

$$\text{arc tan } \frac{i}{2} = \frac{i}{2} \log 3 = 0.54930614i \qquad (7\text{-}14.90)$$

What is the convergence of the approximation (85) as n grows to infinity? The error is proportional to τ, and if τ converges to zero, the approximation converges unlimitedly to $f(z)$. Now in our problem

$$\tau = -\frac{1}{T_0^*(-1/z^2)} \qquad (7\text{-}14.91)$$

The Chebyshev polynomials $T_n^*(x)$ have the property that they grow to infinity as n converges to infinity, at any point x which is outside the interval [0, 1]. Inside that interval they continue to be bounded by ± 1. Hence the convergence will hold at any point z which avoids the condition

$$0 \leq -\frac{1}{z^2} \leq 1 \tag{7-14.92}$$

This shows that the convergence holds everywhere in the complex plane, except along the imaginary axis. But even along the imaginary axis only the points beyond $\pm i$ are excluded. At all points between $-i$ and $+i$ the convergence is still preserved.

This is exactly the convergence behavior which can be predicted on theoretical grounds. The complex ray which connects the points 0 and z must not contain any singular points. Hence a singular point will cast a shadow behind it which reaches out to infinity. This shadow is constructed by continuing the straight line which connects the singular point with the origin. The singular points of the arc tan z function are at $z = \pm i$. Hence the solution must become divergent at all points of the imaginary axis which lie beyond these points.

15. Estimation of the error of the τ method. In the numerical examples of the previous section the fast convergence of the τ expansions was demonstrated in a numerical way. It must be possible, however, to obtain theoretical estimates for the inherent error of these approximations. The relation of the Chebyshev polynomials to the trigonometric functions puts us in the position to develop a simple algebraic method for estimation of the error of the solution of a linear differential equation, obtained by the application of the τ method.

We consider a differential equation of the form

$$A(x)y' + B(x)y + C(x) = \tau T_n^*(x) \tag{7-15.1}$$

While the form of this equation seems very special, it actually covers a wide class of problems. Moreover, in the case of a differential equation of second or higher order it is advisable to introduce surplus functions and transform the given differential equation to a simultaneous system of equations of the form (1). Hence the method we are going to discuss actually has applicability to a larger group of problems which can be reduced to two or more equations of the type (1) and solved by the τ method. For our present purposes it will suffice to consider only one single equation of the form (1).

The right side of (1) has its origin in the fact that we wanted to solve the given differential equation by a finite expansion. Generally we may need more than one τ term on the right side, but we can estimate the error for each τ term separately and then take the absolute sum of these errors. For this reason it will suffice to carry through our investigation for one τ term only.

The exact solution of the given differential equation has no error term on the right side.

$$A(x)y' + B(x)y + C(x) = 0 \tag{7-15.2}$$

The difference between the correct solution $y(x)$ and the approximation $y_n(x)$ represents the error of the solution

$$\eta_n(x) = y(x) - y_n(x) \tag{7-15.3}$$

characterized by the differential equation

$$A(x)\eta_n'(x) + B(x)\eta_n(x) = -\tau T_n^*(x) \tag{7-15.4}$$

It will be our aim to find an *approximate* solution of this differential equation and thus obtain a close estimate of $\eta_n(x)$.

We introduce the angle variable θ as a new variable, replacing x by

$$x = \cos^2 \frac{\theta}{2}, \qquad dx = -\frac{1}{2} \sin \theta \, d\theta \tag{7-15.5}$$

Then we obtain in the new variable a differential equation of the form

$$A_1\eta' + B_1\eta = -\tau \cos n\theta$$

where

$$A_1(\theta) = -\frac{2A \cos^2 (\theta/2)}{\sin \theta} \tag{7-15.6}$$

$$B_1(\theta) = B\left(\cos^2 \frac{\theta}{2}\right) \tag{7-15.7}$$

The right side is replaceable by $-\tau e^{in\theta}$ with the understanding that we are going to use the *real* part of the solution only.

Now the equation

$$A_1\eta' + B_1\eta = -\tau e^{in\theta} \tag{7-15.8}$$

assuming that A_1 and B_1 are constants, can be conceived as the differential equation of a ballistic galvanometer, put into forced vibrations by an external periodic force. The solution of (8) is

$$\eta(\theta) = \frac{-\tau}{inA_1 + B_1} e^{in\theta} \tag{7-15.9}$$

It means that the galvanometer follows the external force, with changed amplitude and phase. In our case A_1 and B_1 are not constants but given functions of θ. Nevertheless, compared with the rapidly changing exponential function $e^{in\theta}$ they still behave *nearly* like constants. Hence we will not go far wrong and obtain a solution for $\eta_n(\theta)$, at least for sufficiently large n, if we utilize the solution (9) for our purposes, although we realize that this solution has now only approximate value. For the exact treatment we should put

$$\eta = -G(\theta)\tau e^{in\theta} \tag{7-15.10}$$

and obtain for the changeable amplitude $G(\theta)$ the differential equation

$$A_1(G' + inG) + B_1G = 1 \tag{7-15.11}$$

However, for sufficiently large n we may consider G' negligible compared with inG. We then return to solution (9).

We see that the error $\eta_n(\theta)$ of our solution will be *periodic* in character. If (9) is written in polar form,

$$\eta_n(\theta) = \frac{-\tau}{\sqrt{B_1^2 + n^2A_1^2}} e^{i(n\theta - \phi)} \tag{7-15.12}$$

with

$$\tan \phi = \frac{B_1}{nA_1} \tag{7-15.13}$$

we obtain, by taking the real part of (12),

$$\eta_n(\theta) = -\frac{\tau}{\sqrt{B_1^2 + n^2A_1^2}} \cos(n\theta - \phi) \tag{7-15.14}$$

This is a harmonic vibration, with variable phase and amplitude.

From the expression (14) several conclusions can be drawn. In the first place we want to know what will be the *maximum error* we may encounter at *any* point of the interval. For this purpose we investigate the quantity

$$B^2(x) + \frac{n^2 A^2(x)}{x(1-x)} = \phi^2(x) \tag{7-15.15}$$

and find its minimum within the interval [0, 1]. Let this minimum be $\phi_{\min}$. Then the maximum possible error $\eta_{\max}$ is estimated by

$$\eta^0 = |\eta_{\max}| \leq \frac{\tau}{\phi_{\min}} \tag{7-15.16}$$

Furthermore, the "method of forced vibration," expressed in (9), will give a sufficiently close estimation of the residual $\eta(\theta)$ at any given point $x = \cos^2(\theta/2)$. We come into difficulty, however, at the two end points $x = 0$ and $x = 1$ of the range. Here the denominator of $A_1(\theta)$ [cf. (6)] becomes zero, and thus $A_1(\theta)$ becomes infinite. The estimated error (9) will thus become zero. This indicates that the error at the points $x = 0$ and $x = 1$ is particularly small. But exactly at the end point $x = 1$ the error is of particular interest because in the parametric method the variable point z becomes the end point of the range.

In many problems the difficulty exists only at the point $x = 1$ since at the point $x = 0$ the function $A(x)$ may become zero, because of the presence of the factor x. Then the error estimate (8) does not fail at the lower limit and requires readjustment only at the upper limit $x = 1$.

We will now avoid the denominator of (6) by multiplying through by $\sin \theta$. Then the right side becomes

$$-\tau \cos n\theta \sin \theta = -\frac{\tau}{2} [\sin (n+1)\theta - \sin (n-1)\theta]$$

and the same method which led to (9) now gives

$$\eta_n = \frac{-\tau}{-2i(n+1)A + B \sin \theta} \frac{e^{i(n+1)\theta}}{2i} \tag{7-15.17}$$
$$+ \frac{\tau}{-2i(n-1)A + B \sin \theta} \frac{e^{i(n-1)\theta}}{2i}$$

At the point $x = 1$, i.e., $\theta = 0$, we obtain

$$\eta_n(1) = \frac{\tau}{2} \left(-\frac{1}{2(n+1)A(1)} + \frac{1}{2(n-1)A(1)} \right) \quad (7\text{-}15.18)$$

$$= \frac{\tau}{2(n^2 - 1)A(1)}$$

This formula shows that while the general order of magnitude of the error is τ/n, at the end point of the range the order of magnitude drops down to τ/n^2.

In the light of these results we will once more examine the examples of the previous section.

Example 1. Here

$$A(x) = -1, \qquad B(x) = 1$$

The upper bound (15.16) for the error at any point of the range becomes

$$\eta_n^0 = \frac{|\tau_n|}{\sqrt{4n^2 + 1}}$$

while the estimated error at the end point $x = 1$ becomes

$$\eta_n(1) = -\frac{\tau_n}{2(n^2 - 1)} \quad (7\text{-}15.19)$$

Now the τ_n of this problem are all positive, and thus it seems that we cannot explain the fact that the e values given in (14.8) approach the limit alternately from above and below. Formula (19) indicates that the error should remain permanently negative. The apparent discrepancy has the following reason. The error estimate (17) gives a definite error at the point $x = 0$, i.e., $\theta = \pi$. This error is $(-1)^{n+1}$ times the error at $x = 1$.

$$\eta_n(0) = (-1)^{n+1}\eta_n(1) \quad (7\text{-}15.20)$$

But in the solution (14.6) we have satisfied the boundary condition (14.3) *exactly*, and this means that we have no error at $x = 0$. For this reason we have to correct our estimated periodic error by a "systematic error," caused by the fact that at the lower boundary we did not make allowance for the natural error which should exist at that point.

In solving the differential equation (4) we did not take into account the solution of the homogeneous differential equation which can always be added, multiplied by an arbitrary constant C. Since in our problem the homogeneous differential equation characterizes the function to be obtained, i.e., e^x, addition of this solution will change $\eta_n(1)$ by an additional term Ce, while $\eta_n(0)$ will change by C. The condition $\eta_n(0) = 0$ determines C to

$$C = (-1)^n \eta_n(1)$$

and thus the resultant error at the point $x = 1$ becomes

$$\eta_n(1) = -\frac{\tau_n}{2(n^2 - 1)} [1 + (-1)^n e] \qquad (7\text{-}15.21)$$

Since e is larger than 1, the sign of $\eta_n(1)$ will alternate, and the error will become *negative* for even n and *positive* for odd n, in accordance with the facts. The estimate (21) is very satisfactory from $n = 3$ on.

Example 2. This example is purely algebraic. The differential operator is missing, with the consequence that the equation (4) becomes exactly solvable. Instead of an *estimate* of the error $\eta_n(x)$ we have an exact algebraic *representation* of the error. The comparison with the error of the Taylor expansion has been discussed before and needs no further elaboration.

Example 3. Here

$$A(x) = x^2, \qquad B(x) = 1 + x$$

The error becomes the largest at $x = 0$; the estimated maximum error is thus

$$\eta_n^0 = |\tau_n|$$

while at the end point $x = 1$ the following estimate holds.

$$\eta_n(1) = \frac{\tau_n}{2(n^2 - 1)} \qquad (7\text{-}15.22)$$

The expression (14.47) for τ_n shows that τ_n is negative for even n and positive for odd n. If we apply (22) to the successive approximations (14.50) of $y(1)$—remembering that τ_n is associated with $y_{n-1}(x)$—we find that the estimate (22) becomes effective from $n = 5$ on. The alternate approach from below and above is not disturbed from that point on.

Example 4. In this example

$$A(x) = 2x, \qquad B(x) = -1$$

and we obtain the estimated upper bound of the error at any point of the range.

$$\eta_n^0 = |\tau_n|$$

and the error estimate at the end point.

$$\eta_n(1) = \frac{\tau_n}{4(n^2 - 1)} \qquad (7\text{-}15.23)$$

The error at the point $x = 1$ is so small compared with the much larger error at $x = 0$ that we lose very little in accuracy if we put $\eta_n(1) = 0$ and determine τ_n from the boundary condition $y(1) = 1$ as we have done in (14.57). However, it is of interest to demonstrate the accuracy of the estimate (23) by taking it into account and obtaining τ from the condition

$$\tau_n \sum_{m=0}^{n} \frac{c_n^m}{2m - 1} = 1 - \frac{\tau_n}{4(n^2 - 1)} \qquad (7\text{-}15.24)$$

If we do so, the quantity $1/(n\tau_n)$ becomes

$$\frac{1}{n\tau_n} = \frac{1}{4n(n^2 - 1)} + \frac{1}{n} \sum_{m=0}^{n} \frac{c_n^m}{2m - 1} \qquad (7\text{-}15.25)$$

According to the error discussion in § 14, this quantity should be theoretically equal to $(-1)^{n+1}\pi$. Earlier we have obtained for the case $n = 6$ (cf. 14.68) the value -3.1427, disregarding the first term on the right side of (25). But if this term is taken into account, we obtain the much closer value -3.141522. The accuracy has increased from $1 : 2800$ to $1 : 45000$.

Example 5. In this example the parametric method was applied, obtaining $y = f(z^2)$ in such fashion that z^2 is considered a mere constant of the given differential equation. Here

$$A(x) = (1 + z^2x)2x, \qquad B(x) = 1 + z^2x$$

The over-all accuracy within the range $[0, 1]$ is here of no importance, since we are going to use the solution solely at the end point $x = 1$, which corresponds in the original variable to the point $x = z$. Hence we get

$$\eta_n(z) = \frac{\tau_n(z)}{4(n^2 - 1)(1 + z^2)}$$

This holds for the function $u(z)$. For the arc tan z function we obtain, according to (14.70),

$$\eta_n(z) = \frac{z\tau_n(z)}{4(n^2 - 1)(1 + z^2)} \qquad (7\text{-}15.26)$$

$$= \frac{-z}{4(n^2 - 1)(1 + z^2)} \frac{1}{T_n^*(-1/z^2)}$$

Let us apply this error estimate to the case $z = 1$, i.e., to the $\pi/4$ values of the table (14.81). The quantity $T_n(-1)$ alternates in sign being positive for even n and negative for odd n. Hence the successive convergents of $\pi/4$ should approach the correct value from below for odd n and from above for even n. We find this verified from $n = 3$ on. On the other hand, the error of (14.87) is not well estimated by (26). If we put $z = 2$, the function $A(x)$ is no longer sufficiently smooth to make the formula (26) applicable for $n = 4$. We would have to go to somewhat larger n for this purpose. Quite different, however, is the case of the substitution $z = i/2$. The surprisingly great accuracy of the approximation (14.88) is explained by the large value of $T_4^*(4)$.

$$T_4^*(4) = 18871$$

Here the estimate (26) gives

$$\eta_4(z) = -\frac{i}{90 \cdot 18817} = -0.59 \cdot 10^{-6} i$$

in close agreement with the actual error.

16. The square root of a complex number. The square root of a complex number is obtainable in the regular algebraic fashion, but the numerical procedure involved is quite laborious. The following method gives the answer much more rapidly.

Let the complex number whose square root is desired be denoted by $A + Bi$. If A is larger in magnitude than B, we put

$$\sqrt{A + Bi} = \sqrt{A}\,\sqrt{1 + Bi/A} \qquad (7\text{-}16.1)$$

If, on the other hand, B is larger in magnitude than A, we put

$$\sqrt{A + Bi} = \sqrt{B}\,\sqrt{A + i} \qquad (7\text{-}16.2)$$

Hence our problem is reducible to the valuation of one of the following functions:

$$y = \sqrt{1 + ix} \qquad (7\text{-}16.3)$$

or

$$y = \sqrt{i + x} \qquad (7\text{-}16.4)$$

where x is limited to the interval $[0, 1]$. By taking out $\sqrt{i}$ as a factor, (4) is reducible to (3). We will agree that x is to vary only between 0 and 1. Hence $\sqrt{1 - ix}$ is not to be obtained by letting x become negative, but by changing i to $-i$.

Our aim will be to obtain a quickly convergent expansion for the function (3) with the help of the τ method. For this purpose we characterize the function (3) by the differential equation

$$\frac{y'}{y} = \frac{i}{2(1 + ix)} \qquad (7\text{-}16.5)$$

We can likewise say, however, that we solve the differential equation

$$\frac{y'}{y} = \frac{1}{2(1 + x)}$$

or

$$2(1 + x)y' - y = 0 \qquad (7\text{-}16.6)$$

along the imaginary axis.

According to the standard procedure we put a τ term on the right side of the differential equation (6). The polynomial $T_n^*(x)$ becomes automatically adjusted to the imaginary axis if we write it in the form $T_n^*(x/i)$. It so happens, however, that in our problem the operator y' is dominant. Hence we obtain better results if we do not use the T_n^* polynomials directly, but their derivative, formulating our problem as follows.

$$2(1 + x)y' - y = \tau T'_{n+1}\left(\frac{x}{i}\right) \qquad (7\text{-}16.7)$$

The Taylor expansion of our problem gives the ordinary binomial expansion with $n = \frac{1}{2}$.

$$S(x) = 1 + \frac{x}{2} - \frac{x^2}{8} + \frac{x^3}{16} - \cdots \qquad (7\text{-}16.8)$$

and the method of the Q polynomials shows that the partial sums of this expansion have to be weighted as follows.

$$y_n(x) = \tau \sum_{m=0}^{n} \frac{c_n^m}{a_m(2m - 1)} S_m(x) \qquad (7\text{-}16.9)$$

Here c_n^m are the coefficients of the polynomial which appears on the right side of (7). Moreover, a_m is the last coefficient of the partial sum $S_m(x)$. Finally, since the variable x is supposed to move along the imaginary axis, x should be replaced by x/i.

As an example, let us choose $n = 2$. Then

$$T_3^*(x) = -1 + 18x - 48x^2 + 32x^3$$

$$T_3^{*'}(x) = 18 - 96x + 96x^2$$

and $$T_3^{*'}(x/i) = 18 + 96ix - 96x^2$$

Moreover: $$a_0 = 1, \qquad a_1 = \frac{1}{2}, \qquad a_2 = -\frac{1}{8}$$

Hence

$$y_2(x) = \tau\left[-18 + 2 \cdot 96i\left(1 + \frac{ix}{2}\right) + \frac{8 \cdot 96}{3}\left(1 + \frac{ix}{2} + \frac{x^2}{8}\right)\right]$$

$$= \tau[238 + 192i + (-96 + 128i)x + 32x^2]$$

The factor τ can be determined from the boundary condition

$$y(0) = 1$$

which gives

$$\tau = \frac{1}{238 + 192i}$$

and the final result becomes

$$\sqrt{1 + ix} = 1 + 0.0184x + 0.0810x^2 + i(0.5201x - 0.0653x^2)$$
$$(7\text{-}16.10)$$

$$= 1 + x(0.0184 + 0.0810x) \quad (\pm 0.002)$$
$$+ ix(0.5201 - 0.0653x) \quad (\pm 0.002i)$$

The accuracy of this quadratic approximation is remarkably high, since the maximum error at any point of the range [0, 1] does not surpass 2 units in the 3rd decimal. This accuracy will suffice for many problems.

If we go one step further and deduce the corresponding formula for the case $n = 3$, we obtain the following cubic approximation, which is already accurate to 2 units of the *fourth* decimal place (in both the real and the imaginary parts).

$$\sqrt{1 + ix} = 1 - 0.00316x + 0.14237x^2 - 0.04079x^3$$
$$+ i(0.50637x - 0.03108x^2 - 0.02020x^3) \quad (7\text{-}16.11)$$

The quadratic approximation for the function $\sqrt{i + x}$ becomes

$$\sqrt{i + x} = 0.7071 + x(0.3807 + 0.0111x)$$
$$+ i[0.7071 - x(0.3548 - 0.1035x)] \quad (7\text{-}16.12)$$

For the sake of convenience we add two further formulas, in order to be prepared for any combination of signs in A and B.

$$\sqrt{-1 + ix} = x(0.5201 - 0.0653x) \qquad (7\text{-}16.13)$$
$$+ i[1 + x(0.0184 + 0.0810x)]$$

$$\sqrt{i - x} = 0.7071 - x(0.3548 - 0.1053x) \qquad (7\text{-}16.14)$$
$$+ i[0.7071 + x(0.3807 + 0.0111x)]$$

As an example, let us evaluate $\sqrt{-3 - 8i}$.

$$\sqrt{-3 - 8i} = \sqrt{8}\,\sqrt{-i - 0.375}$$

Here the formula (14) comes into operation, with $x = 0.375$ and changing i into $-i$.

$$\sqrt{-3 - 8i} = 2.8284[0.7071 - 0.375 \cdot 0.3150$$
$$- i(0.7071 + 0.375 \cdot 0.3849)]$$
$$= 1.6659 - 2.4081i$$

The correct result is

$$\sqrt{-3 - 8i} = 1.66493 - 2.40250i$$

The error is not more than 0.3%.

17. Generalization of the τ method. The method of selected points.
The τ method is directly applicable only if the given linear differential
equation has coefficients which are polynomials of x. For more
general equations we will not succeed with the construction of a
proper error term on the right side of the equation which would
allow a solution in form of a finite power series of strong convergence.
However, we can reformulate the significance of the τ method in a
way which makes it applicable to a much wider class of problems.
Let us return once more to the general idea of the τ method. We
made the linear differential equation

$$Dy = 0 \qquad (7\text{-}17.1)$$

solvable by a finite power expansion through the device that we have
put a properly chosen error term on the right side. Sometimes a term
of the form $\tau T_n^*(x)$ sufficed, but more generally we had to put

$$Dy = T_n^*(x)(\tau_1 + \tau_2 x + \cdots + \tau_m x^{m-1}) \qquad (7\text{-}17.2)$$

By this device we have replaced $y(x)$ by a certain

$$y_p(x) = b_0 + b_1 x + \cdots + b_p x^p \qquad (7\text{-}17.3)$$

which closely approximated $y(x)$ in the range $[0, 1]$. This procedure
is limited, however, to linear differential equations with polynomial
coefficients.

We will now approach the problem from an entirely different
viewpoint. Again we assume that $y(x)$ shall be approximated by a
finite polynomial of the order p. This puts at our disposal the $p + 1$
constants

$$b_0, b_1, \cdots, b_p \qquad (7\text{-}17.4)$$

which are uniquely determined by $p + 1$ independent equations.
This shows immediately that we cannot hope to satisfy the given
differential equation in any *continuous* range of the variable x. If we
satisfy the given differential equation and the given boundary

conditions in only $p + 1$ points, this is generally enough for unique determination of the constants b_i, and any further conditions would lead to overdetermination.

The question is now reduced to the proper selection of the $p + 1$ points in which the given equation is to be satisfied. If initial values or boundary conditions are given, some of the b_i coefficients are already absorbed by these conditions, and the number of free parameters is no longer $p + 1$, but $p + 1 - \nu$, where ν is the number of the given boundary conditions. Our problem is thus reduced to the proper choice of $n = p + 1 - \nu$ points. One possible choice would be to put all these n points exceedingly close to the origin, i.e., to satisfy the given equation at the x values

$$x = 0,\ \varepsilon,\ 2\varepsilon,\ \cdots,\ (n - 1)\varepsilon$$

where ε converges to zero. The expansion thus obtained corresponds to the Taylor expansion of $y = f(x)$, truncated to $p + 1$ terms.

Quite a different distribution of points is suggested by equation (2). The right side of the differential equation (2) is zero at the zeros of the polynomial $T_n^*(x)$. These are the points

$$x_k = \frac{1 + \cos\,[(2k - 1)\pi/2n]}{2} \quad (k = 1, 2, \cdots, n) \quad (7\text{-}17.5)$$

If it is advisable to put $T_{n+1}^{*\prime}(x)$ rather than $T_n^*(x)$ on the right side, the zeros of the "Chebyshev polynomials of the second kind" come into operation, characterized by the conditions

$$x_k = \frac{1 + \cos\,[k\pi/(n + 1)]}{2} \quad (k = 1, 2, \cdots, n) \quad (7\text{-}17.6)$$

These points have a simple geometrical significance. We erect a semicircle with the center $x = \frac{1}{2}$, and the radius $\frac{1}{2}$ above the [0,1] line as diameter, divide it into $n + 1$ equal parts and project these points down on the base line. Excluding the two end points[1] we thus obtain n unequally spaced points which are spread over the entire interval instead of being crowded around the point $x = 0$. With this distribution of the zeros, we will get a polynomial approximation whose

---

[1] We do not lose much in accuracy if we *include* the two end points, replacing $n + 1$ by $n - 1$ in the formula (17.6). The computational scheme is frequently simplified by inclusion of the two end points.

error oscillates practically evenly over the given interval instead of yielding a very small error in the neighborhood of $x = 0$ but a large error in the neighborhood of $x = 1$.

We now see that the inhomogeneous differential equation (2) can be solved in two different ways. One way is to solve the recurrence relations for the coefficients and thus determine the b_i coefficients and the τ_i. The other is to omit the τ_i altogether, perform the differential operator D on the finite expansion (3) and put the resulting expression equal to zero at all those points at which the right-hand side of the differential equation is zero. This, together with the boundary conditions, yields a linear system of algebraic equations for determination of the b_i. The two methods give identical results, but the second method has more universal significance. It is applicable to linear differential equations whose coefficients are not necessarily algebraic in x. It is likewise applicable to integral equations or mixed differential-integral equations as long as they are linear. The general procedure is always the same. We replace $y(x)$ by a finite polynomial expansion, perform the operations indicated by the given differential or integral operator on this expansion, and equate the result of the operation to zero, not for arbitrary values of x, which is impossible, but for a carefully selected discrete set of x values, which will guarantee a satisfactorily even distribution of the error over the given interval. What we obtain is a simultaneous system of ordinary linear algebraic equations for the b_k from which these b_k can be uniquely obtained. If the order p of the approximating polynomial is changed, the entire procedure has to be repeated and an entirely new set of p values obtained.

This method of solving differential or integral equations by well-convergent power expansions is a generalization of the method of "trigonometric interpolation," discussed before in IV, 16. If the functional values of $y = f(x)$ are known, a close polynomial approximation of $f(x)$ can be obtained by fitting the functional values at the points (5) or (6). This amounts to an equidistant trigonometric interpolation of a function of θ with the help of the Fourier cosine functions $\cos k\theta$. If, however, the functional values of $f(x)$ are not known but we possess the basic law which determines $f(x)$ with the help of a linear operator, we can again obtain a well-convergent polynomial approximation of $f(x)$ by interpolating the *operator* instead of the functional values. The procedure is in both cases similar, the only practical difference being that in the latter case we do not have the orthogonality of the resulting set of linear equations which would make them explicitly solvable. Since, however, in many practical applications the order of the approximating polynomial need not go beyond 4 or 5, the linear sets involved are of low order and can be solved by successive eliminations. We thus obtain an elegant numerical procedure which generalizes the benefits of the τ method to a much wider class of problems.[1]

[1] For a practical example of this method of selected points cf. [5], p. 195.

Bibliographical References

[1] BROMWICH, TH. J. I'A, *Introduction to the Theory of Infinite Series* (Macmillan, London, 1926).

[2] JAHNKE, E., and EMDE, F., *Tables of Functions with Formulae and Curves* (Dover Publications, New York, 1945).

[3] SZEGÖ, G., "Orthogonal Polynomials," *Am. Math. Soc. Colloq. Pub.*, 23, 1939.

[4] *Tables of the Chebyshev Polynomials $S_n(x)$ and $C_n(x)$, AMS 9*, National Bureau of Standards (1952).

Articles

[5] LANCZOS, C., "Trigonometric Interpolation of Empirical and Analytical Functions," *J. Math. Phys.*, 17, 123 (1938).

[6] MILLER, J. C. P., "Two numerical applications of Chebyshev polynomials," *Roy. Soc. Edin. Proc.* 62, 204 (1946).

Appendix

NUMERICAL TABLES

Table I

Amplitude of Complex Roots (see p. 36)

r	A(r)	r	A(r)	r	A(r)	r	A(r)
0.80	3.3588	0.90	4.7774	1.00	7.0669	1.10	10.8412
81	4732	91	9594	01	3644	11	11.3353
82	5928	92	5.1505	02	6773	12	8553
83	7179	93	3510	03	8.0062	13	12.4028
84	8490	94	5615	04	3522	14	9790
85	9862	95	7825	05	7161	15	13.5857
86	4.1299	96	6.0146	06	9.0990	16	14.2243
87	2805	97	2585	07	5019	17	8967
88	4383	98	5147	08	9257	18	15.6046
89	6038	99	7839	09	10.3718	19	16.3499
						20	17.1345

Table II

Transformation Matrices B_n (see p. 38)

(Negative numbers underlined)

$$B_1 \quad \begin{bmatrix} \underline{1} & 1 \\ 1 & 1 \end{bmatrix}$$

$$B_2 \quad \begin{bmatrix} 1 & \underline{1} & 1 \\ \underline{2} & 0 & 2 \\ 1 & 1 & 1 \end{bmatrix}$$

$$B_3 \quad \begin{bmatrix} \underline{1} & 1 & 1 & 1 \\ 3 & \underline{1} & \underline{1} & 3 \\ \underline{3} & \underline{1} & 1 & 3 \\ 1 & 1 & 1 & 1 \end{bmatrix}$$

$$B_4 \quad \begin{bmatrix} 1 & \underline{1} & 1 & \underline{1} & 1 \\ \underline{4} & 2 & 0 & \underline{2} & 4 \\ 6 & 0 & \underline{2} & 0 & 6 \\ \underline{4} & \underline{2} & 0 & 2 & 4 \\ 1 & 1 & 1 & 1 & 1 \end{bmatrix}$$

Table II (Cont'd)

B_5

1	1	1	1	1	1
5	3	1	1	3	5
10	2	2	2	2	10
10	2	2	2	2	10
5	3	1	1	3	5
1	1	1	1	1	1

B_6

1	1	1	1	1	1	1
6	4	2	0	2	4	6
15	5	1	3	1	5	15
20	0	4	0	4	0	20
15	5	1	3	1	5	15
6	4	2	0	2	4	6
1	1	1	1	1	1	1

B_7

1	1	1	1	1	1	1	1
7	5	3	1	1	3	5	7
21	9	1	3	3	1	9	21
35	5	5	3	3	5	5	35
35	5	5	3	3	5	5	35
21	9	1	3	3	1	9	21
7	5	3	1	1	3	5	7
1	1	1	1	1	1	1	1

B_8

1	1	1	1	1	1	1	1	1
8	6	4	2	0	2	4	6	8
28	14	4	2	4	2	4	14	28
56	14	4	6	0	6	4	14	56
70	0	10	0	6	0	10	0	70
56	14	4	6	0	6	4	14	56
28	14	4	2	4	2	4	14	28
8	6	4	2	0	2	4	6	8
1	1	1	1	1	1	1	1	1

B_9

1	1	1	1	1	1	1	1	1	1
9	7	5	3	1	1	3	5	7	9
36	20	8	0	4	4	0	8	20	36
84	28	0	8	4	4	8	0	28	84
126	14	14	6	6	6	6	14	14	126
126	14	14	6	6	6	6	14	14	126
84	28	0	8	4	4	8	0	28	84
36	20	8	0	4	4	0	8	20	36
9	7	5	3	1	1	3	5	7	9
1	1	1	1	1	1	1	1	1	1

Table II (Cont'd)

$$B_{10}$$

$$
\begin{bmatrix}
1 & 1 & 1 & 1 & 1 & 1 & 1 & 1 & 1 & 1 & 1 \\
10 & 8 & 6 & 4 & 2 & 0 & 2 & 4 & 6 & 8 & 10 \\
45 & 27 & 13 & 3 & 3 & 5 & 3 & 3 & 13 & 27 & 45 \\
120 & 48 & 8 & 8 & 8 & 0 & 8 & 8 & 8 & 48 & 120 \\
210 & 42 & 14 & 14 & 2 & 10 & 2 & 14 & 14 & 42 & 210 \\
252 & 0 & 28 & 0 & 12 & 0 & 12 & 0 & 28 & 0 & 252 \\
210 & 42 & 14 & 14 & 2 & 10 & 2 & 14 & 14 & 42 & 210 \\
120 & 48 & 8 & 8 & 8 & 0 & 8 & 8 & 8 & 48 & 120 \\
45 & 27 & 13 & 3 & 3 & 5 & 3 & 3 & 13 & 27 & 45 \\
10 & 8 & 6 & 4 & 2 & 0 & 2 & 4 & 6 & 8 & 10 \\
1 & 1 & 1 & 1 & 1 & 1 & 1 & 1 & 1 & 1 & 1
\end{bmatrix}
$$

$$B_{11}$$

$$
\begin{bmatrix}
1 & 1 & 1 & 1 & 1 & 1 & 1 & 1 & 1 & 1 & 1 & 1 \\
11 & 9 & 7 & 5 & 3 & 1 & 1 & 3 & 5 & 7 & 9 & 11 \\
55 & 35 & 19 & 7 & 1 & 5 & 5 & 1 & 7 & 19 & 35 & 55 \\
165 & 75 & 21 & 5 & 11 & 5 & 5 & 11 & 5 & 21 & 75 & 165 \\
330 & 90 & 6 & 22 & 6 & 10 & 10 & 6 & 22 & 6 & 90 & 330 \\
462 & 42 & 42 & 14 & 14 & 10 & 10 & 14 & 14 & 42 & 42 & 462 \\
462 & 42 & 42 & 14 & 14 & 10 & 10 & 14 & 14 & 42 & 42 & 462 \\
330 & 90 & 6 & 22 & 6 & 10 & 10 & 6 & 22 & 6 & 90 & 330 \\
165 & 75 & 21 & 5 & 11 & 5 & 5 & 11 & 5 & 21 & 75 & 165 \\
55 & 35 & 19 & 7 & 1 & 5 & 5 & 1 & 7 & 19 & 35 & 55 \\
11 & 9 & 7 & 5 & 3 & 1 & 1 & 3 & 5 & 7 & 9 & 11 \\
1 & 1 & 1 & 1 & 1 & 1 & 1 & 1 & 1 & 1 & 1 & 1
\end{bmatrix}
$$

Table II (Cont'd)

$$B_{12}$$

1	1	1	1	1	1	1	1	1	1	1	1	1
12	10	8	6	4	2	0	2	4	6	8	10	12
66	44	26	12	2	4	6	4	2	12	26	44	66
220	110	40	2	12	10	0	10	12	2	40	110	220
495	165	15	27	17	5	15	5	17	27	15	165	495
792	132	48	36	8	20	0	20	8	36	48	132	792
924	0	84	0	28	0	20	0	28	0	84	0	924
792	132	48	36	8	20	0	20	8	36	48	132	792
495	165	15	27	17	5	15	5	17	27	15	165	495
220	110	40	2	12	10	0	10	12	2	40	110	220
66	44	26	12	2	4	6	4	2	12	26	44	66
12	10	8	6	4	2	0	2	4	6	8	10	12
1	1	1	1	1	1	1	1	1	1	1	1	1

Table III

Inversion of Eigenvalues (see p. 203). Smallest value of v: $v_0 = \sqrt{3}/\pi = 0.5513288$. For very small u: $u = u_0^* \sqrt{v - v_0}$ where $u_0^* = \sqrt{5\sqrt{3}/\pi} = 1.660314572$. The tabular between $v = 0.56$ and $v = 1$ has to be multiplied by $\sqrt{v - v_0}$. Hence $u^2 = u^{*2}(v - 0.551$. Beyond $v = 1$ the table gives the value of u directly.

v	u^*	v	u^*	v	u^*	v	u^*
0.56	1.6548	0.67	1.5945	0.78	1.5489	0.89	1.5163
57	6485	68	5898	79	5457	90	5138
58	6425	69	5853	80	5427	91	5113
59	6366	70	5810	81	5395	92	5090
60	6309	71	5767	82	5363	93	5067
61	6253	72	5724	83	5331	94	5045
62	6198	73	5684	84	5301	95	5024
63	6145	74	5644	85	5272	96	5004
64	6093	75	5606	86	5243	97	4984
65	6043	76	5568	87	5216	98	4965
66	5993	77	5527	88	5189	99	4947
						1.00	4929

v	u	v	u	v	u	v	u
1.00	1.0000	2.00	2.0000	3.00	3.0000	4.00	4.0000
02	0199	02	0200	02	0200	02	0200
04	0393	04	0397	04	0398	04	0398
06	0584	06	0592	06	0595	06	0596
08	0772	08	0785	08	0790	08	0793

Table III (Cont'd)

v	u	v	u	v	u	v	u
1.10	1.0958	2.10	2.0978	3.10	3.0985	4.10	4.0989
12	1144	12	1169	12	1180	12	1184
14	1336	14	1359	14	1372	14	1379
16	1508	16	1548	16	1564	16	1573
18	1688	18	1737	18	1757	18	1767
20	1867	20	1925	20	1948	20	1960
22	2046	22	2113	22	2139	22	2154
24	2224	24	2300	24	2330	24	2346
26	2405	26	2488	26	2521	26	2539
28	2583	28	2676	28	2712	28	2732
30	2763	30	2863	30	2904	30	2926
32	2944	32	3052	32	3096	32	3119
34	3125	34	3241	34	3287	34	3313
36	3308	36	3431	36	3480	36	3507
38	3492	38	3622	38	3674	38	3702
40	3680	40	3813	40	3868	40	3898
42	3868	42	4007	42	4064	42	4095
44	4059	44	4201	44	4260	44	4292
46	4252	46	4399	46	4456	46	4489
48	4448	48	4595	48	4655	48	4688
50	4647	50	4794	50	4854	50	4887
52	4848	52	4995	52	5055	52	5087
54	5053	54	5198	54	5257	54	5289
56	5260	56	5402	56	5460	56	5491
58	5470	58	5607	58	5663	58	5694
60	5683	60	5814	60	5869	60	5898
62	5898	62	6023	62	6075	62	6103
64	6116	64	6233	64	6282	64	6308
66	6336	66	6444	66	6489	66	6514
68	6557	68	6656	68	6697	68	6721
70	6779	70	6869	70	6906	70	6928
72	7001	72	7082	72	7115	72	7135
74	7226	74	7295	74	7325	74	7342
76	7449	76	7508	76	7534	76	7548
78	7671	78	7721	78	7743	78	7755
80	7892	80	7934	80	7952	80	7963
82	8112	82	8146	82	8160	82	8169
84	8330	84	8357	84	8369	84	8375
86	8547	86	8567	86	8576	86	8580
88	8763	88	8777	88	8782	88	8785
90	8974	90	8984	90	8988	90	8990
92	9183	92	9190	92	9192	92	9194
94	9391	94	9394	94	9396	94	9397
96	9596	96	9598	96	9599	96	9599
98	9800	98	9800	98	9800	98	9800

Table IV

Coefficients of the First 15 Shifted Legendre Polynomials $P_n^*(x)$ (see p. 287).
(Negative numbers underlined; sequence: lowest to highest power;
e.g., $P_3^*(x) = 1 - 12x + 30x^2 - 20x^3$)

$n = 0$:   1

$n = 1$:   1, 2

$n = 2$:   1, 6, 6

$n = 3$:   1, 12, 30, 20

$n = 4$:   1, 20, 90, 140, 70

$n = 5$:   1, 30, 210, 560, 630, 252

$n = 6$:   1, 42, 420, 1680, 3150, 2772, 924

$n = 7$:   1, 56, 756, 4200, 11550, 16632, 12012, 3432

$n = 8$:   1, 72, 1260, 9240, 34650, 72072, 84084, 51480, 12870

$n = 9$:   1, 90, 1980, 18480, 90090, 252252, 420420, 411840, 218790, 48620

$n = 10$:   1, 110, 2970, 34320, 210210, 756756, 1681680, 2333760, 1969110, 92378
184756

$n = 11$:   1, 132, 4290, 60060, 450450, 2018016, 5717712, 10501920, 12471030, 92378(
3879876, 705432

$n = 12$:   1, 156, 6006, 100100, 900900, 4900896, 17153136, 39907296, 62355150, 646646(
42678636, 16224936, 2704156

$n = 13$:   1, 182, 8190, 160160, 1701700, 11027016, 46558512, 133024320, 26189163
355655300, 327202876, 194699232, 97603900, 10400600

$n = 14$:   1, 210, 10920, 247520, 3063060, 23279256, 116396280, 399072960, 9602693)
1636014380, 1963217256, 1622493600, 878850700, 280816200, 40116600

$n = 15$:   1, 240, 14280, 371280, 5290740, 46558512, 271591320, 109745064
3155170590, 6544057520, 9816086280, 10546208400, 7909656300, 39314268(
1163381400, 155117520

Table V

The First 12 Chebyshev Polynomials $T_n(x)$ (see p. 455)

$$T_0(x) = 1 \quad \text{(but } T_0 = \tfrac{1}{2}; \text{ see p. 457)}$$
$$T_1(x) = x$$
$$T_2(x) = 2x^2 - 1$$
$$T_3(x) = 4x^3 - 3x$$
$$T_4(x) = 8x^4 - 8x^2 + 1$$
$$T_5(x) = 16x^5 - 20x^3 + 5x$$
$$T_6(x) = 32x^6 - 48x^4 + 18x^2 - 1$$
$$T_7(x) = 64x^7 - 112x^5 + 56x^3 - 7x$$
$$T_8(x) = 128x^8 - 256x^6 + 160x^4 - 32x^2 + 1$$
$$T_9(x) = 256x^9 - 576x^7 + 432x^5 - 120x^3 + 9x$$
$$T_{10}(x) = 512x^{10} - 1280x^8 + 1120x^6 - 400x^4 + 50x^2 - 1$$
$$T_{11}(x) = 1024x^{11} - 2816x^9 + 2816x^7 - 1232x^5 + 220x^3 - 11x$$
$$T_{12}(x) = 2048x^{12} - 6144x^{10} + 6912x^8 - 3584x^6 + 840x^4 - 72x^2 + 1$$

Table VI

The First 12 Powers $\tfrac{1}{2}(2x)^n$ **Expressed in Chebyshev Polynomials** $T_k(x)$; (with $T_0 = \tfrac{1}{2}$)

	T_0	T_2	T_4	T_6	T_8	T_{10}	T_{12}
$n = 0$:	1						
$n = 2$:	2	1					
$n = 4$:	6	4	1				
$n = 6$:	20	15	6	1			
$n = 8$:	70	56	28	8	1		
$n = 10$:	252	210	120	45	10	1	
$n = 12$:	924	792	495	220	66	12	1

(For example: $32x^6 = 20T_0 + 15T_2 + 6T_4 + T_6$)

	T_1	T_3	T_5	T_7	T_9	T_{11}
$n = 1$:	1					
$n = 3$:	3	1				
$n = 5$:	10	5	1			
$n = 7$:	35	21	7	1		
$n = 9$:	126	84	36	9	1	
$n = 11$:	462	330	165	55	11	1

(For example: $256x^9 = 126T_1 + 84T_3 + 36T_5 + 9T_7 + T_9$)

Table VII

Coefficients of the First 12 Shifted Chebyshev Polynomials $T_n^*(x)$ (see p. 456).
(Negative numbers underlined; sequence: lowest to highest power;
e.g., $T_3^*(x) = -1 + 18x - 48x^2 + 32x^3$)

$T_0^*(x) = 1$ (but $T_0^* = \frac{1}{2}$; see p. 457)

$n = 1$: $\underline{1}$, 2

$n = 2$:   1, $\underline{8}$, 8

$n = 3$: $\underline{1}$, 18, $\underline{48}$, 32

$n = 4$:   1, $\underline{32}$, 160, $\underline{256}$, 128

$n = 5$: $\underline{1}$, 50, $\underline{400}$, 1120, $\underline{1280}$, 512

$n = 6$:   1, $\underline{72}$, 840, $\underline{3584}$, 6912, $\underline{6144}$, 2048

$n = 7$: $\underline{1}$, 98, $\underline{1568}$, 9408, $\underline{26880}$, 39424, $\underline{28672}$, 8192

$n = 8$:   1, $\underline{128}$, 2688, $\underline{21504}$, 84480, $\underline{180224}$, 212992, $\underline{131072}$, 32768

$n = 9$: $\underline{1}$, 162, $\underline{4320}$, 44352, $\underline{228096}$, 658944, $\underline{1118208}$, 1105920, $\underline{589824}$, 131072

$n = 10$:   1, $\underline{200}$, 6600, $\underline{84480}$, 549120, $\underline{2050048}$, 4659200, $\underline{6553600}$, 5570560, $\underline{2621440}$ 524288

$n = 11$: $\underline{1}$, 242, $\underline{9680}$, 151008, $\underline{1208064}$, 5637632, $\underline{16400384}$, 30638080, $\underline{36765696}$ 27394048, $\underline{11534336}$, 2097152

$n = 12$:   1, $\underline{288}$, 13728, $\underline{256256}$, 2471040, $\underline{14057472}$, 50692096, $\underline{120324096}$, 190513152, $\underline{199229440}$, 132120576, $\underline{50331648}$, 8388608

Table VIII

The First 12 Powers $\frac{1}{2}(4x)^n$ Expressed in the Shifted Chebyshev Polynomials $T_k^*(x)$; (with $T_0^* = \frac{1}{2}$), see p. 461

	T_0^*	T_1^*	T_2^*	T_3^*	T_4^*	T_5^*	T_6^*
$n=0$:	1						
$n=1$:	2	1					
$n=2$:	6	4	1				
$n=3$:	20	15	6	1			
$n=4$:	70	56	28	8	1		
$n=5$:	252	210	120	45	10	1	
$n=6$:	924	792	495	220	66	12	1
$n=7$:	3432	3003	2002	1001	364	91	14
$n=8$:	12870	11440	8008	4368	1820	560	120
$n=9$:	48620	43758	31824	18564	8568	3060	816
$n=10$:	184756	167960	125970	77520	38760	15504	4845
$n=11$:	705432	646646	497420	319770	170544	74613	26334
$n=12$:	2704156	2496144	1961256	1307504	735471	346104	134596

	T_7^*	T_8^*	T_9^*	T_{10}^*	T_{11}^*	T_{12}^*
$n=7$:	1					
$n=8$:	16	1				
$n=9$:	153	18	1			
$n=10$:	1140	190	20	1		
$n=11$:	7315	1540	231	22	1	
$n=12$:	42504	10626	2024	276	24	1

(For example: $32768x^8 = 12870T_0^* + 11440T_1^* + 8008T_2^* + 4368T_3^* + 1820T_4^* + 560T_5^* + 120T_6^* + 16T_7^* + T_8^*$)

Appendix

Table IX

Coefficients of the First 13 Shifted Chebyshev Polynomials of the Second Kind, $U_n^*(x)$
(see p. 289). (Negative numbers underlined; sequence: lowest to highest power;
e.g., $U(x) = -4 + 40x - 96x^2 + 64x^3$)

$n = 0$: 1

$n = 1$: $\underline{2}$, 4

$n = 2$: 3, $\underline{16}$, 16

$n = 3$: $\underline{4}$, 40, $\underline{96}$, 64

$n = 4$: 5, $\underline{80}$, 336, $\underline{512}$, 256

$n = 5$: $\underline{6}$, 140, $\underline{896}$, 2304, $\underline{2560}$, 1024

$n = 6$: 7, $\underline{224}$, 2016, $\underline{7680}$, 14080, $\underline{12288}$, 4096

$n = 7$: $\underline{8}$, 336, $\underline{4032}$, 21120, $\underline{56320}$, 79872, $\underline{57344}$, 16384

$n = 8$: 9, $\underline{480}$, 7392, $\underline{50688}$, 183040, $\underline{372736}$, 430080, $\underline{262144}$, 65536

$n = 9$: $\underline{10}$, 660, $\underline{12672}$, 109824, $\underline{512512}$, 1397760, $\underline{2293760}$, 2228224, $\underline{1179648}$, 262144

$n = 10$: 11, $\underline{880}$, 20592, $\underline{219648}$, 1281280, $\underline{4472832}$, 9748480, $\underline{13369344}$, 1120665$\,$
$\underline{5242880}$, 1048576

$n = 11$: $\underline{12}$, 1144, $\underline{32032}$, 411840, $\underline{2928640}$, 12673024, $\underline{35094528}$, 63504384, $\underline{7471104}$
55050240, $\underline{23068672}$, 4194304

$n = 12$: 13, $\underline{1456}$, 48048, $\underline{732160}$, 6223360, $\underline{32587776}$, 111132672, $\underline{254017536}$, 392232960
403701760, 265289728, $\underline{100663296}$, 16777216

$n = 13$: $\underline{14}$, 1820, $\underline{69888}$, 1244672, $\underline{12446720}$, 77395968, $\underline{317521920}$, 889061370
$\underline{1725825024}$, 2321285120, $\underline{2122317824}$, 1258291200, $\underline{436207616}$, 67108864

Table X

Coefficients of the First 11 Laguerre Polynomials $L_n(x)$ (see p. 298). (Negative numbers underlined; sequence: lowest to highest power; e.g., $L_3(x) = 6 - 18x + 9x^2 - x^3$)

$n = 0$:   1

$n = 1$:   1, $\underline{1}$

$n = 2$:   2, $\underline{4}$, 1

$n = 3$:   6, $\underline{18}$, 9, $\underline{1}$

$n = 4$:   24, $\underline{96}$, 72, $\underline{16}$, 1

$n = 5$:   120, $\underline{600}$, 600, $\underline{200}$, 25, $\underline{1}$

$n = 6$:   720, $\underline{4320}$, 5400, $\underline{2400}$, 450, $\underline{36}$, 1

$n = 7$:   5040, $\underline{35280}$, 52920, $\underline{29400}$, 7350, $\underline{882}$, 49, $\underline{1}$

$n = 8$:   40320, $\underline{322560}$, 564480, $\underline{376320}$, 117600, $\underline{18816}$, 1568, $\underline{64}$, 1

$n = 9$:   362880, $\underline{3265920}$, 6531840, $\underline{5080320}$, 1905120, $\underline{381024}$, 42336, $\underline{2592}$, 81, $\underline{1}$

$n = 10$:   3628800, $\underline{36288000}$, 81648000, $\underline{72576000}$, 31752000, $\underline{7620480}$, 1058400, $\underline{86400}$, 4050, $\underline{100}$, 1

$n = 11$:   39916800, $\underline{439084800}$, 1097712000, $\underline{1097712000}$, 548856000, $\underline{153679680}$, 25613280, $\underline{2613600}$, 163350, $\underline{6050}$, 121, $\underline{1}$

Values of the Normalized Laguerre Functions

The following table of the values of the normalized Laguerre functions (in double arguments)

$$\varphi_n(x) = \frac{e^{-x}}{n!} L_n(2x) \qquad (n \leq 20; \text{ see p. 298})$$

is computed under the auspices of the National Bureau of Standards, Washington, D.C., (the original to ten decimal places) and is published by permission of the Bureau. (Negative numbers are underlined.) Recurrence relation:

$$\varphi_{n+1}(x) = \frac{1}{n+1}[(2n + 1 - 2x)\varphi_n(x) - n\varphi_{n-1}(x)]$$

Table X (Cont'd)

n	x = 0	0.1	0.2	0.3	0.4	0.5	1	1.5	2
0	1	.90484	.81873	.74082	.67032	.60653	.36788	.22313	.13534
1	1	72387	49124	29633	13406	0	36788	44626	40601
2	1	56100	22924	01482	18769	30327	36788	11157	13534
3	1	41502	02402	21928	35214	40435	12263	22313	31578
4	1	28478	13231	33974	40505	37908	12263	30680	13534
5	1	16920	24678	39534	38257	28305	26978	18966	11729
6	1	06725	32572	40213	31283	15584	30248	00279	24962
7	1	02207	37478	37349	21730	02455	24409	16655	22040
8	1	09967	39896	32042	11198	09340	13197	24739	08464
9	1	16643	40272	25188	00841	18787	00298	23678	07366
10	1	22318	39000	17508	08548	25410	11370	15620	18666
11	1	27072	36425	09572	16461	29122	19910	04034	22152
12	1	30978	32851	01818	22618	30097	24420	07594	17963
13	1	34108	28541	05423	26909	28682	24827	16576	08569
14	1	36527	23723	11915	29356	25319	21657	21364	02602
15	1	38297	18594	17497	30074	20493	15812	21560	12334
16	1	39478	13321	22074	29244	14687	08355	17701	18375
17	1	40125	08044	25603	27086	08359	00354	10946	19737
18	1	40290	02882	28082	23844	01918	07242	02741	16637
19	1	40023	02070	29544	19769	04284	13676	05464	10197
20	1	39368	06732	30047	15107	09963	18421	12440	02041

n	x = 2.5	3	3.5	4	4.5	5	5.5	6	6.5
0	.08208	.04979	.03020	.01832	.01111	.00674	.00409	.00248	.00150
1	32834	24894	18118	12821	08887	06064	04087	02727	01804
2	28730	34851	34727	31137	26106	20888	16143	12146	08945
3	21889	04979	11072	22589	28883	30770	29561	26523	22652
4	10603	24894	26045	17705	05138	07412	17454	24044	27269
5	25994	18919	01560	14530	23107	23134	16667	06792	03693
6	17158	04979	20664	22019	11984	02321	14545	21169	21493
7	02671	21195	19050	03274	12958	20823	18442	08846	03166
8	18352	19488	00968	16403	20204	10984	03506	15205	19598
9	22095	04979	15857	19313	06441	09967	18730	16310	05896
10	14416	11067	19900	06481	11743	18856	11829	02268	14100
11	00883	19617	10912	09897	18666	09795	06274	16683	15615
12	11891	17647	03693	18313	11013	06673	17117	13214	00087
13	19109	07683	15185	14811	03676	16741	12642	02186	14333
14	18986	04862	18264	03097	14952	14132	01446	14612	14414
15	12543	14626	12615	09489	16505	02276	13535	14520	01997
16	02583	18294	01799	16543	08677	10262	15563	03544	11266
17	07551	15291	09121	15398	03284	16026	07402	09288	15134
18	15024	07356	15887	07472	12939	12566	04829	15215	07857
19	18150	02483	16444	03182	15957	02675	13621	11221	04413
20	16583	11085	11218	12031	11643	08059	14481	00694	13201

Table X (Cont'd)

n	x = 7	7.5	8	8.5	9	9.5	10	11	12
0	.00091	.00055	.00034	.00020	.00012	.00007	.00005	.00002	.00001
1	01185	00774	00503	00326	00210	00135	00086	00035	00014
2	06474	04618	03254	02269	01567	01074	00731	00332	00148
3	18633	14878	11596	08858	06652	04923	03597	01860	00928
4	27752	26292	23650	20443	17117	13963	11143	06726	03835
5	12845	19648	23834	25622	25489	23987	21636	15999	10761
6	16704	08811	00153	08587	15473	20347	23169	23727	20120
7	13397	19359	20363	17055	10796	03120	04623	16792	22394
8	16290	07710	02679	11777	17587	19364	17383	06068	07588
9	06478	15495	18398	15160	07642	01530	09904	18297	14004
0	17900	13137	03108	07567	15064	17427	14654	00028	13831
1	05502	06921	15313	16534	11056	01778	07671	16631	08959
2	12282	16656	11782	01330	09203	15383	15351	01360	13425
3	15471	06424	05978	14443	15160	08741	01177	15038	09302
4	02961	09960	15638	11552	01201	09289	14843	06634	10473
5	11478	15291	07973	04239	13269	14351	07807	10940	12173
6	14972	05954	07186	14539	11907	02055	08548	12373	04492
7	05930	08088	14689	09694	01982	11815	13884	02291	13835
8	07222	14610	08719	04037	13118	12442	03497	13340	04212
9	14360	09254	04280	13433	11240	00595	10024	08361	10225
0	11090	02774	13205	10942	00660	11226	12846	05566	11670

n	x = 13	14	15	16	17	18	19	20	21
0	.00000	.00000	.00000	.00000	.00000	.00000	.00000	.00000	.00000
1	00006	00002	00001	00000	00000	00000	00000	00000	00000
2	00065	00028	00012	00005	00002	00001	00000	00000	00000
3	00450	00213	00099	00045	00020	00009	00004	00002	00001
4	02090	01099	00561	00279	00136	00065	00031	00014	00006
5	06747	04006	02276	01247	00663	00343	00174	00086	00042
6	15126	10433	06740	04132	02428	01376	00757	00405	00212
7	22308	18924	14417	10148	06715	04228	02554	01490	00843
8	17438	21622	21135	17948	13825	09895	06681	04302	02661
9	02391	09606	17713	20893	20144	17130	13319	09668	06641
0	17368	10815	00463	11008	17774	20216	19294	16432	12881
1	05721	15615	15724	07986	02693	11995	17709	19594	18553
2	14490	03407	09596	16080	13825	05537	04450	12695	17568
3	06395	15200	10823	01287	12056	15758	11896	03439	05848
4	13912	02078	11230	14472	06809	04988	13480	14981	10048
5	03187	14048	09353	04095	13522	12379	03016	07777	14166
6	14039	04582	09944	13311	03848	08545	13956	09670	05110
7	02781	11874	10558	03071	12953	10143	01267	11301	10627
8	11868	08945	06459	13084	04354	08634	12970	05994	08959
9	09506	07012	12382	00533	11584	10064	01883	11653	07710
0	05096	12355	00564	12243	07032	06692	12416	05112	09667

Table XI

Conversion of a Harmonic Series into a Polynomial Expansion

The series

$$f(x) = \sum_{k=0}^{N} a_k \cos (2k + 1)\frac{\pi}{2}\, x + \sum_{k=1}^{N} b_k \sin k\pi x$$

is transformed into the series

$$f(x) = \sum_{i=0}^{\infty} c_i T_i(x)$$

(neglecting all $c_i > 13$) by multiplying the a_k by the successive columns of the first matrix, the b_k by the successive columns of the second matrix; see p. 351. (Negative numbers underlined.) (This table was prepared under the auspices of the National Bureau of Standards, Washington, D.C., and is published by permission of the Bureau; N is limited to ≤ 11.)

k	c_0	c_2	c_4	c_6	c_8	c_{10}	c_{12}
0	.472001	.499403	.027992	.000597	.000007	.000000	.00000(
1	265857	292636	740869	205626	024909	001735	000079
2	204268	300940	138934	691873	418338	107419	016207
3	171971	279881	032140	402029	451227	560378	242595
4	151323	259041	097117	211630	465982	121520	567021
5	136676	241238	125502	101019	361305	347427	181560
6	125594	226275	138516	034324	262471	374524	116288
7	116832	213613	144229	007805	185551	337894	256291
8	109679	202772	146188	035480	127870	286247	303192
9	103697	193378	146117	054225	084651	235836	304120
10	098598	185148	144919	067216	051946	191405	284453
11	094184	177867	143093	076366	026880	153695	256880

k	c_1	c_3	c_5	c_7	c_9	c_{11}	c_{13}
1	.569231	.666917	.104282	.006841	.000250	.000006	.00000(
2	424765	058224	745649	315042	058248	006296	00045.
3	353450	167800	294175	589723	503360	170328	03370
4	309062	193132	097432	459239	290219	582676	31808
5	278050	197132	003055	301849	426211	040189	51386
6	254818	194304	047027	189771	385862	240016	29167
7	236577	189152	075527	113651	313247	329561	01067
8	221764	183315	092486	061327	244663	334392	16254
9	209424	177443	102825	024535	187474	306725	24724
10	198938	171797	109159	001931	141479	269136	27834
11	189884	166476	112969	021356	104801	230904	28013

Numerical Tables

Table XII

Curve Fitting of Equidistant Data.

The given $2n + 1$ equidistant ordinates $y_k[k = -n, -(n-1), \cdots, (n-1), n]$ are ted by the infinite expansion

$$f(x) = \sum_{i=0}^{\infty} c_i T_i(x)$$

ich can be truncated at any suitably chosen $i = \nu$ (see p. 351). The last two ordinates are duced to zero (see p. 333). The remaining $2n - 1$ ordinates are divided into the two oups:

$$u_k = y_k + y_{-k} \qquad (k = 0, 1, 2, \cdots, n - 1)$$

d

$$v_k = y_k - y_{-k} \qquad (k = 1, 2, 3, \cdots, n - 1)$$

ie even coefficients c_{2m} are evaluated by multiplying the u_k by the successive columns of one the first group of tables, according to the number of data points (the table extends from to 25 data points, i.e., $n = 2$ to $n = 12$; negative numbers are underlined). The odd efficients c_{2m+1} are similarly evaluated by multiplying the v_k by the successive columns of e of the second group of tables. (These tables were computed under the auspices of the ational Bureau of Standards and are published by permission of the Bureau.)

Even c_i

k	c_0	c_2	c_4	c_6	c_8	c_{10}	c_{12}
				$n = 2$			
0	.051536	.198010	.192215	.051556	.006229	.000434	.000020
1	260872	073103	252040	072489	008804	000614	000028
				$n = 3$			
0	.068402	.081850	.104988	.149683	.073876	.018192	.002715
1	077288	231039	048188	199554	120762	031009	004679
2	201331	064468	265447	046870	061421	017325	002675
				$n = 4$			
0	.029805	.096373	.074723	.062008	.111810	.083692	.032360
1	103761	107491	098060	046475	141858	139541	057575
2	063926	139227	107141	229621	001412	079766	040006
3	170194	116096	197455	150828	047702	029202	019484
				$n = 5$			
0	.038976	.051194	.069490	.070770	.042850	.079105	.082590
1	049960	145764	077724	031293	038520	088787	136363
2	087062	063696	004228	091897	188491	036562	073506
3	055556	086734	155161	140391	135871	091975	020528
4	150219	137968	134730	183270	022172	061958	000980
	c_0	c_2	c_4	c_6	c_8	c_{10}	c_{12}

Table XII (Cont'd)

Even c_i

k	c_0	c_2	c_4	c_6	c_8	c_{10}	c_{12}
				$n = 6$			
0	.021091	.062765	.047450	.050557	.065817	.036969	.053695
1	065054	081523	095971	062839	001731	022276	047351
2	043738	109942	010618	056330	073360	146535	063561
3	076558	033860	045192	116235	096555	108443	114892
4	049724	054366	162047	053758	176085	006155	088080
5	135991	147385	086438	181196	078732	043817	038201
				$n = 7$			
0	.027049	.037636	.05066	.045786	.037667	.058439	.037718
1	037099	104213	066741	059319	053798	014038	002342
2	057015	062020	049309	015222	081722	063230	107341
3	039324	083701	033169	089660	072997	061509	078444
4	069167	013397	067635	096508	007112	150629	029117
5	045377	033142	153880	008840	161891	073419	079610
6	125186	150928	050308	165515	113337	007891	052879
				$n = 8$			
0	.016366	.046282	.035231	.040550	.044555	.030016	.049022
1	047298	064460	079634	057019	033449	044969	013504
2	033458	085966	033788	006173	025525	088291	061550
3	051398	046093	013834	059517	096981	042031	031191
4	035991	064276	058885	089520	032817	018363	119100
5	063597	000980	075649	065946	048602	125873	060961
6	041982	018552	140763	050034	127054	119952	032144
7	116615	151489	023181	145570	131003	027599	045997
				$n = 9$			
0	.020641	.029875	.039438	.034074	.032501	.042584	.026731
1	029584	080574	055967	058507	050576	016780	034104
2	042566	053695	053968	013507	022049	029715	083680
3	030690	071033	006341	030837	061859	089196	024389
4	047188	033442	010528	076840	074948	010331	034580
5	033365	049612	072851	074385	008558	056361	087685
6	059201	011327	076189	036770	075088	078357	110501
7	039239	008150	126688	075785	088391	137623	020381
8	109600	150469	002571	125318	137505	056487	028018
	c_0	c_2	c_4	c_6	c_8	c_{10}	c_{12}

Table XII (Cont'd)

Even c_i

k	c_0	c_2	c_4	c_6	c_8	c_{10}	c_{12}
				$n = 10$			
0	.013392	.036556	.028188	.033378	.033483	.026533	.039264
1	037102	053135	066292	049638	041173	045205	008011
2	027181	069712	037353	026308	000860	038969	032977
3	039031	044121	031191	020473	054546	054572	074808
4	028497	058906	014235	050479	064660	052097	034002
5	043881	023409	026439	077959	041281	051607	052337
6	031230	038319	079522	054432	038873	060942	036059
7	055617	018929	072962	012561	082404	031420	121822
8	036965	000518	113214	091063	052829	136250	062863
9	103719	148599	013294	106345	137101	077852	006458
				$n = 11$			
0	.016656	.024817	.032213	.027288	.028078	.032821	.022764
1	024638	065480	047686	052993	045107	027805	039226
2	033986	046377	050633	023412	004566	006902	047221
3	025270	060263	019820	001567	034004	064401	045676
4	036261	035906	012886	042138	063299	042768	029841
5	026706	049020	028905	057472	050171	009798	069210
6	041192	015398	036391	070649	009167	071840	036492
7	029451	029462	081739	034481	057001	047272	007926
8	052619	024610	067949	006292	078671	006239	109018
9	035040	005211	100895	099390	022613	124033	091632
10	098695	146280	025659	089188	132603	092591	014586
				$n = 12$			
0	.011344	.030160	.023566	.028196	.026858	.023682	.031571
1	030490	045167	056270	043336	041143	041494	018535
2	022928	058339	036051	032318	013014	011452	012403
3	031550	040077	035516	000623	024575	038615	065383
4	023704	052155	004974	021530	050563	060969	021701
5	034014	028939	001160	053580	056613	016386	012406
6	025208	040898	039004	056505	029756	022671	075513
7	038950	008945	042256	059622	016473	074394	007693
8	027940	022402	081269	016535	065292	026587	035796
9	050064	028913	062210	020439	069251	033045	084605
10	033386	009591	189860	103146	002024	106515	107898
11	094339	143741	035405	073931	125799	102028	033242
	c_0	c_2	c_4	c_6	c_8	c_{10}	c_{12}

Table XII (Cont'd)

Odd c_i

k	c_1	c_3	c_5	c_7	c_9	c_{11}
			$n = 2$			
1	.284615	.333458	.052141	.003420	.000125	.000003
			$n = 3$			
1	.041704	.209330	.245354	.092920	.016887	.001819
2	286942	175714	185147	088970	016743	001816
			$n = 4$			
1	.056917	.102788	.152844	.184219	.103589	.031685
2	053945	208679	099615	145721	125778	042581
3	269300	073676	219981	026698	074465	028537
			$n = 5$			
1	.017020	.080262	.109589	.118914	.140971	.102094
2	075576	116690	123542	056363	107482	130114
3	057871	176472	014706	192413	010771	090069
4	251275	012704	196981	107038	050578	037296
			$n = 6$			
1	.023596	.047462	.081095	.103198	.098694	.110050
2	025328	105242	109055	069177	028072	077392
3	082305	106264	065901	046839	154887	035085
4	058729	144187	078069	154339	095037	088995
5	235434	025098	162282	144824	001900	059971
			$n = 7$			
1	.009284	.041395	.059088	.075735	.093361	.086213
2	035218	065979	096903	088815	037615	007724
3	029738	108508	075977	001815	057667	119632
4	084336	090308	018405	090593	107540	076603
5	058302	117106	109225	098261	143562	024587
6	221847	049496	128786	156808	048326	047744
			$n = 8$			
1	.012775	.027172	.048205	.062395	.069469	.082970
2	014824	061249	075895	079106	068921	020267
3	041416	070810	084554	039744	036860	061658
4	032157	103342	039985	049336	077009	067509
5	084285	074452	014588	097211	042814	125327
6	057311	095269	122275	047098	150825	041311
7	210175	065752	099653	155059	081594	021382
	c_1	c_3	c_5	c_7	c_9	c_{11}

Table XII (Cont'd)

Odd c_i

k	c_1	c_3	c_5	c_7	c_9	c_{11}
			$n = 9$			
1	.005862	.025072	.036592	.049810	.061780	.063935
2	020347	041595	067417	072989	061726	051666
3	018311	069600	071071	051153	012115	052059
4	044822	069289	063602	005537	075101	059665
5	033467	095175	009657	071980	058788	003902
6	083256	060284	035879	086055	009673	123219
7	056078	077753	125432	006334	136128	087954
8	200067	076837	075161	146494	102908	007125
			$n = 10$			
1	.007971	.017579	.031591	.041247	.049078	.059205
2	009769	039695	053127	064193	065347	047586
3	025078	048481	069461	056367	022419	007064
4	020577	071807	057395	018085	032006	078029
5	046668	064929	042270	037015	080355	023335
6	034123	086290	013656	077416	028284	038406
7	081767	048083	048853	068189	044280	095626
8	054759	063627	123278	024281	112049	113930
9	191231	084510	054780	134870	115130	032598
			$n = 11$			
1	.004044	.016762	.024808	.034725	.043136	.047442
2	013231	028467	048131	056080	057319	056309
3	012471	047790	056721	056712	038800	001420
4	028127	050967	062849	031079	015812	043909
5	022074	070744	041325	009523	054054	066720
6	047592	059447	023518	055217	066528	016345
7	034379	077647	030769	072744	001351	056299
8	080075	037714	056167	049132	063510	060290
9	053432	052131	118362	046511	085511	123727
10	183434	089860	037839	122190	120867	053386
	c_1	c_3	c_5	c_7	c_9	c_{11}

Table XII (Cont'd)

Odd c_i

k	c_1	c_3	c_5	c_7	c_9	c_{11}
			$n = 12$			
1	.005433	.012301	.022202	.029152	.035916	.043349
2	006939	027708	038677	049996	054521	049668
3	016834	034884	054216	053085	040515	022764
4	014386	051816	053029	039767	006439	036285
5	030123	050903	052570	006559	041103	052154
6	023066	067981	025811	029066	058229	038688
7	047949	053686	008081	063364	044854	045589
8	034381	069626	042940	062861	025028	056327
9	078315	028931	059704	031372	071537	026568
10	052134	042684	112116	062235	059959	122455
11	176496	093587	023715	109503	122242	069418
	c_1	c_3	c_5	c_7	c_9	c_{11}

Table XIII

Zeros and Weights of Gaussian Quadrature (see p. 396)

(Reference: *Applied Math. Series* 37, U.S. Dept. of Commerce, National Bureau of Standards, Washington, D.C., p. 187).

n	$\pm x_i$	w_i
2	0.5773502692	1
3	0	0.8888888889
	0.7745966692	0.5555555556
4	0.3399810435	0.6521451549
	0.8611363116	0.3478548451
5	0	0.5688888889
	0.5384693101	0.4786286705
	0.9061798459	0.2369268851
6	0.2386191861	0.4679139346
	0.6612093865	0.3607615730
	0.9324695142	0.1713244924
7	0	0.4179591837
	0.4058451514	0.3818300505
	0.7415311856	0.2797053915
	0.9491079123	0.1294849662

Table XIII (Cont'd)

n	$\pm x_i$	w_i
8	0.1834346425	0.3626837834
	0.5255324099	0.3137066459
	0.7966664774	0.2238103445
	0.9602898565	0.1012285363
9	0	0.3302393550
	0.3242534234	0.3123470770
	0.6133714327	0.2606106964
	0.8360311073	0.1806481607
	0.9681602395	0.0812743884

Table XIV

Gaussian Quadrature with Rounded-off Zeros (see p. 396)

(This table was computed under the auspices of the National Bureau of Standards and is published by permission of the Bureau.)

n	$\pm x_i$	w_i
2	0.58	1
3	0	0.8755832912
	0.77	0.5622083544
4	0.34	0.6510683761
	0.86	0.3489316239
5	0	0.5652007082
	0.54	0.4860117674
	0.91	0.2313878782
6	0.24	0.4694282398
	0.66	0.3554559718
	0.93	0.1751157884
7	0	0.4087667652
	0.40	0.3816431469
	0.74	0.2855469914
	0.95	0.1284264790
9	0	0.3274777490
	0.32	0.3073797883
	0.61	0.2682840654
	0.84	0.1834544443
	0.97	0.0771428275

Table XV

Quadrature in Terms of End-data (see p. 423)

Range [0, 1]. The numerical coefficients are multiplied by the derivatives at the end-point (symmetrized for the even, anti-symmetrized for the odd derivatives), starting with the lowest order; e.g., the third convergent becomes:

$$A_3 = \frac{1}{2}[f(0) + f(1)] + \frac{3}{28}[f'(0) - f'(1)] + \frac{1}{84}[f''(0) + f''(1)] + \frac{1}{1680}[f'''(0) - f'''(1)$$

$$n = 0: \quad \frac{1}{2}$$

$$n = 1: \quad \frac{6, 1}{12}$$

$$n = 2: \quad \frac{60, 12, 1}{120}$$

$$n = 3: \quad \frac{840, 180, 20, 1}{1680}$$

$$n = 4: \quad \frac{15120, 3360, 420, 30, 1}{30240}$$

$$n = 5: \quad \frac{332640, 75600, 10080, 840, 42, 1}{665280}$$

$$n = 6: \quad \frac{8648640, 1995840, 277200, 25200, 1512, 56, 1}{17297280}$$

$$n = 7: \quad \frac{259459200, 60540480, 8648640, 831600, 55440, 2520, 72, 1}{518918400}$$

$$n = 8: \quad 8821612800, 2075673600, 302702400, 30270240, 2162160,$$
$$110880, 3960, 90, 1 \quad \text{(denominator: } 17643225600)$$

INDEX

A CATALOG OF SELECTED

DOVER BOOKS

IN SCIENCE AND MATHEMATICS

QUALITATIVE THEORY OF DIFFERENTIAL EQUATIONS, V.V. Nemytskii and V.V. Stepanov. Classic graduate-level text by two prominent Soviet mathematicians covers classical differential equations as well as topological dynamics and ergodic theory. Bibliographies. 523pp. 5⅜ × 8½. 65954-2 Pa. $10.95

MATRICES AND LINEAR ALGEBRA, Hans Schneider and George Phillip Barker. Basic textbook covers theory of matrices and its applications to systems of linear equations and related topics such as determinants, eigenvalues and differential equations. Numerous exercises. 432pp. 5⅜ × 8½. 66014-1 Pa. $9.95

QUANTUM THEORY, David Bohm. This advanced undergraduate-level text presents the quantum theory in terms of qualitative and imaginative concepts, followed by specific applications worked out in mathematical detail. Preface. Index. 655pp. 5⅜ × 8½. 65969-0 Pa. $13.95

ATOMIC PHYSICS (8th edition), Max Born. Nobel laureate's lucid treatment of kinetic theory of gases, elementary particles, nuclear atom, wave-corpuscles, atomic structure and spectral lines, much more. Over 40 appendices, bibliography. 495pp. 5⅜ × 8½. 65984-4 Pa. $12.95

ELECTRONIC STRUCTURE AND THE PROPERTIES OF SOLIDS: The Physics of the Chemical Bond, Walter A. Harrison. Innovative text offers basic understanding of the electronic structure of covalent and ionic solids, simple metals, transition metals and their compounds. Problems. 1980 edition. 582pp. 6⅛ × 9¼. 66021-4 Pa. $15.95

BOUNDARY VALUE PROBLEMS OF HEAT CONDUCTION, M. Necati Özisik. Systematic, comprehensive treatment of modern mathematical methods of solving problems in heat conduction and diffusion. Numerous examples and problems. Selected references. Appendices. 505pp. 5⅜ × 8½. 65990-9 Pa. $11.95

A SHORT HISTORY OF CHEMISTRY (3rd edition), J.R. Partington. Classic exposition explores origins of chemistry, alchemy, early medical chemistry, nature of atmosphere, theory of valency, laws and structure of atomic theory, much more. 428pp. 5⅜ × 8½. (Available in U.S. only) 65977-1 Pa. $10.95

A HISTORY OF ASTRONOMY, A. Pannekoek. Well-balanced, carefully reasoned study covers such topics as Ptolemaic theory, work of Copernicus, Kepler, Newton, Eddington's work on stars, much more. Illustrated. References. 521pp. 5⅜ × 8½. 65994-1 Pa. $12.95

PRINCIPLES OF METEOROLOGICAL ANALYSIS, Walter J. Saucier. Highly respected, abundantly illustrated classic reviews atmospheric variables, hydrostatics, static stability, various analyses (scalar, cross-section, isobaric, isentropic, more). For intermediate meteorology students. 454pp. 6⅛ × 9¼. 65979-8 Pa. $14.95

CATALOG OF DOVER BOOKS

ASYMPTOTIC METHODS IN ANALYSIS, N.G. de Bruijn. An inexpensive, comprehensive guide to asymptotic methods—the pioneering work that teaches by explaining worked examples in detail. Index. 224pp. 5⅜ × 8½.   64221-6 Pa. $6.95

OPTICAL RESONANCE AND TWO-LEVEL ATOMS, L. Allen and J.H. Eberly. Clear, comprehensive introduction to basic principles behind all quantum optical resonance phenomena. 53 illustrations. Preface. Index. 256pp. 5⅜ × 8½.
65533-4 Pa. $7.95

COMPLEX VARIABLES, Francis J. Flanigan. Unusual approach, delaying complex algebra till harmonic functions have been analyzed from real variable viewpoint. Includes problems with answers. 364pp. 5⅜ × 8½.   61388-7 Pa. $8.95

ATOMIC SPECTRA AND ATOMIC STRUCTURE, Gerhard Herzberg. One of best introductions; especially for specialist in other fields. Treatment is physical rather than mathematical. 80 illustrations. 257pp. 5⅜ × 8½.   60115-3 Pa. $5.95

APPLIED COMPLEX VARIABLES, John W. Dettman. Step-by-step coverage of fundamentals of analytic function theory—plus lucid exposition of five important applications: Potential Theory; Ordinary Differential Equations; Fourier Transforms; Laplace Transforms; Asymptotic Expansions. 66 figures. Exercises at chapter ends. 512pp. 5⅜ × 8½.   64670-X Pa. $11.95

ULTRASONIC ABSORPTION: An Introduction to the Theory of Sound Absorption and Dispersion in Gases, Liquids and Solids, A.B. Bhatia. Standard reference in the field provides a clear, systematically organized introductory review of fundamental concepts for advanced graduate students, research workers. Numerous diagrams. Bibliography. 440pp. 5⅜ × 8½.   64917-2 Pa. $11.95

UNBOUNDED LINEAR OPERATORS: Theory and Applications, Seymour Goldberg. Classic presents systematic treatment of the theory of unbounded linear operators in normed linear spaces with applications to differential equations. Bibliography. 199pp. 5⅜ × 8½.   64830-3 Pa. $7.95

LIGHT SCATTERING BY SMALL PARTICLES, H.C. van de Hulst. Comprehensive treatment including full range of useful approximation methods for researchers in chemistry, meteorology and astronomy. 44 illustrations. 470pp. 5⅜ × 8½.   64228-3 Pa. $10.95

CONFORMAL MAPPING ON RIEMANN SURFACES, Harvey Cohn. Lucid, insightful book presents ideal coverage of subject. 334 exercises make book perfect for self-study. 55 figures. 352pp. 5⅜ × 8¼.   64025-6 Pa. $9.95

OPTICKS, Sir Isaac Newton. Newton's own experiments with spectroscopy, colors, lenses, reflection, refraction, etc., in language the layman can follow. Foreword by Albert Einstein. 532pp. 5⅜ × 8½.   60205-2 Pa. $9.95

GENERALIZED INTEGRAL TRANSFORMATIONS, A.H. Zemanian. Graduate-level study of recent generalizations of the Laplace, Mellin, Hankel, K. Weierstrass, convolution and other simple transformations. Bibliography. 320pp. 5⅜ × 8½.   65375-7 Pa. $8.95

CATALOG OF DOVER BOOKS

THE ELECTROMAGNETIC FIELD, Albert Shadowitz. Comprehensive undergraduate text covers basics of electric and magnetic fields, builds up to electromagnetic theory. Also related topics, including relativity. Over 900 problems. 768pp. 5⅜ × 8¼.   65660-8 Pa. $18.95

FOURIER SERIES, Georgi P. Tolstov. Translated by Richard A. Silverman. A valuable addition to the literature on the subject, moving clearly from subject to subject and theorem to theorem. 107 problems, answers. 336pp. 5⅜ × 8½.
                                        63317-9 Pa. $8.95

THEORY OF ELECTROMAGNETIC WAVE PROPAGATION, Charles Herach Papas. Graduate-level study discusses the Maxwell field equations, radiation from wire antennas, the Doppler effect and more. xiii + 244pp. 5⅜ × 8½.
                                        65678-0 Pa. $6.95

DISTRIBUTION THEORY AND TRANSFORM ANALYSIS: An Introduction to Generalized Functions, with Applications, A.H. Zemanian. Provides basics of distribution theory, describes generalized Fourier and Laplace transformations. Numerous problems. 384pp. 5⅜ × 8½.   65479-6 Pa. $9.95

THE PHYSICS OF WAVES, William C. Elmore and Mark A. Heald. Unique overview of classical wave theory. Acoustics, optics, electromagnetic radiation, more. Ideal as classroom text or for self-study. Problems. 477pp. 5⅜ × 8½.
                                        64926-1 Pa. $12.95

CALCULUS OF VARIATIONS WITH APPLICATIONS, George M. Ewing. Applications-oriented introduction to variational theory develops insight and promotes understanding of specialized books, research papers. Suitable for advanced undergraduate/graduate students as primary, supplementary text. 352pp. 5⅜ × 8½.   64856-7 Pa. $8.95

A TREATISE ON ELECTRICITY AND MAGNETISM, James Clerk Maxwell. Important foundation work of modern physics. Brings to final form Maxwell's theory of electromagnetism and rigorously derives his general equations of field theory. 1,084pp. 5⅜ × 8½.   60636-8, 60637-6 Pa., Two-vol. set $19.90

AN INTRODUCTION TO THE CALCULUS OF VARIATIONS, Charles Fox. Graduate-level text covers variations of an integral, isoperimetrical problems, least action, special relativity, approximations, more. References. 279pp. 5⅜ × 8½.
                                        65499-0 Pa. $7.95

HYDRODYNAMIC AND HYDROMAGNETIC STABILITY, S. Chandrasekhar. Lucid examination of the Rayleigh-Benard problem; clear coverage of the theory of instabilities causing convection. 704pp. 5⅜ × 8¼.   64071-X Pa. $14.95

CALCULUS OF VARIATIONS, Robert Weinstock. Basic introduction covering isoperimetric problems, theory of elasticity, quantum mechanics, electrostatics, etc. Exercises throughout. 326pp. 5⅜ × 8½.   63069-2 Pa. $7.95

DYNAMICS OF FLUIDS IN POROUS MEDIA, Jacob Bear. For advanced students of ground water hydrology, soil mechanics and physics, drainage and irrigation engineering and more. 335 illustrations. Exercises, with answers. 784pp. 6⅛ × 9¼.   65675-6 Pa. $19.95

NUMERICAL METHODS FOR SCIENTISTS AND ENGINEERS, Richard Hamming. Classic text stresses frequency approach in coverage of algorithms, polynomial approximation, Fourier approximation, exponential approximation, other topics. Revised and enlarged 2nd edition. 721pp. 5⅜ × 8½.
65241-6 Pa. $14.95

THEORETICAL SOLID STATE PHYSICS, Vol. I: Perfect Lattices in Equilibrium; Vol. II: Non-Equilibrium and Disorder, William Jones and Norman H. March. Monumental reference work covers fundamental theory of equilibrium properties of perfect crystalline solids, non-equilibrium properties, defects and disordered systems. Appendices. Problems. Preface. Diagrams. Index. Bibliography. Total of 1,301pp. 5⅜ × 8½. Two volumes. Vol. I 65015-4 Pa. $14.95
Vol. II 65016-2 Pa. $14.95

OPTIMIZATION THEORY WITH APPLICATIONS, Donald A. Pierre. Broad-spectrum approach to important topic. Classical theory of minima and maxima, calculus of variations, simplex technique and linear programming, more. Many problems, examples. 640pp. 5⅜ × 8½. 65205-X Pa. $14.95

THE MODERN THEORY OF SOLIDS, Frederick Seitz. First inexpensive edition of classic work on theory of ionic crystals, free-electron theory of metals and semiconductors, molecular binding, much more. 736pp. 5⅜ × 8½.
65482-6 Pa. $15.95

ESSAYS ON THE THEORY OF NUMBERS, Richard Dedekind. Two classic essays by great German mathematician: on the theory of irrational numbers; and on transfinite numbers and properties of natural numbers. 115pp. 5⅜ × 8½.
21010-3 Pa. $4.95

THE FUNCTIONS OF MATHEMATICAL PHYSICS, Harry Hochstadt. Comprehensive treatment of orthogonal polynomials, hypergeometric functions, Hill's equation, much more. Bibliography. Index. 322pp. 5⅜ × 8½. 65214-9 Pa. $9.95

NUMBER THEORY AND ITS HISTORY, Oystein Ore. Unusually clear, accessible introduction covers counting, properties of numbers, prime numbers, much more. Bibliography. 380pp. 5⅜ × 8½. 65620-9 Pa. $9.95

THE VARIATIONAL PRINCIPLES OF MECHANICS, Cornelius Lanczos. Graduate level coverage of calculus of variations, equations of motion, relativistic mechanics, more. First inexpensive paperbound edition of classic treatise. Index. Bibliography. 418pp. 5⅜ × 8½. 65067-7 Pa. $11.95

MATHEMATICAL TABLES AND FORMULAS, Robert D. Carmichael and Edwin R. Smith. Logarithms, sines, tangents, trig functions, powers, roots, reciprocals, exponential and hyperbolic functions, formulas and theorems. 269pp. 5⅜ × 8½. 60111-0 Pa. $6.95

THEORETICAL PHYSICS, Georg Joos, with Ira M. Freeman. Classic overview covers essential math, mechanics, electromagnetic theory, thermodynamics, quantum mechanics, nuclear physics, other topics. First paperback edition. xxiii + 885pp. 5⅜ × 8½. 65227-0 Pa. $19.95

ROTARY-WING AERODYNAMICS, W.Z. Stepniewski. Clear, concise text covers aerodynamic phenomena of the rotor and offers guidelines for helicopter performance evaluation. Originally prepared for NASA. 537 figures. 640pp. 6⅛ × 9¼.
64647-5 Pa. $15.95

DIFFERENTIAL GEOMETRY, Heinrich W. Guggenheimer. Local differential geometry as an application of advanced calculus and linear algebra. Curvature, transformation groups, surfaces, more. Exercises. 62 figures. 378pp. 5⅜ × 8½.
63433-7 Pa. $8.95

INTRODUCTION TO SPACE DYNAMICS, William Tyrrell Thomson. Comprehensive, classic introduction to space-flight engineering for advanced undergraduate and graduate students. Includes vector algebra, kinematics, transformation of coordinates. Bibliography. Index. 352pp. 5⅜ × 8½. 65113-4 Pa. $8.95

A SURVEY OF MINIMAL SURFACES, Robert Osserman. Up-to-date, in-depth discussion of the field for advanced students. Corrected and enlarged edition covers new developments. Includes numerous problems. 192pp. 5⅜ × 8½.
64998-9 Pa. $8.95

ANALYTICAL MECHANICS OF GEARS, Earle Buckingham. Indispensable reference for modern gear manufacture covers conjugate gear-tooth action, gear-tooth profiles of various gears, many other topics. 263 figures. 102 tables. 546pp. 5⅜ × 8½. 65712-4 Pa. $14.95

SET THEORY AND LOGIC, Robert R. Stoll. Lucid introduction to unified theory of mathematical concepts. Set theory and logic seen as tools for conceptual understanding of real number system. 496pp. 5⅜ × 8¼. 63829-4 Pa. $10.95

A HISTORY OF MECHANICS, René Dugas. Monumental study of mechanical principles from antiquity to quantum mechanics. Contributions of ancient Greeks, Galileo, Leonardo, Kepler, Lagrange, many others. 671pp. 5⅜ × 8½.
65632-2 Pa. $14.95

FAMOUS PROBLEMS OF GEOMETRY AND HOW TO SOLVE THEM, Benjamin Bold. Squaring the circle, trisecting the angle, duplicating the cube: learn their history, why they are impossible to solve, then solve them yourself. 128pp. 5⅜ × 8½. 24297-8 Pa. $4.95

MECHANICAL VIBRATIONS, J.P. Den Hartog. Classic textbook offers lucid explanations and illustrative models, applying theories of vibrations to a variety of practical industrial engineering problems. Numerous figures. 233 problems, solutions. Appendix. Index. Preface. 436pp. 5⅜ × 8½. 64785-4 Pa. $10.95

CURVATURE AND HOMOLOGY, Samuel I. Goldberg. Thorough treatment of specialized branch of differential geometry. Covers Riemannian manifolds, topology of differentiable manifolds, compact Lie groups, other topics. Exercises. 315pp. 5⅜ × 8½. 64314-X Pa. $8.95

HISTORY OF STRENGTH OF MATERIALS, Stephen P. Timoshenko. Excellent historical survey of the strength of materials with many references to the theories of elasticity and structure. 245 figures. 452pp. 5⅜ × 8½. 61187-6 Pa. $11.95

GEOMETRY OF COMPLEX NUMBERS, Hans Schwerdtfeger. Illuminating, widely praised book on analytic geometry of circles, the Moebius transformation, and two-dimensional non-Euclidean geometries. 200pp. 5⅜ × 8¼.
63830-8 Pa. $8.95

MECHANICS, J.P. Den Hartog. A classic introductory text or refresher. Hundreds of applications and design problems illuminate fundamentals of trusses, loaded beams and cables, etc. 334 answered problems. 462pp. 5⅜ × 8½.   60754-2 Pa. $9.95

TOPOLOGY, John G. Hocking and Gail S. Young. Superb one-year course in classical topology. Topological spaces and functions, point-set topology, much more. Examples and problems. Bibliography. Index. 384pp. 5⅜ × 8¼.
65676-4 Pa. $9.95

STRENGTH OF MATERIALS, J.P. Den Hartog. Full, clear treatment of basic material (tension, torsion, bending, etc.) plus advanced material on engineering methods, applications. 350 answered problems. 323pp. 5⅜ × 8½. 60755-0 Pa. $8.95

ELEMENTARY CONCEPTS OF TOPOLOGY, Paul Alexandroff. Elegant, intuitive approach to topology from set-theoretic topology to Betti groups; how concepts of topology are useful in math and physics. 25 figures. 57pp. 5⅜ × 8½.
60747-X Pa. $3.50

ADVANCED STRENGTH OF MATERIALS, J.P. Den Hartog. Superbly written advanced text covers torsion, rotating disks, membrane stresses in shells, much more. Many problems and answers. 388pp. 5⅜ × 8½.   65407-9 Pa. $9.95

COMPUTABILITY AND UNSOLVABILITY, Martin Davis. Classic graduate-level introduction to theory of computability, usually referred to as theory of recurrent functions. New preface and appendix. 288pp. 5⅜ × 8½. 61471-9 Pa. $7.95

GENERAL CHEMISTRY, Linus Pauling. Revised 3rd edition of classic first-year text by Nobel laureate. Atomic and molecular structure, quantum mechanics, statistical mechanics, thermodynamics correlated with descriptive chemistry. Problems. 992pp. 5⅜ × 8½.   65622-5 Pa. $19.95

AN INTRODUCTION TO MATRICES, SETS AND GROUPS FOR SCIENCE STUDENTS, G. Stephenson. Concise, readable text introduces sets, groups, and most importantly, matrices to undergraduate students of physics, chemistry, and engineering. Problems. 164pp. 5⅜ × 8½.   65077-4 Pa. $6.95

THE HISTORICAL BACKGROUND OF CHEMISTRY, Henry M. Leicester. Evolution of ideas, not individual biography. Concentrates on formulation of a coherent set of chemical laws. 260pp. 5⅜ × 8½.   61053-5 Pa. $6.95

THE PHILOSOPHY OF MATHEMATICS: An Introductory Essay, Stephan Körner. Surveys the views of Plato, Aristotle, Leibniz & Kant concerning propositions and theories of applied and pure mathematics. Introduction. Two appendices. Index. 198pp. 5⅜ × 8½.   25048-2 Pa. $7.95

THE DEVELOPMENT OF MODERN CHEMISTRY, Aaron J. Ihde. Authoritative history of chemistry from ancient Greek theory to 20th-century innovation. Covers major chemists and their discoveries. 209 illustrations. 14 tables. Bibliographies. Indices. Appendices. 851pp. 5⅜ × 8½.   64235-6 Pa. $18.95

DE RE METALLICA, Georgius Agricola. The famous Hoover translation of greatest treatise on technological chemistry, engineering, geology, mining of early modern times (1556). All 289 original woodcuts. 638pp. 6¾ × 11.
60006-8 Pa. $18.95

SOME THEORY OF SAMPLING, William Edwards Deming. Analysis of the problems, theory and design of sampling techniques for social scientists, industrial managers and others who find statistics increasingly important in their work. 61 tables. 90 figures. xvii + 602pp. 5⅜ × 8½.
64684-X Pa. $15.95

THE VARIOUS AND INGENIOUS MACHINES OF AGOSTINO RAMELLI: A Classic Sixteenth-Century Illustrated Treatise on Technology, Agostino Ramelli. One of the most widely known and copied works on machinery in the 16th century. 194 detailed plates of water pumps, grain mills, cranes, more. 608pp. 9 × 12.
25497-6 Clothbd. $34.95

LINEAR PROGRAMMING AND ECONOMIC ANALYSIS, Robert Dorfman, Paul A. Samuelson and Robert M. Solow. First comprehensive treatment of linear programming in standard economic analysis. Game theory, modern welfare economics, Leontief input-output, more. 525pp. 5⅜ × 8½.
65491-5 Pa. $14.95

ELEMENTARY DECISION THEORY, Herman Chernoff and Lincoln E. Moses. Clear introduction to statistics and statistical theory covers data processing, probability and random variables, testing hypotheses, much more. Exercises. 364pp. 5⅜ × 8½.
65218-1 Pa. $9.95

THE COMPLEAT STRATEGYST: Being a Primer on the Theory of Games of Strategy, J.D. Williams. Highly entertaining classic describes, with many illustrated examples, how to select best strategies in conflict situations. Prefaces. Appendices. 268pp. 5⅜ × 8½.
25101-2 Pa. $7.95

MATHEMATICAL METHODS OF OPERATIONS RESEARCH, Thomas L. Saaty. Classic graduate-level text covers historical background, classical methods of forming models, optimization, game theory, probability, queueing theory, much more. Exercises. Bibliography. 448pp. 5⅜ × 8¼.
65703-5 Pa. $12.95

CONSTRUCTIONS AND COMBINATORIAL PROBLEMS IN DESIGN OF EXPERIMENTS, Damaraju Raghavarao. In-depth reference work examines orthogonal Latin squares, incomplete block designs, tactical configuration, partial geometry, much more. Abundant explanations, examples. 416pp. 5⅜ × 8¼.
65685-3 Pa. $10.95

THE ABSOLUTE DIFFERENTIAL CALCULUS (CALCULUS OF TENSORS), Tullio Levi-Civita. Great 20th-century mathematician's classic work on material necessary for mathematical grasp of theory of relativity. 452pp. 5⅜ × 8½.
63401-9 Pa. $9.95

VECTOR AND TENSOR ANALYSIS WITH APPLICATIONS, A.I. Borisenko and I.E. Tarapov. Concise introduction. Worked-out problems, solutions, exercises. 257pp. 5⅜ × 8¼.
63833-2 Pa. $7.95

THE FOUR-COLOR PROBLEM: Assaults and Conquest, Thomas L. Saaty and Paul G. Kainen. Engrossing, comprehensive account of the century-old combinatorial topological problem, its history and solution. Bibliographies. Index. 110 figures. 228pp. 5⅜ × 8½. 65092-8 Pa. $6.95

CATALYSIS IN CHEMISTRY AND ENZYMOLOGY, William P. Jencks. Exceptionally clear coverage of mechanisms for catalysis, forces in aqueous solution, carbonyl- and acyl-group reactions, practical kinetics, more. 864pp. 5⅜ × 8½. 65460-5 Pa. $19.95

PROBABILITY: An Introduction, Samuel Goldberg. Excellent basic text covers set theory, probability theory for finite sample spaces, binomial theorem, much more. 360 problems. Bibliographies. 322pp. 5⅜ × 8½. 65252-1 Pa. $8.95

LIGHTNING, Martin A. Uman. Revised, updated edition of classic work on the physics of lightning. Phenomena, terminology, measurement, photography, spectroscopy, thunder, more. Reviews recent research. Bibliography. Indices. 320pp. 5⅜ × 8¼. 64575-4 Pa. $8.95

PROBABILITY THEORY: A Concise Course, Y.A. Rozanov. Highly readable, self-contained introduction covers combination of events, dependent events, Bernoulli trials, etc. Translation by Richard Silverman. 148pp. 5⅜ × 8¼. 63544-9 Pa. $5.95

AN INTRODUCTION TO HAMILTONIAN OPTICS, H. A. Buchdahl. Detailed account of the Hamiltonian treatment of aberration theory in geometrical optics. Many classes of optical systems defined in terms of the symmetries they possess. Problems with detailed solutions. 1970 edition. xv + 360pp. 5⅜ × 8½. 67597-1 Pa. $10.95

STATISTICS MANUAL, Edwin L. Crow, et al. Comprehensive, practical collection of classical and modern methods prepared by U.S. Naval Ordnance Test Station. Stress on use. Basics of statistics assumed. 288pp. 5⅜ × 8½. 60599-X Pa. $6.95

DICTIONARY/OUTLINE OF BASIC STATISTICS, John E. Freund and Frank J. Williams. A clear concise dictionary of over 1,000 statistical terms and an outline of statistical formulas covering probability, nonparametric tests, much more. 208pp. 5⅜ × 8½. 66796-0 Pa. $6.95

STATISTICAL METHOD FROM THE VIEWPOINT OF QUALITY CONTROL, Walter A. Shewhart. Important text explains regulation of variables, uses of statistical control to achieve quality control in industry, agriculture, other areas. 192pp. 5⅜ × 8½. 65232-7 Pa. $7.95

THE INTERPRETATION OF GEOLOGICAL PHASE DIAGRAMS, Ernest G. Ehlers. Clear, concise text emphasizes diagrams of systems under fluid or containing pressure; also coverage of complex binary systems, hydrothermal melting, more. 288pp. 6½ × 9¼. 65389-7 Pa. $10.95

STATISTICAL ADJUSTMENT OF DATA, W. Edwards Deming. Introduction to basic concepts of statistics, curve fitting, least squares solution, conditions without parameter, conditions containing parameters. 26 exercises worked out. 271pp. 5⅜ × 8½. 64685-8 Pa. $8.95

CATALOG OF DOVER BOOKS

TENSOR CALCULUS, J.L. Synge and A. Schild. Widely used introductory text covers spaces and tensors, basic operations in Riemannian space, non-Riemannian spaces, etc. 324pp. 5⅜ × 8¼. 63612-7 Pa. $8.95

A CONCISE HISTORY OF MATHEMATICS, Dirk J. Struik. The best brief history of mathematics. Stresses origins and covers every major figure from ancient Near East to 19th century. 41 illustrations. 195pp. 5⅜ × 8½. 60255-9 Pa. $7.95

A SHORT ACCOUNT OF THE HISTORY OF MATHEMATICS, W.W. Rouse Ball. One of clearest, most authoritative surveys from the Egyptians and Phoenicians through 19th-century figures such as Grassman, Galois, Riemann. Fourth edition. 522pp. 5⅜ × 8½. 20630-0 Pa. $10.95

HISTORY OF MATHEMATICS, David E. Smith. Nontechnical survey from ancient Greece and Orient to late 19th century; evolution of arithmetic, geometry, trigonometry, calculating devices, algebra, the calculus. 362 illustrations. 1,355pp. 5⅜ × 8½. 20429-4, 20430-8 Pa., Two-vol. set $23.90

THE GEOMETRY OF RENÉ DESCARTES, René Descartes. The great work founded analytical geometry. Original French text, Descartes' own diagrams, together with definitive Smith-Latham translation. 244pp. 5⅜ × 8½. 60068-8 Pa. $6.95

THE ORIGINS OF THE INFINITESIMAL CALCULUS, Margaret E. Baron. Only fully detailed and documented account of crucial discipline: origins; development by Galileo, Kepler, Cavalieri; contributions of Newton, Leibniz, more. 304pp. 5⅜ × 8½. (Available in U.S. and Canada only) 65371-4 Pa. $9.95

THE HISTORY OF THE CALCULUS AND ITS CONCEPTUAL DEVELOPMENT, Carl B. Boyer. Origins in antiquity, medieval contributions, work of Newton, Leibniz, rigorous formulation. Treatment is verbal. 346pp. 5⅜ × 8½. 60509-4 Pa. $8.95

THE THIRTEEN BOOKS OF EUCLID'S ELEMENTS, translated with introduction and commentary by Sir Thomas L. Heath. Definitive edition. Textual and linguistic notes, mathematical analysis. 2,500 years of critical commentary. Not abridged. 1,414pp. 5⅜ × 8½. 60088-2, 60089-0, 60090-4 Pa., Three-vol. set $29.85

GAMES AND DECISIONS: Introduction and Critical Survey, R. Duncan Luce and Howard Raiffa. Superb nontechnical introduction to game theory, primarily applied to social sciences. Utility theory, zero-sum games, n-person games, decision-making, much more. Bibliography. 509pp. 5⅜ × 8½. 65943-7 Pa. $12.95

THE HISTORICAL ROOTS OF ELEMENTARY MATHEMATICS, Lucas N.H. Bunt, Phillip S. Jones, and Jack D. Bedient. Fundamental underpinnings of modern arithmetic, algebra, geometry and number systems derived from ancient civilizations. 320pp. 5⅜ × 8½. 25563-8 Pa. $8.95

CALCULUS REFRESHER FOR TECHNICAL PEOPLE, A. Albert Klaf. Covers important aspects of integral and differential calculus via 756 questions. 566 problems, most answered. 431pp. 5⅜ × 8½. 20370-0 Pa. $8.95

CHALLENGING MATHEMATICAL PROBLEMS WITH ELEMENTARY SOLUTIONS, A.M. Yaglom and I.M. Yaglom. Over 170 challenging problems on probability theory, combinatorial analysis, points and lines, topology, convex polygons, many other topics. Solutions. Total of 445pp. 5⅜ × 8½. Two-vol. set.
Vol. I 65536-9 Pa. $7.95
Vol. II 65537-7 Pa. $6.95

FIFTY CHALLENGING PROBLEMS IN PROBABILITY WITH SOLUTIONS, Frederick Mosteller. Remarkable puzzlers, graded in difficulty, illustrate elementary and advanced aspects of probability. Detailed solutions. 88pp. 5⅜ × 8½.
65355-2 Pa. $4.95

EXPERIMENTS IN TOPOLOGY, Stephen Barr. Classic, lively explanation of one of the byways of mathematics. Klein bottles, Moebius strips, projective planes, map coloring, problem of the Koenigsberg bridges, much more, described with clarity and wit. 43 figures. 210pp. 5⅜ × 8½. 25933-1 Pa. $5.95

RELATIVITY IN ILLUSTRATIONS, Jacob T. Schwartz. Clear nontechnical treatment makes relativity more accessible than ever before. Over 60 drawings illustrate concepts more clearly than text alone. Only high school geometry needed. Bibliography. 128pp. 6⅛ × 9¼. 25965-X Pa. $6.95

AN INTRODUCTION TO ORDINARY DIFFERENTIAL EQUATIONS, Earl A. Coddington. A thorough and systematic first course in elementary differential equations for undergraduates in mathematics and science, with many exercises and problems (with answers). Index. 304pp. 5⅜ × 8½. 65942-9 Pa. $8.95

FOURIER SERIES AND ORTHOGONAL FUNCTIONS, Harry F. Davis. An incisive text combining theory and practical example to introduce Fourier series, orthogonal functions and applications of the Fourier method to boundary-value problems. 570 exercises. Answers and notes. 416pp. 5⅜ × 8½. 65973-9 Pa. $9.95

THE THEORY OF BRANCHING PROCESSES, Theodore E. Harris. First systematic, comprehensive treatment of branching (i.e. multiplicative) processes and their applications. Galton-Watson model, Markov branching processes, electron-photon cascade, many other topics. Rigorous proofs. Bibliography. 240pp. 5⅜ × 8½. 65952-6 Pa. $6.95

AN INTRODUCTION TO ALGEBRAIC STRUCTURES, Joseph Landin. Superb self-contained text covers "abstract algebra": sets and numbers, theory of groups, theory of rings, much more. Numerous well-chosen examples, exercises. 247pp. 5⅜ × 8½. 65940-2 Pa. $7.95

Prices subject to change without notice.

Available at your book dealer or write for free Mathematics and Science Catalog to Dept. GI, Dover Publications, Inc., 31 East 2nd St., Mineola, N.Y. 11501. Dover publishes more than 175 books each year on science, elementary and advanced mathematics, biology, music, art, literature, history, social sciences and other areas.